AF905317

Sustainability Sciences in Asia and Africa

Sustainable Aquatic Food Systems

Series Editors
Basanta Kumar Das, ICAR – Central Inland Fisheries Research Institute
Barrackpore, Kolkata, West Bengal, India
Natrah Fatin Mohd Ikhsan, Faculty of Agriculture
Universiti Putra Malaysia
Serdang, Selangor, Malaysia

Sustainability as a scientific endeavor requires broad understanding of the interconnections in our global environment. It is therefore positioned at the crossroads of research in natural sciences, technology and human sciences. This book series supports the global efforts by providing timely coverage of the progress, opportunities, and challenges towards achieving sustainable development goals (SDGs) in Asia and Africa. It fosters the research in transdisciplinary academic fields spanning across agriculture and global food security, health and diseases, access to clean water and affordable energy, climate resilience and effects of global climate change on biodiversity, water and land. This series is planned to narrate the stories of scientific development and progression towards SDGs from the regions of Asia and Africa. The overall aim of the series is to provide a comprehensive publication platform and act as a knowledge engine in the growth of sustainability sciences with a special focus on developing nations. The series constitute of four sub-series each focusing on a specific theme closely linked to a set of SDGs. Each sub-series is leaders and experts in agricultural sciences, environmental sciences, biomedicine and medical sciences respectively. Books in this series are useful for environmentalists, biotechnologists, plant scientists, medical practitioners, conservation biologists, policy makers, NGOs and scientific bodies working in the areas of sustainability sciences in the developing regions.

Aquatic food systems have the potential to nurture human health and support environmental sustainability; however, they are currently threatened by anthropogenic climate change. Providing a growing global population with healthy diets from sustainable aquatic food systems is an immediate challenge. Although global fisheries and aquaculture production surpassed 223 million tonnes, aquaculture contributed to 130 million tonnes. More than 820 million people still have insufficient food, and close to 2 billion people suffer from micronutrient malnutrition, commonly called "hidden hunger."

Aquaculture could be crucial for food and nutritional security with high-quality animal protein (15 %), providing millions of people with livelihood and income generation. It is also expected to contribute towards the Sustainable Development Goals, namely SDG1 – No Poverty, SDG2 – Zero Hunger, and SDG14 – Life Below Water, adopted by all members of the United Nations (UN) in 2015. However, intensive and high-density culture systems could result in suboptimal oxygen levels, accumulation of metabolic waste products and feed leftovers that can create stressful environmental conditions and significantly affect the health status of the aquatic animals. The degraded aquatic environment also leads to decreased growth performance and reduced immune response, which becomes the ideal ground for disease outbreaks. Therefore, technological development to cope with changes in biological characteristics and stabilize production is strongly desired for both capture fisheries and aquaculture sustainability. In this book series, we shift the focus from production to consumption and aim to increase the visibility of aquatic resources as sustainable food that can contribute to "ending poverty, protecting the planet and improving the lives and prospects for everyone, everywhere" (The Sustainable Development Agenda).

Basanta Kumar Das • Vikash Kumar
Suvra Roy
Editors

Bioinformatics Tools for Predictive Ecology and Fisheries

Springer

Editors
Basanta Kumar Das
ICAR–Central Inland Fisheries Research Institute
Barrackpore, West Bengal, India

Vikash Kumar
ICAR–Central Inland Fisheries Research Institute
Barrackpore, West Bengal, India

Suvra Roy
ICAR–Central Inland Fisheries Research Institute
Barrackpore, West Bengal, India

ISSN 2730-6771 ISSN 2730-678X (electronic)
Sustainability Sciences in Asia and Africa
ISSN 3059-4162 ISSN 3059-4170 (electronic)
Sustainable Aquatic Food Systems
ISBN 978-981-95-2572-0 ISBN 978-981-95-2573-7 (eBook)
https://doi.org/10.1007/978-981-95-2573-7

© The Editor(s) (if applicable) and The Author(s), under exclusive license to Springer Nature Singapore Pte Ltd. 2026

This work is subject to copyright. All rights are solely and exclusively licensed by the Publisher, whether the whole or part of the material is concerned, specifically the rights of translation, reprinting, reuse of illustrations, recitation, broadcasting, reproduction on microfilms or in any other physical way, and transmission or information storage and retrieval, electronic adaptation, computer software, or by similar or dissimilar methodology now known or hereafter developed.
The use of general descriptive names, registered names, trademarks, service marks, etc. in this publication does not imply, even in the absence of a specific statement, that such names are exempt from the relevant protective laws and regulations and therefore free for general use.
The publisher, the authors and the editors are safe to assume that the advice and information in this book are believed to be true and accurate at the date of publication. Neither the publisher nor the authors or the editors give a warranty, expressed or implied, with respect to the material contained herein or for any errors or omissions that may have been made. The publisher remains neutral with regard to jurisdictional claims in published maps and institutional affiliations.

This Springer imprint is published by the registered company Springer Nature Singapore Pte Ltd.
The registered company address is: 152 Beach Road, #21-01/04 Gateway East, Singapore 189721, Singapore

If disposing of this product, please recycle the paper.

Preface

Over the past decade, considerable advancements have been made in the development of analytical techniques for molecular biological data. This progress has led to the creation of numerous new algorithms explicitly tailored for data pertaining to biological phenomena, such as gene expression and protein interactions. Conversely, ecological data analysis has continued to predominantly rely on traditional statistical methods, although this trend is gradually evolving with the increasing adoption of cutting-edge techniques. Such methods often entail few assumptions regarding the data and necessitate a more exploratory approach, such as employing Bayesian networks. A publication that provides insights into new bioinformatics tools and their potential cross-disciplinary uses in ecology and fisheries data would be very helpful to showcase the development of models capable of identifying functionally equivalent species across different fish communities to predict functional collapse.

This book, *Bioinformatics Tools for Predictive Ecology and Fisheries*, is a response to this pressing need for integrative approaches. In this context, we are glad to bring out a Handbook on bioinformatics tools covering various aspects of technology and analytical tools that would be beneficial in predictive ecology and fisheries, with 25 chapters. We ensured that each section is detailed, including practical information, step-by-step procedures, and relevant illustrations or diagrams where necessary. Recent updates on bioinformatics tools that incorporate emerging techniques and technologies in the field. This book provides a comprehensive guide, and it is meant for a wide range of readers, including researchers and practitioners in ecology and fisheries science, students interested in how computational tools are used in biology, and policymakers seeking evidence-based approaches to managing resources. By blending scientific rigor with clear explanations, the book aims to be both a go-to reference and a spark for future collaborations across disciplines.

Barrackpore, West Bengal, India

Basanta Kumar Das
Vikash Kumar
Suvra Roy

Acknowledgments

This manual could not have been completed without the tireless help and diligence of Barsha Baisakhi, who formatted the first of many early drafts of the manuscript. We also thank the team members of the Biotechnology division of ICAR-Central Inland Fisheries Research Institute, Kolkata, India, who volunteered to complete this project and continued with numerous revisions that finally evolved into the document presented here.

Contents

Genomics in Fisheries and Aquaculture 1

Basanta Kumar Das and Suvra Roy

1.1 Introduction

In fisheries, genome sequencing has emerged as a vital technology that has significantly enhanced the understanding of fish species and their ecological impacts. This development allows the examination of the genetic composition of various fish species, offering vital insights into their population dynamics, adaptability to changing environmental conditions, and susceptibility to disease (Luo et al. 2020). This cutting-edge technology has transformed fisheries management by supporting more accurate and targeted conservation strategies. Moreover, genome sequencing has facilitated new approaches for selective breeding and genetic improvement in aquaculture.

Fish, which represent half of all modern vertebrates, have served as important tools for understanding vertebrate genome evolution due to their shared evolutionary history that dates back to the origin of vertebrates. In addition to providing insight into the structure of human genes, they also offer surprising insights into how genes work in relation to human disorders. The study of an organism's entire genetic blueprint, known as its genome, is the focus of the field of genomics. DNA sequencing technology, which began with Sanger sequencing in the 1970s, continues to be important today. Following in Sanger's footsteps, next-generation sequencing (NGS) has overcome previous constraints to enable the sequencing of millions of nucleotides in just one experiment (Luo et al. 2020).

B. K. Das (✉) · S. Roy
ICAR-Central Inland Fisheries Research Institute, Barrackpore, Kolkata, India

© The Author(s), under exclusive license to Springer Nature Singapore Pte Ltd. 2026

B. K. Das et al. (eds.), *Bioinformatics Tools for Predictive Ecology and Fisheries*, Sustainable Aquatic Food Systems, https://doi.org/10.1007/978-981-95-2573-7_1

1.1.1 History of Genomics Application in Fisheries and Aquaculture

After James Watson and Francis Crick confirmed DNA as the genetic material in 1953, scientists began exploring its organization within the genome. A significant breakthrough occurred in 1996 with the sequencing of the yeast genome (*Saccharomyces cerevisiae*), followed by the completion of the first multicellular organism's genome, the nematode *Caenorhabditis elegans*, in 1998 ("Genome Sequence of the Nematode Caenorhabditis elegans: A Platform for Investigating Biology"). This milestone set the stage for the Human Genome Project, which led to the publication of the draft human genome sequence in 2001. The limitations of existing sequencing technologies spurred the development of next-generation sequencing, which significantly reduced costs and made extensive genomic data more accessible.

The whole-genome sequence of the Japanese pufferfish, *Fugu rubripes*, was released in 2002 (Aparicio et al. 2002); it is a commercially farmed species in China and Japan. Since the advent of next-generation sequencing around 2005, over 200 fish genomes have been sequenced and made publicly available, greatly improving our understanding of fish diversity. The expansion of fish genomic research is reflected in the National Center for Biotechnology Information (NCBI), where the number of assembled fish genomes increased from around 270 in December 2019 to 594 in December 2020 (Luo et al. 2020). These genomic resources are vital for applied areas such as aquaculture and fisheries, as well as for basic scientific studies like comparative genomics, evolution, and systematics. Although extensive genomic data is now accessible, the main challenge lies in effectively applying this information to real-world issues, including genetic resource management and selective breeding in fisheries and aquaculture. DNA sequencing technologies have played a key role in decoding the genetic code, supporting the sequencing of the human genome and important model fish species like the torafugu (*Fugu rubripes*) (Aparicio et al. 2002), spotted green pufferfish (*Tetraodon nigroviridis*) (Jaillon et al. 2004), medaka (*Oryzias latipes*) (Kasahara et al. 2007), and zebrafish (*Danio rerio*) (Howe et al. 2013).

Genomic sequencing techniques have become much more user-friendly, providing valuable insights into fundamental molecular processes, ranging from evolutionary studies to personalized medical applications. Moreover, these genomic technologies are adaptable across various methodologies and organisms, allowing for the widespread use of genomic techniques and establishing "omics" as essential tools in diverse fields and organisms.

1.1.2 Genomics Application in Fisheries and Aquaculture

Before using publicly available fish genome data, it is essential to evaluate the quality of specific draft genomes carefully, considering the different impacts of various sequencing and analysis methods. Short-read sequencing has been the dominant

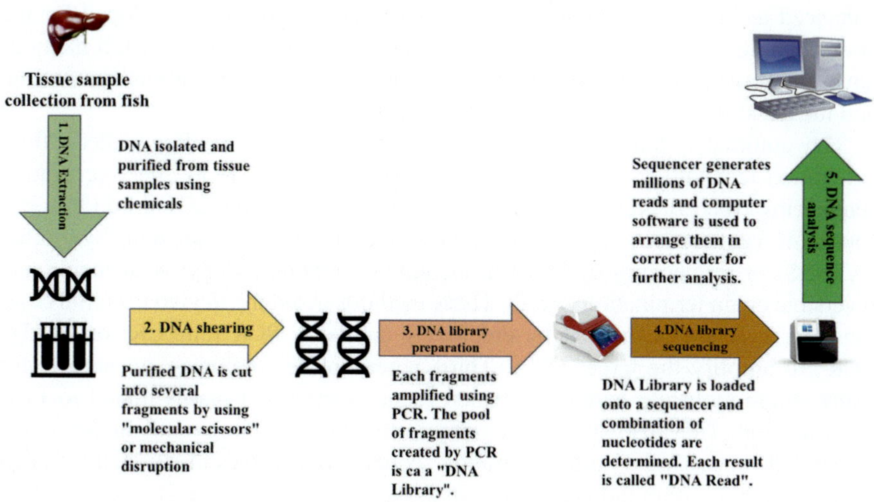

Fig. 1.1 Schematic representation of the genome sequencing process

approach for generating accessible genomes, but its limited read lengths pose challenges for genome assembly. Alternative sequencing technologies can help address this issue, enabling the production of high-quality genomes at the chromosome level. Although draft genomes are available for some important fisheries and aquaculture species like bighead carp (*Hypophthalmichthys nobilis*), silver carp (*H. molitrix*), catla (*Labeo catla*), and rohu (*L. rohita*), many species still lack genome data or need improvements to existing datasets (Luo et al. 2020). Genomic and biotechnological techniques have been employed in critical areas of aquaculture research, such as improving growth, enhancing disease and stress resistance, reducing intramuscular bone, and studying species evolution. Genome sequencing and transcriptome analysis are key tools for generating the necessary resources for these genetic studies (Fig. 1.1).

Successfully sequencing a new genome requires a combination of methods. Illumina sequencing supplies short, highly accurate reads (100–200× coverage), whereas PacBio offers longer reads (50–60× coverage) that aid in assembly. Additionally, to efficiently detect numerous SNPs, researchers often resequence pooled DNA samples from different populations.

1.1.3 Sequencing Platforms

Sequencing involves determining the exact order of nucleotides in a nucleic acid polymer, whether DNA or RNA. For over 20 years, Sanger's DNA sequencing method was the preferred method, successfully sequencing the complete genomes of many species, including humans. However, it has notable drawbacks, including slow speed, high costs, and labor intensity. To address these issues, new and enhanced technologies emerged in 2005. Different terms were developed for the new technologies, starting with next-generation sequencing (NGS). NGS methods have progressed into second-generation sequencing, third-generation sequencing, and the latest fourth-generation sequencing techniques.

The automated Sanger method for DNA sequencing is often called "first-generation technology," while newer methods are referred to as "next-generation sequencing" (NGS). NGS techniques are grouped into second (2G), third (3G), and fourth (4G) generations. Second-generation technologies use sequencing by synthesis (SBS) or sequencing by ligation, including methods like pyrosequencing and reversible chain termination (RCT). These methods avoid the in vivo bacterial cloning step of Sanger's method by employing emulsion PCR (emPCR) or "bridge PCR" to amplify the target DNA. Third-generation technologies are notable for using single-molecule templates, which reduce errors and biases linked to PCR amplification. They also skip cyclic array techniques, enabling much higher levels of parallel processing. Fourth-generation sequencing involves in-situ methods that sequence DNA within individual cells in tissue sections, utilizing 2G NGS chemistry to analyze nucleic acids directly in fixed cells and tissues.

1.1.4 Genome Sequencing

Whole genome data provides insights into gene structure and organization, improving our understanding of their functions and roles within biological systems. Genome resequencing helps identify and interpret genetic variations, aiding the development of genome-wide markers to measure genetic diversity and link it to specific traits. Additionally, comparative genomics offers clues about the evolutionary history and ancestry of genes across different species. The completion of the Human Genome Project around 2000, combined with advances in high-throughput DNA sequencing and powerful computational tools, has transformed genomics. Initially, human genome sequencing depended on Sanger sequencing, which provided accurate results but was slow and costly. In the early 2000s, second-generation sequencing technologies like Roche 454 pyrosequencing and Illumina's bridge amplification facilitated large-scale genomic studies in fisheries. Subsequently, third-generation sequencing methods, such as Pacific Biosciences and Oxford Nanopore, which generate long reads, further revolutionized high-resolution fish genome research. Fish genome sequencing projects began in the early 2000s, with the Japanese pufferfish, *Fugu rubripes*, being the first to have its complete genome sequenced in 2002. In 2009, the Genome 10 K Project was initiated by a consortium of scientists in Santa Cruz, California, to sequence 10,000 vertebrate genomes, including 4000 fish species, to explore how complex animal life evolved through changes in DNA. The goal was to sequence at least one individual from each of the roughly 66,000 vertebrate species. The publication of the zebrafish (*Danio rerio*) genome in 2013 marked a major milestone in fish genomics. In 2020, the Fish10K (10,000 Fish Genomes) Project was launched as part of the Earth BioGenome Project, focusing solely on sequencing 10,000 fish species to explore the evolution and development of vertebrates through extensive genomic data. Unlike previous draft assemblies, the current approach emphasizes using a combination of short-reads, long-reads, and Hi-C technologies to generate chromosomal-level genome assemblies. So far, about 867 fish genomes have been sequenced and assembled at various levels, including more than 86 aquaculture species. The smallest assembled genome belongs to the silver spinyfish, *Diretmus argenteus* (302.36 Mb), while the largest is from the cloudy catshark, *Scyliorhinus storazame* (4470.98 Mb).

1.1.5 Transcriptome Sequencing

If the genome represents the static molecular component, then the transcriptome, comprising all mRNA, rRNA, tRNA, and ncRNA, can be seen as its dynamic counterpart. Transcriptome sequencing, also known as RNA-Seq, is a crucial technique that utilizes deep sequencing methods to elucidate gene functions and identify large-scale expression patterns. This approach can enhance our understanding of the relationship between genotypes and phenotypes, considering various biological and environmental factors, and explain how gene expression profiles influence specific traits.

The Fish-T1K project, initiated in 2013 by the Beijing Institute of Genomics (BGI) and the China National GeneBank (CNGB), aims to produce large-scale transcriptome sequences for 1000 diverse fish species, facilitating evolutionary studies. Over 98,400 RNA-Seq reads related to fish species are available in the NCBI database. This initiative has enhanced the mechanisms for tolerance to biotic (bacterial and fungal) and abiotic stresses (hypoxia, ammonia, salinity, temperature, etc.), as well as provided insights into biological processes, including growth, reproduction, and disease resistance. In India, transcriptome sequencing efforts have focused on identifying genes that govern key production traits, such as growth, disease resistance, and reproduction, as well as marker discovery. The species studied include Indian major carps (*Labeo rohita* and *L. catla*), hilsa shad (*Tenualosa ilisha*), magur (*Clarias magur*), Nile tilapia (*Oreochromis niloticus*), common carp (*Cyprinus carpio*), Indian white shrimp (*Fenneropenaeus indicus*), and spotted snakehead (*Channa punctata*).

1.2 Application of Genomics

1.2.1 Genome-Wide Association Studies, Linkage Maps, and QTLs

Genome-wide association studies (GWAS) are an emerging method in fisheries science that aims to pinpoint genes linked to performance and production traits. This method entails scanning markers, primarily SNPs, across genomes to detect genomic variations tied to commercially significant traits like growth and disease resistance. Although promising findings have been achieved in terrestrial livestock and human medical fields, the application of GWAS in aquaculture species remains limited (Tsai et al. 2015a). However, the declining costs of next-generation sequencing (NGS) and advancements in genotyping technology have spurred GWAS efforts in various aquaculture species. For instance, four quantitative trait loci (QTL) linked to columnaris disease resistance have been identified in catfish (Geng et al. 2015), and nine QTL associated with resistance to *Vibrio anguillarum* have been discovered in the Japanese flounder (*Paralichthys olivaceus*) genome (Shao et al. 2015). GWAS has also revealed conserved genetic loci connected to migration tendencies across different geographical areas (Hecht et al. 2013) and specific regions of the genome associated with phenotypes or selection related to sea ages (the ages at which they return from marine migration to spawn) (Johnston et al. 2014). Additionally, researchers have utilized GWAS to unveil the polygenic genetic architecture underlying growth-related traits in Atlantic salmon.

A high-density genetic linkage map underpins genomic comparisons among related nonmodel species, illuminating genome organization and evolution through comparative studies. For example, the comparison of meiotic maps in the spotted gar (*Lepisosteus oculatus*) demonstrated that this lineage branched off from teleosts prior to the teleost genome duplication (TGD), with its genome bearing more resemblance to humans than to teleosts (Amores et al. 2011). Additionally, comparative genomic analysis has revealed a highly conserved synteny among teleosts (Kai et al. 2014) and significant inter- and intrachromosomal rearrangements in catfish (Zhang et al. 2013). Linkage maps have also played a role in pinpointing sex-specific recombination sites, identifying departures from Mendelian inheritance based on sex (Kakioka et al. 2013), and discovering sex-specific markers. Furthermore, high-resolution linkage maps have been created for genetic breeding programs, utilizing marker- and gene-assisted selection for orange-spotted grouper (*Epinephelus coioides*), mapping ecologically significant traits in cichlid fishes (Henning and Meyer 2014), and revealing the genetic basis for phenotypic adaptation in the Mexican tetra, *Astyanax mexicanus*.

QTL refers to chromosomal regions, which can be single genes or clusters, that influence quantitative traits (Geldermann 1975). Research utilizing NGS-based QTL analysis has aimed to pinpoint the genetic regions linked to migratory behaviors in rainbow trout, Oncorhynchus mykiss (Hecht et al. 2012; Hale et al. 2013). These investigations indicated a complex polygenic structure underlying migration-related traits in this species. Development rate is a crucial characteristic that shows variability among natural populations of aquaculture species. A QTL analysis comparing two clonal lines of rainbow trout, Clearwater (Cl) and Swanson (Sw), from geographically and genetically distinct populations, identified a genomic region associated with the rapid development observed in the Sw line. Notably, this region coincided with that found in the Cl line (Miller et al. 2012). The findings led the authors to propose that the parallel rapid development traits may have evolved through the recurring use of a pre-existing haplotype containing the key mutation(s). In a separate study, quantitative genetic analysis of retinal degeneration in *Astyanax mexicanus* revealed four QTLs associated with degeneration across specific retinal layers (O'Quin et al., 2013). The study's results suggested that mutations leading to retinal degeneration in Astyanax might also impact the evolution of eye size in the species. Furthermore, Eshel et al. (2014) conducted a comprehensive transcriptome search in Nile tilapia (*Oreochromis niloticus*) during early embryonic development, identifying the QTL responsible for determining sex. Performance and quality traits are the most frequently researched in fisheries due to their significant economic impact on the aquaculture sector. Tsai et al. (2015b) explored the genetic underpinnings of various growth and fillet traits within a large farmed salmon cohort, identifying multiple QTLs linked to these traits. The authors inferred a polygenic structure for the traits studied and noted the pleiotropic nature of the identified QTL. Most studies mentioned above utilized lower-density SNP arrays for GWAS, linkage mapping, and QTL analysis. To create a high-resolution genetic map and effectively discover QTL, a high-density SNP array is essential. The employment of lower-density SNP arrays could lead to reduced linkage disequilibrium. Thus, future research should prioritize the use of higher-density SNP arrays to identify markers related to traits, thereby aiding functional genomics and selective breeding programs.

1.2.2 Studies on Population Genetic Structure

Next-generation sequencing (NGS) technologies have been utilized to investigate the genetic effects of stocking on natural brook charr (*Salvelinus fontinalis*) populations. Analyzing SNP markers from NGS RNA-seq in 336 individuals across nine lakes in Quebec's Portneuf Wildlife Reserve revealed a positive link between stocking intensity and genetic diversity within groups, along with decreased population differentiation (Lamaze et al. 2012). Another study by Fernandez-Silva et al. (2013) used 12 microsatellite markers, developed from shallow shotgun pyrosequencing of genomic DNA, to investigate the genetic structure of yellow-striped goatfish (*Mulloidichthys flavolineatus*). The microsatellite data showed no significant deviations from Hardy-Weinberg equilibrium at the sampling sites (Johnston Atoll and Okinawa), suggesting a single, panmictic population in the regions studied. Bourret et al. (2013) conducted a detailed analysis of Atlantic salmon (*Salmo salar*) population structure using an SNP array based on expressed sequence tags (ESTs) and genomic sequencing. Principal component analysis revealed notable genetic differences between European and North American populations, with three main regional groups within Europe: the Atlantic, Baltic Sea, and Barents-White Sea regions. The application of NGS methods like restriction-site associated DNA sequencing (RAD-seq) allowed researchers to uncover historical population structures and recent colonization events in stickleback samples from marine and freshwater environments along thc coast, including the Willamette Basin and central Oregon (Catchen et al. 2013). RAD-seq was also used to examine the population genetic structure of sea mullet (*Mugil cephalus*) in Australian waters, indicating a multispecies fishery stock (Krück et al. 2013). Additionally 5, 5985 new SNPs identified through RAD sequencing helped estimate genome-wide divergence between two Baltic Sea herring (*Clupea harengus*) populations separated by 387 km. This SNP analysis revealed significant divergence (Fst = 0.128), challenging the notion of genetic uniformity in Baltic Sea herring (Corander et al. 2013). Larson et al. (2014) applied RAD-sequencing to examine population structures in five Chinook salmon (*Oncorhynchus tshawytscha*) populations from western Alaska (Tubutulik River, Anvik River, Kogrukluk River, Koktuli River, and Big Salmon River). Both individual-based clustering methods and assignment tests confirmed genetic structure among the populations of the Anvik River, Kogrukluk River, and Koktuli River. Furthermore, Pujolar et al. (2014) adopted a population genomics approach to examine genetic diversity in European eel (*Anguilla anguilla*). Their Fst-based analysis of 259 individuals from eight locations across the species' range revealed very low genetic differentiation (Fst = 0.0007), suggesting genomic panmixia and no significant population sub-structuring. These findings highlight the power of next-generation sequencing techniques not only for identifying genetic markers but also for uncovering historical population structures and recent colonization patterns. Moreover, NGS has proven helpful in revealing subtle population structures even across smaller geographical distances (387 km) (Corander et al. 2013). Nonetheless, some of the aforementioned studies have utilized relatively few samples to explore population structures. Additionally, many studies did not employ effective sampling strategies to uncover population structures, historical processes, and patterns of interstock and interpopulation connectivity. Since fish species may exhibit spatial and temporal variability, historical conclusions require suitable temporal sampling designs. Thus, future research should focus on employing larger sample sizes and incorporating both spatial and temporal sampling to enhance confidence in historical inferences, levels of genetic variability, and population differentiation.

1.2.3 Studies on Traceability

Traceability entails the ability to identify species and track their population origins, which is crucial for food conservation and quality certification. Different approaches, such as conventional tagging, labeling, and DNA-based techniques, are used for this purpose. Among them, DNA-based methods offer higher reliability and precision. Furthermore, DNA markers can be utilized in processed fish products that have undergone freezing, salting, cooking, and canning. Recent advancements in next-generation sequencing (NGS) technologies have led to the creation of species-specific panels of SNP markers, effectively assigning individuals to their population of origin. For instance, using transcriptome-derived SNPs, high precision (93–100%) was achieved in assigning individuals of Atlantic cod (*Gadus morhua*), Atlantic herring (*Clupea harengus*), sole (*Soleasolea*), and European hake (*Merluccius merluccius*) to their populations. Similarly, 60–70% of European anchovy (*Engraulis encrasicolus*) individuals were accurately identified based on species-specific SNPs (Montes et al. 2014). Additionally, De Battisti et al. (2013) devised a new DNA-based method using pyrosequencing technology, which can uniquely identify raw and processed fish products from 25 species within the Clupeiformes and Pleuronectiformes groups. Another study employed a DNA metabarcoding approach, leveraging NGS to determine whether the fish species present in Egyptian aqua-feed samples are from endangered populations (Galal-Khallaf et al. 2016). Thirteen fish species from four orders (Clupeiformes, Perciformes, Aulopiformes, and Siluriformes) were distinctly identified, revealing that 46% of all species detected were either overexploited or experiencing significant decline. Moreover, a combination of simulated and empirical SNP datasets was employed to differentiate between captive-bred and natural marine fish populations (Bylemans et al. 2016), aiming to protect the ecological and genetic integrity of wild fish from the increasing mixing of these populations. The accessibility of NGS technologies has enhanced our ability to determine the stock or population of origin for individual fish, thereby improving food safety, combating illegal fishing, and promoting sustainable fish stocks globally. However, solely testing fish identity will not suffice to curb the decline in fish resources; instead, traceability methods must be integrated with sustainable fishing practices, reduced bycatch, and ecosystem-based management policies.

1.2.4 Studies on Local Adaptation, Evolution, and Phylogenetics

Understanding how local adaptations occur is a key focus for evolutionary biologists. However, the exploration of the molecular foundations that drive phenotypic changes and reproductive isolation has faced significant challenges due to the limited availability of DNA sequences and genomic resources. Recent advances in sequencing technologies enable whole transcriptome sequencing, providing valuable insights into gene expression divergence and the mechanisms underlying phenotypic variation. The diverse environmental habitats and natal homing behaviors of Salmonid fishes create conditions conducive to local adaptation (Quinn 2005). Gene expression analyses indicate that the differences in genotypic and phenotypic traits between closely related species or populations of the same species often result from regulatory modifications in genes associated with biosynthesis, metabolic functions, immunity, DNA replication and repair, and development (Elmer et al. 2010; Jeukens et al. 2010). Additionally, next-generation sequencing (NGS) studies of adaptations in alewife (*Alosa pseudoharengus*) to freshwater and marine environments have shown that phenotypic differences between anadromous and lake alewives stem from extensive regulatory modifications rather than changes in coding sequences (Czesny et al. 2012). The conservation of specific genes and haplotypes also contributes to the local adaptation of fish. For example, researchers identified a conserved haplotype through a combination of genetic crosses and NGS technology, which correlated with the parallel adaptation of rapid development rates in two geographically and genetically distinct populations of rainbow trout (*Oncorhynchus mykiss*) (Miller et al. 2012). Genome-wide variation data from two oceanic and three freshwater populations of three-spined sticklebacks (*Gasterosteus aculeatus*) show that oceanic populations frequently give rise to diverse freshwater populations phenotypically (Hohenlohe et al. 2010). Additionally, local adaptation in three-spined sticklebacks has been linked to copy-number variation (Feulner et al. 2013), changes in gene regulation patterns (Lenz et al. 2013), and differences in genomic regions unique to specific populations (Feulner et al. 2015). Based on observed similarities in genetic variability and differentiation patterns between ten marine populations and previous studies of marine-freshwater divergence in three-spined sticklebacks, Guo et al. (2015) suggested that similar genetic processes and loci are involved in adaptation to freshwater and brackish water habitats. However, environmental heterogeneity, such as salinity and temperature variations, influences genetic differentiation and drives local adaptation in fish species (Guo et al. 2015).

Cichlid fishes are recognized for their wide range of phenotypes and extensive species diversity. However, phylogenetic analyses of NGS data have revealed a monophyletic group among 16 cichlid species from Lake Victoria (Wagner et al. 2013). Using the same approach, Jones et al. (2013) examined the phylogenetic relationships among all 26 species of swordtail and Platyfish (genus *Xiphophorus*), dividing them into three major clades. They also suggested that at least two species (*X. clemenciae* and *X. monticolus*) arose through hybridization events. A study on thick- and thin-lipped Midas cichlids revealed that traits associated with foraging behavior evolved in parallel. In contrast, traits related to diet and food processing followed non-parallel evolutionary paths (Manousaki et al. 2013). This indicates a mosaic pattern of both parallel and nonparallel evolution in these cichlids. The European eel and the American eel (*A. rostrata*), two Atlantic eel species, are partially isolated sister species that naturally hybridize. The existence of hybrids suggests incomplete reproductive isolation, caused by cytonuclear incompatibility that impairs ATP synthase function (Gagnaire et al. 2012). When exploring parallel evolution, Pavey et al. (2013) studied major histocompatibility complex (MHC) II diversity and bacterial infections in five sympatric pairs of dwarf and normal lake whitefish (*Coregonus clupeaformis*). Their results imply that pathogens influencing MHC II evolution do not directly cause parallel phenotypic evolution in whitefish. To investigate the genetic and evolutionary aspects of optic adaptation, the genome sequence of Pacific bluefin tuna (*Thunnus orientalis*) was determined (Nakamura et al. 2013). Phylogenetic analysis of five visual pigment genes revealed that Pacific bluefin tuna experienced evolutionary changes in three genes (RH1, RH2, and SWS2). These changes could improve their ability to detect blue-green contrasts and estimate prey distance in the blue-pelagic ocean. To investigate the origins and evolution of B chromosomes, Valente et al. (2014) sequenced complete genomes of cichlid fish with and without B chromosomes, including microdissected B chromosomes. They proposed a model based on sequence analysis, suggesting that the B chromosome originated early in Lake Victoria cichlids from a small fragment of an autosome. Recently, Meyer et al. (2015) examined the phylogenetic relationships among East African Lake Tanganyika (LT) cichlid lineages. Using an NGS dataset of 42 nuclear loci, they identified LT mouthbrooding cichlids, known as the "H-lineage" (Nishida 1991), as a sister group to the Lamprologini lineage. Additionally, Martin et al. (2015) uncovered a complex history of speciation involving gene flow, with both allopatric and sympatric phases among four Cameroon Crater Lake cichlids.

1.2.5 Studies on Fish Immunology

Next-generation sequencing (NGS) technologies have proven valuable in identifying genes linked to the innate immune response and innate host defense against pathogens. For instance, Ordas et al. (2011) used NGS and microarray data to annotate infection-responsive genes in zebrafish embryos exposed to *Salmonella typhimurium*. Additionally, the authors discovered innate host defense genes expressed both in the presence and absence of a fully developed adaptive immune system. In another investigation, Zhang et al. (2011) applied whole-genome sequencing and RNA-seq data to pinpoint innate immune genes in common carp (*Cyprinus carpio*). Moreover, NGS technology has been employed to assess the efficacy of an oral vaccine against pathogens responsible for fish diseases (Sarropoulou et al. 2012). It has also been utilized to measure genetic variability in antigen-presenting genes of the major histocompatibility complex (MHC) and their potential interactions with microbiota (Wegner et al. 2012). Findings indicated that marine fish populations exhibit relatively low immunogenetic diversity. In yet another study, NGS technology played a crucial role in elucidating how dietary and pathogenic factors influence the transcription of the mucin gene (Perez-Sanchez et al. 2013). Gene expression analysis revealed that nutritional factors have a significant influence on intestinal mucin.

1.2.6 Ecotoxicological Studies

Toxic metal contaminants in aquatic ecosystems pose a global threat to both environments and human health because they accumulate in the food chain, affecting species at all trophic levels. Therefore, having practical tools to measure and predict their effects on natural populations, communities, and ecosystems is essential. Transcriptome analysis has become a valuable approach to understanding the molecular basis of biological responses to toxic substances. While microarrays remain widely used for transcriptome analysis, recent years have seen increased reliance on next-generation sequencing (NGS) technologies to quantify transcripts from experimental samples. For example, the toxic effects of cadmium (Cd) and copper (Cu) on yellow perch (*Perca flavescens*) were studied using 454-sequencing technology (Pierron et al. 2011). This study found a downregulation of genes involved in protein biosynthesis, immune response, and lipid and energy metabolism. The researchers suggested that reduced gene expression might result from disruptions in bile acid metabolism and epigenetic modifications affecting histones and DNA, leading to gene silencing. Jenny et al. (2012) used NGS in combination with microarray and real-time PCR to examine the short-term effects of 2,3,7,8-tetrachlorodibenzo-p-dioxin (TCDD) on miRNA expression in zebrafish embryos. They concluded that TCDD exposure could alter hematopoiesis and cardiovascular development by changing miRNA levels. In another study, van Aerle et al. (2013) employed NGS to investigate the molecular toxicity mechanisms of silver nanoparticles in zebrafish embryos, observing significant changes in gene expression, particularly related to oxidative phosphorylation and protein synthesis. Gene expression studies related to persistent organic pollutants (POPs) in burbot (*Lota lota*) indicated that POP exposure induces oxidative stress in fish (Olsvik et al. 2013). Additionally, Sun et al. (2014) employed NGS in conjunction with liver histology to investigate the mechanisms of nitrite toxicity in *Megalobrama amblycephala*. The results showed that nitrites can cause liver damage through oxidative stress, histotoxic hypoxia, and cell death. A recent investigation employed NGS to identify pathways involved in xenobiotic biodegradation and metabolism in *Tigriopus japonicus* (Kim et al. 2015). This research demonstrated that NGS effectively detects gene expression changes caused by toxicants and provides critical mechanistic insights into toxicity within the context of whole-animal physiology.

1.2.7 Gene Expression Studies in Different Tissues, Developmental Stages and/or Sex

In recent years, RNA-seq analyses have become increasingly popular for studying gene expression in fish because they directly measure mRNA levels across the entire transcriptome. Additionally, transcriptomes are closely associated with genes or functional regions within the genome, which are often the main focus of aquaculture research. Le Cam et al. (2012) conducted transcriptome analyses on the testis, ovary, brain, and gill of rainbow trout (*Oncorhynchus mykiss*), identifying genes that are predominantly expressed in each tissue. This information can help identify gene activity under various physiological states. Tissue-specific expression profiling also allows for the identification of genes critical for certain functions. For example, genes important for myogenesis and muscle growth were identified through transcriptome analysis of the fast skeletal muscle in gilthead sea bream (*Sparus aurata*) (Garcia de la Serrana et al. 2012). Micallef et al. (2012) highlighted the importance of NGS in transcriptome analysis, reporting the first skin-specific mRNA sequences from skin and epidermal tissue samples of salmon (*Salmo salar*). Another study by Vieira et al. (2013) indicated that the teleost skeleton acts as both a calcium store and an endocrine tissue. Moreover, NGS technology has enabled the development of a transcriptome database, supporting comprehensive studies of individual genes and their proteins (Calduch-Giner et al. 2013). MicroRNAs (miRNAs) are essential regulators of gene expression (Krol et al. 2010). Transcriptome analyses of gonads have shown that miRNAs significantly influence sexual development in teleosts (Bizuayehu et al. 2012; Gu et al. 2014). Additionally, miRNAs are critical for regulating genes during early embryonic stages in fish. For instance, transcriptome analysis across four stages of early zebrafish development revealed different expression patterns of miRNAs and piRNAs (Wei et al. 2012). Gene expression studies have shown that temperature changes during early development (embryonic and larval stages) can modify miRNA profiles, affecting short- and long-term gene regulation (Bizuayehu et al. 2015). These results suggest that climate-induced temperature increases could influence fish life histories. Recently, Wongwarangkana et al. (2015) analyzed miRNA expression in nine tissues of the

Japanese puffer (*Takifugu rubripes*), including fast and slow muscles, heart, eye, brain, intestine, liver, ovaries, and testes. The authors identified 1420 known miRNAs with expression patterns comparable to those in other animals, indicating that vertebrate miRNAs likely share standard evolutionary functions. Consequently, *T. rubripes* can serve as a model species for studying gene expression patterns in other animals.

1.3 Conclusion and Future Perspectives

The rapid evolution and decreasing costs of genomics technologies have revolutionized genomics and related fields. Genomics has led to a remarkable array of scientific breakthroughs, shifting the focus from gene-level to genome-wide research in fisheries and aquaculture. Fish biologists have leveraged genomics to uncover new genetic markers for population and traceability studies. Furthermore, genomics technologies have been utilized in ecotoxicology, enabling genome-wide characterizations to identify genomic regions associated with commercially significant traits, and for profiling messenger RNAs and microRNAs to investigate the regulation of biological processes alongside inquiries into evolutionary issues. Recently, significant strides have been made in sequencing the genomes of key aquaculture species, revealing tremendous potential for transformative impacts in genetic and biological research within fisheries and aquaculture. However, genomic data for many fish species remains scarce, and some genomics studies have been published without adequate attention to sampling or experimental design. Future research should extend to various nonmodel species, ensuring that sampling methods align with the specific goals of such studies. Only then can we fully comprehend the genetic and biological importance of the species examined for fisheries and aquaculture.

References

Amores A, Catchen J, Ferrara A, Fontenot Q, Postlethwait JH (2011) Genome evolution and meiotic maps by massively parallel DNA sequencing: spotted gar, an outgroup for the teleost genome duplication. Genetics 188(4):799–808

Aparicio S, Chapman J, Stupka E, Putnam N, Chia JM, Dehal P et al (2002) Whole-genome shotgun assembly and analysis of the genome of *Fugu rubripes*. Science 297(5585):1301–1310

Bizuayehu TT, Babiak J, Norberg B, Fernandes JM, Johansen SD, Babiak I (2012) Sex-biased miRNA expression in Atlantic halibut (*Hippoglossus hippoglossus*) brain and gonads. Sex Dev 6(5):257–266

Bizuayehu TT, Johansen SD, Puvanendran V, Toften H, Babiak I (2015) Temperature during early development has long-term effects on microRNA expression in Atlantic cod. BMC Genomics 16:1–12

Bourret V, Kent MP, Primmer CR, Vasemägi A, Karlsson S, Hindar K et al (2013) SNP-array reveals genome-wide patterns of geographical and potential adaptive divergence across the natural range of Atlantic salmon (*Salmo salar*). Mol Ecol 22(3):532–551

Bylemans J, Maes GE, Diopere E, Cariani A, Senn H, Taylor MI et al (2016) Evaluating genetic traceability methods for captive-bred marine fish and their applications in fisheries management and wildlife forensics. Aquac Environ Interact 8:131–145

Calduch-Giner JA, Bermejo-Nogales A, Benedito-Palos L, Estensoro I, Ballester-Lozano G, Sitjà-Bobadilla A, Pérez-Sánchez J (2013) Deep sequencing for de novo construction of a marine fish (*Sparus aurata*) transcriptome database with an extensive coverage of protein-coding transcripts. BMC Genomics 14:1–11

Catchen J, Bassham S, Wilson T, Currey M, O'Brien C, Yeates Q, Cresko WA (2013) The population structure and recent colonization history of O regon three spines stickle back determined using restriction-site associated DNA-sequencing. Mol Ecol 22(11):2864–2883

Corander J, Majander KK, Cheng L, Merilä J (2013) High degree of cryptic population differentiation in the Baltic Sea herring *Clupea harengus*. Mol Ecol 22(11):2931–2940

Czesny S, Epifanio J, Michalak P (2012) Genetic divergence between freshwater and marine morphs of alewife (*Alosa pseudoharengus*): a 'next-generation'sequencing analysis. PLoS One 7(3):e31803

De Battisti C, Salomoni A, Ormelli S, Monne I, Capua I, Cattoli G (2013) Rapid pathotyping of Newcastle Disease Virus by pyrosequencing. J Virol Methods 188(1–2):13–20

Elmer KR, Fan S, Gunter HM, Jones JC, Boekhoff S, Kuraku S, Meyer A (2010) Rapid evolution and selection inferred from the transcriptomes of sympatric crater lake cichlid fishes. Mol Ecol 19:197–211

Eshel O, Shirak A, Dor L, Band M, Zak T, Markovich-Gordon M et al (2014) Identification of male-specific amh duplication, sexually differentially expressed genes and microRNAs at early embryonic development of Nile tilapia (*Oreochromis niloticus*). BMC Genomics 15:1–18

Fernandez-Silva I, Snelgrove BN, Bowen BW (2013) Twelve microsatellite DNA markers to resolve population structure of the Yellow-Striped Goatfish (family Mullidae). Conserv Genet Resour 5:565–568

Feulner PG, Chain FJ, Panchal M, Eizaguirre C, Kalbe M, Lenz TL et al (2013) Genome-wide patterns of standing genetic variation in a marine population of three-spined sticklebacks. Mol Ecol 22(3):635–649

Feulner PG, Chain FJ, Panchal M, Huang Y, Eizaguirre C, Kalbe M et al (2015) Genomics of divergence along a continuum of parapatric population differentiation. PLoS Genet 11(2):e1004966

Gagnaire PA, Normandeau E, Bernatchez L (2012) Comparative genomics reveals adaptive protein evolution and a possible cytonuclear incompatibility between European and American eels. Mol Biol Evol 29(10):2909–2919

Galal-Khallaf A, Osman AG, Carleos CE, Garcia-Vazquez E, Borrell YJ (2016) A case study for assessing fish traceability in Egyptian aquafeed formulations using pyrosequencing and metabarcoding. Fish Res 174:143–150

Garcia de la Serrana D, Vieira VL, Andree KB, Darias M, Estévez A, Gisbert E, Johnston IA (2012) Development temperature has persistent effects on muscle growth responses in gilthead sea bream. PLoS One 7(12):e51884

Geldermann H (1975) Investigations on inheritance of quantitative characters in animals by gene markers I. Methods. Theoretical and Applied Genetics, 46(7):319–330

Geng X, Sha J, Liu S, Bao L, Zhang J, Wang R et al (2015) A genome-wide association study in catfish reveals the presence of functional hubs of related genes within QTLs for columnaris disease resistance. BMC Genomics 16:1–12

Gu Y, Zhang L, Chen X (2014) Differential expression analysis of *Paralichthyso livaceus* microRNAs in adult ovary and testis by deep sequencing. Gen Comp Endocrinol 204:181–184

Guo B, DeFaveri J, Sotelo G, Nair A, Merilä J (2015) Population genomic evidence for adaptive differentiation in Baltic Sea three-spined sticklebacks. BMC Biol 13:1–18

Hale MC, Thrower FP, Berntson EA, Miller MR, Nichols KM (2013) Evaluating adaptive divergence between migratory and nonmigratory ecotypes of a salmonid fish, *Oncorhynchus mykiss*. G3 Genes|Genomes|Genetics 3(8):1273–1285

Hecht BC, Thrower FP, Hale MC, Miller MR, Nichols KM (2012) Genetic architecture of migration-related traits in rainbow and steelhead trout, *Oncorhynchus mykiss*. G3 Genes|Genomes|Genetics 2(9):1113–1127

Hecht BC, Campbell NR, Holecek DE, Narum SR (2013) Genome-wide association reveals genetic basis for the propensity to migrate in wild populations of rainbow and steelhead trout. Mol Ecol 22(11):3061–3076

Henning F, Meyer A (2014) The evolutionary genomics of cichlid fishes: explosive speciation and adaptation in the postgenomic era. Annu Rev Genomics Hum Genet 15(1):417–441

Hohenlohe PA, Bassham S, Etter PD, Stiffler N, Johnson EA, Cresko WA (2010) Population genomics of parallel adaptation in threespine stickleback using sequenced RAD tags. PLoS Genet 6(2):e1000862

Howe K, Clark MD, Torroja CF, Torrance J, Berthelot C, Muffato M et al (2013) The zebrafish reference genome sequence and its relationship to the human genome. Nature 496(7446):498–503

Jaillon O, Aury JM, Brunet F, Petit JL, Stange-Thomann N, Mauceli E et al (2004) Genome duplication in the teleost fish *Tetraodon nigroviridis* reveals the early vertebrate proto-karyotype. Nature 431(7011):946–957

Jenny MJ, Aluru N, Hahn ME (2012) Effects of short-term exposure to 2, 3, 7, 8-tetrachlorodibenzo-p-dioxin on microRNA expression in zebrafish embryos. Toxicol Appl Pharmacol 264(2):262–273

Jeukens J, Renaut S, St-Cyr JÉRÔME, Nolte AW, Bernatchez L (2010) The transcriptomics of sympatric dwarf and normal lake whitefish (Coregonus clupeaformis spp., Salmonidae) divergence as revealed by next-generation sequencing. Mol Ecol 19(24):5389–5403

Johnston SE, Orell P, Pritchard VL, Kent MP, Lien S, Niemelä E et al (2014) Genome-wide SNP analysis reveals a genetic basis for sea-age variation in a wild population of Atlantic salmon (*Salmo salar*). Mol Ecol 23(14):3452–3468

Jones, J.C., Fan, S., Franchini, P., Schartl, M. and Meyer, A., (2013) The evolutionary history of X iphophorus fish and their sexually selected sword: a genome-wide approach using restriction site-associated DNA sequencing. Molecular ecology, 22(11):2986–3001

Kai W, Nomura K, Fujiwara A, Nakamura Y, Yasuike M, Ojima N et al (2014) A ddRAD-based genetic map and its integration with the genome assembly of Japanese eel (*Anguilla japonica*) provides insights into genome evolution after the teleost-specific genome duplication. BMC Genomics 15:1–16

Kakioka R, Kokita T, Kumada H, Watanabe K, Okuda N (2013) A RAD-based linkage map and comparative genomics in the gudgeons (genus Gnathopogon, Cyprinidae). BMC Genomics 14:1–11

Kasahara M, Naruse K, Sasaki S, Nakatani Y, Qu W, Ahsan B et al (2007) The medaka draft genome and insights into vertebrate genome evolution. Nature 447(7145):714–719

Kim HS, Lee BY, Won EJ, Han J, Hwang DS, Park HG, Lee JS (2015) Identification of xenobiotic biodegradation and metabolism-related genes in the copepod *Tigriopus japonicus* whole transcriptome analysis. Mar Genomics 24:207–208

Krol J, Loedige I, Filipowicz W (2010) The widespread regulation of microRNA biogenesis, function and decay. Nat Rev Genet 11(9):597–610

Krück NC, Tibbetts IR, Ward RD, Johnson JW, Loh WK, Ovenden JR (2013) Multi-gene barcoding to discriminate sibling species within a morphologically difficult fish genus (Sillago). Fish Res 143:39–46

Lamaze FC, Sauvage C, Marie A, Garant D, Bernatchez L (2012) Dynamics of introgressive hybridization assessed by SNP population genomics of coding genes in stocked brook charr (*Salvelinus fontinalis*). Mol Ecol 21(12):2877–2895

Larson WA, Seeb LW, Everett MV, Waples RK, Templin WD, Seeb JE (2014) Genotyping by sequencing resolves shallow population structure to inform conservation of Chinook salmon (*Oncorhynchus tshawytscha*). Evol Appl 7(3):355–369

Le Cam A, Bobe J, Bouchez O, Cabau C, Kah O, Klopp C et al (2012) Characterization of rainbow trout gonad, brain and gill deep cDNA repertoires using a Roche 454-Titanium sequencing approach. Gene 500(1):32–39

Lenz TL, Eizaguirre C, Rotter B, Kalbe M, Milinski M (2013) Exploring local immunological adaptation of two stickleback ecotypes by experimental infection and transcriptome-wide digital gene expression analysis. Mol Ecol 22(3):774–786

Luo J, Chai J, Wen Y, Tao M, Lin G, Liu X et al (2020) From asymmetrical to balanced genomic diversification during rediploidization: subgenomic evolution in allotetraploid fish. Sci Adv 6(22):eaaz7677

Manousaki T, Hull PM, Kusche H, Machado-Schiaffino G, Franchini P, Harrod C et al (2013) Parsing parallel evolution: ecological divergence and differential gene expression in the adaptive radiations of thick-lipped M idas cichlid fishes from N icaragua. Mol Ecol 22(3):650–669

Martin CH, Cutler JS, Friel JP, Dening Touokong C, Coop G, Wainwright PC (2015) Complex histories of repeated gene flow in Cameroon crater lake cichlids cast doubt on one of the clearest examples of sympatric speciation. Evolution 69(6):1406–1422

Meyer BS, Indermaur A, Ehrensperger X, Egger B, Banyankimbona G, Snoeks J, Salzburger W (2015) Back to Tanganyika: a case of recent trans-species-flock dispersal in East African haplochromine cichlid fishes. R Soc Open Sci 2(3):140498

Micallef G, Bickerdike R, Reiff C, Fernandes JM, Bowman AS, Martin SA (2012) Exploring the transcriptome of Atlantic salmon (*Salmo salar*) skin, a major defense organ. Mar Biotechnol 14:559–569

Miller MR, Brunelli JP, Wheeler PA, Liu S, Rexroad CE III, Palti Y et al (2012) A conserved haplotype controls parallel adaptation in geographically distant salmonid populations. Mol Ecol 21(2):237–249

Montes JM, Technow F, Martin M, Becker K (2014) Genetic diversity in Jatropha curcas L. assessed with SSR and SNP markers. Diversity 6(3):551–566

Nakamura Y, Mori K, Saitoh K, Oshima K, Mekuchi M, Sugaya T et al (2013) Evolutionary changes of multiple visual pigment genes in the complete genome of Pacific bluefin tuna. Proc Natl Acad Sci 110(27):11061–11066

Nishida M. (1991) Lake Tanganyika as an evolutionary reservoir of old lineages of east African cichlid fishes: inferences from allozyme data. Experientia. 47:974–9

Olsvik PA, Berg V, Lyche JL (2013) Transcriptional profiling in burbot (*Lota lota*) from Lake Mjøsa—a Norwegian Lake contaminated by several organic pollutants. Ecotoxicol Environ Saf 92:94–103

Ordas A, Hegedus Z, Henkel CV, Stockhammer OW, Butler D, Jansen HJ et al (2011) Deep sequencing of the innate immune transcriptomic response of zebrafish embryos to Salmonella infection. Fish Shellfish Immunol 31(5):716–724

O'Quin, K.E., Yoshizawa, M., Doshi, P. and Jeffery, W.R., (2013) Quantitative genetic analysis of retinal degeneration in the blind cavefish Astyanax mexicanus. PloS one, 8(2), p.e57281

Pavey SA, Sevellec M, Adam W, Normandeau E, Lamaze FC, Gagnaire PA et al (2013) Nonparallelism in MHCII β diversity accompanies nonparallelism in pathogen infection of lake whitefish (C oregonus clupeaformis) species pairs as revealed by next-generation sequencing. Mol Ecol 22(14):3833–3849

Perez-Sanchez J, Estensoro I, Redondo MJ, Calduch-Giner JA, Kaushik S, Sitja-Bobadilla A (2013) Mucins as diagnostic and prognostic biomarkers in a fish-parasite model: transcriptional and functional analysis. PLoS One 8(6):e65457

Pierron, F., Normandeau, E., Defo, M.A., Campbell, P.G., Bernatchez, L. and Couture, P., (2011) Effects of chronic metal exposure on wild fish populations revealed by high-throughput cDNA sequencing. Ecotoxicology, 20(6):1388–1399

Pujolar JM, Jacobsen MW, Als TD, Frydenberg J, Magnussen E, Jónsson B et al (2014) Assessing patterns of hybridization between North Atlantic eels using diagnostic single-nucleotide polymorphisms. Heredity 112(6):627–637

Quinn TP (2005) Comment: sperm competition in salmon hatcheries the need to institutionalize genetically benign spawning protocols. Trans Am Fish Soc 134(6):1490–1494

Sarropoulou E, Galindo-Villegas J, García-Alcázar A, Kasapidis P, Mulero V (2012) Characterization of European sea bass transcripts by RNA SEQ after oral vaccine against V. anguillarum. Mar Biotechnol 14:634–642

Shao C, Niu Y, Rastas P, Liu Y, Xie Z, Li H et al (2015) Genome-wide SNP identification for the construction of a high-resolution genetic map of Japanese flounder (*Paralichthys olivaceus*): applications to QTL mapping of *Vibrio anguillarum* disease resistance and comparative genomic analysis. DNA Res 22(2):161–170

Sun S, Zhu J, Jiang X, Li B, Ge X (2014) Molecular cloning, tissue distribution and expression analysis of a manganese superoxide dismutase in blunt snout bream *Megalobrama amblycephala*. Fish Shellfish Immunol 38(2):340–347

Tsai HY, Hamilton A, Guy DR, Tinch AE, Bishop SC, Houston RD (2015a) The genetic architecture of growth and fillet traits in farmed Atlantic salmon (*Salmo salar*). BMC Genet 16:1–11

Tsai HY, Hamilton A, Tinch AE, Guy DR, Gharbi K, Stear MJ et al (2015b) Genome wide association and genomic prediction for growth traits in juvenile farmed Atlantic salmon using a high-density SNP array. BMC Genomics 16:1–9

Valente GT, Conte MA, Fantinatti BE, Cabral-de-Mello DC, Carvalho RF, Vicari MR et al (2014) Origin and evolution of B chromosomes in the cichlid fish *Astatotilapia latifasciata* based on integrated genomic analyses. Mol Biol Evol 31(8):2061–2072

van Aerle R, Lange A, Moorhouse A, Paszkiewicz K, Ball K, Johnston BD et al (2013) Molecular mechanisms of toxicity of silver nanoparticles in zebrafish embryos. Environ Sci Technol 47(14):8005–8014

Vieira FA, Thorne MA, Stueber K, Darias M, Reinhardt R, Clark MS et al (2013) Comparative analysis of a teleost skeleton transcriptome provides insight into its regulation. Gen Comp Endocrinol 191:45–58

Wagner, C.E., Keller, I., Wittwer, S., Selz, O.M., Mwaiko, S., Greuter, L., Sivasundar, A. and Seehausen, O., (2013) Genome-wide RAD sequence data provide unprecedented resolution of species boundaries and relationships in the L ake V ictoria cichlid adaptive radiation. Molecular ecology, 22(3):787–798

Wegner KM, Shama LN, Kellnreitner F, Pockberger M (2012) Diversity of immune genes and associated gill microbes of European plaice Pleuronectes platessa. Estuar Coast Shelf Sci 108:87–96

Wei C, Salichos L, Wittgrove CM, Rokas A, Patton JG (2012) Transcriptome-wide analysis of small RNA expression in early zebrafish development. RNA 18(5):915–929

Wongwarangkana C, Fujimori KE, Akiba M, Kinoshita S, Teruya M, Nezuo M et al (2015) Deep sequencing, profiling and detailed annotation of microRNAs in Takifugu rubripes. BMC Genomics 16:1–14

Zhang Y, Xu P, Lu C, Kuang Y, Zhang X, Cao D et al (2011) Genetic linkage mapping and analysis of muscle fiber-related QTLs in common carp (*Cyprinus carpio L.*). Mar Biotechnol 13:376–392

Zhang J, Liu S, Rajendran KV, Sun L, Zhang Y, Sun F et al (2013) Pathogen recognition receptors in channel catfish: III Phylogeny and expression analysis of Toll-like receptors. Dev Comp Immunol 40(2):185–194

Advanced Bioinformatics Techniques for Ecosystem Monitoring

2

Smruti Priyambada Pradhan, Ayushman Gadnayak, Sukanta Kumar Pradhan, Venkatarao Epari, and Subhasmita Behera

2.1 Introduction

The accelerating rates of environmental crises such as habitat loss, pollution, climate change, species extinction, and genetic resource depletion are among the most devastating factors affecting the ecosystem. As the vital functions of the ecosystem are diminished, these issues contribute to enormous economic repercussions (Sparrow et al. 2020). The term "Ecosystem Monitoring" is the systematic collection of data on the environment to track changes in habitat conditions, species populations, and other factors using live creatures. It involves regular analysis and data interpretation to better understand how ecosystems like forests, oceans, and wetlands function. This includes tracing key elements such as the diversity of plant and animal species, population health, water quality, and the condition of soils. This process has become an essential part of ecological management and research due to the growing number of threats to ecosystems caused by factors such as anthropogenic activities, altered habitats, and fluctuation in species distribution and abundance (such as the introduction of invasive species, the extinction of large carnivores,

S. P. Pradhan (✉)
ICAR-Central Inland Fisheries Research Institute, Barrackpore, Kolkata, India

Community Medicine, Siksha 'O' Anusandhan Deemed to be University Institute of Medical Sciences and SUM Hospital, Bhubaneswar, India

A. Gadnayak · S. Behera
ICAR-Central Inland Fisheries Research Institute, Barrackpore, Kolkata, India

S. K. Pradhan
Bioinformatics, Odisha University of Agriculture and Technology, Bhubaneswar, India

V. Epari
Community Medicine, Siksha 'O' Anusandhan Deemed to be University Institute of Medical Sciences and SUM Hospital, Bhubaneswar, India

© The Author(s), under exclusive license to Springer Nature Singapore Pte Ltd. 2026

B. K. Das et al. (eds.), *Bioinformatics Tools for Predictive Ecology and Fisheries*, Sustainable Aquatic Food Systems, https://doi.org/10.1007/978-981-95-2573-7_2

or an excess of herbivores) (Yaccoz et al. 2012). Continuous observation over time enables the detection of variations caused by natural dynamics or human activities, including pollution, deforestation, and climate change. Advances in *Bioinformatics* are transforming the processing and interpretation of ecological data. For instance, metagenomics, metabarcoding, genome sequencing, environmental DNA, now allow researchers to measure gene expression across thousands of species simultaneously, while artificial intelligence (AI) and machine learning (ML), and tools like remote sensing can predict ecosystem shifts caused by anthropogenic activities or climate variability. From the population genomics of endangered species to the microbial communities in degraded soils, these tools are essential for identifying patterns in large datasets. Using the collected data, we can make more informed decisions about environmental management to guarantee that ecosystems can continue to offer vital services to coming generations. Hence, ecosystem monitoring is a vital tool for understanding how ecosystems work, identifying potential threats, and guiding conservation efforts to maintain the balance of our planet's natural systems (Biggs et al. 2015).

2.2 Bioinformatics in Ecosystem Monitoring

Ecosystem monitoring involves time-consuming field surveys and direct observations, which are often limited by geography and the sheer effort required. However, in recent years, advances in bioinformatics have revolutionized the study approach towards ecosystems. By diving into the molecular and genomic details of ecosystems, bioinformatics has introduced more precise, scalable, and noninvasive methods that allow us to understand ecosystem dynamics on a deeper level (Cordier et al., 2021). At its core, bioinformatics blends biology, computer science, and statistics to create powerful tools for analyzing complex biological data, offering new ways to monitor and interpret ecosystems (Vazquez-Baeza et al. 2013). This shift has opened up exciting new possibilities, particularly when it comes to studying the genetic material in the environment itself. One of the most promising breakthroughs has been the use of environmental DNA (eDNA) and metagenomics. These techniques analyze the genetic material found in the environment, whether in water, soil, or air, without the need to observe or capture the species directly. This means that researchers can track biodiversity and ecosystem changes across large areas without disturbing the organisms of that habitat. It also provides a holistic picture of ecosystem health, including species interactions and functions that traditional methods often miss. For example, instead of relying on visual sightings of animals, eDNA can reveal the presence of species that are hard to detect even in remote areas.

These advanced techniques are particularly useful for overcoming some of the major limitations of traditional monitoring methods. Field surveys, while invaluable, can be costly, biased by where and how they are conducted, and limited in their ability to detect all species, especially those that are rare, microscopic, or transient (Díaz et al. 2015). With eDNA and metagenomics, scientists can gather a broader range of data, providing a more accurate understanding of an ecosystem. Therefore,

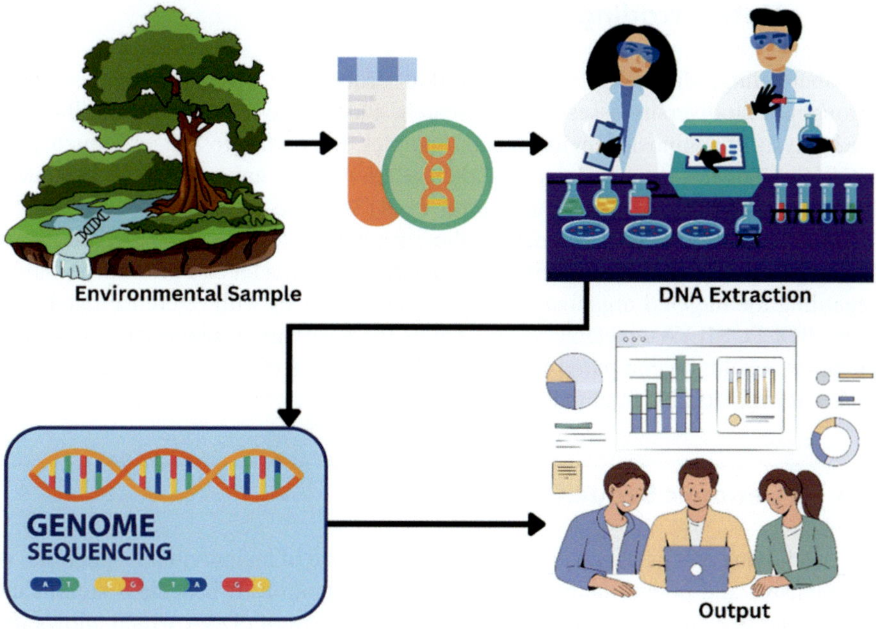

Fig. 2.1 Bioinformatics techniques for ecosystem monitoring by combining DNA sequencing, species tracking, and environmental data analysis

bioinformatics has opened a new era for ecosystem monitoring. By combining these pioneering tools with traditional methods, we're now able to monitor ecosystems in ways that were once unimaginable, offering us a clearer picture of biodiversity and ecological health critical information for protecting our planet's fragile environments (Fig. 2.1).

2.3 Metagenomics

Metagenomics (meta: beyond; beyond the genome study) has developed recently as a hopeful area that examines the intricate genomes held within a microbial niche (Gilbert et al., 2011). Therefore, rather than harvesting live microorganisms from a microbial community to be cultured or viewed in the laboratory, isolating DNA can give intricate details associated to the microbial diversity that resides in a given area and can comprehensively enlist information related to their functions and biological roles. As it makes use of high-throughput sequencing techniques, it offers an affordable and potentially useful way to assess the ecological interactions and biological diversity in complex ecosystems, including microbes, plants, and animals. These advanced bioinformatics protocols have a wide range of applications and benefits across various fields, such as microbial community analysis, environmental health assessment, bioremediation, and biodiversity monitoring (Wang et al. 2013).

2.4 Metabarcoding

Metabarcoding, driven by next-generation sequencing (NGS), provides high-throughput biodiversity assessment, research, and effective biomonitoring through the analysis of composite environmental samples. In contrast to Sanger sequencing, NGS provides fast, low-cost species identification using universal DNA barcodes, but with shorter reads and increased error rates (Stoeck et al., 2024).

Metabarcoding technique enables the simultaneous analysis of several taxonomic groups with the use of universal barcodes without a lot of a prior information regarding the targeted organisms. By aligning sequences with reference databases (e.g., BOLD, SILVA), metabarcoding avoids the necessity of taxonomic skills and enhances resolution compared to traditional approaches (Deiner et al. 2017), which makes it irreplaceable in global biomonitoring.

2.5 Genomic Sequencing

This technique entails deciphering the DNA sequence of an individual's genome, is an intensive and potent method of genomic investigation. It became a significant tool to monitor the ecosystem as it brought the concept of eDNA which allows for examining species diversity and assess environmental stressors causing considerable disturbance to the environment.

2.6 Environmental DNA

The modern science of environmental DNA (eDNA) is an innovative approach in environmental genomics to detect and monitor sequenced nucleic acid data from environmental samples. It acts as a high-tech "fingerprint" of life hidden in air, water, and soil. It is like dipping a test tube in murky water and revealing the entire ecosystem. This method acts like nature's surveillance camera by exposing harmful bacteria in contaminated water, giving details on predator-prey dynamics through DNA traces in feces or water. Sequence-based eDNA data is all about detecting organisms or function-specific genes through presence/absence or relative abundance measurements. For example, it is used to measure nitrogen cycle genes in soils to monitor how land-use changes affect them. Generally, eDNA data has focused on observing specific indicator organisms. By analyzing the presence or absence, or even the relative abundance, of these organisms, researchers can infer the health and function of ecosystems. This valuable data then helps in making informed decisions and taking appropriate actions. While taxonomic observations provide valuable insights, they do not always tell us everything about how an ecosystem functions. However, this approach can be challenging, especially with microorganisms. Microbes can lose genes or gain new ones through processes like horizontal gene transfer, which can change their metabolic capabilities. Also, a single microorganism can have different metabolic abilities depending on its environment (Beale et al. 2022).

With advancements in biotechnology, environmental DNA (eDNA) has become a widely used tool in aquatic research. It helps identify whether specific species are present or absent, estimate their population sizes, and assess the biological and genetic diversity of ecosystems in both freshwater and marine environments. This approach is making it easier to understand and safeguard aquatic biodiversity.

2.7 Advanced Technologies for Next-Generation Ecosystem Monitoring

The field of bioinformatics, which is defined by the use of computational tools to analyze massive biological data, has undergone a radical change in the past decade. Decoding the massive amounts of data produced by modern technology into useful knowledge continues to be one of computational biology's most urgent problems. In this situation, the incorporation of artificial intelligence (AI) plays a pivotal role, acting as a powerful engine for revolutionizing biological research approaches and the onset of a new era of innovation (Fig. 2.2) (Jamialahmadi et al. 2024). Machine learning (ML) has also emerged to drive this transformation, enabling scientists to

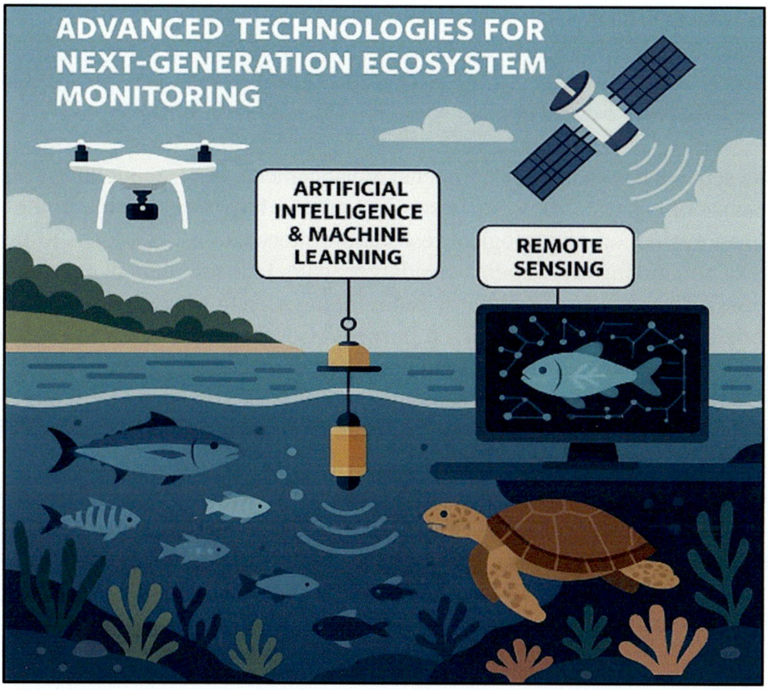

Fig. 2.2 Advanced technologies for next-generation ecosystem monitoring

uncover intricate patterns, predict biological interactions, and accelerate discoveries (Larranaga et al. 2006).

2.8 Artificial Intelligence and Machine Learning

In recent periods, the integration of AI and ML into environmental sciences has led to substantial advancements in environmental monitoring and data processing. By providing real-time monitoring, more accurate forecasts, and the capacity to analyze enormous datasets that are inefficiently handled by conventional techniques, these technologies provide powerful tools to tackle difficult environmental issues (Alotaibi and Nassif 2024).

AI-driven technologies have been used to detect and track environmental challenges such as climate change, deforestation, ocean pollution, and climate change impacts, such as the rise in global temperatures, increased frequency of extreme weather events, and sea level rise.

For example, AI models can process large volumes of data from sources such as satellite imagery, sensor networks, and historical datasets, offering insights that were previously unattainable. Additionally, biodiversity loss is being addressed through AI applications in automated species identification and ecosystem monitoring, providing crucial data for conservation efforts. For instance, machine learning algorithms have been used to predict air pollution levels, identify changes in land use, and model climate change impacts with high precision.

2.9 Remote Sensing

Remote sensing has enabled phenomenal advancements in the modelling, mapping, and science of ecosystems. Remote sensing sensors have a prime role in the monitoring and evaluation of ecological restoration interventions as it leads to the possibility of conducting assessments on a larger scale. In the domain of environmental detection, remote sensing technology is now being used to significant extents around the world. Rapid advancements in remote sensing technology from the 1970s to the 1990s led to an explosion in the use of satellite remote sensing for ecological restoration monitoring. Indicators for ecological restoration remote sensing monitoring overseas have also been developed to evaluate the growth and advancement of ecological restoration initiatives. These cutting-edge instruments are able to image and record information on the Earth's surface remotely from the atmosphere and give scientists and environmental managers timely information regarding ecosystem status and alterations. Whether it is studying the changes in vegetation cover, following up on the recovery process of devastated areas, or evaluating the long-term effects of ecological engineering, remote sensing technology provides a cost-effective, accurate, and non-destructive method (Wang et al. 2024).

2.10 Applications

Biodiversity Monitoring It is possible to discover and monitor biodiversity via the use of metagenomics, which includes species that are both uncommon and impossible to cultivate (Datta et al. 2020). The monitoring of changes in species composition and abundance over time is especially valuable for this since it provides insights into the consequences of environmental changes and actions taken by humans (Techtmann and Hazen 2016).

Biodiversity and Ecosystem Monitoring In order to monitor biodiversity and evaluate the wellness of ecosystems, artificial intelligence methods, especially those requiring picture identification and processing, are used. For instance, the identification and classification of species from camera capture images can be facilitated by automated systems that employ deep learning, which aids in the monitoring of wildlife populations and biodiversity changes over time.

Aquatic Ecosystem Monitoring It is the systematic observation, measurement, and analysis of biological, chemical, and physical components within aquatic environments, such as rivers, lakes, wetlands, and oceans. It aims to assess ecosystem health, detect changes over time, and identify potential threats to ecological integrity. The biological aspect focuses on studying the composition, abundance, and interactions of species, including indicator species, to evaluate ecosystem biodiversity. The chemical aspect involves analyzing water quality parameters such as pH, dissolved oxygen, nutrient concentrations, and pollutant levels, which influence the viability of aquatic life. The physical aspect examines factors like water flow, temperature, and sediment transport, which affect ecosystem processes and stability. By integrating these components, monitoring programs provide data essential for tracking environmental changes, guiding conservation strategies, and informing policy decisions. This scientific approach supports evidence-based management to maintain ecosystem functionality and resilience in the face of natural and anthropogenic pressures.

Microbial Community Analysis Microbial populations in many environments, including soil, water, and air, may be located and studied using metagenomics (Techtmann and Hazen 2016). Insights into the functions of different microorganisms in nutrient cycling, organic matter decomposition, and other ecosystem processes are enhanced by this (Flinkstrom et al. 2024).

Environmental Health Assessment The identification of infections, pollutants, and other contaminants may be accomplished via the use of metagenomics, which involves the examination of the genetic material found in environmental samples. Identifying possible threats to human and environmental health is made easier with the use of this information, which is important for evaluating the state of ecosystems.

Bioremediation The study of the microbial communities that are engaged in the degradation of pollutants and toxins is something that may be done with the help of metagenomics. The use of this information may be utilized to design bioremediation systems that utilize natural microbial processes in order to clean up habitats that have become polluted (Techtmann and Hazen 2016).

Conservation Genetics At the present time, numerous factors are endangering natural populations: climate change, habitat loss, overharvesting, illnesses, and invading species. Ecosystems have declined by around 50% from their pristine levels, and 25% of species are in risk of extinction, according to the latest Intergovernmental Science-Policy Platform on Biodiversity and Ecosystem Services (IPBES). The term "conservation genetics" refers to a multifaceted discipline that employs genetic ideas and techniques in order to describe and enhance the preservation of biodiversity. Evolutionary biology provides a solid theoretical basis for conservation genetics (Kardos 2021).

Laser Radar In order to detect, identify, and measure objects, laser radar technology is a remote sensing method that makes use of the idea of electromagnetic wave reflection at its core. With the rapid advancement of technology in recent times, laser radar has the ability to provide high-resolution images that may be used for the precise identification of plants. Dense data can be gathered with numerous measurement points per square meter because of its strong penetration capabilities. The ability to properly measure distances to different components of vegetation, such as the highest and lowest regions, as well as other structures, enables the production of high-precision three-dimensional models.

Normalized Difference Vegetation Index (NDVI) The Normalized Difference Vegetation Index (NDVI) is a reliable metric based on the absorption and reflection characteristics of plant chlorophyll. Changes and growth in plant diversity may be tracked with its help. When it comes to vegetation monitoring on a worldwide scale, NDVI is a versatile tool with many potential uses. Also, remote sensing data collected from satellites and airplanes offers a wide range of climatic and geographical variables because of its enormous coverage.

Hyperspectral Remote Sensing By monitoring the reflection or radiation of objects throughout a wide range of tiny spectral bands, hyperspectral remote sensing technology is a kind of remote sensing that was developed to collect information. A number of water quality indicators, such as chlorophyll, turbidity, dissolved organic matter, and pollution concentrations, may be detected with great accuracy using hyperspectral remote sensing. It helps with comprehending the dynamics of waterbodies by differentiating between freshwater, saltwater, algal blooms, and sediments, among other kinds of water.

Satellite Remote Sensing The technique known as satellite remote sensing uses sensors on board satellites to gather data about the Earth's surface. By processing and interpreting the electromagnetic radiation that satellite sensors gather from the Earth, it is able to provide data on the environment and surface characteristics. Numerous disciplines, including geology, meteorology, environmental monitoring, and resource management, make substantial use of this technique. Large and small bodies of water can both be monitored using satellite remote sensing, which can cover wide geographical regions. Time-series data on water bodies may be obtained by satellites taking frequent pictures of the same region. For tracking seasonal variations, water level swings, climate effects, and other things, this is quite beneficial.

Synthetic Aperture Radar (SAR) Using microwave frequencies and the reflected signals, the synthetic aperture radar (SAR) technology generates pictures. Similar to LIDAR technology, SAR images are two-dimensional and may be captured in a range of weather and time conditions. Research on soil moisture retrieval often uses Synthetic Aperture Radar because of its high resolution, polarization capabilities, all-weather and all-time observation skills, and significant penetration abilities, particularly its sensitivity to soil moisture content.

Multispectral Remote Sensing To collect a variety of data about the Earth's surface, multispectral remote sensing equipment usually uses many distinct spectral bands, often in the visible and near-infrared spectral range. Though the number of bands is generally small, often ranging from 3 to 15 bands in standard experimental settings, these bands usually span a wide range. Basic information on land cover and land use classification is provided by multispectral images. Multispectral remote sensing technology, for example, is often used in investigations pertaining to soil moisture retrieval in bare soil because of its affordability. Additionally, this technology meets a variety of critical demands by streamlining processes, lowering effort, extending the scope of studies, and enabling quick detection. The limits of conventional detection techniques are addressed by multispectral remote sensing technology, which also shows encouraging development trends (Jiao 2024).

Air and Water Quality Monitoring The utilization of AI and ML methodologies is employed to attain the prediction and monitoring of air and water quality levels. For instance, machine learning models can predict future pollution levels by examining historical air pollution data and meteorological variables, as well as water quality parameters such as pH, dissolved oxygen, and contaminant levels. These models are capable of processing data from remote sensing technologies and sensor networks, thereby enabling the provision of real-time assessments that are indispensable for resource management.

Disaster Management Artificial intelligence and machine learning are very important in the field of disaster management and response. Predictive models have the ability to foresee disasters such as hurricanes, floods, and wildfires, which enables immediate actions and the allocation of resources. When it comes to assessing the effect of disasters and coordinating response operations, artificial intelligence may also aid in the analysis of data from social media platforms.

2.11 Conclusion and Future Prospects

Ecosystem monitoring has been transformed by the combination of bioinformatics, artificial intelligence, and remote sensing technologies, which allow for previously unheard of levels of precision in monitoring biodiversity, evaluating the condition of the environment, and forecasting ecological changes. By overcoming conventional constraints, methods such as metagenomics, eDNA analysis, and machine learning have provided scalable, noninvasive answers to conservation problems. In order to tackle climate change and biodiversity loss, the field needs to concentrate on standardizing data protocols, improving computational tools for multiomics integration, and encouraging international cooperation. By combining state-of-the-art bioinformatics with real-time sensor networks and predictive modelling, scientists can gain a better understanding of ecosystem dynamics, which will ultimately help them make sustainable management and policy choices for a resilient planet in the future.

References

Alotaibi E, Nassif N (2024) Artificial intelligence in environmental monitoring: in-depth analysis. Discov Artif Intell 4(1):84

Beale DJ, Jones OA, Bose U, Broadbent JA, Walsh TK, van de Kamp J, Bissett A (2022) Omics-based ecosurveillance for the assessment of ecosystem function, health, and resilience. Emerg Top Life Sci 6(2):185–199

Biggs J, Ewald N, Valentini A, Gaboriaud C, Dejean T, Griffiths RA et al (2015) Using eDNA to develop a national citizen science-based monitoring programme for the great crested newt (Triturus cristatus). Biol Conserv 183:19–28

Cordier T, Alonso-Sáez L, Apothéloz-Perret-Gentil L, Aylagas E, Bohan DA, Bouchez A & Lanzén A et al (2021) Ecosystems monitoring powered by environmental genomics: a review of current strategies with an implementation roadmap. Mol Ecol 30(13):2937–2958

Díaz S, Demissew S, Carabias J, Joly C, Lonsdale M, Ash N et al (2015) The IPBES conceptual framework—connecting nature and people. Curr Opin Environ Sustain 14:1–16

Deiner K, Bik HM, Mächler E, Seymour M, Lacoursière-Roussel A, Altermatt F & Bernatchez L et al (2017) Environmental DNA metabarcoding: Transforming how we survey animal and plant communities. Mol Ecol 26(21): 5872–5895

Eckert IM, Littlefair JE, Zhang GK, Chain FJ, Crease TJ, Cristescu ME (2018) Bioinformatics for biomonitoring: species detection and diversity estimates across next-generation sequencing platforms. In: Advances in ecological research, vol 59. Academic Press, New York, pp 1–32

Flinkstrom Z, Bryson S, Candry P, Winkler MKH (2024) Metagenomic clustering links specific metabolic functions to globally relevant ecosystems. MSystems 9(8):e00573–e00524

Gilbert JA, Dupont CL et al (2011) Microbial metagenomics: beyond the genome. Annu Rev Mar Sci 3(1):347–371

Jamialahmadi H, Khalili-Tanha G, Nazari E, Rezaei-Tavirani M (2024) Artificial intelligence and bioinformatics: a journey from traditional techniques to smart approaches. Gastroenterol Hepatol Bed Bench 17(3):241

Jiao Z (2024) The application of remote sensing techniques in ecological environment monitoring. Highl Sci Eng Technol 81:449–455

Kardos M (2021) Conservation genetics. Curr Biol 31(19):R1185–R1190

Larranaga P, Calvo B, Santana R, Bielza C, Galdiano J, Inza I et al (2006) Machine learning in bioinformatics. Brief Bioinform 7(1):86–112

Li J, Pei Y, Zhao S, Xiao R, Sang X, Zhang C (2020) A review of remote sensing for environmental monitoring in China. Remote Sens 12(7):1130

National Research Council, Board on Science, & Technology for International Development (1992) Conserving biodiversity: a research agenda for development agencies. National Academies Press, Washington, DC

Saptashwa D, Narayanan RK, Samuel MS, Arivalagan P, Ethiraj S (2020) Metagenomic applications in microbial diversity, bioremediation, pollution monitoring, enzyme and drug discovery. A review. Environ Chem Lett 18(4):1229–1241

Sparrow BD, Edwards W, Munroe SE, Wardle GM, Guerin GR, Bastin JF et al (2020) Effective ecosystem monitoring requires a multi-scaled approach. Biol Rev 95(6):1706–1719

Stoeck T, Katzenmeier SN, Breiner HW, Rubel V et al (2024) Nanopore duplex sequencing as an alternative to Illumina MiSeq sequencing for eDNA-based biomonitoring of coastal aquaculture impacts. Metabarcoding and Metagenomics 8: e121817

Techtmann SM, Hazen TC (2016) Metagenomic applications in environmental monitoring and bioremediation. J Ind Microbiol Biotechnol 43(10):1345–1354

Vázquez-Baeza Y, Pirrung M, Gonzalez A, Knight R et al (2013) EMPeror: a tool for visualizing high-throughput microbial community data. Gigascience 2(1):2047–217X

Wang J, Mc Lenachan PA, Biggs PJ, Winder LH, Schoenfeld BI, Narayan VV et al (2013) Environmental bio-monitoring with high-throughput sequencing. Brief Bioinform 14(5):575–588

Wang R, Sun Y, Zong J, Wang Y, Cao X, Wang Y et al (2024) Remote sensing application in ecological restoration monitoring: a systematic review. Remote Sens 16(12):2204

Yoccoz NG et al (2012) The future of environmental DNA in ecology. Mol Ecol 21(8):2031–2038

Machine Learning Application in Aquatic Health Prediction

3

Babita Pruseth, Subhashree Subhasmita Raut, Soumya Prasad Panda, and Basanta Kumar Das

3.1 Introduction

Aquatic ecosystems are among the most valuable and ecologically diverse components of the Earth's biosphere (Irfan and Alatawi 2019; Wanjari et al. 2025). They provide essential ecosystem services such as regulating climate, supporting biodiversity, ensuring freshwater availability, and sustaining the livelihoods of millions worldwide. From coral reefs and coastal wetlands to rivers and deep-sea habitats, these ecosystems are intricately connected and provide vital resources for human communities, from fisheries to tourism. However, these critical systems face unprecedented threats due to anthropogenic pressures such as climate change, pollution, habitat destruction, and overexploitation (Mukherjee et al. 2023). The degradation of aquatic ecosystems is not only a direct threat to biodiversity but also undermines the resources on which humans depend, leading to food security risks, the loss of livelihoods, and the collapse of local economies (Adla et al. 2022).

As global environmental challenges intensify, the need for innovative, scalable, and efficient monitoring solutions has never been more significant. Traditional methods of assessing aquatic health have predominantly relied on manual sampling, laboratory analyses, and time-intensive fieldwork. While effective in specific contexts, these conventional approaches are limited in their ability to address the dynamic and complex nature of aquatic ecosystems. They cannot often scale across large, remote, or poorly monitored areas and fall short in providing real-time, continuous data-critical features needed to make timely and informed decisions in managing aquatic health (Chen et al. 2024). For instance, manual water sampling can miss rapid changes in water quality, such as harmful algal blooms, or overlook

B. Pruseth · S. S. Raut · S. P. Panda · B. K. Das (✉)
ICAR-Central Inland Fisheries Research Institute, Barrackpore, Kolkata, India

© The Author(s), under exclusive license to Springer Nature Singapore Pte Ltd. 2026

B. K. Das et al. (eds.), *Bioinformatics Tools for Predictive Ecology and Fisheries*, Sustainable Aquatic Food Systems, https://doi.org/10.1007/978-981-95-2573-7_3

subtle trends in biodiversity loss that may not be immediately visible through conventional survey techniques.

In response to these challenges, machine learning (ML) has emerged as a groundbreaking tool in environmental and aquatic health research. ML offers a transformative alternative to traditional monitoring and analysis methods by harnessing the power of big data and computational algorithms. Unlike static statistical models or predefined rules, ML algorithms are dynamic and adaptive, capable of learning from vast and complex datasets to detect patterns, make predictions, and optimize decision-making processes. These algorithms excel in managing the voluminous and diverse data that aquatic ecosystems generate from satellite imagery and remote sensing data to sensor networks and ecological surveys. ML models can detect subtle changes in environmental variables that might be imperceptible to human observers and provide insights into long-term trends, making them indispensable for real-time and predictive monitoring. This chapter delves into integrating machine learning techniques in aquatic health management, exploring their applications, the challenges of adopting these technologies, and their future potential. It examines the role of smart technologies in aquaculture, the use of predictive models for water quality and ecosystem dynamics, and how ML can enhance environmental sustainability and ecosystem resilience. Through this exploration, we aim to highlight how ML is revolutionizing aquatic health monitoring and paving the way for more sustainable and adaptive strategies to safeguard the future of aquatic ecosystems worldwide.

3.2 Role of Machine Learning in Aquatic Health

Given the critical importance of aquatic ecosystems and the challenges in monitoring them, integrating advanced technologies like machine learning has become essential. This section explores how ML transforms aquatic health management by overcoming traditional monitoring limitations and enabling more efficient, real-time, and large-scale analysis.

3.2.1 How Machine Learning Is Transforming Environmental Science in Aquatic Health

Machine learning (ML) is revolutionizing the way we understand and manage environmental systems, particularly in the context of aquatic ecosystems. Traditional methods of monitoring aquatic health, such as manual sampling, lab analysis, and fieldwork, have limitations in scalability, speed, and the ability to process large volumes of data. Machine learning, with its ability to analyze vast and complex datasets, can provide previously difficult or impossible insights.ML algorithms enable researchers and environmental scientists to process data from diverse sources, such as satellite imagery, sensor networks, and ecological surveys. By learning from this data, ML models can identify patterns, predict future trends, and provide real-time insights into the state of aquatic ecosystems. This capability is particularly important in the face of rapid environmental changes, such as climate change and pollution, where traditional methods may fall short. For instance, satellite data and sensor networks can produce massive amounts of water quality, temperature, and ecosystem health data. ML algorithms, especially deep learning models, can analyze this data more efficiently, detecting subtle patterns and making accurate predictions without human intervention. This allows for more proactive and informed decision-making when managing aquatic resources and protecting biodiversity (Li et al. 2021).

3.2.2 The Challenges of Traditional Methods and How ML Addresses Them

Conventional methods for monitoring aquatic ecosystems include physical sampling (collecting water samples), lab analysis (testing for pollutants, bacteria, or other contaminants), and field surveys (measuring environmental variables like temperature, salinity, and pH). While these methods have been valuable, they come with several challenges:

Scalability Aquatic ecosystems cover vast areas, including oceans, rivers, lakes, and wetlands, which are often difficult and expensive to monitor comprehensively using traditional techniques. Collecting samples at a large scale is resource-intensive and time-consuming. For example, field surveys to monitor biodiversity or pollutants across large regions can take months or even years. In contrast, ML can analyze data from satellite imagery or sensor networks, providing a broader, more comprehensive view of aquatic health with minimal human intervention (Deng et al. 2024). For example, instead of manually sampling water quality across an entire river system, ML can analyze data from a network of sensors placed throughout the system, providing continuous, real-time updates.

Real-Time Monitoring Traditional methods struggle to provide real-time insights, especially in remote or inaccessible areas. For example, detecting and responding to a harmful algal bloom or pollution event requires rapid data collection and analysis. ML models, however, can analyze data continuously in real-time, offering immediate insights and predictions (Chen et al. 2023). This capability enables quicker response times to environmental threats, such as identifying pollution hotspots or detecting early signs of ecosystem degradation.

Complexity of Aquatic Ecosystems Aquatic ecosystems are dynamic and influenced by numerous variables, such as water temperature, pH, salinity, nutrient levels, pollutants, and biological factors like species interactions. Traditional methods may be too simplistic to capture the complexity of these interactions. ML, on the other hand, excels at handling complex, nonlinear relationships within large datasets. By recognizing hidden patterns and correlations in the data, ML can provide a more holistic and accurate understanding of aquatic health. For example, machine learning models have been used to predict the onset of coral bleaching based on multiple environmental factors, helping conservationists take proactive measures to protect vulnerable reefs.

Data Overload The volume of data collected from sensors, remote sensing technologies, and ecological surveys is enormous. Manually processing this data can be overwhelming and time-consuming. ML automates data analysis, reducing the burden on scientists and enabling them to focus on interpreting results and making decisions. Additionally, ML can handle unstructured data, such as images, sensor data, or audio recordings, which traditional methods might struggle with (Brown et al. 2024). Machine learning is rapidly transforming the field of aquatic health management by overcoming the limitations of traditional methods. Its ability to handle large datasets, provide real-time analysis, and understand complex relationships within aquatic ecosystems makes it an invaluable tool in addressing environmental challenges. As the volume and complexity of environmental data continue to grow, the role of ML in aquatic health will only become more critical in ensuring the sustainability and resilience of our planet's aquatic ecosystems.

3.3 Different Types of Machine Learning for Aquatic Health

Machine learning (ML) is a subset of artificial intelligence (AI), and its applications are revolutionizing how researchers approach ecological data. The three primary types of ML techniques used in aquatic health and environmental monitoring include the following:

Supervised Learning Supervised learning is employed when the model is trained on labeled data, where both inputs and outcomes are known (Zhou 2018). This method is particularly useful for predictive tasks, such as forecasting water quality based on sensor data or predicting the presence of certain species based on environmental parameters (Arrighi and Castelli 2023). Supervised learning models can predict indicators like water quality indices or oxygen levels in the context of aquatic health, assisting in the early detection of ecosystem stressors.

Unsupervised Learning In unsupervised learning, the algorithm works with unlabeled data to identify hidden patterns and groupings within the dataset (Naeem et al. 2023). This technique is essential when researchers do not have predefined outcomes but seek to discover relationships or anomalies in the data. For example, unsupervised learning can help identify areas of a water body experiencing elevated levels of pollutants based on chemical or biological profiles.

Reinforcement Learning Reinforcement learning is focused on decision-making and optimal strategies (Li 2023). In this approach, an algorithm learns to make decisions based on rewards and penalties from actions taken. It is particularly useful in optimizing real-time management strategies, such as controlling water pollution levels or managing aquaculture systems for disease prevention. By continuously learning from feedback, reinforcement learning can enhance environmental management practices. The details of different ML types, the algorithms used, their approaches, applications in aquatic science, and their advantages and disadvantages are presented in Table 3.1.

Workflow

Creating a machine learning workflow involves several essential steps to ensure a structured approach. Begin by defining the problem clearly, identifying objectives, and determining whether it's a classification, regression, or clustering task. Next, the data will be collected and preprocessed by cleaning missing values, outliers, duplicates, or inconsistent values, performing exploratory data analysis (EDA) to uncover patterns and relationships, and scaling features. Feature engineering is critical to selecting relevant variables or creating new ones that enhance model performance. Choose suitable algorithms based on the problem, train the models, and evaluate their performance using accuracy, precision, recall, or F1-score metrics. Simplify the process using tools like Scikit-learn for pipelines, TensorFlow for deep learning, or ML flow for workflow automation. The workflow is shown in Fig. 3.1.

3.4 Machine Learning Applications in Aquatic Health Prediction

Aquatic health prediction is essential for monitoring the quality of water bodies like rivers, lakes, and oceans. Machine learning (ML) offers a powerful alternative using data-driven models to analyze large datasets, detect patterns, and predict water quality issues in real time. Below, we explore how ML is applied in different areas of aquatic health prediction, shown in Fig. 3.2.

3.4.1 Water Quality Monitoring and Prediction

Water quality monitoring is essential for assessing the health of aquatic ecosystems. ML models help predict key parameters like temperature, pH, turbidity, dissolved oxygen, and pollutant levels by analyzing real-time sensor data and historical trends. Regression models forecast temperature fluctuations and pH changes based on environmental factors, while neural networks analyze complex interactions between multiple water quality parameters. Time-series forecasting predicts future pollution levels, enabling early intervention. For example, ML models can process satellite and sensor data to anticipate toxic algal blooms before they occur, allowing authorities to take preventive action.

Table 3.1 Machine learning types and algorithms in aquatic science

Machine learning type	Algorithm	Approach	Applications in aquatic science	Advantages	Disadvantages
Supervised learning	Random forest	Ensemble learning	Predicting water quality levels, detecting marine pollution, and classifying ecosystems	Robust to overfitting, handles missing data well	Computationally expensive
	Support vector machine (SVM)	Classification	Classifying marine pollutants, identifying marine diseases, species identification	High accuracy, especially with high-dimensional data	Requires careful tuning of hyperparameters
	K-nearest neighbors (KNN)	Classification/regression	Monitoring fish health, predicting species distribution, estimating water quality parameters	Simple to implement, intuitive	Computationally expensive with large datasets
	Decision trees	Classification/regression	Classifying water conditions and predicting aquatic health based on environmental factors	Easy to interpret, suitable for rule-based decisions	Prone to overfitting if not pruned
Unsupervised learning	K-Means	Clustering	Identifying patterns in water quality, classifying marine ecosystems, detecting anomalies	No need for labeled data, it is good for exploratory analysis	Difficult to interpret results
Reinforcement learning	Q-learning, deep Q-networks	Decision making	Optimizing fishery management, controlling environmental variables to improve aquatic health	Can adapt over time, suitable for dynamic systems	Requires a large number of simulations, complex setup
Deep learning	Neural networks	Neural networks	Analyzing satellite imagery for coral reef health, detecting marine animal behavior from video data	High accuracy with complex data, feature extraction	Requires large datasets, computationally intensive

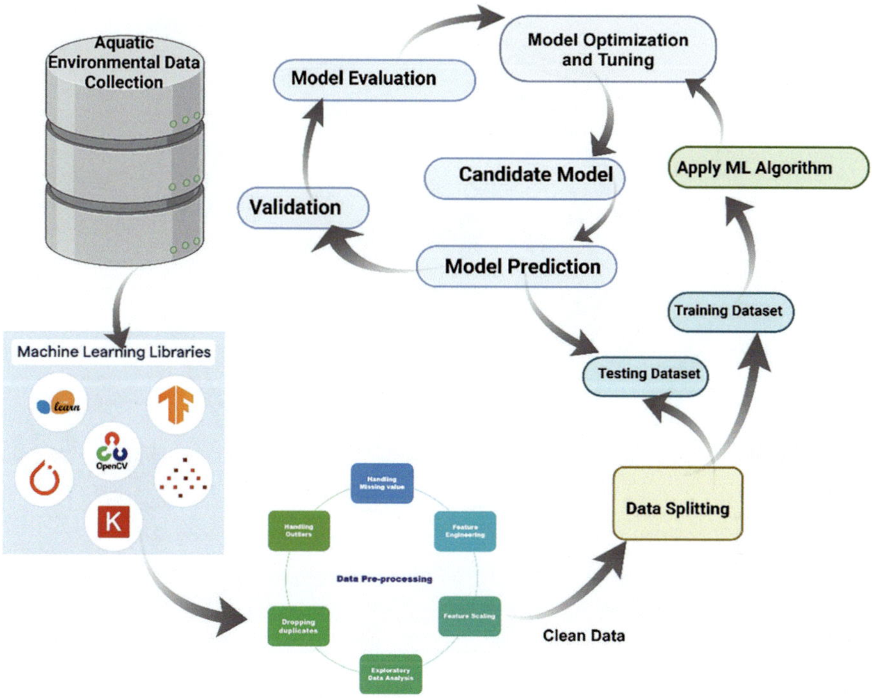

Fig. 3.1 Machine learning workflow for aquatic health prediction

3.4.2 Detection of Water Pollution Sources

Water pollution originates from various sources, such as industrial waste, agricultural runoff, sewage discharge, and microplastics. Identifying these sources manually is challenging, but ML improves efficiency by automating detection processes. Classification algorithms help determine whether industrial discharge, agricultural activities, or urban waste cause pollution. Anomaly detection models identify sudden pollution spikes, while remote sensing techniques analyze satellite images to detect oil spills, chemical leaks, or excessive sedimentation. For example, ML-powered image processing can pinpoint pollution hotspots in rivers and coastal waters by detecting color changes in satellite imagery.

3.4.3 Early Warning Systems for Aquatic Life Protection

Aquatic organisms are highly sensitive to water quality changes. ML plays a crucial role in predicting environmental risks before they become severe. Deep learning models analyze underwater images to assess fish health and population dynamics,

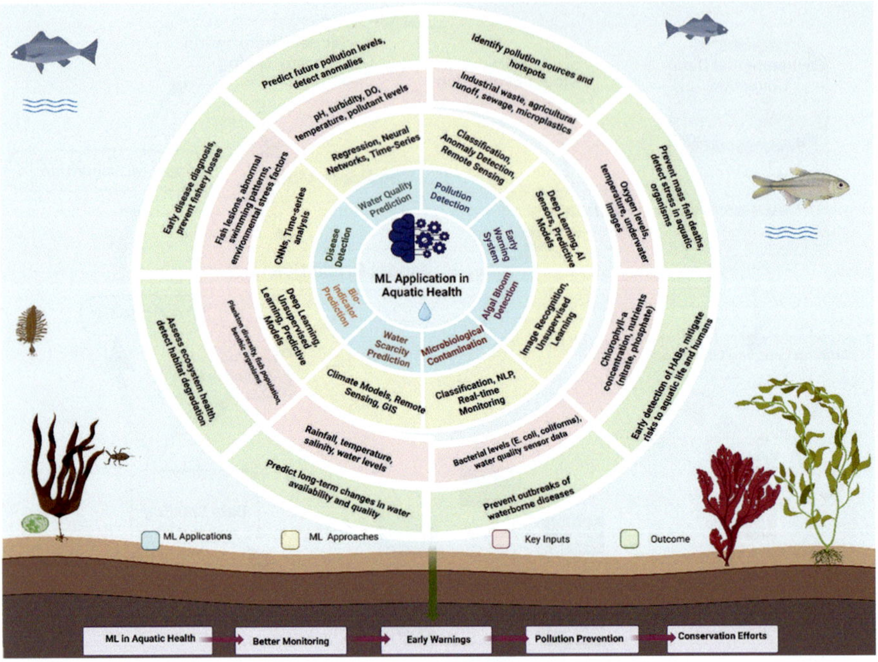

Fig. 3.2 Machine learning applications in aquatic health prediction

while AI-driven sensors monitor oxygen levels and send alerts if conditions become hazardous for marine life. Predictive analytics also help forecast mass fish mortality events caused by oxygen depletion or toxic pollutants. For example, an AI-based system can analyze water temperature and dissolved oxygen levels to predict fish kill events in lakes and reservoirs.

3.4.4 Identifying Harmful Algal Blooms (HABs)

Harmful algal blooms (HABs) occur when excess nutrients, such as nitrates and phosphates, lead to the rapid growth of toxic algae. These blooms can severely impact aquatic ecosystems and human health. ML models help detect and predict HABs by analyzing satellite images, water quality data, and environmental factors. Image recognition algorithms assess chlorophyll-a concentrations, while supervised learning models predict HAB outbreaks based on nutrient levels and temperature variations. Unsupervised clustering identifies patterns in water quality changes that signal bloom formation. For instance, ML models trained on historical satellite data and weather conditions can predict when and where a harmful algal bloom is likely to occur.

3.4.5 Microbiological Contamination Prediction

Microbiological contamination from bacteria such as *E. coli* and coliforms poses serious public health risks. Traditional water testing methods are slow, but ML enables faster detection and prediction. Classification models assess whether water is safe or contaminated based on real-time sensor readings. At the same time, NLP (natural language processing) techniques analyze online reports and social media to track waterborne disease outbreaks. ML-based monitoring systems also predict bacterial growth patterns in drinking water sources. For example, deep learning models trained on water quality sensor data can forecast *E. coli* outbreaks before contamination spreads, allowing timely intervention.

3.4.6 Predicting Bioindicators of Aquatic Ecosystems

Bioindicators, including fish species, plankton diversity, microbial communities, and aquatic plants, provide valuable insights into the health of water bodies. ML models help predict bioindicator presence and abundance based on environmental parameters such as nutrient levels, pH, dissolved oxygen, and temperature. Deep learning techniques analyze underwater images to monitor biodiversity, while unsupervised clustering identifies species tolerance to pollution. Predictive models assess ecosystem stability under climate change scenarios. For example, ML models trained on historical biodiversity data can forecast declining fish populations due to increasing pollution levels, enabling conservation efforts.

3.4.7 Predicting Water Scarcity and Climate Change Impact

Climate change alters aquatic ecosystems by affecting rainfall patterns, temperature, and salinity levels. ML models help predict long-term changes in water availability and quality. Climate forecasting models use ML to predict droughts, floods, and seasonal water quality variations, while remote sensing techniques track changes in water levels, ice cover, and sedimentation over time. AI-powered GIS (geographic information systems) analyzes land use patterns to evaluate human impacts on water resources. For example, AI-driven climate models can predict freshwater shortages in certain regions based on temperature and precipitation trends, aiding in water resource management and policy planning.

3.4.8 Disease Detection in Aquatic Organisms

Aquatic organisms are vulnerable to various diseases caused by bacteria, viruses, parasites, and environmental stressors. Early disease detection is vital for maintaining healthy ecosystems and preventing large-scale outbreaks. ML models assist in diagnosing diseases by analyzing images, sensor data, and behavioral patterns.

Convolutional neural networks (CNNs) can identify symptoms of infections in fish through automated image analysis, while time-series models detect abnormal swimming behavior or respiratory stress. AI-powered monitoring systems analyze water parameters to predict conditions conducive to disease outbreaks. For example, ML models trained on fish health data can predict viral infections in aquaculture farms, allowing preventive measures to be taken before significant losses occur.

3.5 Challenges

Despite its transformative potential, the application of machine learning in aquatic health faces several critical challenges. One of the primary obstacles is data availability, as high-quality, labeled datasets are often scarce, particularly in remote or under-monitored aquatic regions. This limitation hinders the development and validation of accurate models. Additionally, the computational costs associated with processing large datasets, especially for deep learning algorithms, can be prohibitive, requiring substantial computational power and storage (L'heureux et al. 2017). Another significant issue is model interpretability, where complex models, such as deep neural networks, operate as "black boxes," making it difficult for stakeholders and decision-makers to understand or trust their predictions. The lack of transparency in model decision-making processes can hinder the adoption of these technologies in critical real-world applications, such as water quality monitoring or species conservation (Taherdoost 2023). Furthermore, ethical considerations arise due to potential biases in data collection methods, which may lead to the underrepresentation of certain ecosystems or species, ultimately impacting the fairness and inclusivity of conservation efforts (Pagano et al. 2022). Biases in training data may lead to skewed predictions, which could disproportionately affect regions or species that are already underrepresented in the datasets. Addressing these challenges is essential to maximizing the effectiveness and reliability of machine learning in safeguarding aquatic ecosystems.

3.6 Interdisciplinary Collaboration in Aquatic Health ML Projects

Machine learning (ML) in aquatic health is not a standalone effort; it thrives on interdisciplinary collaboration among experts from diverse fields. Such collaboration ensures that ML models are both technically sound and environmentally relevant. Ecologists and marine biologists bring critical domain expertise on species behavior, water quality, and ecosystem dynamics, providing the foundation for ecological validity. Data scientists contribute by developing and refining ML models through data preprocessing, feature engineering, and algorithm optimization, ensuring the models are robust and applicable to real-world challenges. Policymakers and regulators play a vital role in translating ML-driven insights into actionable strategies, enabling informed decision-making for conservation and sustainable resource management. By integrating these diverse perspectives, ML projects can effectively address complex challenges in aquatic health, ensuring accurate, impactful, and socially responsible solutions.

3.7 Conclusion and Future Perspective

Machine learning is transforming aquatic health prediction by making it more accurate, efficient, and scalable. It helps forecast water quality, detect pollution, and conserve biodiversity in real time. However, challenges like data quality and ethical concerns must be fully addressed to maximize its potential. Adopting ML responsibly can improve aquatic health monitoring and create a sustainable future. The future of aquatic ecosystem management will undoubtedly be shaped by the continued advancements in machine learning and the integration of cutting-edge technologies like explainable AI, federated learning, genomic studies, and real-time monitoring. By leveraging these emerging tools, scientists, policymakers, and practitioners can move toward a more data-driven, efficient, and adaptable approach to managing aquatic health. Explainable AI will allow for greater transparency in decision-making, fostering trust and improving the interpretability of ML models used in aquatic health management. Federated learning could enable the secure sharing of data across institutions without compromising privacy, further enhancing collaboration in ecosystem monitoring. The application of genomic studies will deepen our understanding of species behavior and resilience in the face of environmental stressors, while real-time monitoring technologies will provide continuous, actionable insights into the health of aquatic systems. These technologies not only have the potential to address current challenges but also open up new opportunities for proactive and sustainable management practices, ultimately ensuring the health and resilience of the world's aquatic ecosystems for generations to come.

> The future of life on Earth depends on our ability to work with nature and apply science and technology in new ways to preserve it.—David Suzuki.

References

Adla K, Dejan K, Neira D, Dragana Š (2022) Degradation of ecosystems and loss of ecosystem services. In: One health. Academic Press, pp 281–327. https://doi.org/10.1016/B978-0-12-822794-7.00008-3

Arrighi C, Castelli F (2023) Prediction of ecological status of surface water bodies with supervised machine learning classifiers. Sci Total Environ 857:159655

Brown, K.E., Yan, C., Li, Z., Zhang, X., Collins, B.X., Chen, Y., Clayton, E.W., Kantarcioglu, M., Vorobeychik, Y. and Malin, B.A., (2024) Not the models you are looking for: Traditional ML outperforms LLMs in clinical prediction tasks. medRxiv

Chen W, Milosevic Z, Rabhi FA, Berry A (2023) Real-time analytics: concepts, architectures, and ML/AI considerations. IEEE Access 11:71634–71657

Chen W, Xu D, Pan B, Zhao Y, Song Y (2024) Machine learning-based water quality classification assessment. Water 16(20):2951

Deng Y, Zhang Y, Pan D, Yang SX, Gharabaghi B (2024) Review of recent advances in remote sensing and machine learning methods for lake water quality management. Remote Sens 16(22):4196

Irfan, S. and Alatawi, A.M.M., (2019) Aquatic ecosystem and biodiversity: a review. Open Journal of Ecology, 9(1), pp.1–13

Li, F., Zhang, Y., Altermatt, F. and Zhang, X., (2021) Consideration of multitrophic biodiversity and ecosystem functions improves indices on river ecological status. Environmental science & technology, 55(24), pp.16434–16444

L'heureux A, Grolinger K, Elyamany HF, Capretz MA (2017) Machine learning with big data: challenges and approaches. IEEE Access 5:7776–7797

Li SE (2023) Reinforcement learning for sequential decision and optimal control. Springer Nature, Singapore

Mukherjee S, Rizvi SS, Biswas G, Paswan AK, Vaiphei SP, Warsi T, Mitran T (2023) Aquatic ecosystems under influence of climate change and anthropogenic activities: potential threats and its mitigation strategies. In: Hydrogeochemistry of aquatic ecosystems, pp 307–331. https://doi.org/10.1002/9781119870562.ch14

Naeem S, Ali A, Anam S, Ahmed MM (2023) An unsupervised machine learning algorithm: comprehensive review. Int J Comput Digit Syst 13(1):911–921

Pagano TP, Loureiro RB, Lisboa FVN, Cruz GOR, Peixoto RM, Guimarães GADS et al (2022) Bias and unfairness in machine learning models: a systematic literature review. arXiv preprint arXiv:2202.08176

Taherdoost H (2023) Deep learning and neural networks: decision-making implications. Symmetry 15(9):1723

Wanjari RN, Hamid I, Abass Z, Magloo AH, Ahmad I (2025) Ecosystem services of aquatic biodiversity for food production and livelihoods. In: Food security, nutrition and sustainability through aquaculture technologies. Springer Nature Switzerland, Cham, pp 207–225

Zhou ZH (2018) A brief introduction to weakly supervised learning. Natl Sci Rev 5(1):44–53

Comparative Genomics for Aquaculture Improvement

4

Sanjeev Kumar Singh, Nikita Sahu, Anand Kumar, Basanta Kumar Das, and Mukunda Goswami

4.1 Introduction

Presently, aquaculture is the fastest-growing sector in food production worldwide, taking care of food security challenges arising from population growth, climate change, and resource constraints. Although the global human population could grow to 9.7 billion by 2050, it is projected to peak at around 10.4 billion people in the 2080s and remain at that level until 2100 (Ngo 2023). It has been a challenge to feed the expanding population, so food production needs to rise significantly from current levels, with an increase of up to 70 percent anticipated by the year 2050. However, this high demand for fish protein has put a significant strain on wild fish populations because of near- or above-sustainable harvest exploitation. Aquaculture promises to fill this gap as a fast-growing contributor to the world's food supply, with global consumption expected to reach 25.5 kg per capita by 2050 (FAO 2022). The world consumption of aquaculture products and fisheries will continue from the 1990s to 2022, with a growth that has doubled from 81.6 to 164.6 million metric tons per year (Zhu et al. 2024). Scientific breakthroughs and technological

S. K. Singh · N. Sahu
ICAR-Central Institute of Fisheries Education, Mumbai, India

ICAR-Central Inland Fisheries Research Institute, Barrackpore, Kolkata, India

A. Kumar
ICAR-Central Inland Fisheries Research Institute, Barrackpore, Kolkata, India

College of Fisheries, Kishanganj, Bihar, India

B. K. Das
ICAR-Central Inland Fisheries Research Institute, Barrackpore, Kolkata, India

M. Goswami (✉)
ICAR-Central Institute of Fisheries Education, Mumbai, India

© The Author(s), under exclusive license to Springer Nature Singapore Pte Ltd. 2026

B. K. Das et al. (eds.), *Bioinformatics Tools for Predictive Ecology and Fisheries*, Sustainable Aquatic Food Systems, https://doi.org/10.1007/978-981-95-2573-7_4

innovations in aquaculture are urgently needed to relieve the pressure on food systems globally (Wang et al. 2025). To effectively address these global concerns, it is clear that we must secure the widespread availability of high-quality, nutrient-dense food. However, the aquaculture industry, trying to accommodate rising human demand, has been impeded by several concerns ranging from ecological sustainability to production efficiency. In addition to crops and livestock, fish are indispensable sources of protein for humans, especially in low- and middle-income countries (Andersen et al. 2025). Advances in genomic technologies have revolutionized our understanding of aquatic species, and comparative genomics has now become a powerful tool for improving aquaculture species through precise genetic interventions (Houston 2017). Comparative genomics is a field where evolutionarily diverged species are compared to indicate differences or similarities in their genome sequences so as to identify desirable genes or their regulatory elements. The approach opens up a multitude of opportunities for reproduction on the evolutionary relationship, the exploration of genetic diversity, and the study of the mechanism of growth, disease resistance, and stress tolerance in cultured fish and shellfish (Robinson et al. 2023). With the aid of comparative genomics, scientists could develop breeding programs to improve productivity, resilience, and sustainability. New developments of genome manipulation tools, including RNAi, gene transfer and genome editing, have been applied to aquaculture to create heritable variations at the DNA or RNA level, allowing dramatic increases in the production or performance of fishes and crustaceans (Wang et al. 2025). The implementation of next-generation sequencing (NGS) technologies, together with the fields of bioinformatics and functional genomics, has started to unify and integrate the extensive functional analysis of important aquaculture species. This multidisciplinary approach enables the identification of genes associated with key traits such as growth, disease resistance, and stress tolerance, thereby enhancing selective breeding programs and overall increasing aquaculture productivity (Rather et al. 2024). Moreover, the genomic data continues to expand, and the integration of machine learning and artificial intelligence (AI) into comparative genomics research is becoming more prevalent (Zhang et al. 2012). The modern approaches encompassing genome-wide association studies (GWAS), transcriptomic profiling, and marker-assisted selection (MAS) permit the identification of some key genetic determinants for improved performance traits (Yáñez et al. 2020). Moreover, the study of model organisms along with their genomic counterparts with commercially important aquaculture species is generating insights into gene functions amenable to application in selective breeding and genetic engineering (Robledo et al. 2018). This chapter discusses the principles and applications of comparative genomics in aquaculture improvements concerning its breeding strategies, different applications in aquaculture improvement, and stress adaptation. Recent breakthroughs, challenges, and future perspectives in utilizing genomic information for the sustainable development of aquaculture are also covered (Yáñez et al. 2020). Such a comparative genomic approach provides scientists and stakeholders in the industry with a comprehensive capability to sustain a greater genetic performance on these species, thus assuring long-term productivity while preserving environmental sustainability (FAO 2022; Houston 2017).

4.2 Key Concepts in Comparative Genomics for Enhancing Aquaculture

Comparative genomics in aquaculture is defined as a comparative analysis done on the genomes of various species to identify genetic factors that underlie desirable traits in aquaculture species, including growth, disease resistance, and environmental tolerance. Understanding these factors provides a basis for improving aquaculture productivity and sustainability.

4.2.1 Genome Sequencing and Annotation

Genome sequencing and annotation are the first critical steps in comparative genomics. This nonlethal nature of these methods allowed their use in fish, making genome sequencing now more affordable through different NGS techniques (Lien et al. 2016; Yáñez et al. 2020). Hence, the reference genome provides important insights into understanding the biology, evolution, and potential for genetic improvement through nutrient management practices, thus further making it possible to identify traits in economic species concerning resistance to diseases and growth rates.

4.2.2 Genetic Marker Discovery and Applications

Another significant application of comparative genomics in aquaculture is the identification of molecular markers that will help to select fish that have the traits that are important economically. That way, through GWAS, scientists can locate single-nucleotide polymorphisms (SNPs) or other markers that are close to traits such as resistance to diseases, growth rate, and feed conversion efficiency (Robledo et al. 2018). These markers can subsequently be applied in breeding programs for genetic improvement of aquaculture species through marker-assisted selection (MAS) (Houston 2017).

4.2.3 Integrating Comparative Genomics into Breeding Programs

Diverse breeding programs under aquaculture allow fish to be selected with better traits such as fast growth, disease resistance, and enhanced adaptation to changing environments. The genomic data helps to improve the breeding plans to develop a new aquaculture strain for economic efficiency and also sustainable development (Yue and Wang 2017). Genomic selection would be a tool for the breeders, with which they can make better-informed decisions in less time, using fewer resources for conventional breeding methods (Robledo et al. 2018).

4.2.4 Functional Genomics and Transcriptomics

Functional genomics plays an important role in understanding the biological mechanisms of important traits in aquaculture species. Transcriptomic analyses that assess gene expression profiles in different conditions offer insights into key processes: stress response, immune function, and metabolic regulation. There are genes that are now well characterized for growth regulation and stress tolerance, which would help in developing strong aquaculture species. Gene-editing technologies have exponentially changed functional genomics in aquaculture. The employment of CRISPR/Cas9 technology in fish species helped create desirable transgenic changes designed to improve growth and disease resistance. CRISPR-Cas9-generated modifications in the group of genes known to be related to growth regulation in tilapia lead to improved growth performance and feed conversion efficiency (Liang et al. 2024). Moreover, not only are the changes in the gut microbiota of native fish controlled by food supplementation, but it is also the changes that maintain their health and stability (Wu et al. 2018). A recent study on rainbow trout (*Oncorhynchus mykiss*) was conducted, where fish meal was substituted with insect-based meals, and timely changes in gut microbial flora were observed to take place, resulting in improved gut health and nutrient absorption. In addition to this, RNA analysis on crustaceans has shown that the constitutive mechanisms of the immune system and metabolic pathways, which are essential for stress relief, are activated on stress exposure (Li and Xiang 2013).

4.2.5 Integrating Comparative Genomics into Breeding Programs

Comparative genomics in breeding programs is a potential technique for the selection of fish with superior traits like faster growth, resistance to diseases, and better adaptation to different environmental conditions. By the introduction of genomic data to breeding programs, a more rapid improvement of aquaculture strains that are more efficient and sustainable is obtained (Yue and Wang 2017). Besides, genetic prediction tools drastically change the way breeders make decisions by making these choices more informative, thus saving time and resources from the past (Robledo et al. 2018) (Table 4.1).

Table 4.1 Genetic improvement strategies in aquaculture

Strategy	Target traits	Species applied
Selective breeding	Growth rate, disease resistance	Salmon, Tilapia, Carp
Marker-assisted selection (MAS)	Disease resistance, stress tolerance	Atlantic Salmon, Catfish
Genomic selection	Feed conversion efficiency, growth rate	Tilapia, Shrimp
CRISPR gene editing	Disease resistance, faster growth	Carp, Tilapia, Catfish
Hybridization	Hybrid vigor (heterosis)	Rohu × Catla, Tilapia hybrids

4.3 Comparative Genomics: Application in Aquaculture

Aquaculture is one of the most crucial sectors in global food production, providing nearly half of the world's seafood supply. It is also the fastest-growing food industry, playing a vital role in meeting the increasing global demand for sustainable and nutritious food sources (FAO 2018). However, despite its rapid expansion and economic significance, aquaculture faces numerous challenges that hinder its efficiency and productivity. These challenges include disease outbreaks, slow growth rates, environmental stressors, limited genetic diversity, and the risk of genetic constrictions, all of which pose serious threats to the stability and profitability of the industry (Houston 2017). To overcome these issues, modern scientific advancements have increasingly turned to comparative genomics, a field that studies the similarities and differences in the genetic makeup of different species. Comparative genomics is a powerful tool for improving aquaculture by providing insights into the genetic basis of important traits such as disease resistance, rapid growth, stress tolerance, and reproductive efficiency (Yue 2014). It allows analyzing the genetic composition of multiple aquaculture species, which helps to understand their evolutionary relationships and identify functional elements within their genomes in several ways:

4.3.1 Enhancing Disease Resistance

Disease is one of the most emerging issues in aquaculture, which leads to significant economic losses. Infectious diseases caused by bacteria, viruses, fungi, and parasites often spread rapidly in densely populated aquaculture systems. Comparative genomics enables us to identify disease-resistant genes and understand host-pathogen interactions, leading to the development of disease-resistant strains through selective breeding or genetic modification. For example, genomic studies in salmon have helped identify markers associated with resistance to sea lice and viral infections, contributing to the production of more resilient fish stocks (Houston 2017). Comparative genomics involves analyzing the genetic makeup of different species to identify conserved and unique genes associated with immune responses. Key applications include the following:

4.3.1.1 Genome-Wide Association Studies (GWAS)

Genome-wide association studies (GWAS) play a vital role in identifying genetic markers associated with disease resistance in aquaculture species. By comparing the genomes of resistant and susceptible organisms, researchers can find genetic loci related to enhanced immune responses, such as resistance to sea lice in salmon (Robledo et al. 2019) and *Streptococcus agalactiae* in Nile tilapia (Houston 2017), which aids in marker-assisted selection (MAS) for developing disease-resistant strains.

4.3.1.2 Transcriptomics

It further enriches our understanding by examining RNA expression patterns to uncover immune-response pathways. This method identifies crucial immune components such as pattern recognition receptors (PRRs) and antiviral peptides in shrimp, contributing to genetic enhancements for viral resistance (Xu et al. 2014). Similarly, comparative transcriptomics reveals differences in gene expression in carp, connecting interferon signaling and antiviral pathways to natural resistance against Koi Herpes virus (Yáñez et al. 2015).

4.3.1.3 The Groundbreaking CRISPR-Cas9

This technology propels genetic progress by accurately altering genes associated with disease resistance, growth performance, and adaptability to environmental changes. Unlike conventional breeding methods, CRISPR allows for targeted modifications, significantly shortening the time needed for genetic enhancements. Innovations in base editing, prime editing, and epigenome editing further broaden its usage in aquaculture genetics (Gaudelli et al. 2017). Importantly, CRISPR-based editing of the toll-like receptor gene tlr22 has boosted antiviral immunity in fish, while genetic markers related to muscle development and feed efficiency have been discovered in multiple species. The knockout of the myostatin (MSTN) gene, known for inhibiting muscle growth, has successfully increased muscle mass in catfish and tilapia (Wang et al. 2025). With its affordability and high precision, CRISPR is an invaluable asset in aquaculture breeding programs, enhancing disease resistance and productivity.

4.3.2 Next-Generation Sequencing (NGS)

New technologies greatly influenced our ability to monitor the behavior of fish by tracing marathon runners from start to finish. The revolution of fish and aquaculture biotechnologies started with the Next-Generation Sequencing (NGS) technique, which is the method used for rapid, high-throughput sequencing of whole genomes and mutation detection, through which genetic diversity and functional genomics can be analyzed. Technologies such as RNA sequencing (RNA-Seq), Whole-genome sequencing, and Single-Cell Sequencing facilitate the identification of gene expression patterns, genetic variations, and adaptive traits that contribute to aquaculture success. Besides, whole-genome sequencing (WGS) has paved the way for high-density single-nucleotide polymorphism (SNP) array development, which would also represent an advancement for genomic selection strategies in aquaculture breeding programs (Yáñez et al. 2015). By integrating next-generation sequencing with bioinformatics tools and machine learning models, predictive breeding is enhanced, resulting in the formulation of genetically superior stocks by aquaculture industries with super efficiency and resilience in climate change. Genome-wide association studies (GWAS) and quantitative trait loci mapping have become essential tools for linking genetic markers to economically significant traits, such as growth rate, disease resistance, and environmental tolerance. These approaches have led to the identification of genetic loci associated with resistance to bacterial and viral pathogens in key aquaculture species, including Atlantic salmon (*Salmo salar*) (Houston 2017). Even with these advancements, there are still challenges to overcome, including the high costs of sequencing, the complexity of bioinformatics, and the necessity for reference genomes that are specific to each species. Tackling these issues will help maximize the potential of NGS in enhancing genetic improvement strategies for sustainable aquaculture.

4.3.3 Vaccine Development Using Genomic Insights

Comparative genomics is crucial in the development of vaccines as it uncovers important immune-related genes and the interactions between pathogens and hosts, resulting in more effective immunization approaches. For instance, genomics research on the antiviral gene responses in tilapia has guided the creation of DNA vaccines for Tilapia Lake Virus (TiLV), significantly enhancing the survival rates of vaccinated fish (Dong et al. 2020). In the case of shrimp, genomic discoveries regarding viral invasion methods have paved the way for RNA interference (RNAi)-based vaccines that increase disease resistance. These developments illustrate how comparative genomics propels vaccine innovation, bolstering the health and productivity of aquaculture.

4.3.4 Precision Nutrition for Immune Enhancement

Comparative genomics has shed light on how certain dietary components affect the expression of genes related to the immune system. These investigations have pinpointed essential nutrients, probiotics, and immunostimulants that can boost immune function by influencing gene expression in the host. Functional feeds that are fortified with these bioactive substances are being developed to enhance health and resistance to diseases in both humans and animals. Probiotic strains like *Lactobacillus* and *Bifidobacterium* have demonstrated the ability to increase the expression of immune-related genes, thereby strengthening the host's defense mechanisms. Likewise, immunostimulants such as β-glucans and polyphenols derived from plants activate mechanisms that enhance both innate and adaptive immunity. Progress in the field of nutrigenomics allows for the precise customization of diets to maximize immune responses through the interactions between genes and diet. In aquaculture, the use of genomic information in designing functional feeds has led to improvements in disease resistance and growth rates among farmed fish species. Incorporating genomics into feed formulation promotes sustainable production practices by decreasing reliance on antibiotics (Dawood 2021). Continuous research is revealing new bioactive compounds that may further bolster immune resilience.

4.3.5 Improving Growth Rates and Feed Efficiency

Growth rate is a crucial factor in aquaculture profitability. Comparative genomics helps to identify genes that influence growth performance, muscle development, and feed conversion efficiency. Growth and feed conversion efficiency are complex traits influenced by multiple genes and environmental factors. The *growth hormone (GH) insulin-like growth factor (IGF) axis* is one of the primary regulators of growth in vertebrates, including fish. Comparative genomics demonstrate that variations in *GH, IGF-1, IGF-2, and their receptor genes (GHR, IGFR)* influence growth performance in many aquaculture species like the following:

- *Atlantic Salmon (Salmo salar)*: Comparative genomics studies between fast-growing and slow-growing strains have identified variations in GH, IGF-1, and IGF-2 genes that contribute to enhanced growth rates (Fuentes et al. 2017). The GH gene has been found to regulate both muscle growth and feed conversion efficiency, making it a key target for selective breeding.
- *Tilapia* (*Oreochromis niloticus*): Genetic studies have identified variations in GH and IGF-1 genes, with some alleles being strongly associated with improved feed efficiency and increased muscle growth (Macqueen et al. 2017).

Some other key genetic traits studied in comparative genomics for aquaculture are listed in Table 4.2.

Table 4.2 Key genetic traits studied in comparative genomics for aquaculture

Trait	Species	Gene(s) identified	Function
Disease resistance	Atlantic salmon	*MHC, TLR, IL*	Immune response
Growth rate	Nile tilapia	*GH, IGF*	Growth hormone regulation
Stress tolerance	Pacific oyster	*Hsp70, SOD*	Heat shock and oxidative stress
Salinity adaptation	Eel (*Anguilla*)	*NKA, CFTR*	Ion transport
Fillet quality	Rainbow trout	Myostatin	Muscle growth

4.3.6 Genetic Diversity and Conservation of Wild Populations

The genetic variation of cultivated species is vital for enduring sustainability; however, intensive breeding practices frequently result in genetic bottlenecks, diminishing diversity and heightening susceptibility to diseases and environmental shifts. Comparative genomics facilitates the evaluation of genetic structures in both wild and domestic populations, helping to devise strategies that maintain diversity while enhancing desirable characteristics. It also contributes to conservation by pinpointing critical genetic variations necessary for the survival and adaptation of species. For example, genomic research on Atlantic salmon and rainbow trout has advanced selective breeding efforts while safeguarding genetic diversity. These findings bolster sustainable aquaculture and conservation initiatives (Houston 2017).

4.3.7 Enhancing Stress Tolerance and Environmental Adaptation

The adaptation of aquaculture species to environmental stress is influenced by various genes and molecular pathways, with comparative genomics assisting in the identification of significant genetic mechanisms related to thermal tolerance, osmoregulation, hypoxia resistance, and detoxification of pollutants. For example, Heat Shock Proteins (HSP70, HSP90) are essential for temperature stress resilience in oysters (Zhang et al. 2012). Furthermore, hypoxia-inducible factors (HIF) and erythropoietin (EPO) assist species such as catfish in coping with low-oxygen conditions. The detoxification of pollutants is facilitated by genes like Cytochrome P450 (CYP) and glutathione S-transferase (GST), which provide protection against harmful substances. Given the increasing environmental pressures caused by climate change, comparative genomics allows for the development of more effective breeding approaches for aquaculture species that are resilient (Gjedrem et al. 2012).

Comparative genomics is crucial for improving aquaculture by tackling significant issues such as resistance to diseases, efficiency of growth, genetic variability, and adaptation to environmental changes. Researchers utilize methods such as genome-wide association studies (GWAS), transcriptomics, and CRISPR-Cas9 to pinpoint genetic markers associated with disease resistance, paving the way for the creation of hardier aquaculture species. The use of next-generation sequencing (NGS) boosts breeding techniques by allowing for extensive genomic analysis at a high throughput, while insights from genomics support the creation of vaccines against prevalent aquatic diseases. Precision nutrition, guided by nutrigenomics, enhances immune response and feeding efficiency, leading to better growth performance. Furthermore, comparative genomics contributes to the preservation of genetic diversity, promoting sustainability by avoiding genetic bottlenecks. It also supports research on stress tolerance by identifying genes linked to resilience against temperature fluctuations, salinity changes, and detoxification of pollutants, which are vital for the aquaculture sector to cope with climate change. By

Table 4.3 Comparative genomics studies in major aquaculture species

Species	Genomic study	Key findings
Atlantic salmon (*Salmo salar*)	Whole-genome sequencing	Evidence of a whole-genome duplication event
Nile tilapia (*Oreochromis niloticus*)	Genomic selection	Identification of growth and disease-resistance markers
Pacific oyster (*Crassostrea gigas*)	Transcriptomics	Genes involved in stress response and shell formation
Common carp (*Cyprinus carpio*)	Functional genomics	Adaptation mechanisms in different environments
Whiteleg shrimp (*Penaeus vannamei*)	Metagenomics	Gut microbiota influencing disease resistance

incorporating genomic technologies, the industry can improve productivity, sustainability, and resilience, ensuring a reliable seafood supply for the future (Table 4.3).

4.4 Challenges and Limitations

In the life sciences, decoding the genomic sequences of model organisms and humans represents a major milestone. While individual genome sequencing offers valuable structural information, it provides limited insights into gene function (Lander 2001). The primary objective of the next phase of genomic research is to identify functional DNA elements and determine their roles. One of the key applications of comparative genomics is the functional annotation of genomes. Genomic sequences undergoing purifying (negative) selection evolve at a slower rate compared to nonfunctional sequences, whereas those under Darwinian (positive) selection change more rapidly (Lawrie and Petrov 2014). These evolutionary patterns can be identified by comparing genetic sequences, allowing researchers to distinguish functional elements within the genome.

In aquaculture, comparative genomics has become a crucial technology for enhancing desirable traits, contributing to sustainable aquaculture by guiding practices that reduce environmental impact and improve resource efficiency. It plays a vital role in strengthening disease resistance and supports breeding and conservation efforts by providing insights into the evolutionary processes that have shaped the genomes of aquaculture species (Ma et al. 2021). By analyzing and comparing the genomes of different species or strains, researchers can identify genetic variations and evolutionary patterns that inform selective breeding programs and disease management strategies.

Comparative genomics has an immense amount of potential; however, there are a lot of obstacles and restrictions to its use in aquaculture. When compared to model organisms, the genetic complexity of many aquaculture species, including genome duplications, polyploidy, and high genetic diversity, makes genomic analysis more challenging (Abdelrahman et al. 2017). The precision and usefulness of genomic data are restricted by the absence of functional annotations and high-quality reference genomes for numerous commercially significant species. Predicting and improving desired features is made more difficult by environmental factors that further complicate genotype-phenotype connections, such as changing water conditions, pathogen exposure, and nutritional variability. Furthermore, the availability of comparative genomic studies in aquaculture is limited by computational and technological obstacles, such as the demand for sophisticated bioinformatics tools, expensive sequencing, and substantial data processing needs (Abdelrahman et al. 2017). The broad use of genomic technologies is also impacted by ethical and legal issues surrounding genome editing and genetic alteration in aquaculture.

4.4.1 Interactions Between Genotype and Environment

Aquaculture species frequently display unique genetic traits and phenotypic manifestations that are impacted by a range of environmental variables, including water temperature, salinity, and nutritional availability. These genotype-by-environment (G × E) interactions make it difficult to predict phenotypic performance in various aquaculture settings, which makes it difficult to create breeding programs that are universally successful (Olesen et al. 2000). When aquaculture species are raised in varying environmental settings, G × E interactions can result in differences in growth rates, survival, and disease resistance (Houston 2017). For genomic selection to be successfully applied in a number of aquaculture settings, it is essential to comprehend and control these interactions. Selective breeding techniques may produce inconsistent outcomes if G × E interactions are not properly taken into account, which would limit genetic gains and overall production (Olesen et al. 2000). The accuracy of selection can be improved by taking into consideration G × E interactions thanks to advancements in genomic technologies like genome-wide association studies (GWAS) and genomic estimated breeding values (GEBVs). Furthermore, studies indicate that by identifying resilient genotypes that consistently perform under a variety of conditions, the use of reaction norms mathematical models that explain how genotype performance varies across environments can increase the effectiveness of breeding programs. Research on species like tilapia (*Oreochromis niloticus*), Atlantic salmon (*Salmo salar*), and shrimp (*Penaeus vannamei*) highlights the importance of environment-specific breeding strategies by showing the substantial influence of environmental factors on genetic expression (Yáñez et al. 2014).

4.4.2 Methods of Genome Manipulation

Aquaculture species may benefit from improved traits because of cutting-edge genome editing techniques like gene transfer and RNA interference (RNAi). Technical difficulties, legal restrictions, and moral dilemmas, however, restrict their use. These genome modification approaches have potential, but a number of obstacles prevent their widespread use. Their successful application is severely hampered by technical issues such ineffective gene delivery systems, unwanted genetic alterations, and the requirement for refined gene-targeting techniques (Wang et al. 2025). Furthermore, in many areas, governmental approvals and public acceptance of the use of genetically modified organisms (GMOs) in aquaculture are restricted by ethical and legal considerations. Alternative genome-editing techniques, including base editing and prime editing, which provide more accurate and regulated alterations with fewer off-target consequences, are being investigated by researchers in order to overcome these difficulties. The creation of frameworks that are ethically sound, legally compliant, and scientifically sound will determine the future of genome editing in aquaculture. These frameworks will guarantee the appropriate application of these revolutionary technologies while optimizing their advantages for the production of seafood worldwide. These complex problems must be resolved in order to create genome editing techniques that are both effective and palatable (Wang et al. 2025).

4.4.3 Evolutionary Constraints in Polyploid Species

Many aquaculture species exhibit polyploidy, which poses unique evolutionary difficulties. Triploid species, for example, frequently struggle with meiosis, which lowers fertility and presents problems for breeding efforts. Aquaculture species, such as Salmonids (*Salmo salar*, *Oncorhynchus* spp.), Sturgeons (Acipenseridae), and Oysters (*Crassostrea gigas*), exhibit polyploidy, which is the state of having more than two complete sets of chromosomes (Zhou and Gui 2017). Polyploidy introduces evolutionary restrictions even if it might offer benefits like greater genetic diversity and flexibility. Breeding programs are complicated by the lower fertility that results from meiotic disturbances in triploid and higher-order polyploid species (Arai 2023). Polyploid species experience significant genomic rearrangements, such as chromosomal fusions, gene deletions, and epigenetic changes that affect gene expression, according to comparative genomic investigations (Robertson et al. 2017). The way polyploid organisms adjust to environmental stresses and aquaculture settings is greatly influenced by these genetic alterations. Comparative genomics offers important insights into the evolutionary processes that allow polyploid species to survive and flourish, despite the difficulties posed by polyploidy. A characteristic of polyploidy, whole-genome duplication (WGD) events have been demonstrated to provide functional redundancy by enabling the subfunctionalization and neo-functionalization of duplicated genes (Lien et al. 2016). In salmonids, for example, comparative genomic studies have shown how maintained duplicate genes have evolved to serve specialized activities in osmoregulation, immunological response, and metabolism (Berthelot et al. 2014). Because they offer molecular targets for selective breeding and genome editing initiatives meant to improve desired aquaculture traits, these results are essential for genetic improvement projects. Researchers can minimize the evolutionary restrictions associated with genome duplication and optimize polyploid organisms for commercial production by using comparative genomics to develop more accurate breeding tactics. Effective genetic improvement tactics require an understanding of the evolutionary dynamics of polyploid species.

4.4.4 Species Diversity in Aquaculture

Hundreds of species, such as mollusks, finfish, and crabs, are cultivated in aquaculture. Numerous variables, including market demands, ecological compatibility, and regional preferences, contribute to this diversity. For example, although they are all essential to aquaculture worldwide, species such as Pacific white shrimp (*Litopenaeus vannamei*), Nile tilapia (*Oreochromis niloticus*), and Atlantic salmon (*Salmo salar*) have very different genetic compositions (Zhou and Gui 2017). Every species has a different genome in terms of size, organization, and structure. For instance, the genome size of Atlantic salmon is over 2.97 billion base pairs. Because of these variations, genome sequencing and annotation techniques must be species-specific. High-quality reference genomes, which are necessary for comparative genomics and the creation of molecular markers, are lacking in many aquaculture species. The implementation of breeding programs and genomic selection across a variety of species is made difficult by the lack of these reference points. Aquaculture species' genetic diversity helps the industry to be resilient and adaptable, but it also presents major obstacles to the creation of universal genomic tools. A concentrated effort in species-specific genomic research and the adaptation of tactics to fit the distinct genetic profiles of each species are needed to address these issues.

4.4.5 Ethical and Regulatory Concerns

Ethical and regulatory issues have a big impact on the application of comparative genomics in aquaculture. Although genome editing technologies like CRISPR/Cas9 have the potential to enhance characteristics like growth efficiency and disease resistance, they also bring up moral questions about their effects on the environment, animal welfare, and consumer acceptance (Robinson et al. 2024). Comparative genomics is essential for finding genetic variants that can improve the resilience of aquaculture species, but converting these discoveries into gene-editing applications necessitates carefully weighing the dangers. For instance, if genetically modified fish escape, unanticipated genetic alterations may disturb wild populations, with unexpected ecological repercussions. The implementation of comparative genomics in aquaculture is severely hampered by ethical and regulatory issues. Though they raise ethical concerns about ecological effect, animal welfare, and consumer acceptance, genome editing technologies like CRISPR/Cas9 have the potential to improve features like disease resistance and growth efficiency (Robinson et al. 2024). Global regulatory frameworks for genetically modified and gene-edited organisms also differ, with some areas enforcing stringent limitations that impede commercialization and research. The confidence of the public and regulatory approval for genetic breakthroughs in aquaculture can be increased by using an open, scientific method that incorporates risk-benefit analyses, stakeholder involvement, and ethical standards (FAO 2022). By combining comparative genomics with prudent regulatory processes, the industry can capitalize on the potential of genetic innovations while maintaining consumer confidence and environmental sustainability. The broad use of genomic technology in aquaculture is hampered by these regulatory difficulties, which call for careful balancing of innovation with moral and legal obligations.

4.4.6 The Resource-Intensive Nature

The genomic analyses are one of the major constraints. Smaller research institutes or aquaculture businesses may not have the financial or computational capacity to do whole-genome sequencing, annotation, and comparison investigations. Genotyping-by-sequencing (GBS), which reduces sequencing costs and simplifies genome complexity while still identifying important genetic markers, is one of the affordable solutions that researchers are using to address these issues (Robledo et al. 2019). Furthermore, improvements in computer tools and bioinformatics have increased the effectiveness of data processing, opening up genomic investigations to institutions with minimal resources (Yáñez et al. 2020).

4.4.7 Adaptation of Bioinformatics Tools

Although bioinformatics is essential for genome-based biotechnology in aquaculture, it faces difficulties since biologists have little training in informatics, and informatics specialists lack biological knowledge. Bioinformatics analyses require high-performance computing (HPC), yet this technology is difficult to use and expensive, particularly in developing nations. Furthermore, several bioinformatics platforms are less accessible due to their command-line competency requirements. Although cloud computing presents a viable remedy, it still necessitates a significant investment in IT infrastructure. These difficulties underscore the need for more accessible bioinformatics tools and transdisciplinary training (Rana et al. 2020).

4.4.8 Lack of Use of AI Technologies in Aquaculture

There is great potential for increasing sustainability and efficiency in aquaculture with the application of artificial intelligence (AI). However, a number of obstacles prevent its broad use. The main challenges are data availability and quality; unstructured, noisy, and non-standardized data require extensive preprocessing, and model development and collaboration are hampered by the scarcity of high-quality labeled datasets (Yang et al. 2025). Additionally, because specialized hardware and processing infrastructure are costly, small-scale farms are particularly discouraged from implementing AI technologies due to high implementation costs and limited resources. These difficulties are made worse by a lack of technological know-how, which makes it challenging for aquaculture operations to successfully deploy and manage AI systems. There are a number of tactics that can be used to get past these obstacles. More robust AI model training can be facilitated by improving data quality and accessibility through the development of standardized data collection techniques and the promotion of data-sharing collaborations (Rather et al. 2024). Smaller businesses can adopt technology more easily by investing in cloud-based services and reasonably priced AI solutions, which can eliminate the need for substantial on-site infrastructure (FNB Tech 2025). Additionally, offering focused training programs can help employees get the technical skills they need to adopt and manage AI systems in aquaculture settings.

4.5 Conclusion and Future Perspectives

A comprehensive knowledge of biological systems is made possible by the integration of multi-omics data, such as transcriptomics, proteomics, metabolomics, and genomes. By exposing complex gene connections and regulatory networks that affect phenotypic traits, this all-encompassing method improves breeding accuracy (Cembrowska-Lech et al. 2023). Through the identification of conserved and divergent genetic components across species, comparative genomics plays a critical role in this process and provides important insights into the evolution and adaptation of traits in aquaculture species. Researchers can identify functionally important genes and regulatory sequences that support growth efficiency, disease resistance, and environmental adaptability through comparative analysis, which will ultimately enhance selective breeding tactics. The use of cutting-edge technologies like CRISPR genome editing, multi-omics integration, third-generation sequencing, and AI-driven analytics is revolutionizing aquaculture genetics. These developments enhance the precision of genetic changes for desired phenotypes, functional annotations, and genome assembly accuracy (Andersen et al. 2025). By offering a framework for comprehending genetic variation both within and between species, comparative genomics reinforces these developments by making it easier to identify target genes for genetic improvement. Furthermore, the field of comparative genomics is growing in scope due to developments in high-throughput sequencing and machine learning-driven genomic analyses, which make it possible to quickly identify genetic variations and adaptive features in aquaculture species. Additionally, by accelerating the development of aquaculture species that are resistant to disease and climate change, genomic selection is transforming breeding operations. In order to maintain output in the face of shifting environmental conditions, comparative genomic investigations make it possible to identify genetic markers linked to immune responses and stress tolerance (Robledo et al. 2018). Even so, there are still issues with aquaculture genetics, such as high expenses, a lack of genomic data for nonmodel species, and moral dilemmas with genome editing. The use of genetic technology in aquaculture is also impacted by differing public acceptance and regulatory difficulties. Genetic selection and trait prediction will be improved in the future by ongoing developments in sequencing technology, multiomics methodologies, and AI-powered analytics, which will increase the effectiveness and precision of breeding operations. Because it helps us better understand evolutionary adaptations and makes informed genetic alterations possible, comparative genomics will continue to be at the forefront of these advancements. The full potential of genetic advances for a more robust and productive aquaculture industry would require uniform regulatory frameworks, cooperative research initiatives, and sustainable breeding techniques to overcome existing constraints. Long-term sustainability and global food security will be ensured by optimizing aquaculture species for future challenges through the combination of comparative genomic approaches with functional genomics and systems biology.

Comparative genomics has considerably benefited aquaculture by revealing the genetic basis of critical features including growth rate, disease resistance, stress tolerance, and environmental adaptability. These findings are critical as the global need for sustainable and efficient aquaculture operations increases. However, obstacles remain, such as the complexity of aquatic genomes, limited genetic resources for nonmodel species, and computational limitations. Addressing these challenges is critical to the continued development and implementation of genomic technology in aquaculture.

Rapid advances in sequencing technology, multiomics integration, artificial intelligence, and genome editing will determine the future of comparative genomics in aquaculture. High-throughput sequencing approaches, such as third-generation sequencing (PacBio, Oxford Nanopore), may allow for more accurate reference genome assembly, particularly for complex polyploid species, hence improving trait selection and breeding programs (Houston 2017). Furthermore, combining genomes with transcriptomics, proteomics, and metabolomics will provide a comprehensive understanding of the molecular pathways driving essential features such as growth, disease resistance, and environmental adaptability. Additionally, Machine learning and artificial intelligence will enhance genotype-phenotype predictions, expediting selective breeding efforts and increasing genome annotation and variation discovery. Furthermore, CRISPR/Cas9 genome editing is projected to revolutionize trait improvement in aquaculture species by enabling the generation of disease-resistant and climate-resilient strains while lowering antibiotic use (Hollenbeck and Johnston 2018). As climate change presents additional problems, genetic selection will be vital in breeding species that can tolerate changing climatic conditions (Yáñez et al. 2020). However, the successful implementation of these technologies will be contingent on increasing ethical concerns and regulatory frameworks that enable responsible genetic alteration and public acceptability. By tackling these issues, comparative genomics will continue to push sustainable and efficient aquaculture techniques, influencing the future of food security and environmental resilience.

Concluding, the field of comparative genomics presents significant opportunities to enhance aquaculture by increasing disease resistance, improving selective breeding practices, and promoting sustainability. By pinpointing essential genetic differences linked to growth, immune response, and resilience to stress, scientists can create aquaculture species that are stronger and more productive. These genomic findings are vital for satisfying the rising global need for sustainable seafood while reducing environmental effects. Nonetheless, several challenges impede the broad implementation of comparative genomics, such as the intricate genetic make-up of aquatic organisms, environmental factors that influence gene expression, a lack of genomic information for non-model species, and ethical and regulatory issues related to genome editing.

Furthermore, the steep expenses associated with whole-genome sequencing, along with the substantial computational requirements and the necessity for advanced bioinformatics tools, create obstacles to accessibility, especially for smaller research institutions and aquaculture businesses. The limited access to high-quality reference genomes and functional annotations adds further complexity to genetic analysis and breeding initiatives. In addition, the complex relationships between genotype and environmental conditions, such as water temperature, salinity, and pathogen exposure, hinder the ability to predict and enhance desirable traits in various aquaculture environments. Ethical issues surrounding genome editing technologies, like CRISPR/Cas9, alongside the regulatory framework for genetically modified organisms (GMOs), also present significant challenges to the widespread implementation of genomic innovations in aquaculture.

Overcoming these challenges necessitates ongoing improvements in genome annotation, the establishment of strong regulatory frameworks, and the adoption of AI-driven bioinformatics solutions. The outlook for comparative genomics in aquaculture is bright, with advancements in high-throughput sequencing, multi-omics strategies, artificial intelligence, and genome editing propelling progress in trait enhancement and sustainability. Next-generation sequencing methods, like PacBio and Oxford Nanopore, provide more precise and thorough genome assemblies, especially for intricate polyploid species. Merging genomics with transcriptomics, proteomics, and metabolomics enables a more profound insight into the molecular processes that regulate growth, disease resistance, and adaptability to environmental changes. Machine learning and artificial intelligence will further refine genotype-phenotype predictions, enhance genome annotation, and speed up the identification of genetic variations associated with crucial aquaculture attributes.

Additionally, technologies like CRISPR/Cas9 and newer genome-editing methods, including base editing and prime editing, are set to transform aquaculture by allowing for the targeted modification of genes linked to disease resistance, stress adaptability, and growth efficiency. These innovations are expected to minimize dependency on antibiotics, boost food security, and promote aquaculture practices that are resilient to climate change. As climate change brings forth more challenges, genetic selection will be essential in breeding species that can prosper in changing environmental conditions. Nonetheless, the effective application of these technologies relies on tackling ethical issues, enhancing regulatory frameworks, and encouraging public support for genomic changes in aquaculture. Ensuring clarity in research, conducting risk-benefit assessments, and engaging stakeholders will be vital for the responsible implementation of genetic advancements. By confronting these obstacles, comparative genomics will further advance sustainable and effective aquaculture practices, impacting global food security and environmental sustainability. With ongoing investment in research, technology, and ethical oversight, comparative genomics has the capacity to transform aquaculture, guaranteeing the production of high-quality seafood in an environmentally sound and economically sustainable way.

References

Abdelrahman H, ElHady M, Alcivar-Warren A, Allen S and Al-Tobasei R (2017) LSU Scholarly Repository. Aquac 2:20

Agriculture Organization of the United Nations. Fisheries Department (2018) The state of world fisheries and aquaculture. Food and Agriculture Organization of the United Nations

Andersen LK, Thompson NF, Abernathy JW, Ahmed RO, Ali A, Al-Tobasei R, Reading BJ (2025) Advancing genetic improvement in the omics era: status and priorities for United States aquaculture. BMC Genomics 26(1):1–24

Arai K (2023) Meiosis and gametogenesis in hybrid, polyploid, and clonal fishes: case studies in the dojo loach *Misgurnus anguillicaudatus*. Fish Sci 89(5):537–570

Berthelot C, Brunet F, Chalopin D, Juanchich A, Bernard M, Noël B, Guiguen Y (2014) The rainbow trout genome provides novel insights into evolution after whole-genome duplication in vertebrates. Nat Commun 5(1):3657

Cembrowska-Lech D, Krzemińska A, Miller T, Nowakowska A, Adamski C, Radaczyńska M, Mikiciuk M (2023) An integrated multi-omics and artificial intelligence framework for advance plant phenotyping in horticulture. Biology 12(10):1298

Dawood MA (2021) Nutritional immunity of fish intestines: important insights for sustainable aquaculture. Rev Aquac 13(1):642–663

Dong HT, Senapin S, Gangnonngiw W, Nguyen VV, Rodkhum C, Debnath PP, Mohan CV (2020) Experimental infection reveals transmission of tilapia lake virus (TiLV) from tilapia broodstock to their reproductive organs and fertilized eggs. Aquaculture 515:734541

Fuentes EN, Martin SA, Johnston IA, Macqueen DJ (2017) Divergent regulation of insulin-like growth factor binding protein genes in cultured Atlantic salmon myotubes under different models of catabolism and anabolism. Gen Comp Endocrinol 247:53–65

FNB Tech (2025) Aquaculture Technology 2.0: Automation and AI in Fish Farming. Retrieved from https://fnb.tech/aquaculture-technology/

Food and Agriculture Organization of the United Nations. (2022). The state of world fisheries and aquaculture 2022: Towards blue transformation. FAO. https://www.fao.org/3/cc0461en/online/sofia/2022/fisheries-and-aquaculture-projections.html

Gaudelli NM, Komor AC, Rees HA, Packer MS, Badran AH, Bryson DI, Liu DR (2017) Programmable base editing of A• T to G• C in genomic DNA without DNA cleavage. Nature 551(7681):464–471

Gjedrem T, Robinson N, Rye M (2012) The importance of selective breeding in aquaculture to meet future demands for animal protein: a review. Aquaculture 350:117–129

Hollenbeck CM, Johnston IA (2018) Genomic tools and selective breeding in molluscs. Front Genet 9:253

Houston RD (2017) Future directions in breeding for disease resistance in aquaculture species. Rev Bras Zootec 46:545–551

Lander ES (2001) Initial sequencing and analysis of the human genome. Nature 409:860–921

Lawrie DS, Petrov DA (2014) Comparative population genomics: power and principles for the inference of functionality. Trends Genet 30(4):133–139

Liang C, Lu Y, Mai R (2024) CRISPR-Cas9 mediated gene editing in tilapia: enhancing growth rates and disease resistance. Genomics Appl Biol 15

Li F and Xiang J (2013) Signalling pathways regulating innate immune responses in shrimp. Fish and shellfish Immunol 34(4):973–980

Lien S, Koop BF, Sandve SR, Miller JR, Kent MP, Nome T, Davidson WS (2016) The Atlantic salmon genome provides insights into rediploidization. Nature 533(7602):200–205

Ma X, Su B, Bangs M, Alston V, Backenstose NJ, Simora RM, Dunham RA (2021) Comparative genomic and transcriptomic analyses revealed twenty-six candidate genes involved in the air-breathing development and function of the bighead catfish Clarias macrocephalus. Mar Biotechnol 23:90–105

Macqueen DJ, Primmer CR, Houston RD, Nowak BF, Bernatchez L, Bergseth S, FAASG Consortium Craig R. Primmer Daniel J. Macqueen Ross D. Houston Barbara F. Nowak Willie S. Davidson Sigbjørn Lien Ben F. Koop (2017) Functional Annotation of All Salmonid Genomes (FAASG): an international initiative supporting future salmonid research, conservation and aquaculture. BMC Genomics 18:1–9
Ngo TD (2023) Demographic trends and population health: tackling inequality in a world of eight billion people. BMJ Glob Health 8(4):e012137
Olesen I, Groen AF, Gjerde B (2000) Definition of animal breeding goals for sustainable production systems. J Anim Sci 78(3):570–582
Rana KMS, Ahammad K, Salam MA (2020) Bioinformatics: scope and challenges in aquaculture research of Bangladesh – a review. Int J Agric Res Innov Technol 10(2):137–145
Rather MA, Ahmad I, Shah A, Hajam YA, Amin A, Khursheed S, Rasool S (2024) Exploring opportunities of Artificial Intelligence in aquaculture to meet increasing food demand. Food Chem: X 22:101309
Robertson FM, Gundappa MK, Grammes F, Hvidsten TR, Redmond AK, Lien S, Macqueen DJ (2017) Lineage-specific rediploidization is a mechanism to explain time-lags between genome duplication and evolutionary diversification. Genome Biol 18:1–14
Robinson NA, Robledo D, Sveen L, Daniels RR, Krasnov A, Coates A, Houston RD (2023) Applying genetic technologies to combat infectious diseases in aquaculture. Rev Aquac 15(2):491–535
Robinson NA, Østbye TKK, Kettunen AH, Coates A, Barrett LT, Robledo D and Dempster T (2024) A guide to assess the use of gene editing in aquaculture. Rev Aquac 16(2):775–784
Robledo D, Matika O, Hamilton A and Houston RD (2018) Genome-wide association and genomic selection for resistance to amoebic gill disease in Atlantic salmon. G3: Genes, Genomes, Genetics, 8(4):1195–1203
Robledo D, Gutiérrez AP, Barría A, Lhorente JP, Houston RD, Yáñez JM (2019) Discovery and functional annotation of quantitative trait loci affecting resistance to sea lice in Atlantic salmon. Front Genet 10:56
Wang J, Cheng Y, Su B, Dunham RA (2025) Genome manipulation advances in selected aquaculture organisms. Rev Aquacult 17(1):e12988
Wu ZB, Gatesoupe FJ, Li TT, Wang XH, Zhang QQ, Feng DY and Li AH (2018) Significant improvement of intestinal microbiota of Gibel carp (Carassius auratus gibelio) after traditional Chinese medicine feeding. J Appl Microbiol 124(3):829–841
Xu P, Zhang X, Wang X, Li J, Liu G, Kuang Y, Sun X (2014) Genome sequence and genetic diversity of the common carp, Cyprinus carpio. Nat Genet 46(11):1212–1219
Yáñez JM, Houston RD, Newman S (2014) Genetics and genomics of disease resistance in salmonid species. Front Genet 5:415
Yáñez JM, Houston RD and Newman S (2015) Genomics in aquaculture to better understand species biology and accelerate genetic progress. Trends Genet 31(9):493–504
Yáñez JM, Houston RD and Newman S (2020) Genetics and genomics of disease resistance in salmonid species. Front Genet 11:576
Yang H, Feng Q, Xia S, Wu Z, Zhang Y (2025) AI-driven aquaculture: a review of technological innovations and their sustainable impacts. Artif Intell Agric 15:508
Yue GH (2014) Recent advances of genome mapping and marker-assisted selection in aquaculture. Fish Fish 15(3):376–396
Yue GH, Wang L (2017) Current status of genome sequencing and its applications in aquaculture. Aquaculture 468:337–347
Zhang G, Fang X, Guo X, Li LI, Luo R, Xu F, Wang J (2012) The oyster genome reveals stress adaptation and complexity of shell formation. Nature 490(7418):49–54
Zhou L, Gui J (2017) Natural and artificial polyploids in aquaculture. Aquacult Fish 2(3):103–111
Zhu M, Sumana SL, Abdulla teef MM, Falayi OC, Shui Y, Zhang C, Su S (2024) CRISPR/Cas9 technology for enhancing desirable traits of fish species in aquaculture. Int J Mol Sci 25(17):9299

RNA-Seq Analysis in Aquatic Organisms

5

Vikash Kumar, Timothy J. Bruce, Vikas Kumar, and Basanta Kumar Das

5.1 Introduction

Transcriptomics is the analysis of the entire set of transcripts from a specific cell, tissue, or organism during a particular developmental stage or physiological situation. This whole set is collectively referred to as the transcriptome, encompassing both protein-coding messenger RNA (mRNA) and noncoding RNA, including ribosomal RNA (rRNA), transfer RNA (tRNA), and other noncoding RNAs (Lindberg and Lundeberg 2010). In contrast to the relatively static genome, the transcriptome changes in response to developmental stages, physiological states, and external factors. Analysing the transcriptome is an effective way to investigate how genotype relates to phenotype, thereby improving our understanding of the pathways and mechanisms that control cell fate, development, and disease progression. The objectives of transcriptomics extend beyond merely quantifying gene expression changes across different transcriptome samples; they also encompass mapping and annotating the transcriptome, as well as determining the functional context of each gene within the genome (Costa et al. 2010).

The intricate nature of the transcriptome necessitates the use of high-throughput tools for analysing the transcriptome. Over time, various technologies, both hybridization- and sequence-based, have emerged to conduct large-scale surveys of the transcriptome. Hybridization-based technologies typically involve incubating

V. Kumar · B. K. Das (✉)
ICAR-Central Inland Fisheries Research Institute, Barrackpore, Kolkata, India

T. J. Bruce
School of Fisheries, Aquaculture & Aquatic Sciences, Auburn University, Auburn, AL, USA

V. Kumar
Washington University, St. Louis, MO, USA

© The Author(s), under exclusive license to Springer Nature Singapore Pte Ltd. 2026

B. K. Das et al. (eds.), *Bioinformatics Tools for Predictive Ecology and Fisheries*, Sustainable Aquatic Food Systems, https://doi.org/10.1007/978-981-95-2573-7_5

fluorescently labelled cDNA with probes affixed to a solid substrate, such as a microarray. Since their initial implementation in 1995 (Schena et al. 1995), microarrays have become extensively utilized in transcriptomics. The tiling microarray, an advanced version featuring high-density probes that represent the genome, can analyse transcriptomes with considerable resolution and also identify novel transcripts. The application of microarrays in fish biological research has spanned over a decade and is well-documented elsewhere. However, microarray technology has some inherent limitations, including dependence on existing genomic data, signal saturation for very abundant transcripts, and higher background noise caused by nonspecific hybridization. Additionally, there are specific challenges for fish transcriptomic studies using microarrays: limited commercial platforms exist, primarily designed for model species like zebrafish (*Danio rerio*); creating customized microarrays or tiling arrays requires prior knowledge of EST sequences or genomic data, which is often unavailable for many fish species.

Unlike microarray technologies, sequence-based methods directly determine the cDNA sequence. They produce specific EST tags that represent transcript fragments in a sample, which are then concatenated prior to cloning and sequencing (Fullwood et al. 2009). Several tag-based sequencing techniques have been developed, such as serial analysis of gene expression (SAGE), polony multiplex analysis of gene expression (PMAGE), and massively parallel signature sequencing (MPSS). SAGE has been utilized in three transcriptomic studies involving zebrafish, focusing on oogenesis, sexual dimorphism, and responses to toxicant stress. Nevertheless, many tag-based sequencing methods rely on costly Sanger sequencing; these approaches demand significant labour and time for the cloning process. Precise annotation of the tags is often unfeasible without complete genome information; only a fraction of transcripts is analysed, and distinguishing differential isoform and allelic expressions is typically challenging (Wang et al. 2009a, b).

The recent advent of affordable next-generation sequencing (NGS) technologies has revolutionized transcriptome mapping and measurement through RNA-seq. This method surpasses almost all the limitations of earlier techniques (see Table 5.1). Despite being relatively new, RNA-seq is already transforming transcriptomics, improving our knowledge of genome expression and regulation. Notably, when combined with advanced omics technologies, RNA-seq can analyse comprehensive personal omics profiles to assess disease risk and track disease progression for personalized treatment. Furthermore, RNA-seq has been extensively used in numerous studies related to fish biology (see Table 5.2). This chapter provides an overview of the RNA-seq method, covering the entire process from data generation to bioinformatics analysis. We address the challenges associated with RNA-seq and then examine the biological insights obtained from it across various fish species. We extend our apologies to the researchers whose work was not referenced due to space limitations.

Table 5.1 Comparing RNA-Seq with alternative transcriptome survey methods

	Tilling microarray	Tag-based sequencing	RNA-Seq
Principle	Hybridization	Sanger sequencing	Parallelized high-throughput sequencing
Resolution of data	Several to 100 bp	Single base	Single base
Sensitivity	Low	Moderate	High
Throughput	High	Low	High
Turnaround time	Long	Long	Short
Required amount of RNA samples	High	High	Low
Cost per sample (excluding equipment)	High	High	Relatively low
Reliance on existing genomic sequence	Yes	No	No
Linear dynamic range of expression levels	<2 orders of magnitude	Not practical	Limited only by sequencing depth
Discovery of unknown transcribed regions	Limited	Yes	Yes
Detection of differences in isoformic and allelic expressions	Limited	Limited	Yes
Detection of mutations	Limited	Yes	Yes
Determination of splicing sites	Limited	Yes	Yes
Identification of UTRs	Limited	Limited	Yes

5.2 Overview of RNA Sequencing

RNA-Seq (RNA sequencing) is a high-throughput sequencing technology that enables precise quantification and characterization of transcriptomes, playing a transformative role in transcriptomics by offering unparalleled precision in gene expression analysis. It involves RNA extraction, fragmentation, reverse transcription into cDNA, library preparation, sequencing, and bioinformatic analysis. Unlike traditional transcriptome profiling techniques, such as microarrays and quantitative PCR (qPCR), which rely on predesigned probes and have limited ability to detect novel transcripts, RNA-Seq offers a comprehensive and unbiased approach to characterizing the entire transcriptome. This technology surpasses conventional methods by detecting low-abundance transcripts, alternative splicing events, and non-coding RNAs, thereby offering deeper insights into gene expression, regulation, and functional genomics, revolutionizing aquatic organism research in disease resistance and immune responses. By leveraging next-generation sequencing (NGS) technologies, RNA-Seq facilitates the identification of gene expression patterns, alternative splicing events, and regulatory elements, offering deeper insights into cellular functions and biological responses. The RNA-Seq workflow consists of several critical stages, each contributing to the accuracy and reproducibility of the results. The process begins with RNA extraction, ensuring high-quality,

Table 5.2 Publications with fish transcriptomes analysed by RNA-Seq

Major application	Fish species	Sequencing platform
Transcriptome mapping and genome annotation	*Poecilia reticulate*	Roche 454
	Ictalurus punctatus	Illumina
	Danio rerio	Illumina
	Ictalurus punctatus	Illumina
	Nothobranchius furzeri	Illumina
Novel transcript discovery	*Labeorohita*, Hamilton	Illumina
	Danio rerio	Illumina
	Oncorhynchus mykiss	Illumina
	Salmo salar	Illumina
Detection of RNA splicing	*Danio rerio*	AB SOLiD
SNP discovery	*Ictalurus furcatus*	Illumina
	Cyprinus carpio	Illumina
	Oncorhynchus mykiss	Illumina
Quantification of gene expression	*Danio rerio*	AB SOLiD
	Lates calcarifer	Roche 454
	Lateolabrax japonicus	Illumina
	Fundulus heteroclitus	Roche 454
	Salmo trutta	Illumina
	Fundulus grandis	Illumina
	Perca flavescens	Roche 454
	Melanotaenia duboulayi	Illumina
	Coregonus clupeaformis spp., Salmonidae	Roche 454
	Gasterosteus aculeatus	Illumina
	Astyanax mexicanus	Roche 454

contaminant-free RNA, followed by library preparation, where RNA is fragmented, reverse-transcribed into cDNA, and adapted for sequencing (Zhao et al. 2019). Various sequencing platforms have been employed, including Illumina, which provides high-throughput, short-read sequencing ideal for differential gene expression analysis (Kukurba and Montgomery 2015), PacBio single-molecule real-time (SMRT) sequencing, known for generating long-read sequences that enhance isoform resolution and alternative splicing detection (Rhoads and Au 2015), and Oxford Nanopore technology, which enables real-time, portable sequencing with the ability to directly detect RNA modifications. The bioinformatics pipeline for RNA-Seq analysis involves multiple steps, including quality control, read alignment to reference genomes or transcriptomes, quantification of transcript abundance, differential expression analysis, and functional annotation. However, despite its advantages, RNA-Seq presents several challenges, such as sequence biases, mapping errors in non-model species, and the complexity of de novo transcriptome assembly in organisms lacking reference genomes. Solutions such as improved statistical models, robust normalization techniques, and the integration of multi-omics approaches have been developed to enhance data interpretation and biological relevance. RNA-seq utilizes next-generation sequencing (NGS) technologies to sequence complementary DNA (cDNA) directly from a chosen RNA sample. The transcriptome analysis via RNA-seq involves three main steps: constructing the library, sequencing on a designated NGS platform, and performing bioinformatics analysis. In the following section, we will delve into each step of a standard RNA-seq experiment, addressing the associated challenges and potential solutions.

5.3 RNA-Seq Workflow

The RNA sequencing (RNA-Seq) workflow involves multiple critical steps, including experimental design, library construction, sequencing, and bioinformatics analysis for data interpretation. Each step is pivotal in ensuring reliable and biologically meaningful transcriptome analysis. A well-structured experimental design is essential, as poor planning can lead to inconclusive results, loss of resources, and misinterpretation of biological processes. Key considerations include sequencing depth, transcriptome coverage, and the number of biological and technical replicates. The selection of the biological sample, whether tissue-specific or single-cell, is equally important, depending on the research objective. Traditional RNA-Seq experiments typically involve bulk tissue samples, which may contain heterogeneous cell populations, potentially confounding transcriptomic signals (Islam et al. 2011). To address this limitation, single-cell RNA sequencing (scRNA-Seq) has emerged as a powerful tool, enabling the investigation of cell-type-specific transcriptomes (Shapiro et al. 2013). However, scRNA-Seq presents challenges, particularly the low RNA concentration available for analysis. Despite these challenges, transcriptome analysis of complex tissues remains valuable for uncovering novel pathways and understanding disease mechanisms at the cellular level. Following experimental design, RNA preparation and library construction are performed, beginning with total RNA extraction from the chosen tissue or cell type. The integrity and quality of RNA must be assessed before proceeding, typically using the Agilent Bioanalyzer, which assigns an RNA Integrity Number (RIN) ranging from 1 to 10. A high RIN score indicates minimal degradation, whereas values below 6 can yield misleading results. Total RNA consists of ribosomal RNA (rRNA), pre-mRNA, messenger RNA (mRNA), and various noncoding RNAs (ncRNAs). Depending on the research focus, specific RNA species must be isolated. For mRNA sequencing, poly(A) enrichment is used, while ribo-depletion allows for sequencing of mRNA, pre-mRNA, and ncRNAs. Small RNA sequencing often employs size selection. Once the RNA species of interest are isolated, they are either fragmented before reverse transcription into double-stranded cDNA or reverse-transcribed first, followed by fragmentation. Fragmentation is crucial for improving transcript coverage during sequencing. The cDNA fragments are then ligated with sequencing adapters, completing the library construction step (Kukurba and Montgomery 2015). Advancements in next-generation sequencing (NGS) technologies have significantly enhanced the efficiency and affordability of transcriptomic studies. Platforms such as Illumina, Ion Torrent, and PacBio are commonly utilized, with Illumina dominating due to its high accuracy and scalability. Each platform employs distinct sequencing chemistries and adapter sequences, influencing read length and quality. Long-read sequencing (>1000 bp) is advantageous for de novo transcriptome assembly and the identification of alternative splicing events, while short-read sequencing (50–100 bp) provides higher coverage with lower error rates. The sequencing output is stored in FASTQ format and subsequently analysed using bioinformatics pipelines. The bioinformatics workflow of RNA-Seq begins with preprocessing, which includes removing adapter sequences, quality trimming, and filtering based on K-mer coverage. If a reference genome is available, the reads are aligned to it; otherwise, de novo assembly is performed to reconstruct the transcriptome. Alignment tools such as STAR, TOPHAT, GSNAP, and MAPslice are commonly used, while transcript assembly and quantification are performed using Cufflinks, iReckon, and FluxCapacitor. Differential gene expression analysis, crucial for understanding disease mechanisms and immune system complexity, is conducted using tools such as DESeq, Cuffdiff, EdgeR, and BaySeq. With continuous advancements in bioinformatics, RNA-Seq pipelines are constantly evolving to enhance accuracy, scalability, and computational efficiency, making them indispensable for transcriptomic studies in aquatic organisms (Kvam et al. 2012) (Fig. 5.1).

5.3.1 Library Construction

Library preparation is crucial for RNA-seq, as it significantly influences the accuracy of sequencing data in reflecting the original transcriptome. The initial procedure involves gathering suitable samples, typically tissues, for analysis. A key challenge is the inherently heterogeneous nature of tissues, which often consist of tens or even hundreds of distinct cell types. This diversity complicates the transcriptomic analysis of tissue, making it difficult to accurately capture the true profiles of individual cell types. One solution is to analyse single cells instead of entire cell populations. Various single-cell isolation methods have been developed and successfully combined with RNA-seq technology (Shapiro et al. 2013). Advances in single-cell RNA-seq now enable many new applications in fundamental and clinical research, such as studying early differentiation in embryogenesis, investigating tumour heterogeneity, and analysing rare, transient adult stem cells, among others.

After collecting samples, total RNA is typically extracted through organic methods and/or by binding to silica membranes in spin columns. Subsequently, RNA is transformed into a library of cDNA fragments. While total RNA can be utilized directly, it is generally necessary to fractionate it because ribosomal RNA (rRNA) constitutes over 80% of total cellular RNA and is typically not the primary focus of research. Its presence significantly limits the transcript coverage available in the sequencing process. Therefore, total RNA samples are either subjected to direct poly(A) RNA selection or to ribo-depletion, where rRNA is selectively removed (Costa et al. 2010). The widely used oligo(dT)-based mRNA purification technique in eukaryotes utilizes the poly(A) tail located at the 3′ end of eukaryotic

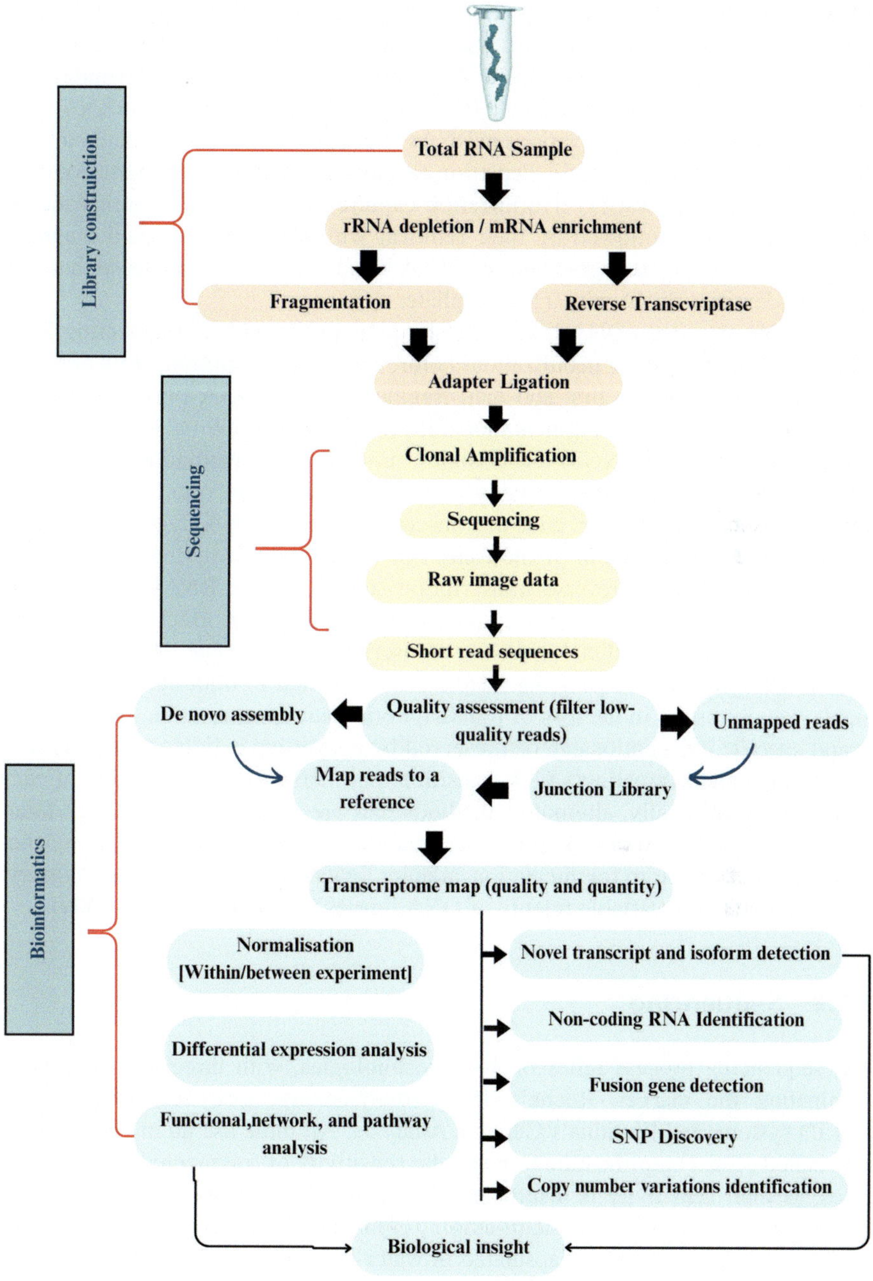

Fig. 5.1 Diagram of the RNA-Seq data analysis pipeline

mRNA. However, a significant portion of nonribosomal RNAs, both coding and noncoding, in eukaryotes do not possess a poly(A) tail and are thus often overlooked. In comparison to poly(A) RNA selection, ribo-depletion is favoured as it enriches all types of nonribosomal RNA, including non-poly (A) mRNA, preprocessed RNA, tRNA, and various ncRNAs with known or yet-to-be-defined functions. Although multiple rRNA depletion techniques exist, the two most common methods are as follows: (1) hybridization capture of rRNA using biotin-labelled anti-rRNA probes, followed by their removal with streptavidin-coated magnetic beads; and (2) targeted degradation of rRNA through a 5′-3′ exonuclease that specifically identifies rRNA with a 5′ phosphate.

A double-stranded cDNA library can be prepared by either fragmenting RNA (using RNA hydrolysis or nebulization) before reverse transcription, or by performing reverse transcription first and then fragmenting the cDNA (through DNase I treatment or sonication). Fragmentation aims to achieve the optimal length for next-generation sequencing (NGS) technologies; however, each method introduces different biases in the results. Notably, cDNA fragmentation often results in the underrepresentation of the 5′ ends of transcripts, whereas RNA fragmentation provides a good representation of the transcript body while depleting the transcript ends. Additionally, the reverse transcription step can make RNA-seq data more complex because reverse transcriptase might generate incorrect second-strand cDNA and produce artificial chimeric transcripts via template switching. Once cDNA is fragmented, sequencing adapters are attached to both ends of the fragments, which results in the loss of transcript orientation information. Nonetheless, strand-specific information can be preserved by converting cytidine to uridine with sodium bisulphate, creating a C–T transition that marks the coding strand of each transcript. Additionally, alternative methods that preserve strand specificity focus on ligating adapters to cDNA fragments, and detailed reviews of these approaches are available. Similar to fragmentation, adapter ligation can also cause non-uniform coverage due to the variable fidelity of DNA ligases (Faulhammer et al. 2000).

5.3.2 Sequencing

The sequencing process relies on NGS technologies, with three main platforms dominating the market: Roche's 454-pyrosequencing, Life Technologies' AB SOLiD system, and Illumina's Genome Analyzer. All three use an in vitro cloning step called clonal amplification to boost the sensitivity of fragmented cDNA molecules, as their sensitivity is insufficient for single-molecule sequencing (Metzker 2009). The 454 and SOLiD systems employ emulsion PCR, where cDNA fragments are attached to beads and encapsulated in water-in-oil emulsion droplets, ensuring each droplet contains only one DNA molecule. This method allows the separation of template fragments within these tiny droplets. In contrast, the Illumina Genome Analyser employs bridge PCR amplification, which involves immobilizing adapter-linked, single-stranded cDNA fragments on a glass slide through oligonucleotide hybridization, followed by clonal PCR amplification. This clonal amplification results in a collection of identical templates, each of which undergoes the subsequent sequencing reaction. However, PCR artefacts may introduce biases in RNA-seq results. To identify these artefacts, conducting biological replicates can help determine if the same short reads appear across different replicates (Wang et al. 2009a, b).

NGS platforms use diverse sequencing strategies. Roche 454 utilizes pyrosequencing, a bioluminescence-based technique that does not rely on electrophoresis. Here, each bead contains a clonally amplified cDNA template and is placed in a ~ well about 29 μm wide on a Picotiter Plate, a fiberoptic chip. Enzymes like DNA polymerase, ATP sulfurylase, and luciferase are included in each well. During sequencing, four DNA nucleotides are introduced in a specific sequence. When they bind to the complementary strand, a light signal is emitted, captured by the instrument's camera. The signal's intensity relates to the number of nucleotides incorporated. The SOLiD system employs sequencing by ligation (SBL). After clonal amplification, a sequencing primer hybridizes to the template, with its 3′ end at position 'n'. Fluorescently labelled di-base probes, 8 bp long, compete to ligate to the primer. The best match is connected via DNA ligase, and unligated probes are washed away. Fluorescence imaging identifies the ligated probe. Multiple ligation cycles are performed, then the extension product is removed, and a new primer is used for additional cycles. This process analyses each base twice with different primers, making SOLiD slower than other approaches. Finally, the Illumina platform employs cyclic reversible termination (CRT), a sequencing-by-synthesis method that uses reversible terminator nucleotides and DNA polymerase, differing from SBL. After incubation, unincorporated nucleotides are washed out. Imaging detects the incorporated base, then the dye and 3′ blocker are chemically removed for the next cycle. Ion semiconductor sequencing (Life Technologies) uses detection of released hydrogen ions during DNA polymerization, enabling fast and low-cost sequencing (Rothberg et al. 2011). Literature contains various reviews that discuss these mechanisms and compare their advantages and disadvantages. Despite the growth of NGS technologies, third-generation sequencing methods, or single-molecule sequencing, are also emerging. These techniques, such as Heliscope sequencing and single-molecule real-time (SMRT) (Pacific Biosciences), do not require clonal amplification and allow for real-time signal capture. Heliscope sequencing has already been utilized in an RNA-seq study. Furthermore, Heliscope's direct RNA sequencing (DRS) technology is evolving, enabling the direct sequencing of RNA molecules in a massively parallel manner, thereby avoiding biases introduced by RNA reverse transcription, ligation, and clonal amplification. However, these technologies are still in development and face challenges, including low single-pass accuracy (81%–83%), low sequencing efficiency, and low throughput.

5.3.3 Bioinformatic Analysis

After signal processing, NGS platforms generate millions of short sequences called reads, each associated with quality scores that indicate the confidence in each base call. These reads range from 25 to 450 base pairs in length, depending on the

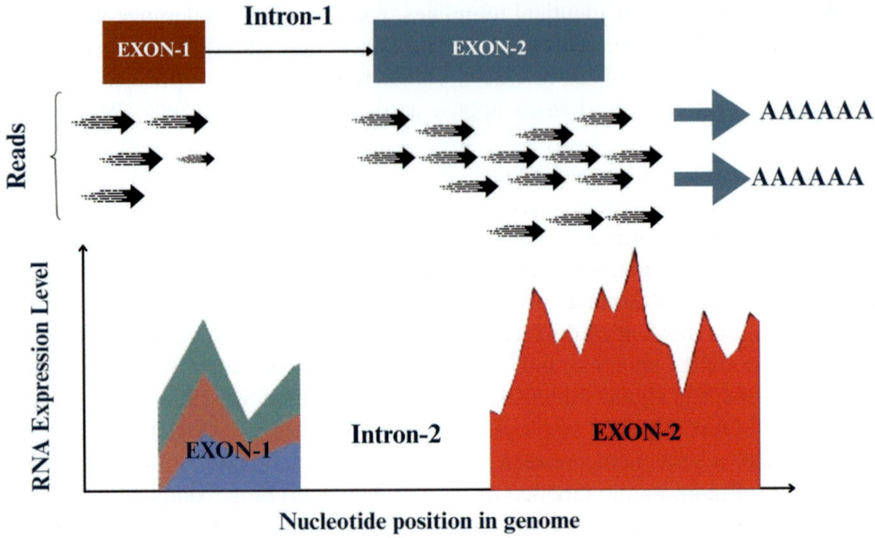

Fig. 5.2 Mapping short reads and quantifying the digital signal. (Adapted from Wang et al. 2009b). The aligned reads match a reference genome and are classified into three categories: exonic reads (green), junction reads (red-brown), and poly(A) end reads (blue). These reads are used to generate an expression profile with single-base accuracy

specific NGS platform. They are classified into three types: exonic reads, exon–intron junction reads, and poly(A) reads (Fig. 5.2). While NGS technology in RNA-seq simplifies access to transcriptomic sequences, analysing vast amounts of RNA-seq data remains a significant challenge in fully understanding the transcriptome. Fortunately, a range of bioinformatics tools for RNA-seq data analysis has emerged over the past few years. Notably, Bioconductor deserves particular attention. Bioconductor is a free, open-source software project that is developed collaboratively, mainly using the R programming language for analysing and annotating RNA-seq and other high-throughput genomic data (Gentleman et al. 2004). Most of its components are structured as R packages, and both these packages and other RNA-seq data analysis tools have been extensively reviewed in scientific literature. Today, researchers can effectively integrate these tools to formulate the most suitable analysis pipeline for their specific research objectives.

Figure 5.1 illustrates a typical RNA-seq data analysis pipeline. The first step is quality assessment; ensuring coherent outcomes requires the elimination of low-quality sequences, over-represented sequences, and adapter sequences, which can be accomplished using Bioconductor packages such as ShortRead and Biostrings (Morgan et al. 2009). After obtaining high-quality reads, the next step is aligning these short reads to a reference genome or transcriptome. This alignment process is typically performed using software that is not part of Bioconductor. Unlike traditional alignment algorithms, these tools utilize indexing strategies that enable the efficient alignment of millions of short reads within a reasonable timeframe (Langmead et al. 2009). Generally, these aligners can be categorized into two groups: those that utilize the Burrows-Wheeler transform algorithm, such as Bowtie and BWA, and those based on the Needleman-Wunsch or Smith-Waterman algorithms, including GNUMAP, BFAST, and SHRiMP (Rumble et al. 2009). The former category is known for its speed, while the latter, though slower, tends to be more sensitive and produces more accurately aligned reads. Because many reads span exon–exon junctions and cannot be aligned directly, specialized spliced aligners like Erange and IsoformEx have been created to isolate junction reads and align these read fragments separately. Literature provides comprehensive reviews on methods for analysing RNA splicing from short RNA-seq data. When a reference genome or transcriptome is not available, de novo transcriptome assembly can be used (Robertson et al. 2010). This approach is especially useful in fish studies that use RNA-seq, since genomic data is only available for a few fish species. De novo assembly tools such as ABySS and Velvet use a de Bruijn graph method, aligning user-defined sequence overlaps (k-mers) across reads to assemble contigs. Challenges in transcriptome de novo assembly stem from repetitive elements within extensive short reads, alternatively spliced transcripts, and uneven coverage of the transcriptome. To optimize transcriptome assembly, various strategies, including the use of different k-mer lengths rather than a single length, have been implemented. Furthermore, employing a reference proteome or genome from related species can aid de novo assembly processes. After all reads have been properly filtered and either mapped or assembled, they are counted, which enables the inference of gene expression levels based on the total read counts linked to the exons of specific genes. This process produces an expression score for each base, creating a genome-wide transcriptome map that indicates both quality and quantity (see Fig. 5.2). The single-base (digital) resolution of RNA-seq allows for the detection of gene expression at isoform and allelic levels, aiding in the discovery of previously unknown genes. RNA-seq not only measures gene expression levels within individual samples but also identifies differential expression (DE) between different treatments or conditions, which is often the main goal of many research studies. However, normalization must be performed in the DE analysis of RNA-seq data to account for differences between samples, such as library size and inherent gene-specific characteristics, including GC content and gene length. Due to the relatively recent introduction of RNA-seq technology, no standardized method exists for detecting DE. Popular tools for DE analysis within Bioconductor include edgeR, DESeq, baySeq, and tweeDESeq, as well as non-Bioconductor methods such as Shrink Bayes and TSPM. Detailed comparisons of these DE analysis tools and their gene ranking performances have been conducted. Furthermore, RNA-seq analysis supports the detection of single-nucleotide polymorphisms (SNPs), fusion genes, and post-transcriptional gene regulation processes, including RNA editing, degradation, and translation.

5.4 New Biological Insights from Fish Research Utilizing RNA-Seq

The advent of RNA-seq technology is revolutionizing fish transcriptomics. In recent years, it has been widely used across numerous studies involving different fish species. This includes not only model organisms, such as zebrafish, but also commercially significant and ecologically relevant species, such as channel catfish (*Ictalurus punctatus*), European sea bass (*Dicentrarchuslabrax*), and rainbow trout (*Oncorhynchus mykiss*) (c). Although the use of RNA-seq in fish transcriptomics is still in its early stages, a PubMed literature search using the keywords 'fish' and 'RNA-seq' reveals a notable rise in publications within this field over the past three years. To our knowledge, no comprehensive review has yet been conducted on this topic. Thus, the purpose of this section is to analyse the existing literature on fish transcriptomic studies employing RNA-seq.

5.4.1 Transcriptome Mapping and Genome Annotation

The high-throughput RNA-seq technology, which does not require prior knowledge, enables effective transcriptome annotation. This includes identifying transcript start and end sites, as well as discovering novel transcripts. Liu and colleagues sequenced and de novo assembled the transcriptome of the doubled haploid channel catfish using RNA-seq, producing 370,798 nonredundant transcript-derived contigs (Liu et al. 2012). Their functional annotation revealed 25,144 unique protein-coding transcripts. Of these, more than 14,000 were full-length transcripts with a complete open reading frame, and nearly 90% had both 5′ and 3′ ends fully identified. The 5′ UTRs were found to be significantly shorter, averaging approximately 254 bp, compared to the 3′ UTRs, which averaged around 1096 bp. Similarly, reference transcriptomes have been established for various fish species, such as zebrafish, guppy (*Poecilia reticulata*), and turquoise killifish (*Nothobranchius furzeri*). Creating a detailed reference transcriptome using RNA-seq offers essential information for genome annotation, including gene and exon boundaries, as well as the discovery of new transcribed regions. For example, in an effort to improve zebrafish genome annotation, RNA-seq data analyzed with an optimized pipeline was used to fine-tune intron/exon boundaries of specific genes, confirm their expression, and increase the coverage of the genes' 3′ untranslated regions (Collins et al. 2012). This enhanced analysis pipeline shows potential for advancing genome annotation in other organisms.

5.4.2 Novel Transcript Discovery

RNA-seq has been instrumental in identifying new transcribed regions within the genome. Findings from RNA-seq indicate a significant number of previously unknown transcribed areas across various fish species, including zebrafish, rohu carp (*Labeo rohita*), and rainbow trout. Notably, among these undiscovered transcripts, many noncoding RNAs (ncRNAs), aside from rRNA and tRNA, are particularly intriguing due to their crucial functions in various cellular processes, such as translation, RNA splicing, gene regulation, and genome defence. These ncRNAs can be categorized into long non-coding RNAs (lncRNAs, >200 nucleotides) and small RNAs (sRNAs, <200 nucleotides). Research has identified over 550 distinct, long intervening, noncoding RNAs (lincRNAs, a subtype of lncRNAs) in zebrafish embryos, while 224 unique mature microRNAs (miRNAs, a subtype of sRNAs) have been found in Atlantic salmon (*Salmo salar*). The discovery of these transcripts will enhance genome annotation and pave the way for future functional studies.

5.4.3 Interrogation of Post-transcriptional Modification (RNA Splicing)

The mature mRNA sequence can vary considerably from the genome sequence. This difference occurs because precursor mRNA undergoes three main post-transcriptional modifications—5′ capping, 3′ polyadenylation, and RNA splicing—in the cell nucleus before translation. Studies using RNA-seq have investigated RNA splicing in fish. Specifically, an analysis of the zebrafish transcriptome dynamics during the transition from maternal to zygotic stages revealed that the rate of alternative splicing in zebrafish embryos ranges from 50% to 60%. This figure, while significantly high, is much lower than the estimated splicing frequency in humans, which ranges from 92% to 95%. Another RNA-seq investigation in zebrafish embryos identified 3532 transcripts as novel isoforms and partial transcripts of known RefSeq genes, thereby enhancing the existing exon–intron structures of many zebrafish genes. These diverse splicing variants, alongside numerous newly transcribed regions, provide compelling evidence of previously overlooked transcriptome complexity. Furthermore, RNA-seq has also been applied to explore the regulation of alternative splicing in zebrafish. The genome-wide and target-specific functions of the U1C protein in recognizing and selecting 5′ splice sites have been characterized, thereby adding this protein to the expanding roster of known splicing regulators (Rösel et al. 2011).

5.4.4 SNP Discovery

SNP, or single-nucleotide polymorphism, refers to a variation at a specific location in the genome that differs among members of a biological species. These SNPs are distributed throughout the genome and are widely utilized as markers in various genomics and genetics applications. RNA-seq, renowned for its high sensitivity and ability to resolve single-nucleotide differences, has proven to be an effective method for identifying gene-associated single-nucleotide polymorphisms (SNPs) on a genome-wide scale. Numerous studies using RNA-seq to identify SNPs in fish have been published. In a study focused on creating SNP arrays for catfish, Liu and colleagues conducted RNA-seq on samples from both channel catfish and blue catfish (*Ictalurus furcatus*). They used the SNP calling feature in CLC Genomics Workbench (CLC bio, Aarhus, Denmark), combined with SNP quality screening process, they identify 342,104 intraspecific SNPs in channel catfish, 366,269 in blue catfish, and 420,727 interspecific SNPs between the two species. These SNPs were distributed across 16,562 unique genes in channel catfish and 17,423 unique genes in blue catfish. Another study used RNA-seq to detect gene-associated SNPs in four common carp (*Cyprinus carpio*) strains. BWA and SAMtools were used to align reads to a reference transcriptome and identify SNPs, revealing 712,042 intrastrain and 53,893 interstrain SNPs. These SNPs offer a solid basis for future genetic research on these fish and will support the creation of a high-throughput SNP genotyping platform.

5.4.5 Developmental Biology

Zebrafish is a crucial model organism, especially in developmental biology. Most RNA sequencing (RNA-seq) studies on fish development have focused on zebrafish. A comprehensive analysis used RNA-seq to compare transcriptome profiles across four early developmental stages: the 1-cell, 16-cell, 512-cell stages, and at 50% epiboly. This study found only 177 genes were developmentally regulated, with most gene transcripts remaining stable; a notable change in gene regulation and transcriptional activity occurred between the 512-cell and 50% epiboly stages. To explore the role of the vitamin D receptor (VDR) in embryogenesis, RNA-seq analyzed transcriptome changes in embryos and larvae treated with 1α,25(OH)2D3 (the active vitamin D3 form) or a control over different periods (Craig et al. 2012). Genes related to transcription factors, peptide hormones, receptor-activator of NFκB ligand (RANKL), and pathways involved in fatty acid, amino acid, and xenobiotic metabolism showed significant changes, suggesting that 1α,25(OH)2D3 affects multiple pathways during zebrafish development. Additionally, RNA-seq research on zebrafish retino-genesis identified Id2a protein as a key internal regulator that controls proliferation and differentiation by regulating Notch pathway gene expression.

5.4.6 Immunology

Comparing fish transcriptomes before and after immune challenges uncovers immune-related genes and pathways involved in immune responses. This research aids in developing immune-based therapies for fish diseases, selecting disease-resistant brood stocks, and understanding the origins and evolution of the immune system. RNA-seq has been used in studies involving live fish, fish embryos, and primary fish cells. Most RNA-seq studies on fish immune responses to pathogens focus on economically significant species. For example, a transcriptomic analysis of the channel catfish gill in response to *Flavobacterium columnare*, a Gram-negative bacterium linked to worldwide fish disease outbreaks, utilized Illumina sequencing. Using a 1.5-fold change cutoff, researchers identified 2605 uniquely annotated genes involved in pathogen recognition, cytoskeletal dynamics, junction integrity, oxidative stress responses, apoptosis, lysosomal activity, and both pro- and anti-inflammatory pathways, all exhibiting significant differential expression patterns. Quantitative PCR (qPCR) confirmed 15 differentially expressed genes identified by RNA-seq, including a rhamnose-binding lectin (RBL) gene, which exhibited a remarkable 10^5-fold increase in expression levels. This discovery prompted further research by the same team, revealing that RBL ligands, L-rhamnose, and D-galactose provided strong, dose-dependent protection for channel catfish against columnaris disease (Beck et al. 2012). Additionally, this group examined the transcriptomic changes in the intestinal epithelium of channel catfish after challenge with *Edwardsiella ictaluri*, identifying 1633 differentially expressed genes involved in actin cytoskeletal polymerization and remodelling, as well as junction regulation, both of which are crucial for pathogen entry and subsequent inflammatory responses. Similar investigations were conducted in Asian seabass (*Lates calcarifer*) and Japanese sea bass (*Lateolabrax japonicus*), both of which were subjected to *Vibrio harveyi* challenges.

5.4.7 Aquatic Toxicology

Aquatic toxicology examines the effects of toxic chemicals on aquatic life at multiple levels, ranging from subcellular interactions to individual organisms and communities, and even to entire ecosystems. Fish are among the primary subjects of research in this field. Utilizing RNA-seq, researchers can analyse the transcriptomic responses of these organisms to environmental pollutants. This foundational knowledge ultimately aims to foresee and mitigate the adverse effects of aquatic contaminants on ecosystems. For instance, Oleksiak and colleagues (2011) employed RNA-Seq in conjunction with previous microarray data to investigate how embryos and larvae from both polychlorinated biphenyl (PCB)-sensitive and PCB-resistant populations of Atlantic killifish (*Fundulus heteroclitus*) respond to polychlorinated biphenyl (PCB) exposure. Their differential expression analysis uncovered a substantial array of PCB-responsive genes in the sensitive population, a significantly smaller set in the resistant fish, and few overlapping genes between the two groups. The RNA-seq findings validated most of the results from the microarray and uncovered novel transcripts that had not been identified previously. Similar discrepancies between microarray and RNA-seq results have also been observed in other studies, which may be addressed through the use of appropriate statistical methods for RNA-seq to assess the significance of observed changes accurately (Esnaola et al. 2013). In a separate study, the effects of perfluorooctane sulfonate, a widely recognized persistent organic pollutant, on *Oryzias melastigma* embryos were explored using RNA-seq (Huang et al. 2012). Out of the 145,394 genes detected, 325 were significantly upregulated and 349 significantly downregulated. These differentially expressed genes were associated with neurobehavioral issues, mitochondrial dysfunction, and the metabolism of proteins and lipids. Recently, RNA-seq identified specific miRNAs in the muscle tissue of Atlantic salmon that could serve as potential biomarkers for toxicological stress (Kure et al. 2013). A total of 18 miRNAs were notably differentially expressed in response to exposure to acidic aluminium-rich water, with 4 miRNAs downregulated and 14 upregulated. These identified miRNAs might also serve as biomarkers for other fish species. Similar RNA-seq studies have been conducted on brown trout (*Salmo trutta*), Gulf killifish (*Fundulus grandis*, and yellow perch (*Perca flavescens*) (Pierron et al. 2011), each subjected to stressors such as a mixture of copper and zinc, Deepwater Horizon oil, and a mix of cadmium and copper, respectively. Overall, these investigations demonstrate that RNA-Seq is an effective tool for studying fish ecotoxicological responses in polluted environments.

5.4.8 Physiological Processes

Stress induces specific physiological reactions and plays a crucial role in the process of natural selection. In the wild, the unpredictable aquatic environment exposes fish to various stressors, including insufficient food, extreme salinity, temperature fluctuations, pathogens, toxic substances, and other environmental factors. These physiological reactions are reflected in changes in gene expression. Therefore, identifying differentially expressed genes and pathways during stressful conditions is vital for understanding fish stress responses. While previous sections have covered stress responses to pathogens and toxicants, we will now examine other types of stressors.

The liver transcriptomic response of the crimson-spotted rainbowfish (*Melanotaenia duboulayi*) to increased temperatures (21 °C compared to 33 °C) was analysed using RNA sequencing (Smith et al. 2013). Out of 107,749 assembled transcripts, 4251 were differentially expressed, with over 1000 annotated. Beyond the well-known temperature-responsive genes related to apoptosis protection and protein structure maintenance, gene ontology analysis revealed many novel temperature-responsive genes associated with catabolism, lipid metabolism, and oxidoreductase activity, indicating an increase in metabolic activity to manage elevated temperatures and associated hypoxic conditions. Similarly, RNA sequencing was used to evaluate the heat stress-induced gene expression profile in channel catfish (25 °C versus 36 °C), yielding findings consistent with those in crimson-spotted rainbowfish. These investigations indicate that fish exhibit considerable adaptability to variations in environmental temperature, supporting the development of heat-tolerant fish strains for aquaculture. Additionally, RNA sequencing was employed to investigate how temperature fluctuations during embryonic development impact adult thermal adaptation capabilities in zebrafish. Besides heat stress, transcriptomic responses to salinity and fasting have been investigated in Asian seabass.

5.4.9 Evolution Biology

Understanding the molecular foundation of population divergence and speciation presents a significant challenge in evolutionary biology. Gene expression divergence is believed to play a role in phenotypic variation. Recently, RNA-seq has emerged as a crucial method for examining phenotypic divergence, enabling whole-transcriptome sequencing. For example, RNA-seq analysis was utilized to investigate adaptive transcriptomic divergence between dwarf and normal lake whitefish (*Coregonus clupeaformis* spp.) (Jeukens et al. 2010). The findings indicated that normal whitefish show higher expression of genes involved in protein synthesis, whereas dwarf whitefish exhibited elevated expression of genes linked to immunity, DNA replication and repair, and energy metabolism. Additionally, a positive correlation was observed between the RNA-seq findings and prior microarray data. RNA-seq was also applied in studies of the genetic factors contributing to phenotypic differentiation between siscowet and lean lake trout (*Salvelinus namaycush*) (Goetz et al. 2010), the contrasting pigmentation patterns in marine and freshwater sticklebacks (*Gasterosteus aculeatus*), the adaptation of Mexican tetra (*Astyanax mexicanus*) to cave habitats, and the evolutionary relationship between zebrafish swim bladders and mammalian lungs. These transcriptomic investigations revealed numerous differentially expressed genes, setting the stage for future functional genomic research aimed at uncovering the heritable genetic modifications that drive phenotypic divergence during evolution (Qian et al. 2014).

5.5 Disease Mechanisms in Aquatic Organisms

5.5.1 Insights into Fish Immune Responses Against Bacterial Pathogens

The application of RNA-Seq technology has significantly enhanced our understanding of fish immune responses to bacterial infections, elucidating complex host-pathogen interactions and immune evasion strategies. Among the model organisms, the zebrafish (*Danio rerio*) remains extensively utilized due to its genetic tractability and transparent embryos, which facilitate real-time infection studies. Research on zebrafish embryos infected with *Salmonella typhimurium* has highlighted the critical role of Traf6 in innate immunity, while transcriptomic analyses have revealed dynamic changes in gene expression during infection (Ordas et al. 2011). RNA-Seq has enhanced our understanding of pathogen-host interactions, identifying immune functions in zebrafish muscle tissue during *Mycobacterium marinum* infection (Rougeot et al. 2019) and linking TIR domain variants to mycobacterial infections in common carp (*Cyprinus carpio*) (Zhang et al. 2011). In channel catfish (*Ictalurus punctatus*), *Edwardsiella ictaluri* infections revealed roles for actin polymerization, NLRs, Rab GTPase, and TLRs in pathogen entry and immune responses. RNA-Seq also identified biomarkers for disease resistance (Zhu et al. 2017). Transcriptomic studies on *Flavobacterium columnare* and *Aeromonas hydrophila* infections have uncovered mucosal immune shifts, bacterial adhesion mechanisms, and vaccine-induced responses in various fish species. In *Streptococcus agalactiae*-infected Nile tilapia, RNA-Seq identified TLR-mediated pathways and miRNA-driven immune responses. In contrast, global warming was found to impair immune gene function in Mozambique tilapia (*Oreochromis mossambicus*), exacerbating infections (Wang et al. 2010). RNA-Seq has been pivotal in studying *Vibrio* spp. infections in fish, uncovering key immune pathways. In the half-smooth tongue sole (*Cynoglossus semilaevis*) and Japanese sea bass (*Lateolabrax japonicus*), which were challenged with *Vibrio anguillarum*, numerous immune-related genes were identified. Transcriptomic studies on Japanese sea bass during *Vibrio harveyi* infection revealed novel fish-specific immune transcripts. Similarly, *Epinephelus coioides* exhibited rapid immune activation against *Vibrio alginolyticus*, including complement and antimicrobial peptide responses. *Epinephelus lanceolatus* showed TLR5-mediated cytokine regulation (Wang et al. 2010). RNA-Seq analysis of Asian sea bass (*Lates calcarifer*) revealed immune adaptability to Vibrio challenges and environmental stressors (Xia et al. 2013). Additionally, zebrafish vaccinated against *Edwardsiella tarda* exhibited differential MHC-I and MHC-II regulation (Yang et al. 2012). Similarly, oral vaccination of European sea bass (*Dicentrarchus labrax*) against *Vibrio anguillarum* resulted in the enhanced expression of immune-related rare transcripts, such as leukocyte immune-type receptors (Sarropoulou et al. 2012). *Yersinia ruckeri*, the causative agent of enteric redmouth disease, has been studied using RNA-Seq to elucidate its virulence factors (Kumar et al. 2016, 2017). Transcriptomic analyses of Amur sturgeon (*Acipenser schrenckii*) have identified immune-related pathways in response to *Y. ruckeri* infection (Li et al. 2017). Meanwhile, studies on the *Y. ruckeri* flagellar master operon flhD have highlighted its role in bacterial virulence. RNA-Seq technology continues to uncover novel insights into fish immune responses against bacterial pathogens, advancing our understanding of host-pathogen interactions, resistance mechanisms, and vaccine development strategies.

5.5.2 Insights into Fish Immune Responses to Viral Pathogens

Viral diseases pose a significant challenge to the aquaculture industry, causing severe economic losses and mortality in various fish species. During infection, once a virus surpasses initial host defences and establishes itself within host cells, a cascade of nonspecific and specific immune mechanisms is activated to limit viral spread and minimize cellular damage. Transcriptomic analyses using RNA-Seq technology have provided valuable insights into these immune responses, unravelling host-pathogen interactions and immune regulatory networks in infected fish. The grass carp (*Ctenopharyngodon idella*), a vital aquaculture species, is highly susceptible to grass carp reovirus (GCRV), a double-stranded RNA virus in the genus Aquareovirus (Lenoir et al. 1975). GCRV causes haemorrhagic disease with high mortality in early developmental stages. RNA-Seq analyses revealed that GCRV disrupts lipid and carbohydrate metabolism, with the spleen exhibiting stronger immune responses than the head kidney. Alternative splicing plays a crucial role in immune regulation, and complement and coagulation cascades were identified as key antiviral pathways. Similarly, viral haemorrhagic septicaemia virus (VHSV), an OIE-listed pathogen that infects over 80 fish species, poses a threat to salmonid aquaculture. RNA-Seq studies in rainbow trout (*Oncorhynchus mykiss*) uncovered genetic factors governing resistance and susceptibility to VHSV. The study found that VHSV suppresses metabolic functions, including ATP synthesis and antioxidant responses, ultimately leading to cellular necrosis (Genet et al. 2014). Similarly, the infectious salmon anaemia virus (ISAV), a segmented, single-stranded RNA virus of the Orthomyxoviridae family, causes systemic infections in the endothelial cells of salmonid blood vessels, leading to disease outbreaks. RNA-Seq analysis of *Salmo salar* during ISAV infection revealed key immune-related genes involved in viral replication and host-pathogen interactions, highlighting potential molecular targets for disease intervention (Dettleff et al. 2017).

Piscine Orthoreo Virus (PRV), a double-stranded RNA virus belonging to the Reoviridae family, causes heart and skeletal muscle inflammation (HSMI) in Atlantic salmon (*Salmo salar*) but not in sockeye salmon (*Oncorhynchus nerk*a), despite its active replication (Garver et al. 2016). Infectious hematopoietic necrosis virus (IHNV), a Rhabdoviridae virus, threatens fish farming worldwide. Transcriptomic studies in sockeye salmon kidneys revealed that PRV replication remains unaffected despite a robust antiviral response induced by IHNV coinfection. Salmonid alphavirus (SAV) causes severe pancreatic and muscular necrosis in *S. salar*, with RNA-Seq confirming a strong type I interferon response in infected macrophage-like cells. Additionally, cyprinid herpesvirus-3 (CyHV-3) causes high mortality in Cyprinus carpio, with spleen transcriptomics revealing novel immune-related antiviral. In European sea bass (*Dicentrarchus labrax*), transcriptomics studies have revealed that NNV evades cell-mediated cytotoxicity, leading to a fatal infection. Moreover, RNA-Seq data have been leveraged to develop customized microarrays for VNN studies in Atlantic cod (*Gadus morhua*). These studies highlight the power of RNA-Seq in dissecting fish immune responses, providing insights into host-virus interactions and viral immune evasion strategies. Such transcriptomic data provide crucial knowledge for developing antiviral strategies and enhancing disease resistance in aquaculture species.

5.5.3 Insights into Fish Immune Responses to Parasitic Infections

Parasitic infections pose significant challenges to both wild and cultured fish populations, affecting their health, survival, and economic viability. Fish are hosts to a diverse array of parasites, including protozoa, myxozoans, helminthic worms, and crustaceans, each of which employs sophisticated immune evasion strategies. These mechanisms include avoiding immune surveillance, adopting indirect life cycles, molecular mimicry, immune suppression, immune modulation, and even manipulating host endocrine and neural systems to establish infection.

Recent advances in RNA sequencing (RNA-Seq) have provided valuable insights into the molecular interactions between fish hosts and their parasites, offering new avenues for understanding immune responses and identifying potential targets for disease management. One extensively studied ectoparasite is *Cryptocaryon irritans*, which causes cryptocaryoniasis or 'white spot disease' in marine fish, leading to significant economic losses. RNA-Seq studies on infected large yellow croaker (*Larimichthys crocea*) have revealed profound immune responses involving the Toll-like receptor (TLR) pathway, chemokine signalling, complement system activation, and coagulation cascades. Notably, single-sequence repeats (SSRs) and single-nucleotide polymorphisms (SNPs) identified in these studies provide crucial genetic markers for future linkage mapping and quantitative trait locus (QTL) analysis. The alternative complement pathway plays a pivotal role in the early stages of *C. irritans* infection, facilitating pathogen recognition and immune activation. Interestingly, even low, nonlethal infections trigger significant immune responses, whereas severe infections result in an immunosuppression-like state in large yellow croaker (Yin et al. 2016). Moreover, transcriptomic analyses of *C. irritans*-infected orange-spotted grouper (*Epinephelus coioides*) have highlighted local immune responses characterized by leukocyte migration and TLR-associated pathogen recognition. In an effort to develop control strategies, RNA-Seq studies have been conducted on different developmental stages of *C. irritans* to identify potential vaccine candidates. Transcriptome profiling of the tomont stage under low temperatures has identified several genes critical for dormancy and parasite survival in adverse conditions, providing potential targets for diagnostic and therapeutic interventions. Additionally, transcriptomic data have revealed nine putative immobilization antigen transcripts and 161 protease transcripts, which could serve as key targets for vaccine development and drug design (Mo et al. 2016). Similarly, transcriptomic analyses of the salmon louse (*Caligus rogercresseyi*) across six developmental stages and adult sexes offer a valuable genomic resource for understanding host-parasite interactions and designing targeted control measures.

Beyond ectoparasites, RNA-Seq studies have shed light on immune responses to trematodes like *Diplostomum pseudospathaceum*. In three-spined stickleback (*Gasterosteus aculeatus*), infection with different *D. pseudospathaceum* genotypes alters immune gene expression, revealing genetic variation in host immunity (Haase et al. 2014). Further research shows that host adaptive immunity prioritizes broad-spectrum responses over strain-specific defences (Haase et al. 2016). Myxozoan parasites also elicit complex immune responses. In large yellow croaker, *Enteromyxum scophthalmi* infection triggers a strong local immune response in the pyloric caeca, while adaptive immunity remains subdued (Robledo et al. 2014). In turbot, *E. scophthalmi* activates interferon-mediated pathways in early infection (Ronza et al. 2016). *Tetracapsuloides bryosalmonae*, the causative agent of proliferative kidney disease, completes its life cycle in brown trout (*Salmo trutta*) but not rainbow trout (Oncorhynchus mykiss), indicating species-specific resistance (Kumar et al. 2015). The lack of transcriptomic data on *T. bryosalmonae* and its invertebrate host, *Fredericella sultana* remains.

Overall, RNA-Seq has significantly advanced our understanding of fish immune responses to parasitic infections by unravelling key molecular pathways involved in host defence. These findings provide a foundation for developing novel strategies to mitigate parasitic diseases in aquaculture, including the development of vaccines, targeted therapeutics, and selective breeding programs for disease-resistant fish strains. Future research should focus on integrating multi-omics approaches, including proteomics and metabolomics, to achieve a more comprehensive understanding of host-parasite interactions in fish and to identify biomarkers for early disease detection and intervention.

5.5.4 Insights into the Immune Architecture of Healthy Fish

Fish represent an evolutionarily ancient lineage of vertebrates, possessing a highly complex immune system that operates continuously to defend against pathogens. RNA-Seq-based transcriptome analysis has become a valuable tool for understanding the immune system in healthy fish, offering insights into immune mechanisms and their evolutionary significance even in the absence of infections. Unlike higher vertebrates, fish lack lymph nodes and bone marrow; however, most of the fundamental immune organs found in mammals, such as the kidney, spleen, thymus, and gut-associated lymphoid tissue (GALT), are present in fish. Among these, the kidney and spleen are most commonly used in RNA-Seq studies to decipher immune responses and host-pathogen interactions. In addition to primary immune organs, RNA-Seq has uncovered the expression of immune-related genes in various tissues, including the gills, swim bladder, skin, brain, muscle, gonads, heart, liver, and pancreas. Transcriptomic studies have demonstrated evolutionary conservation between fish and vertebrate immune systems. RNA-Seq analysis of the rainbow trout (*Oncorhynchus mykiss*) spleen highlighted homology between fish innates and adaptive immunity and that of higher vertebrates (Ali et al. 2014). Studies in Japanese flounder, naked carp, schizothoracine fish, top mouthculter, and grass carp have further expanded knowledge of immune-related genes and pathways. RNA-Seq has also uncovered evolutionary links between fish swim bladders and mammalian lungs, with *Takifugu rubripes* exhibiting immune-related pathway activity. In tilapia, adaptive evolution has driven the selection of immune genes Notch2 and nfatc3b (Xiao et al. 2015). Additionally, transcriptomic studies in European sea bass, elephant shark, and ayu fish have revealed tissue-specific immune cell populations and specialized immune responses (Venkatesh et al. 2014).

Single-cell RNA sequencing (scRNA-Seq) has also emerged as a powerful tool for studying haematopoiesis and the evolution of immune cells in fish. scRNA-Seq analysis of kidney-derived blood cells in zebrafish (*Danio rerio*) demonstrated a striking degree of conservation in haematopoietic processes between fish and higher vertebrates (Athanasiadis et al. 2017). Moreover, scRNA-Seq investigations of zebrafish spleen tissue have provided insights into the evolution of immune cells in lower vertebrates, offering a detailed view of immune system complexity in aquatic organisms (Carmona et al. 2017). These findings underscore the immense potential of RNA-Seq technology in elucidating the immune architecture of healthy fish. By providing a deeper understanding of immune gene expression, cellular responses, and evolutionary adaptations, RNA-Seq has expanded our knowledge of vertebrate immunology and paved the way for future research aimed at enhancing fish health, improving aquaculture sustainability, and developing disease resistance strategies. As transcriptomic studies continue to evolve, they will further unravel the complexities of fish immune systems, providing new insights into immune defence mechanisms and their implications for both natural populations and aquaculture settings.

5.6 Conclusion and Future Prospects

RNA-Seq has revolutionized aquatic health research by offering unprecedented insights into gene expression dynamics, host-pathogen interactions, and immune responses in diverse aquatic organisms. This technology has facilitated the identification of key molecular players involved in disease resistance, stress adaptation, and immune modulation, thereby advancing the understanding of transcriptomic complexity in fish and other aquatic species. Through its ability to capture both qualitative and quantitative changes in the transcriptome, RNA-Seq has paved the way for significant breakthroughs in fish immunogenetics, vaccine development, and disease management. The continuous refinement of sequencing protocols and bioinformatics tools is expected to further enhance the precision and applicability of this approach. Beyond disease research, RNA-Seq holds immense potential in promoting sustainable aquaculture practices and conservation strategies. By enabling the identification of genetic markers linked to disease resistance, growth performance, and environmental adaptability, this technology supports selective breeding programs aimed at developing robust and resilient fish strains. Moreover, RNA-Seq-based transcriptomic profiling contributes to assessing ecological stressors, pollution impacts, and climate change-induced shifts in aquatic ecosystems, thereby guiding conservation efforts and habitat restoration initiatives. The integration of transcriptomics with ecological studies can provide a deeper understanding of species responses to environmental changes, aiding in the formulation of effective management strategies for aquatic biodiversity.

Looking ahead, transcriptomics is set to play a transformative role in combating aquatic diseases and ensuring ecosystem health. Advances in high-throughput sequencing, single-cell RNA-Seq, and multiomics approaches will enable a more comprehensive analysis of host-pathogen interactions at the cellular and molecular levels. The growing interest in exploring non-coding RNAs, epigenetic modifications, and gene regulatory networks will further expand the scope of transcriptomic studies, offering novel targets for disease prevention and therapeutic interventions. As the cost of sequencing continues to decline and analytical pipelines become more sophisticated, RNA-Seq is poised to become an indispensable tool for aquatic health research, driving innovations in disease diagnostics, treatment strategies, and conservation genomics. By leveraging these advancements, researchers can unlock new frontiers in the health of aquatic organisms, ensuring the sustainability of aquaculture and the resilience of aquatic ecosystems in the face of emerging challenges.

References

Ali A, Rexroad CE, Thorgaard GH, Yao J, Salem M (2014) Characterization of the rainbow trout spleen transcriptome and identification of immune-related genes. Front Genet 5:348

Athanasiadis EI, Botthof JG, Andres H, Ferreira L, Lio P, Cvejic A (2017) Single-cell RNA-sequencing uncovers transcriptional states and fate decisions in haematopoiesis. Nat Commun 8(1):2045

Beck BH, Farmer BD, Straus DL, Li C, Peatman E (2012) Putative roles for a rhamnose binding lectin in Flavobacterium columnare pathogenesis in channel catfish Ictalurus punctatus. Fish Shellfish Immunol 33(4):1008–1015

Carmona SJ, Teichmann SA, Ferreira L, Macaulay IC, Stubbington MJ, Cvejic A, Gfeller D (2017) Single-cell transcriptome analysis of fish immune cells provides insight into the evolution of vertebrate immune cell types. Genome Res 27(3):451–461

Craig TA, Zhang Y, McNulty MS, Middha S, Ketha H, Singh RJ, Magis AT, Funk C, Price ND, Ekker SC and Kumar R (2012) Research resource: whole transcriptome RNA sequencing detects multiple 1α, 25- dihydroxyvitamin D3-sensitive metabolic pathways in developing zebrafish. Mol Endocrinol 26(9):1630–1642

Collins JE, White S, Searle SM, Stemple DL (2012) Incorporating RNA-seq data into the zebrafish. Genome Res 22(10):2067–2078

Costa PR, Acencio ML, Lemke N (2010) A machine learning approach for genome-wide prediction of morbid and druggable human genes based on systems-level data. BMC Genomics 11:1

Dettleff P, Moen T, Santi N and Martinez V (2017) Transcriptomic analysis of spleen infected with infectious salmon anemia virus reveals distinct pattern of viral replication on resistant and susceptible Atlantic salmon (Salmo salar). Fish & Shellfish Immunol 61:187–193

Esnaola M, Puig P, Gonzalez D, Castelo R, Gonzalez JR (2013) A flexible count data model to fit the wide diversity of expression profiles arising from extensively replicated RNA-seq experiments. BMC Bioinformatics 14:1–22

Fullwood MJ, Wei CL, Liu ET, Ruan Y (2009) Next-generation DNA sequencing of paired-end tags (PET) for transcriptome and genome analyses. Genome Res 19(4):521–532

Faulhammer D, Lipton RJ and Landweber LF (2000) Fidelity of enzymatic ligation for DNA computing. J Comput Biol 7(6):839–848

Garver KA, Johnson SC, Polinski MP, Bradshaw JC, Marty GD, Snyman HN, et al. (2016) Piscine orthoreovirus from western North America is transmissible to Atlantic salmon and Sockeye salmon but fails to cause heart and skeletal muscle inflammation. PLoS One 11:e0146229

Gentleman RC, Carey VJ, Bates DM, Bolstad B, Dettling M, Dudoit S, Zhang J (2004) Bioconductor: open software development for computational biology and bioinformatics. Genome Biol 5:1–16

Genet C, Verrier ER, Ciobotaru C, Klopp C, Esquerre D, Laloë D, Boudinot P and Quillet E (2014) RNA-Seq analysis of transcriptome response to VHS-V infection in two target tissues of resistant vs susceptible trout clonal lines. In Proceedings of the 10th World Congress on Genetics Applied to Livestock Production, Vancouver, BC, Canada (pp. 17–22)

Goetz F, Rosauer D, Sitar S, Goetz G, Simchick C, Roberts S, Mackenzie S (2010) A genetic basis for the phenotypic differentiation between siscowet and lean lake trout (Salvelinus namaycush). Mol Ecol 19:176–196

Haase D, Rieger JK, Witten A, Stoll M, Bornberg-Bauer E, Kalbe M, Reusch TB (2014) Specific gene expression responses to parasite genotypes reveal redundancy of innate immunity in vertebrates. PLoS One 9(9):e108001

Haase D, Rieger JK, Witten A, Stoll M, Bornberg-Bauer E, Kalbe M, Reusch TB (2016) Immunity comes first: the effect of parasite genotypes on adaptive immunity and immunization in three-spined sticklebacks. Dev Compar Immunol 54(1):137–144

Hayatsu H (2008) Discovery of bisulfite-mediated cytosine conversion to uracil, the key reaction for DNA methylation analysis—a personal account. Proceedings of the Japan Academy, Series B, 84(8):321–330

Huang Q, Fang C, Chen Y, Wu X, Ye T, Lin Y, Dong S (2012) Embryonic exposure to low concentration of bisphenol A affects the development of *Oryziasmelastigma* larvae. Environ Sci Pollut Res 19:2506–2514

Islam S, Kjällquist U, Moliner A, Zajac P, Fan JB, Lönnerberg P, Linnarsson S (2011) Characterization of the single-cell transcriptional landscape by highly multiplex RNA-seq. Genome Res 21(7):1160–1167

Jeukens J, Renaut S, St-Cyr J, Nolte AW, Bernatchez L (2010) The transcriptomics of sympatric dwarf and normal lake whitefish (Coregonus clupeaformis spp., Salmonidae) divergence as revealed by next-generation sequencing. Mol Ecol 19(24):5389–5403

Kukurba KR, Montgomery SB (2015) RNA sequencing and analysis. Cold Spring Harb Protoc 2015(11):pdb.top084970

Kumar G, Abd-Elfattah A, El-Matbouli M (2015) Identification of differentially expressed genes of brown trout (Salmo trutta) and rainbow trout (Oncorhynchus mykiss) in response to Tetracapsuloidesbryosalmonae (Myxozoa). Parasitol Res 114:929–939

Kumar G, Hummel K, Ahrens M, Menanteau-Ledouble S, Welch TJ, Eisenacher M, El-Matbouli M (2016) Shotgun proteomic analysis of Yersinia ruckeri strains under normal and iron-limited conditions. Vet Res 47:1–13

Kumar G, Hummel K, Welch TJ, Razzazi-Fazeli E, El-Matbouli M (2017) Global proteomic profiling of Yersinia ruckeri strains. Vet Res 48:1–11

Kure EH, Sæbø M, Stangeland AM, Hamfjord J, Hytterød S, Heggenes J, Lydersen E (2013) Molecular responses to toxicological stressors: profiling microRNAs in wild Atlantic salmon (Salmo salar) exposed to acidic aluminum-rich water. Aquat Toxicol 138:98–104

Kvam VM, Liu P, Si Y (2012) A comparison of statistical methods for detecting differentially expressed genes from RNA-seq data. Am J Bot 99(2):248–256

Langmead B, Trapnell C, Pop M, Salzberg SL (2009) Ultrafast and memory-efficient alignment of short DNA sequences to the human genome. Genome Biol 10:1–10

Lenoir G and De Kinkelin P (1975) Fish rhabdoviruses: comparative study of protein structure. J Virol 16(2):259–262

Li S, Zhang Y, Cao Y, Wang D, Liu H, Lu T (2017) Trancriptome profiles of Amur sturgeon spleen in response to Yersinia ruckeri infection. Fish Shellfish Immunol 70:451–460

Lindberg J, Lundeberg J (2010) The plasticity of the mammalian transcriptome. Genomics 95(1):1–6

Liu S, Zhang Y, Zhou Z, Waldbieser G, Sun F, Lu J, Liu Z (2012) Efficient assembly and annotation of the transcriptome of catfish by RNA-Seq analysis of a doubled haploid homozygote. BMC Genomics 13:1–18

Metzker ML (2009) Sequencing in real time. Nat Biotechnol 27(2):150–151

Mo ZQ, Li YW, Wang HQ, Wang JL, Ni LY, Yang M, Dan XM (2016) Comparative transcriptional profile of the fish parasite *Cryptocaryon irritans*. Parasit Vectors 9:1–12

Morgan M, Anders S, Lawrence M, Aboyoun P, Pagès H and Gentleman R (2009) ShortRead: a bioconductor package for input, quality assessment and exploration of high-throughput sequence data. Bioinformatics 25(19):2607–2608

Ordas A, Hegedus Z, Henkel CV, Stockhammer OW, Butler D, Jansen HJ, Meijer AH (2011) Deep sequencing of the innate immune transcriptomic response of zebrafish embryos to Salmonella infection. Fish Shellfish Immunol 31(5):716–724

Pierron F, Normandeau E, Defo MA, Campbell PG, Bernatchez L, Couture P (2011) Effects of chronic metal exposure on wild fish populations revealed by high-throughput cDNA sequencing. Ecotoxicology 20:1388–1399

Qian X, Ba Y, Zhuang Q, Zhong G (2014) RNA-Seq technology and its application in fish transcriptomics. Omics 18(2):98–110

Rhoads A, Au KF (2015) PacBio sequencing and its applications. Genomics Proteomics Bioinformatics 13(5):278–289

Robertson G, Schein J, Chiu R, Corbett R, Field M, Jackman SD, Birol I (2010) De novo assembly and analysis of RNA-seq data. Nat Methods 7(11):909–912

Robledo D, Ronza P, Harrison PW, Losada AP, Bermúdez R, Pardo BG, José Redondo M, Sitjà Bobadilla A, Quiroga MI and Martínez P (2014) RNA-seq analysis reveals significant transcriptome changes in turbot (Scophthalmus maximus) suffering severe enteromyxosis. Bmc Genomics 15(1):1149

Ronza P, Robledo D, Bermúdez R, Losada AP, Pardo BG, Sitjà-Bobadilla A, Martínez P (2016) RNA-seq analysis of early enteromyxosis in turbot (Scophthalmus maximus): new insights into parasite invasion and immune evasion strategies. Int J Parasitol 46(8):507–517

Rougeot J, Torraca V, Zakrzewska A, Kanwal Z, Jansen HJ, Sommer F, Spaink HP and Meijer AH (2019) RNAseq profiling of leukocyte populations in zebrafish larvae reveals a cxcl11 chemokine gene as a marker of macrophage polarization during mycobacterial infection. Front Immunol 10:832

Rösel TD, Hung LH, Medenbach J, Donde K, Starke S, Benes V, Bindereif A (2011) RNA-Seq analysis in mutant zebrafish reveals role of U1C protein in alternative splicing regulation. EMBO J 30(10):1965–1976

Rothberg JM, Hinz W, Rearick TM, Schultz J, Mileski W, Davey M, Bustillo J (2011) An integrated semiconductor device enabling non-optical genome sequencing. Nature 475(7356):348–352

Rumble SM, Lacroute P, Dalca AV, Fiume M, Sidow A, Brudno M (2009) SHRiMP: accurate mapping of short color-space reads. PLoS Comput Biol 5(5):e1000386

Sarropoulou E, Galindo-Villegas J, García-Alcázar A, Kasapidis P, Mulero V (2012) Characterization of European sea bass transcripts by RNA SEQ after oral vaccine against V. anguillarum. Mar Biotechnol 14:634–642

Schena M, Shalon D, Davis RW, Brown PO (1995) Quantitative monitoring of gene expression patterns with a complementary DNA microarray. Science 270(5235):467–470

Shapiro E, Biezuner T, Linnarsson S (2013) Single-cell sequencing-based technologies will revolutionize whole-organism science. Nat Rev Genet 14(9):618–630

Smith S, Bernatchez L, Beheregaray LB (2013) RNA-seq analysis reveals extensive transcriptional plasticity to temperature stress in a freshwater fish species. BMC Genomics 14:1–12

Venkatesh B, Lee AP, Swann JB, Ohta Y, Flajnik MF, Kasahara M, Warren WC (2014) Venkatesh et al. reply. Nature 511(7508):E9–E10

Wang SJ, Wong G, de Heer AM, Xia W, Bourguignon LY (2009a) CD44 variant isoforms in head and neck squamous cell carcinoma progression. Laryngoscope 119(8):1518–1530

Wang C, Xin X, Xiang R, Ramos FJ, Liu M, Lee HJ, Dong LQ (2009b) Yin-Yang regulation of adiponectin signaling by APPL isoforms in muscle cells. J Biol Chem 284(46):31608–31615

Wang Y, Chen P, Cui R, Si W, Zhang Y, Ji W (2010) Heavy metal concentrations in water, sediment, and tissues of two fish species (*Triplohysapappenheimi, Gobio hwanghensis*) from the Lanzhou section of the Yellow River, China. Environ Monit Assess 165:97–102

Xia JH, Lin G, He X, Liu P, Liu F, Sun F, Yue GH (2013) Whole genome scanning and association mapping identified a significant association between growth and a SNP in the IFABP-a gene of the Asian seabass. BMC Genomics 14:1–12

Xiao J, Zhong H, Liu Z, Yu F, Luo Y, Gan X, Zhou Y (2015) Transcriptome analysis revealed positive selection of immune-related genes in tilapia. Fish Shellfish Immunol 44(1):60–65

Yang M, Lv Y, Xiao J, Wu H, Zheng H, Liu Q, Wang Q (2012) Edwardsiella comparative phylogenomics reveal the new intra/inter-species taxonomic relationships, virulence evolution and niche adaptation mechanisms. PLoS One 7(5):e36987

Yin F, Gao Q, Tang B, Sun P, Han K, Huang W (2016) Transcriptome and analysis on the complement and coagulation cascades pathway of large yellow croaker (Larimichthyscrocea) to ciliate ectoparasite Cryptocaryon irritans infection. Fish Shellfish Immunol 50:127–141

Zhang Y, Stupka E, Henkel CV, Jansen HJ, Spaink HP, Verbeek FJ (2011) Identification of common carp innate immune genes with whole-genome sequencing and RNA-Seq data. J Integr Bioinform 8(2):165–175

Zhao L, Zhang H, Kohnen MV, Prasad KV, Gu L, Reddy AS (2019) Analysis of transcriptome and epitranscriptome in plants using PacBio Iso-Seq and nanopore-based direct RNA sequencing. Front Genet 10:253

Zhu L, Ni W, Liu S, Cai B, Xing H, Wang S (2017) Transcriptomics analysis of apple leaves in response to Alternaria alternata apple pathotype infection. Front Plant Sci 8:22

6 Innovations in CRISPR Technology for Aquaculture and Fisheries

Suvra Roy, Vikash Kumar, and Basanta Kumar Das

6.1 Introduction

Food security has emerged as a significant global issue due to rising population and demand for animal protein. Aquaculture serves as a vital source of high-quality animal protein for human nutrition, playing a crucial role in enhancing global food security (FAO 2020). Gene editing introduces innovative options, and its application in aquaculture represents an effective strategy to realize its full potential, making nutrient-rich foods more accessible worldwide (Hallerman 2021). Initially, genome editing utilized zinc finger nucleases (ZFNs) and transcription activator-like endonucleases (TALEN). Currently, the most advanced genome editing technique, the clustered regularly interspaced palindromic repeats (CRISPR) system, is globally dominant (Jinek et al. 2012; Doudna and Charpentier 2014) and offers a cost-effective and efficient alternative for targeted genome editing compared to ZFN and TALEN. Researchers have rapidly applied the CRISPR/Cas system to various plant, animal, and fish species, with no indication of slowing down. As a powerful and user-friendly tool, CRISPR/Cas editing represents a promising new breeding technology for aquaculture, having recently targeted traits in a few aquaculture species; however, significant challenges remain before it can be widely adopted in commercial aquaculture. This discussion will cover the current understanding of CRISPR/Cas genome editing technology, its present status, challenges, and future prospects in aquaculture applications (Roy et al. 2022; Momose and Concordet 2016).

S. Roy · V. Kumar · B. K. Das (✉)
ICAR-Central Inland Fisheries Research Institute, Barrackpore, Kolkata, India

© The Author(s), under exclusive license to Springer Nature Singapore Pte Ltd. 2026

B. K. Das et al. (eds.), *Bioinformatics Tools for Predictive Ecology and Fisheries*, Sustainable Aquatic Food Systems, https://doi.org/10.1007/978-981-95-2573-7_6

6.2 Fundamental Characteristics of Natural CRISPR Systems and the CRISPR/Cas9 Toolkit for Genome Editing

The CRISPR/Cas9 genome editing toolkit was developed based on the discovery of the CRISPR locus's role in bacterial adaptive immunity. This locus protects bacteria and archaea from viruses and plasmids by cleaving and degrading foreign DNA. It produces numerous CRISPR RNAs (crRNA), which are short RNAs matching 20-base pair segments of invader DNA, along with transactivating CRISPR RNA (tracrRNA) and Cas proteins. In class I systems, such as those in *Escherichia coli*, multiple Cas proteins work together to enable DNA cleavage. Conversely, class II systems utilize a single protein for both target identification and DNA cleavage. The first-class II all-in-one effectors were Cas9 proteins derived from *Streptococcus pyogenes* and *Streptococcus thermophilus* (Jinek et al. 2012; Lander 2016).

Cas9 interacts with guide RNAs formed by the partial hybridization of crRNA and tracrRNA. In genome editing, a single-guide RNA (sgRNA), which combines crRNA and tracrRNA, is often used. DNA sequences are only cleaved if they contain a specific protospacer-adjacent motif (PAM) located 3′ to the target. For *S. pyogenes* Cas9, the PAM is NGG, whereas for *S. thermophilus*, it is NGGNG. Importantly, the CRISPR locus lacks PAM motifs near crRNA sequences, preventing self-cleavage in the bacterial genome. *S. pyogenes* Cas9 is more frequently employed in genome editing because it is easier to find target sequences compatible with its PAM than those for *S. thermophilus* Cas9.

Research into biochemistry, structure, and biophysics has deepened our understanding of how Cas9 cleaves DNA, which is essential for genome editing applications (Fig. 6.1) (Sternberg and Doudna 2015). Initially, Cas9 binds to the guide composed of crRNA and tracrRNA and searches the genome for PAM motifs. Upon finding a PAM, Cas9 unwinds the DNA upstream, allowing the crRNA to potentially anneal in a 3′ to 5′ direction. If annealing occurs, Cas9 cuts both DNA strands 3 base pairs upstream of the PAM. It has two catalytic domains, each cleaving one DNA strand. This makes DNA cleavage specificity dependent on both the PAM recognition by Cas9, especially from species other than *S. pyogenes*, which tend to recognize longer PAMS for higher specificity, and the crRNA sequence. For example, Cas9 from *S. aureus*, which is smaller than SpCas9, enables easier integration into AAV vectors for use in mammalian models. Initially, SpCas9 was thought to have low specificity, tolerating up to five mismatches in the guide RNA during experiments (Jinek et al. 2012). In cultured cells, some guide RNAs caused off-target effects with similar mismatches. It is now clear that the guide RNA sequence heavily influences the DNA cleavage specificity. Genome-wide studies revealed that SpCas9's off-target sites always correspond to closely related sequences, aligning with the guide target through a few mismatches. This off-target issue can thus be addressed through bioinformatics (Haeussler and Concordet 2016). The best approach is to avoid using guides that might target off-site sequences that could be cleaved significantly, typically those with fewer than three mismatches to the target (Kim et al. 2016) or when the mutations might be problematic, such as those located in exons or on the same chromosome as the target.

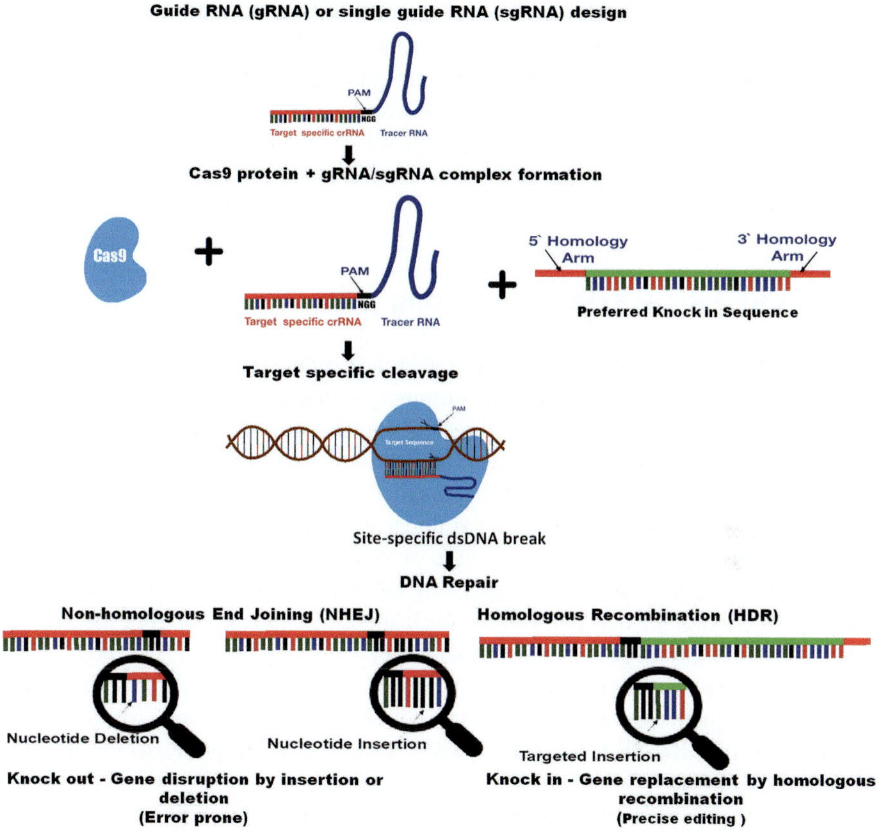

Fig. 6.1 This schematic illustrates the CRISPR/Cas-based genome editing technique. The guide RNA (gRNA) or single-guide RNA (sgRNA) consists of a CRISPR RNA (crRNA) sequence that precisely binds to DNA, and a transactivating crRNA (tracrRNA) segment that interacts with the Cas9 protein. The target site within the gRNA or sgRNA contains a species-specific short protospacer adjacent motif (PAM) sequence, specifically NGG. The gRNA/sgRNA guides the Cas9 endonuclease to a specific genomic location, where the CRISPR/Cas9 complex binds and cuts the DNA. The site-specific DNA cleavage by Cas9 occurs through complementary base pairing between the gRNA or sgRNA and the target sequence, contingent upon the presence of a PAM sequence on the opposite strand. The Cas9 enzyme cuts both strands of genomic DNA, leading to a double-strand break (DSB). This CRISPR-Cas-induced double-strand break (DSB) is rectified through either nonhomologous end-joining (NHEJ) or homologous recombination (HRR). While NHEJ is generally more efficient, it is prone to errors that can create unpredictable insertions or deletions (indels), whereas HDR enables precise gene modifications

In species without a complete genome sequence, off-targets cannot be effectively reduced through bioinformatics. To rigorously eliminate concerns of off-target effects, confirming phenotypes in multiple independent founder animals and employing two unrelated guide RNAs to create mutant animals is essential (Iyer et al. 2015). Using highly specific variants like SpCas9-HF1 and eSpCas9(1.1) can reduce off-target concerns because these newer variants block DNA cleavage even with just one mismatch, eliminating off-target effects seen with wild-type Cas9. Moreover, positively charged amino acids that help Cas9 interact nonspecifically with the DNA strand complementary to the guide RNA or the nontarget strand (Slaymaker et al. 2016) were changed to Alanine to prevent unintended cleavage. The on-target activity of the examined guides was only slightly affected, but further investigation is necessary to determine if this consistency holds with more guides tested.

6.3 CRISPR/Cas9 Genome Editing

Genome editing with CRISPR/Cas9 builds on the fundamental principle first shown with the I-SceI nuclease in human cell cultures. In that case, efficient DNA modifications happen during repair after targeted double-strand breaks (Carroll 2014). This concept originated from the understanding that DNA breaks can trigger homologous recombination, especially during meiosis and yeast mating-type switching. Early programmable nucleases like zinc finger nucleases (ZFNs) and TALE nucleases (TALENs) faced the main challenge of designing nucleases that specifically target the desired gene. In contrast, targeting DNA cleavage with CRISPR/Cas9 has become much simpler and more efficient; it merely requires changing the crRNA sequence to complement a 20-nucleotide target sequence plus a PAM motif (Mali et al. 2013). The current challenges lie in (i) selecting the optimal strategy to ensure the desired sequence modification occurs post-DNA cleavage, and (ii) optimizing the delivery of CRISPR/Cas9 components, specifically the guide RNAs and the Cas9 endonuclease.

DNA repair of double-strand breaks occurs through several molecular pathways, including nonhomologous end-joining (NHEJ) and homologous recombination (HR), as well as several lesser-known but significant pathways such as microhomology-mediated end-joining (MMEJ) (Sfeir and Symington 2015; Symington and Gautier 2011) (Fig. 6.2a). As indicated by their titles, these pathways differ primarily in the role and extent of sequence homology involved in the repair process. NHEJ operates without sequence homology, and directly ligating DNA ends typically restores the original sequence accurately (Bétermier et al. 2014). However, during genome editing, subsequent cleavage can lead to small deletions or insertions during the repair process. In homologous recombination (HR), resection exposes single-stranded DNA on both sides of the break, which then searches for a homologous DNA template within the nucleus. In genome editing, donor DNA with homologous arms flanking the break can be introduced to guide HR, allowing precise modification of the target sequence to match the donor. The mutations introduced may include point mutations or insertions of up to 10 kilobases, as observed in human cultured cells and common animal models (mouse, rat, and *Drosophila melanogaster*). Typically, homology arms range from several hundred to about 1000 base pairs. In MMEJ, 2–10 base pairs of identical sequences on either side of the break anneal after DNA resection, followed by processing of the nonannealed flaps, before ligation completes the repair. While NHEJ and MMEJ efficiently

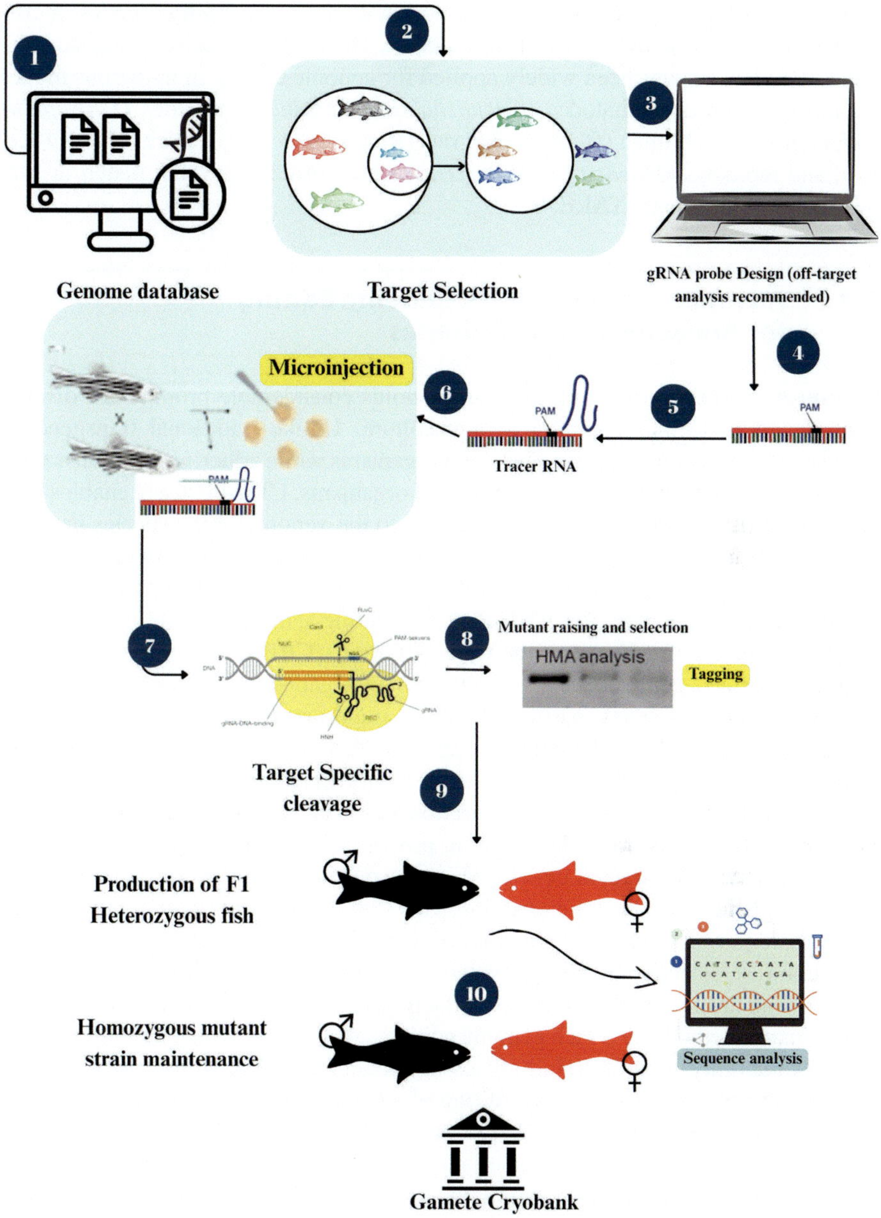

Fig. 6.2 A straightforward roadmap outlining the general methodology for CRISPR/Cas genome editing in aquaculture and fisheries. First, the target gene must be selected by searching the genome database of relevant species. Next, sgRNA should be designed using sgRNA-designing tools, followed by synthesizing the sgRNA oligo. For target-specific cleavage, deliver the sgRNA and Cas9 mixture to the newly fertilized embryo at the one-cell stage via microinjection or similar techniques. The final step involves assessing the genome-editing outcomes, which includes mutagenesis analysis, selecting mutants, crossing with wild populations, and producing a specific mutant line. Additionally, evaluate the phenotypes associated with CRISPR-induced mutations and establish new varieties with enhanced value in aquaculture.

generate small mutations, usually deletions that disrupt open reading frames or create knockouts (KOs) for gene function studies, HR enables precise sequence editing. CRISPR/Cas9 has been widely applied for genome editing in numerous model organisms, such as nematodes (*Caenorhabditis elegans*), fruit flies (*Drosophila melanogaster*), zebrafish (*Danio rerio*), rats (*Rattus norvegicus*), mice (*Mus musculus*), and monkeys (*Macaca fascicularis*), reaching efficiencies that match or surpass those of ZFNs and TALENs.

6.4 Advances in CRISPR/Cas Genome Editing for Aquaculture and Fisheries

CRISPR/Cas9 genome editing technology holds considerable promise for disease control and molecular breeding in aquaculture. Unlike traditional transgenesis, which entails transferring a gene between organisms while adhering to public safety regulations concerning genetically modified organisms, CRISPR/Cas9 enables precise, targeted, and often minor modifications to the genome of the species in question without introducing foreign DNA. This approach effectively addresses major public safety concerns surrounding GMOs (Okoli et al. 2021). The in vitro synthesis of sgRNA facilitates multiplex CRISPR/Cas9 editing at multiple gene sites, increasing the likelihood of achieving gene knockout events. CRISPR/Cas9 has fundamentally transformed genetics across several model organisms and is now utilized in various nonmodel species, particularly those important to aquaculture. This technology represents a novel breeding approach that enhances sustainable genetic improvement in aquaculture, owing to its high efficiency and flexibility in gene targeting without external gene integration (Blix et al. 2021; Hallerman 2021). Genome editing accelerates the breeding process with pinpoint accuracy, enabling changes to occur in targeted genes and increasing the likelihood of expressing the desired traits in subsequent generations. This trait transmission probability is high, typically requiring only 1 to 3 generations to establish a new breed.

In contrast, selective breeding involves a random process with a low probability of achieving desired traits, often taking many generations to produce a stable breed. The primary methods by which genome editing can enhance aquaculture production and sustainability include fixing alleles at existing quantitative trait loci (QTLs), targeted introgression through the editing of alleles from various strains or species, and generating de novo variants based on trait knowledge (Gratacap et al. 2019). The life history and culture characteristics of many aquaculture species, including high fecundity, short generation times, external egg fertilization, and embryonic and larval development outside the mother, along with established methods for artificial spawning and larval rearing, offer a significant advantage for genome editing research compared to farmed terrestrial species animals.

Conducting CRISPR/Cas9 genome editing experiments in farmed fish involves several crucial steps (Fig. 6.2). Initially, the target gene must be identified through genome databases of potential species. Next, sgRNA must be designed using sgRNA-design tools, followed by the synthesis of sgRNA oligos. Various methods for genome editing, such as recombinant vectors (which contain sgRNA tailored to the target DNA sequence along with the Cas9 protein sequence) or a cloning-free PCR technique that amplifies the gRNA template for transcription, can be directly transfected into cells. Alternatively, the lentivirus CRISPR system can be utilized for in vitro gene editing. For in vivo targeting, the sgRNA and Cas9 mixture must be delivered to newly fertilized embryos at the one-cell stage via techniques such as microinjection or electroporation. The preferred method for inducing in vivo mutations in aquaculture species is microinjecting the CRISPR/Cas9 complex, consisting of SgRNAs and Cas9, into one-cell-stage embryos. For heritable modifications through gene editing to occur, they must happen very early in development, ideally at the one-cell stage, to ensure incorporation into the germline. This necessitates specialized techniques, such as microinjection, for the precise delivery of the CRISPR/Cas complex. Other methods for delivering the CRISPR/Cas9 complex into one-cell fertilized eggs or cultured cells include lentivirus-mediated approaches, electroporation, and ribonucleoprotein complexes (Okoli et al. 2021; Yang et al. 2022). Thorough screening of mutants through sequence analysis is crucial, along with assessing off-target effects. In the F0 generation, mutants exhibit various mosaic mutation patterns, which can complicate phenotypic identification. As a result, off-target-free candidate founder mutants should be crossed with wild populations, and a heterozygous mutant population should be maintained. Crossing F1 heterozygotes produces homozygous F2 offspring. The final stage involves selecting phenotypes with CRISPR-induced mutations and developing new varieties with improved traits for aquaculture (Luo et al. 2022). All phenotypic assessments should be performed using at least the F3 homozygous generation progeny.

6.5 The Present State of CRISPR/Cas Genome Editing in Aquaculture and Fisheries

The CRISPR/Cas technology offers new possibilities, with research in aquaculture showing that this system is user-friendly, highly effective, and has reduced off-target effects. It can also delete long DNA segments. Recent studies on the CRISPR/Cas9 system have explored gene functions and breeding in over 20 key aquaculture species, such as Atlantic salmon, Nile tilapia, red sea bream, various catfish, rainbow trout, blunt snout bream, and farmed carps like Rohu, grass, and common carp. Other species include Pacific oyster, Atlantic killifish, fighting fish, Chinese tongue sole, olive flounder, and several prawn species (Blix et al. 2021). So far, extensive CRISPR research in aquaculture has complemented traditional genetic improvement efforts, targeting traits like growth, disease resistance, reproduction, sterility, and coloration. Genes frequently targeted include myostatin (mstn), a key regulator of skeletal muscle growth and fertility; sex-related genes such as dead end (dnd) and dmrt; as well as slc45a2, involved in pigment synthesis, and immunity-related genes like Toll-like receptor (TLR22) (Hallerman 2021).

6.5.1 Growth and Productivity

Genetic enhancement programs often aim at different breeding goals, with an emphasis on improving growth and achieving a rapid growth rate, which are valued traits. The CRISPR genome-editing technique has been used to enhance growth by targeting the myostatin (mstn) gene in various cultured fish species. Myostatin, part of the transforming growth factor β (TGFβ) superfamily, is a key regulator of skeletal muscle development (Sun et al. 2020b). The mstn gene's expression is linked to the double-muscled phenotype observed in cattle and other animals. CRISPR technologies have been employed to enhance growth by deactivating the mstn gene in several fish species, including the common carp (*Cyprinus carpio*), channel catfish (*Ictalurus punctatus*), red sea bream (*Pagrus major*), olive flounder (*Paralichthys olivaceus*), pacific oyster (*Crassostrea gigas*), blunt snout bream (*Megalobrama amblycephala*), and the Nile tilapia (*Oreochromis niloticus*), developed by AquaBounty, a U.S.-based company. The Mstn knockout red sea bream has been noted for significantly enhancing its muscle mass (a 16% increase in skeletal muscle), compact body length, and improved feed efficiency, resulting in enhanced overall growth not observed in other fish species. Recently, the CRISPR/Cas9 technique was applied to disrupt the mstn gene, resulting in the first growth-enhanced mud loach (*Misgurnus anguillicaudatus*) (Tao et al. 2021). Besides the mstn gene, CRISPR is also employed to modify growth-related traits in the tiger puffer (*Takifugu rubripes*) by disrupting the leptin receptor gene, which regulates appetite, prompting the fish to consume more. Japan has recently approved the sale of two CRISPR-edited fish species, tiger puffer and red sea bream, both of which display accelerated growth compared to their non-CRISPR-edited counterparts. Growth hormone (GH) and insulin-like growth factor (IGF) positively regulate growth, while IGF-binding proteins (IGFBPs) inhibit IGF and function as negative regulators of somatic growth. CRISPR/Cas9 was used to disable the IGFBP-2b gene in rainbow trout (*Oncorhynchus mykiss*) (Cleveland et al. 2018). To improve productivity and reduce cannibalism, we created AVTR (arginine vasotocin receptor) V1a2 knockout chub mackerel (*Scomber japonicus*) (Ohga and Matsuyama 2021). These fish not only showed the targeted traits but also became more docile and required less oxygen. This innovative method of enhancing productivity through genome editing has the potential to contribute to greater sustainability in aquaculture in the future.

6.5.2 Disease Resistance

Disease resistance is a crucial trait for commercial production, making its genetic enhancement a top breeding focus. Genome-editing studies have been conducted to improve immunity and disease resistance. Specifically, a targeted disruption of the Toll-like receptor TLR22 gene was performed in farmed Indian major carp, Rohu (*Labeo rohita*). This model helps investigate TLR22's role against pathogenic dsRNA viruses, bacteria, and parasites, including fish lice (Chakrapani et al. 2016).

As a strategy to combat hemorrhagic disease caused by grass carp reovirus (GCRV), CRISPR/Cas9 was used to knock out the Junctional adhesion molecule-A (JAM-A) gene in *Ctenopharyngodon idella*, the grass carp, which was then evaluated for in vitro resistance to GCRV (Ma et al. 2018). In channel catfish (*Ictalurus punctatus*), researchers targeted the altered immunity gene TICAM1, related to the TLR3 signaling pathway, as well as the disease susceptibility gene rhamnose-binding lectin (RBL) (Elaswad et al. 2018). Sun et al. (2020a, b) highlighted the role of the carotenoid isomerase oxygenase (EcNinaB-X1) gene in the immune defense of *Exopalaemon carinicauda*, using CRISPR/Cas9 for gene mutagenesis. Prawns with EcNinaB-X1 knocked out showcased lower mortality rates than wild-type individuals following bacterial challenges. CRISPR/Cas9 HDR was also utilized to insert an exogenous alligator cathelicidin gene, known for its broad-spectrum antimicrobial properties, into a specific noncoding region of the channel catfish genome (Simora et al. 2020). Because multiple genes influence complex traits such as disease resistance and growth, several genes may need to be edited simultaneously to achieve the desired phenotypes, leading to challenges in genome editing. Extensive research will be necessary to apply CRISPR effectively and identify the appropriate genes and versions targeted for editing.

6.5.3 Sterility and Reproductive Confinement

The escape of fish from aquaculture facilities can lead to environmental, ecological, and genetic issues, particularly when non-native species are involved. Producing sterile animals in aquaculture is crucial for the industry, as it prevents interbreeding with wild populations, thus conserving genetic integrity and avoiding early maturation-related production challenges. Furthermore, reproductive confinement within genetic improvement programs safeguards breeders' investments. In this regard, CRISPR/Cas 9 has been utilized to create new sterility models in Atlantic salmon (*Salmo salar*) by knocking out the germ cell-specific dead end (dnd) gene, resulting in germ cell-free salmon (Wargelius et al. 2016). These germ-cell-free Atlantic salmon remained immature and did not reach puberty. Another study produced sterile channel catfish (*Ictalurus punctatus*) by editing the luteinizing hormone (LH) gene using modified zinc finger nucleases (ZNF) technology combined with electroporation (Qin et al. 2016). In addition to using genome editing for creating sterile fish in breeding programs, developing a knock-in technique is essential for implementing an inducible "on-off" sterility system. However, previous experiments with model fish species, such as medaka and zebrafish, indicate that restoring fertility in breeding stocks may be achievable (Nagasawa et al. 2019). In a related study, we at Kyushu University employed platinum TALEN to knock out the LH gene in Japanese anchovy (*Engraulis japonicus*) and were able to reverse the LH-KO induced "immature phenotype" through gonadotropin [human chorionic gonadotropin (HCG)] administration, even at the F 0 generation (unpublished data). A strategy is needed to restore reproductive capabilities in salmon lacking germ cells for effective breeding and aquaculture. A rescue approach has been reported to produce germ cells in dnd knockout fish (Güralp et al. 2020). Sterlet (*Acipenser ruthenus*), known for its rapid reproductive cycle, can serve as a surrogate for critically endangered and late-maturing sturgeon species. CRISPR/Cas9 has been employed to knock out the dnd1 gene, thereby preparing a sterile Sterlet host for the surrogate production of large, late-maturing sturgeon species (Chen et al. 2018).

6.5.4 Omega-3 Production

The health benefits of omega-3 long-chain polyunsaturated fatty acids (n-3 LC-PUFAs), particularly EPA (eicosapentaenoic acid, 20:5n-3) and DHA (docosahexaenoic acid, 22:6n-3), are well recognized as beneficial to human health. Due to the limited ability of humans to synthesize LC-PUFAs (EPA and DHA) endogenously, dietary supplements are the most effective means to fulfill these needs. Fish, including Atlantic salmon, serve as the primary sources of omega-3 long-chain polyunsaturated fatty acids (LC-PUFAs) in our diet. The ability of any species to synthesize LC-PUFAs relies on the combined functions of fatty acyl desaturases (Fads2) and elongases of very-long-chain fatty acids (Elovls). A comprehensive understanding of endogenous synthesis and the regulatory mechanisms of LC-PUFAs is essential for in vivo functional studies. In 2019, Datsomor et al. (2019a, b) applied CRISPR/Cas9 editing to $\Delta 5$ and $\Delta 6$ desaturases in Atlantic salmon (*Salmo salar*), which hindered $\Delta 8$-desaturation and the production of docosahexaenoic acid. In another study that same year, they used CRISPR/Cas9 to edit elovl2, inhibiting PUFA elongation, which emphasized the crucial role of elovl2 in PUFA synthesis and indicated that sterol regulatory element binding protein-1 (Srebp-1) is a significant regulator of endogenous PUFA synthesis in Atlantic salmon.

6.5.5 Pigmentation

Fish coloration and patterns are significant factors in aquaculture, especially in the ornamental trade. Six types of pigment cells have been identified in fish (Fujii 2000), though different species may possess distinct types of these cells. Numerous genes that influence fish pigmentation also impact the development of color patterns. With the advent of genome-editing tools like CRISPR/Cas9, research on pigmentation genes has advanced our knowledge, leading to the creation of more colorful mutant fish (Kimura et al. 2014). In Nile tilapia (*Oreochromis niloticus*), 25 pigmentation-related genes associated with melanogenesis, pteridine metabolism, and carotenoid absorption have been edited using CRISPR/Cas9, with 13 showing phenotypes in both F0 and F2 generations. Four pigment cell types are present in wild-type tilapia, which, along with various natural and artificial color mutants, makes tilapia a model for studying teleost fish color patterns (Wang et al. 2021). Additionally, gene removal for pigmentation can serve as a strategy to track successful genome modifications, as modified individuals often lack pigment, resulting in a white or yellow appearance. The CRISPR/Cas9 system was employed to knock out two pigmentation-related genes, tyrosinase (tyr) and solute carrier family 45, member 2 (slc45a2), in Atlantic salmon, resulting in a spectrum of phenotypes ranging from complete albinism to partial loss of pigmentation (Edvardsen et al. 2014). In economically important Asian fish, such as the loach (*Paramisgurnus dabryanus*) and white crucian carp (*Carassius auratus cuvieri*) (Liu et al. 2019), the CRISPR/Cas9 system was applied to the tyr gene, revealing varying levels of pigmentation, ranging from normal to an albino state. Disruption of the HDL receptor/Scavenger receptor B1 (Scarb1) and Gch1 (GTP cyclohydrolase 1) genes in ornamental common carp led to regions of red skin fading to white (Du et al. 2021). Another study disrupted two Agouti signaling protein genes (ASIP 1 and ASIP 2) via CRISPR/Cas9, eliminating black patches in Oujiang color common carp. Recently, the market demand for red tilapia, which are hybrids of various tilapia species or red variants of highly inbred Nile tilapia, has increased. This red phenotype is unstable, resulting in inconsistent coloration or dark red blotches, which negatively impact market value. To address this, CRISPR/Cas9 targeted the slc45a2 gene involved in melanin biosynthesis, resulting in the generation of an albino variant of Nile tilapia with red skin and eyes (Segev-Hadar et al. 2021). Furthermore, researchers are using slc family genes to enhance the visual confirmation of knockout fish strains; Pandey et al. (2021) demonstrated that slc2415 mutants exhibited reduced melanin pigmentation in the retina and bodies of Kawakawa (*Euthynus affinis*), indicating a future where genome-edited stocks can be visually recognized more easily.

6.6 Optimization of CRISPR/Cas9 Experimental Design

The selection of the target sequence is influenced by the objectives of the genome editing experiment. In gene knockout (KO) scenarios, the first coding exon is often chosen to reduce the risk of creating a truncated protein with unforeseen functions. However, if the domains and activities of the protein are well understood, targeting exons that code for these functional domains can lead to higher KO rates, as even in-frame deletions may disrupt function. For gene knock-in (KI), the target sequence can be defined with precision. For instance, when tagging with GFP, the insertion of the GFP cDNA typically occurs at either the ATG or the stop codon. In these instances, it is advisable to keep the DNA cleavage site by Cas9 close to the insertion point, ideally within 20–50 base pairs.

6.7 Guide RNA Selection

The NGG PAM motif is prevalent, which often results in numerous potential guide RNAs for the target sequence of interest. In preliminary experiments, selecting the most efficient guide RNAs is typically prioritized due to the significant variability in cutting efficiency among different guides (Haeussler and Concordet 2016). For instance, in *C. elegans*, only half of the tested guides were sufficiently active to enable the easy isolation of mutants (Kim et al. 2014). Therefore, it is crucial to evaluate several guide RNAs for a single target gene and choose the most effective ones for genome editing. Specifically, when seeking to perform knock-in (KI) at a precise genomic location, as observed in gene tagging experiments, only guides that induce cleavage with a 30-bp overhang will be beneficial. Additionally, low activity of appropriately positioned guides may diminish the feasibility of effective genome editing. The mutation rates caused by guides can be assessed through various procedures. Target sequences can be amplified via PCR, with mutation rates estimated using Tracking of Indels by Decomposition (https://tide.nki.nl). This method is simple, fast, and cost-effective, relying on analyzing Sanger sequencing chromatograms from treated versus control DNA with online software (Brinkman et al. 2014). Although other techniques are available, they tend to be more time-consuming such as endonuclease assays with T7 or Surveyor enzymes, sequencing from individual colonies after PCR cloning, or Mi-Seq and PGM sequencing or require specialized equipment like high-resolution melt analysis (HRMA), heteroduplex melting analysis (HMA), and indel detection by amplicon analysis (IDAA) (Haeussler and Concordet 2016).

When numerous potential guides are available in the target sequence, online tools can help select those predicted to exhibit the highest activity (Doench et al. 2016). Although on-target predictions lack robustness, they remain statistically significant when analyzed using large datasets, making them more reliable than random chance. We recently evaluated the various on-target predictions. For guide RNAs expressed from exogenous promoters, the Fusi/sgRNA designer score (Doench et al. 2016) is most suitable, whereas for guide RNAs produced via in vitro transcription, the Moreno-Mateos/CRISPR scan score (Moreno-Mateos et al. 2015) is preferred (the sgRNA designer score is available at broad institute. org and the Moreno-Mateos score at crisprscan.org). Bioinformatic methods can identify potential off-target guides, which can be minimized through careful guide selection, necessitating a complete genome sequence assembly. Particularly, off-targets with fewer than three mismatches should be avoided. Cas-off finder is a valuable web tool for identifying potential off-targets (Bae et al. 2014). The cutting frequency determination score helps predict off-target mutation rates by considering the position and nature of mismatches; it can be accessed at broadinstitute.org. On-target and off-target predictions can be generated using several web tools mentioned above. Additionally, we created crispor.org, which offers easy access to this information for over 100 known genomes and includes tools for guide cloning and primer design (Haeussler et al. 2016). More genomes can be added upon request, and the software is available for local use on preliminary genome assemblies. While having a complete genome assembly helps reduce off-target effects, it is not essential for applying the technology in new species. PCR amplification of target sequences is sufficient for guide design and mutation detection. For example, we recently contributed to genome editing in a crop pest moth without an available genome sequence (Koutroumpa et al. 2016). In such cases, verifying that the mutant phenotype is linked to the DNA mutation is crucial, and validating with a second independent guide can be helpful. In *Paryhale hawaiensis*, four Hox genes were inactivated using CRISPR/Cas9, and the resulting limb phenotypes matched those obtained with a second guide, effectively ruling out off-target effects (Martin et al. 2016).

6.8 Cas9 Expression System

Cas9 can be expressed in various formats, including DNA, RNA, or protein. The simplest and most efficient option, particularly for new species, is to acquire SpCas9 protein and synthetic guide RNA from commercial sources and then combine them to create the nuclease complex before introducing this ribonucleoprotein complex into the cells. Initially viewed as an expensive solution, costs are rapidly decreasing due to increased demand and a greater number of suppliers. Crude preparations of recombinant Cas9 can also be utilized after extraction using His tags from *E. coli* (Gagnon et al. 2014). Comparisons show that Cas9 protein typically outperforms Cas9 mRNA or plasmid-based methods (Square et al. 2015). One reason for this is that concentrations of guide RNA may diminish before the translation of Cas9 mRNA, highlighting the superior efficiency of Cas9 protein. In certain organisms, enhancing the translation of Cas9 mRNA may require adjustments to the coding sequence to align better with specific codon usage. Moreover, while plasmids containing expression cassettes for Cas9 and guide RNA are relatively straightforward to produce and use, the regulatory elements, such as promoters, must be tested and optimized for each organism, particularly to ensure the effective expression of the small guide RNAs. Additionally, expression from plasmids might be uneven and inadequate for consistently modifying the germ line. Ultimately, the success of Cas9/sgRNA delivery is greatly influenced by the specific species involved.

6.9 Challenges in Fisheries and Aquaculture Applications

The CRISPR/Cas technology holds great promise for aquaculture, enabling the assessment of gene function and enhancing critical commercial traits (Fig. 6.3). Nevertheless, technical challenges and ethical concerns surrounding its use in breeding remain. For CRISPR products to be commercialized, risk assessments, regulatory approvals, and public/consumer acceptance are essential. Effective CRISPR editing demands a thorough understanding of genetic backgrounds and genomic sequences. Although over 70 significant fish and aquatic species have been sequenced (Wargelius 2019), this number is minor compared to the more than 600 aquaculture species identified. Furthermore, many genome sequences for these species are poorly assembled and lack proper annotations, leaving much of the gene functionality undefined (Sundaram et al. 2017). In aquaculture, only a limited number of genes associated with key traits, such as growth, disease resistance, and robustness, have been identified, restricting the applications of CRISPR/Cas in the field. Therefore, enhancing our understanding of genome sequences and trait-related genes will facilitate the design of gRNAs that target specific regions effectively.

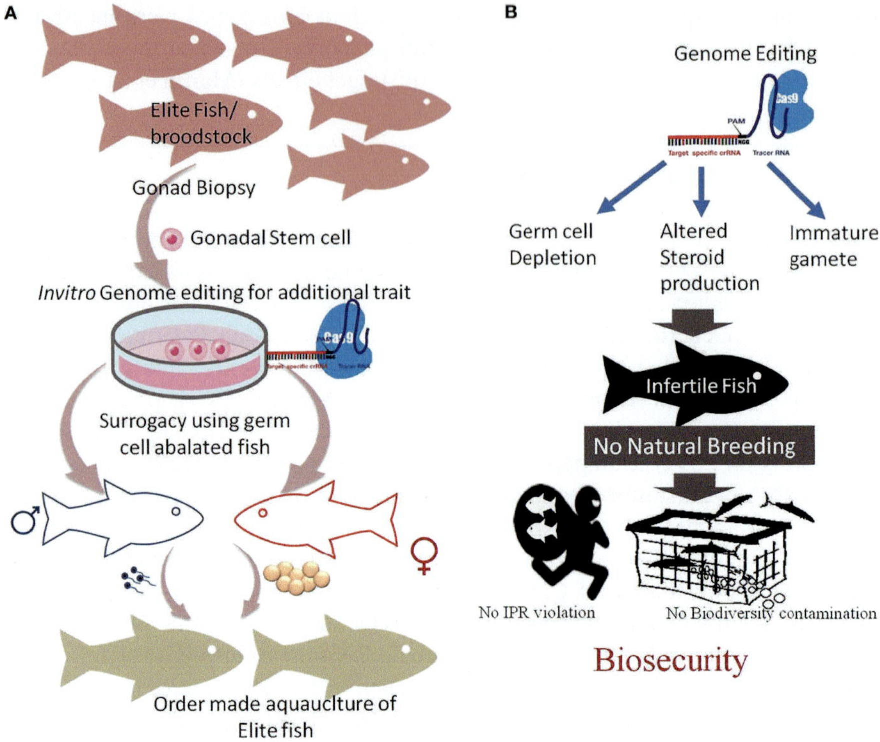

Fig. 6.3 Depicts the future of genome editing in aquaculture and fisheries. It is expected that genome editing will become widely used in aquaculture to enhance sustainability. To create a genome-edited stock, a germline mutation is crucial. (**a**) The right panel illustrates the collection, cultivation, and modification of gonadal stem cells (germ cell precursor cells or gamete-producing cells) from various elite fish. When these in vitro genome-edited germ cell precursors, which contain the mutation, are transplanted into sterilized hosts (lacking germ cells), they will produce male and female gametes, depending on the sex of the host. Combining these gametes will yield a distinct superior strain. (**b**) An illustration demonstrating the application of GSC-based custom strain production in producing infertile fish, managing controlled aquaculture, and ensuring biosecurity

To quickly identify optimal gRNAs, it is preferable to test CRISPR/Cas constructs in cell cultures before in vivo editing (Gratacap et al. 2020; Yang et al. 2022). However, aquaculture research faces a challenge due to a lack of well-characterized cell lines for many target species. Furthermore, many crustacean and molluscan species lack established immortalized cell lines. Important aspects of in vitro genome editing must also be optimized for aquaculture species. Thus, building these platforms will enhance the feasibility of genome-wide screening in key aquaculture species. For fish, CRISPR/Cas constructs should be delivered to fertilized eggs at the one-cell stage (Yang et al. 2022). The time from fertilization to first cell division varies among fish species, making it crucial to know this timing to avoid missing the critical period for genome editing. Microinjection, although effective, is labor-intensive and requires an expensive microinjection setup, which complicates the process of injecting many eggs quickly. Additionally, for certain shrimp species, obtaining newly fertilized embryos can be challenging. Therefore, alternative CRISPR/Cas9 delivery methods must be developed and tested for breeding, including electroporation, sperm-mediated transfer, microinjection of unfertilized ova, and editing of primordial germ cells. The egg membrane in oviparous fish results in low microinjection success rates, whereas in ovoviviparous fish, no established gene editing platform exists. Moreover, the hardness of the egg membrane varies among fish species (e.g., soft in some and very hard in others, such as tilapia), making the standardization of needle types and injection techniques crucial (Okoli et al. 2021). Teleost fish have undergone a whole-genome duplication (WGD), with salmon experiencing an additional, salmon-specific fourth round (Ss 4R) of duplication. This duplication can reduce CRISPR editing efficiency; comparing genes with various copies within the genome could help clarify this issue (Gratacap et al. 2019). In aquatic species where CRISPR/Cas 9 is in use, methods need to be refined to enhance editing efficiency, minimize off-target effects, and address mosaicism. Off-target modifications can lead to unintended genome alterations, potentially causing detrimental effects on organisms and necessitating a thorough assessment of both genomes off-targeting and associated food safety risks. Postediting screening for unintended mutations is essential; however, natural genetic variations among different families and strains complicate detection efforts (Blix et al. 2021).

6.10 Regulatory and Consumer Acceptance

Government regulation and public acceptance present major hurdles for CRISPR/Cas genome editing in aquaculture. Although CRISPR holds the potential to improve aquaculture productivity, securing public approval and regulatory clearance is essential for realizing this potential (Okoli et al. 2021). Currently, most CRISPR/Cas applications are confined to research, with limited adoption in aquaculture breeding programs. This reluctance mainly stems from public mistrust of genome-edited products and an uncertain global regulatory environment. Unlike GMOs, where foreign genes are inserted into the genome, CRISPR/cas editing involves making precise, small-scale modifications without introducing foreign DNA. It improves traits by removing less favorable alleles, rather than adding new genetic material (Yang et al. 2022). This suggests that GMO fish are organisms that would not naturally exist, whereas CRISPR-edited fish display alterations to their original genes, akin to selective breeding, effectively shortening the evolutionary process that occurs over many generations. Despite this, an ongoing debate exists about whether genome-editing techniques should be classified separately from GMOs. To date, there are no specific, harmonized guidelines for the regulatory acceptance of genome-edited products. However, various regulatory authorities globally, including those in the EU, Argentina, Brazil, Australia, New Zealand, Canada, the United States, and Norway, have initiated discussions regarding the regulation of products developed using the new CRISPR editing methods

(Eckerstorfer et al. 2019). Regulatory acceptance and guidelines may vary significantly by country; in some places, genome-edited products are not categorized as GMOs, thereby bypassing biosafety screenings, while others will mandate such evaluations. CRISPR genome editing products are likely to gain consumer acceptance more readily than GMOs, as products derived from CRISPR/Cas technology are referred to as "genome edited" instead of "genetically modified." This change in terminology can help mitigate the negative reactions previously directed at GMOs (Fig. 6.4). To enhance consumer acceptance, it is essential to effectively communicate how and why CRISPR/Cas has been utilized, emphasizing its benefits, such as producing healthy, high-quality food and improving animal welfare, while also highlighting its safety and sustainability aspects (Hallerman 2021).

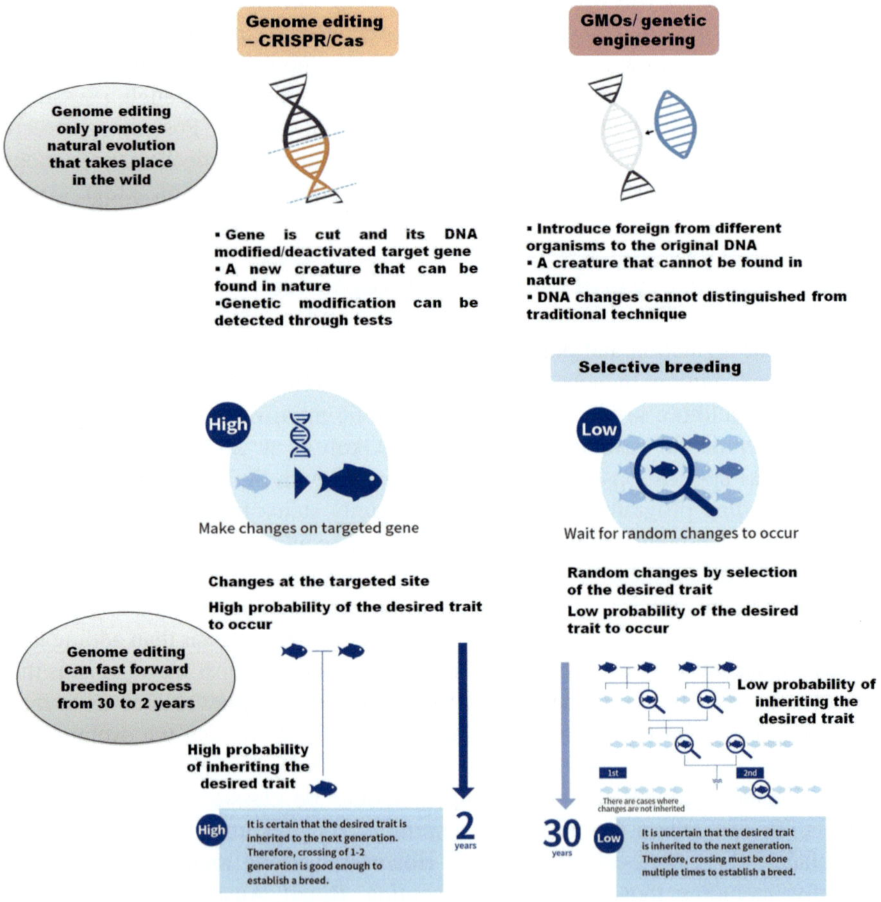

Fig. 6.4 CRISPR/Cas as a rapid new breeding technology in aquaculture: CRISPR/Cas genome editing facilitates the natural evolutionary processes occurring in the wild, accelerating genetic improvement with minimal targeted alterations in the genome

6.11 Conclusion and Future Prospective

The CRISPR/Cas9 revolution is transforming all areas of life sciences. The review demonstrates that techniques developed in human cultured cells and standard laboratory organisms are easily applicable to nonmodel organisms. Thus, the primary hurdles in advancing genome editing involve rearing nonmodel organisms in the lab and the transfer of CRISPR/Cas9 reagents. The results of CRISPR/Cas genome editing hold promise for accelerating the aquaculture breeding process. It is clear that most of the anticipated genome-edited aquaculture products for commercial use will arise from CRISPR/Cas technology. However, the use of this technology for genetic enhancement in aquaculture remains in its early stages. Currently, significant research efforts in aquaculture focus on traditional genetic improvement strategies involving growth, production, and reproduction. Looking ahead, more precise and sophisticated genome editing will need to be conducted across multiple loci or chromosomal locations to enhance complex traits. Additionally, incorporating other economically significant traits and expanding the diversity of aquaculture species is necessary. The ongoing advancement of the CRISPR/Cas9 technique and its implementation as a new breeding tool in aquaculture has the potential to revolutionize the industry by enhancing both production and quality. Furthermore, innovative solutions in aquaculture, such as disease resistance, sterile breeding, and targeted therapies, are expected to emerge soon options that traditional methods cannot effectively address. For CRISPR/Cas to assume a larger role in aquaculture as a new breeding technology, identifying and functionally annotating genes related to traits and signaling pathways is essential. Notably, significant progress has been made with CRISPR/Cas editing in Atlantic salmon and tilapia, and insights gained from these species can be applied to others. As such, both Atlantic salmon and tilapia can serve as model organisms in aquaculture to initiate the optimization of aquatic protocols and evaluate off-target effects. It is also crucial to guarantee that the genome editing process is sustainable, considering environmental impacts and animal welfare. Ultimately, the success of CRISPR/Cas genome editing in aquaculture hinges on favorable public perceptions, regulatory frameworks, and consumer acceptance.

References

Bae S, Park J, Kim JS (2014) Cas-OFFinder: a fast and versatile algorithm that searches for potential off-target sites of Cas9 RNA-guided endonucleases. Bioinformatics 30(10):1473–1475

Bétermier M, Bertrand P, Lopez BS (2014) Is non-homologous end-joining really an inherently error-prone process? PLoS Genet 10(1):e1004086

Blix TB, Dalmo RA, Wargelius A (2021) Genome editing in aquaculture – current status and future challenges. Front Genet 12:622614

Brinkman EK, Chen T, Amendola M, Van Steensel B (2014) Easy quantitative assessment of genome editing by sequence trace decomposition. Nucleic Acids Res 42(22):e168–e168

Carroll D (2014) Genome engineering with targetable nucleases. Annu Rev Biochem 83:409–439

Chakrapani V, Kumar R, Kumar D, Behera BK, Das BK (2016) Role of TLR22 in fish immunity: a review. Fish Shellfish Immunol 55:339–347

Chen Y, Zhu J, Liu X (2018) Creating sterile sterlet (*Acipenser ruthenus*) via CRISPR/Cas9 knock-out of dnd1 for surrogate production. Aquacult Rep 10:7–13

Cleveland B. M.Yamaguchi G.Radler L. M.Shimizu M.(2018) Editing the Duplicated Insulin-Like Growth Factor Binding Protein-2b Gene in Rainbow Trout (Oncorhynchus Mykiss). Sci. Rep. 8. https://doi.org/10.1038/s41598-018-34326-6

Datsomor AK, Skugor A, Gjøen T, Krasnov A (2019a) CRISPR/Cas9-mediated knock-out of Δ5 and Δ6 desaturase genes in Atlantic salmon impairs long-chain polyunsaturated fatty acid biosynthesis. Sci Rep 9:16806

Datsomor AK, Skugor A, Gjøen T, Krasnov A (2019b) Functional genomics of lipid metabolism in Atlantic salmon: CRISPR/Cas9 editing of elovl2. Mar Biotechnol 21:636–650

Doench JG, Fusi N, Sullender M, Hegde M, Vaimberg EW, Donovan KF, Root DE (2016) Optimized sgRNA design to maximize activity and minimize off-target effects of CRISPR-Cas9. Nat Biotechnol 34(2):184–191

Doudna JA, Charpentier E (2014) The new frontier of genome engineering with CRISPR-Cas9. Science 346(6213):1258096

Du SJ, Zhang Q, Lin H, Xu D, Zhou Y (2021) Genome editing in ornamental fish: approaches and applications for color and pattern modification. Dev Comp Immunol 122:104106

Eckerstorfer MF, Engelhard M, Heissenberger A, Simon S, Teichmann H (2019) Plants developed by new genetic modification techniques – comparison of existing regulatory frameworks in the EU and non-EU countries. Front Bioeng Biotechnol 7:26

Edvardsen RB, Leininger S, Kleppe L, Skaftnesmo KO, Wargelius A, von Schalburg KR (2014) Targeted mutagenesis in Atlantic salmon using the CRISPR/Cas9 system induces complete knockout individuals in the F0 generation. PLoS One 9(9):e108622

Elaswad A, Khalil K, Emam M, Liu Z (2018) Disruption of TICAM1 and RBL genes in channel catfish using CRISPR/Cas9: toward enhanced disease resistance. Aquac Res 49(12):3824–3835

FAO (2020) The State of World Fsheries and Aquaculture 2020: Sustainability in Action (Rome: FAO). Available at: 10.4060/ca9229en

Fujii R (2000) The regulation of motile activity in fish chromatophores. Pigment Cell Res 13(5):300–319

Gagnon JA, Valen E, Thyme SB, Huang P, Ahkmetova L, Pauli A, Schier AF (2014) Efficient mutagenesis by Cas9 protein-mediated oligonucleotide insertion and large-scale assessment of single-guide RNAs. PLoS One 9(5):e98186

Gratacap RL, Wargelius A, Edvardsen RB, Houston RD (2019) Potential of genome editing to improve aquaculture breeding and production. Trends Genet 35(9):672–684

Gratacap R. L.Regan T.Dehler C. E.Martin S. A. M.Boudinot P.Collet B.et al. (2020) Efficient CRISPR/Cas9 Genome Editing in a Salmonid Fish Cell Line Using a Lentivirus Delivery System. BMC Biotechnol.20. https://doi.org/10.1186/s12896-020-00626-x

Güralp H.Skaftnesmo K. O.Kjærner-Semb E.Straume A. H.Kleppe L.Schulz R. W.et al . (2020) Rescue of Germ Cells in Dnd Crispant Embryos Opens the Possibility to Produce Inherited Sterility in Atlantic Salmon. Sci. Rep.10, 1–12. https://doi.org/10.1038/s41598-020-74876-2

Haeussler M, Concordet JP (2016) Genome-scale prediction of off-target effects for CRISPR-Cas9. Nucleic Acids Res 44(13):e57

Haeussler M, Schönig K, Eckert H, Eschstruth A, Mianné J, Renaud JB, Concordet JP (2016) Evaluation of off-target and on-target scoring algorithms and integration into the guide RNA selection tool CRISPOR. Genome Biol 17:1–12

Hallerman EM (2021) Perspectives on the application of genome editing in aquaculture. Front Genet 12:730825

Iyer V, Shen B, Zhang W, Hodgkins A, Keane TM, Huang X, Skarnes WC (2015) Off-target mutations are rare in Cas9-modified mice. Nat Methods 12(6):479–479

Jinek M, Chylinski K, Fonfara I, Hauer M, Doudna JA, Charpentier E (2012) A programmable dual-RNA–guided DNA endonuclease in adaptive bacterial immunity. Science 337(6096):816–821

Kim H, Ishidate T, Ghanta KS, Seth M, Conte D Jr, Shirayama M, Mello CC (2014) A co-CRISPR strategy for efficient genome editing in Caenorhabditis elegans. Genetics 197(4):1069–1080

Kim D, Bae S, Park J, Kim E, Kim S, Yu HR, Kim JS (2016) Digenome-seq: genome-wide profiling of CRISPR-Cas9 off-target effects in human cells. Nat Methods 12(3):237–243

Kimura Y, Hisano Y, Kawahara A, Higashijima S (2014) Efficient generation of albino and transparent medaka using CRISPR/Cas9 system. Genes Cells 19(7):554–564

Koutroumpa FA, Monsempes C, François MC, de Cian A, Royer C, Concordet JP, Jacquin-Joly E (2016) Heritable genome editing with CRISPR/Cas9 induces anosmia in a crop pest moth. Sci Rep 6(1):29620

Lander ES (2016) The heroes of CRISPR. Cell 164(1–2):18–28

Liu Q, Zhao Y, Wang X, Wang H, Zhang W (2019) CRISPR/Cas9-mediated gene editing in loach and crucian carp: implications for ornamental aquaculture. Fish Physiol Biochem 45(6):2071–2081

Luo M, Wang J, Dong Z, Wang C, Lu G (2022) CRISPR-Cas9 sgRNA Design and Outcome Assessment: Bioinformatics Tools and Aquaculture Applications. Aquac. Fish. 7:121–130. https://doi.org/10.1016/j.aaf.2021.10.002

Ma J, Zhang Q, Chen Y, Wu Z, Li Y, Wang X (2018) CRISPR/Cas9-mediated knockout of JAM-A in grass carp (*Ctenopharyngodonidella*) enhances resistance to grass carp reovirus. Fish Shellfish Immunol 82:48–57

Martin A, Serano JM, Jarvis E, Bruce HS, Wang J, Ray S, Patel NH (2016) CRISPR/Cas9 mutagenesis reveals versatile roles of Hox genes in crustacean limb specification and evolution. Curr Biol 26(1):14–26

Mali P, Yang L, Esvelt K, Aach J, Guell M, DiCarlo J et al (2013) RNA-Guided Human Genome Engineering via Cas9. Sci. (80-.).339, 823–826. https://doi.org/10.1126/science.1232033

Momose T, Concordet JP (2016) Diving into marine genomics with CRISPR/Cas9 systems. Mar Genomics 30:55–65

Moreno-Mateos MA, Vejnar CE, Beaudoin JD, Fernandez JP, Mis EK, Khokha MK, Giraldez AJ (2015) CRISPRscan: designing highly efficient sgRNAs for CRISPR-Cas9 targeting in vivo. Nat Methods 12(10):982–988

Nagasawa K, Ishida M, Octavera A, Kusano K, Kezuka F, Kitano T et al (2019) Novel Method for Mass Producing Genetically Sterile Fish From Surrogate Broodstock via Spermatogonial Transplantation. Biol. Reprod.100, 535–546. https://doi.org/10.1093/biolre/ioy204

Ohga H, Matsuyama M (2021) Effects of LPXR Famide Peptides on Chub Mackerel Gonadotropin Secretion†. Biol. Reprod.105, 1179–1188. https://doi.org/10.1093/biolre/ioab130

Okoli AS, Eding EH, Janssen HJ, Verdegem MCJ, Schneider O (2021) CRISPR genome editing: implications for sustainable aquaculture. Front Genet 12:622782

Pandey PK, Jain A, Singh RK (2021) slc2415 mutation reduces retinal melanin pigmentation in Kawakawa (*Euthynnusaffinis*). Mar Genomics 58:100823

Qin G et al (2016) Sterility induction in channel catfish through LH gene knockout using zinc finger nucleases. BMC Genomics 17:600. https://doi.org/10.1186/s12864-016-2954-5

Roy S, Kumar V, Behera BK, Parhi J, Mohapatra S, Chakraborty T, Das BK (2022) CRISPR/Cas genome editing – can it become a game changer in the future fisheries sector? Front Mar Sci 9:924475

Segev-Hadar A, Manor R, Ron M (2021) Enhancing the commercial value of red tilapia using CRISPR-mediated pigmentation gene knockout. Aquac Rep 20:100682

Sfeir A, Symington LS (2015) Microhomology-mediated end joining: a back-up survival mechanism or dedicated pathway? Trends Biochem Sci 40(11):701–714

Simora RM, Vallejo RL, Liu S, Robison BD, Peatman E (2020) Disease-resistant channel catfish generated by CRISPR/Cas9 knock-in of an alligator cathelicidin gene. Sci Rep 10:14230

Slaymaker IM, Gao L, Zetsche B, Scott DA, Yan WX, Zhang F (2016) Rationally engineered Cas9 nucleases with improved specificity. Science 351(6268):84–88

Square T, Romášek M, Jandzik D, Cattell MV, Klymkowsky M, Medeiros DM (2015) CRISPR/Cas9-mediated mutagenesis in the sea lamprey Petromyzon marinus: a powerful tool for understanding ancestral gene functions in vertebrates. Development 142(23):4180–4187

Sternberg SH, Doudna JA (2015) Expanding the biologist's toolkit with CRISPR-Cas9. Mol Cell 58(4):568–574

Sun Y, Zhang Q, Li Y, Wang C, Liu Y, Zhou Z (2020a) CRISPR/Cas9 editing of myostatin gene improves growth and muscle quality in fish. Aquac Rep 18:100491
Sun X, Zhu J, Zhang J (2020b) CRISPR/Cas9-mediated mutagenesis of EcNinaB-X1 gene reduces mortality in prawns following bacterial challenge. Fish Shellfish Immunol 104:364–372
Sundaram A, Tengs T, Grimholt U (2017) Issues with RNA-seq analysis in non-model organisms: a salmonid example. Dev Compar Immunol 75:38–47
Symington LS, Gautier J (2011) Double-strand break end resection and repair pathway choice. Annu Rev Genet 45:247–271
Tao W, Wang D, Sun L, Wang C, Zhou L, Zhang X (2021) CRISPR/Cas9-mediated mutagenesis of the myostatin gene improves growth performance in mud loach. Aquaculture 530:735949
Wang X, Li Y, Chen H, Liu Y (2021) Genome editing of pigment-related genes in *Nile tilapia* using CRISPR/Cas9 reveals color pattern regulation. Aquaculture 530:735890
Wargelius A, Leininger S, Skaftnesmo KO, Kleppe L, Andersson E, Taranger GL et al (2016) Dnd Knockout Ablates Germ Cells and Demonstrates Germ Cell Independent Sex Differentiation in Atlantic Salmon. Sci Rep 6:1–8. https://doi.org/10.1038/srep21284
Wargelius A (2019) Application of Genome Editing in Aquatic Farm Animals: Atlantic Salmon. Transgenic Res.28, 101–105. https://doi.org/10.1007/s11248-019-00163-0
Yang C, Wang C, Xie H, Liu Y (2022) Advances in CRISPR/Cas9 technology for genome editing in aquaculture. Aquac Rep 23:101087

Environmental RNA: A Molecular Revolution in Ecological Surveillance and Biodiversity Monitoring

7

Biswajit Mandal, Anupam Kumar Sahu, and Basanta Kumar Das

7.1 Introduction

The advent of molecular techniques has transformed the landscape of ecological monitoring, with environmental DNA (eDNA) emerging as a widely used tool for detecting biodiversity (Cristescu 2019). However, as technological advancements continue to push the boundaries of ecological surveillance, environmental RNA (eRNA) has surfaced as a promising yet underexplored frontier. Unlike eDNA, which reflects historical and cumulative biological presence, eRNA provides a dynamic snapshot of active biological processes, offering unprecedented insights into species presence, physiological states, and ecosystem health (Wood et al. 2019). Despite its inherent instability, the rapid degradation of eRNA is paradoxically advantageous, as it ensures that detected signals are temporally relevant, allowing researchers to monitor recent biological activity with higher precision than eDNA (Cristescu 2019). The emergence of environmental RNA (eRNA) as a molecular tool for ecological surveillance heralds a transformative era in biodiversity monitoring and ecosystem assessment. Unlike DNA-based methods that offer static snapshots of species presence, eRNA provides a dynamic, real-time perspective on the metabolic activity and physiological state of organisms within their environments. However, despite its immense potential, several critical challenges must be addressed to harness eRNA's full capabilities.

One of the foremost questions in eRNA research is whether it can be reliably quantified to reflect the biological activity of organisms in a given habitat. Unlike DNA, which remains relatively stable over time, RNA exhibits inherent variability due to its transient nature and differential expression across tissues, developmental

B. Mandal · A. K. Sahu · B. K. Das (✉)
Aquatic Environmental Biotechnology (AEB) Division, ICAR-Central Inland Fisheries Research Institute (CIFRI), Barrackpore, Kolkata, India

© The Author(s), under exclusive license to Springer Nature Singapore Pte Ltd. 2026

B. K. Das et al. (eds.), *Bioinformatics Tools for Predictive Ecology and Fisheries*, Sustainable Aquatic Food Systems, https://doi.org/10.1007/978-981-95-2573-7_7

stages, and environmental conditions. This raises crucial methodological considerations do eRNA profiles represent a balanced snapshot of total RNA sources in an ecosystem, or are they skewed toward specific phenotypes, tissue types, or highly expressed genes? Deciphering these nuances will be fundamental to refining eRNA-based ecological assessments. Environmental RNA enters ecosystems through the same biological shedding processes as eDNA via mucus, skin cells, feces, and other organic materials (Wang et al. 2009). However, its transient nature poses methodological challenges in collection, preservation, and analysis. While eDNA-based approaches focus on species detection through PCR and metabarcoding, eRNA expands the scope by enabling real-time assessment of gene expression, thereby providing functional insights into ecosystem dynamics (Glover et al. 2025). For instance, eRNA can differentiate between living and dead organisms, detect metabolically active species, and identify molecular responses to environmental stressors, thus refining the ecological and evolutionary narratives derived from molecular monitoring (Wood et al. 2019). Given these attributes, eRNA has the potential to revolutionize ecological surveillance by complementing and, in some cases, surpassing eDNA in its resolution and applicability. Recent studies have demonstrated its utility in identifying active meiofaunal assemblages, tracking invasive species, and assessing the physiological health of aquatic organisms through transcriptomic analyses (Von Ammon et al. 2019; Kitahashi et al. 2020). Notably, the detection of tissue-specific mRNA markers in zebrafish water samples has provided a glimpse into the future applications of eRNA, suggesting that it could be leveraged for health assessments at both individual and population levels (Tsuri et al. 2021). The integration of eRNA into biomonitoring frameworks holds promise for advancing ecological forecasting, environmental conservation, and wildlife management. Despite its promise, several challenges must be addressed before eRNA can be widely adopted in ecological research. Its detectability is influenced by environmental factors such as temperature, pH, and microbial activity, all of which impact RNA degradation rates (Goldberg et al. 2016). Additionally, the interpretation of gene expression data in ecological contexts requires rigorous validation, as mRNA expression is highly sensitive to external stimuli and confounding variables such as age, reproductive status, and seasonal changes (Van Cleef et al. 2000; Dragun et al. 2009). The development of standardized protocols for eRNA sampling, preservation, and analysis will be critical in ensuring the reliability and reproducibility of results across studies.

One particularly promising aspect of eRNA research lies in its potential for tracking organismal health through stress-responsive biomarkers. For decades, molecular ecologists have relied on mRNA profiles to assess physiological responses to environmental perturbations, with well-established biomarkers including heat shock proteins (Iwama et al. 1998), cytochrome P450 enzymes (Oris and Roberts 2007), and vitellogenin (Matozzo et al. 2008). More recently, microRNAs (miRNAs) short, noncoding RNAs involved in post-transcriptional gene regulation have emerged as potential indicators of environmental stress. Studies have shown that miRNAs respond to stressors such as temperature fluctuations (Ikert and Craig

2020), hypoxia, and pollutant exposure (Bertucci et al. 2019). The detection of these molecules in environmental samples suggests that miRNA-based eRNA assays could serve as noninvasive tools for assessing organismal health at a molecular level (Ikert and Craig 2020). The ability of eRNA to provide high-resolution biological data has significant implications for ecosystem monitoring and management. Unlike eDNA, which primarily enables species identification, eRNA can facilitate deeper ecological inquiries, such as tracking metabolic activity, monitoring life-history stages, and detecting phenotypic variations within species (Cristescu 2019). By capturing real-time transcriptional responses to environmental changes, eRNA may enhance our ability to predict ecosystem shifts before they manifest as large-scale biodiversity losses. Furthermore, its application in conservation biology could help assess the impact of climate change, pollution, and habitat degradation on sensitive species and ecosystems (Glover et al. 2025). As the field of eRNA research advances, there is a pressing need for empirical studies that explore its full potential across diverse ecosystems. Future research should focus on optimizing methodological frameworks, improving RNA preservation techniques, and developing high-specificity assays that can distinguish among different physiological states and stress responses (Sohel 2016). Integrating eRNA into multiomics approaches combining genomic, transcriptomic, and metabolomic data could further enhance ecological monitoring by providing a holistic view of ecosystem function and resilience (Huang et al. 2011).

Environmental RNA (eRNA) is revolutionizing ecological surveillance by capturing real-time snapshots of biodiversity and ecosystem dynamics. Unlike eDNA, eRNA's rapid degradation offers unparalleled temporal resolution, enabling the detection of short-term species activity and physiological states. However, preserving RNA integrity from collection to sequencing remains a critical challenge. Standardizing methodologies and advancing bioinformatics will be key to unlocking eRNA's full potential. As techniques refine, eRNA promises to redefine conservation biology, offering a powerful molecular lens into ecosystem resilience and environmental change.

This chapter explores the revolutionary potential of environmental RNA (eRNA) in ecological surveillance, offering unparalleled molecular precision in biodiversity assessments. We first examine eRNA dynamics, addressing its persistence and degradation in environmental systems. We then highlight its advantages over eDNA for high-resolution organism detection, including life-history stages and physiological states. We discuss methodological advancements, from extraction to sequencing, and the role of environmental transcriptomics in monitoring ecosystem responses. Beyond community composition, eRNA enables insights into evolutionary and ecological processes. Finally, we address challenges, research gaps, and future prospects, emphasizing the integration of eRNA with emerging technologies for dynamic, real-time ecosystem monitoring. This paradigm shift heralds a new era in biomonitoring and conservation science (Table 7.1).

Table 7.1 Notable innovations in environmental RNA (eRNA) research, from foundational studies to recent advancements

Reference	Innovation	Key contribution	Scientific impact
Pace et al. (1986)	Introduced rRNA analysis as a tool for microbial diversity	Pioneering use of rRNA sequences to explore microbial populations in their natural environments, laying the foundation for using RNA as an indicator of active life	Established rRNA as a powerful biomarker for microbial ecological studies
Olsen et al. (1986)	Theoretical framework for rRNA in microbial phylogeny	Presented the rRNA approach for studying microbial phylogeny, influencing later eRNA applications in ecological research	Provided the scientific basis for the growing field of eRNA
DeLong et al. (1989)	Development of rRNA-based probes for single-cell identification	Introduced ribosomal RNA-targeted probes, enabling the detection of individual microbial cells in environmental samples	Advanced eRNA's use for identifying and quantifying specific microorganisms in complex environments
Giovannoni et al. (1990)	RNA-based tools for marine microbial diversity	Applied RNA-based methods to study microbial diversity in marine environments, demonstrating eRNA's use for monitoring marine ecosystems	Expanded the scope of eRNA applications to marine biodiversity monitoring
Urich et al. (2008)	Metatranscriptomics for simultaneous analysis of microbial community structure and function	Introduced the concept of metatranscriptomics to assess both the taxonomic composition and functional profiles of microbial communities	Enabled a more comprehensive understanding of microbial ecology and function in environmental samples
Blazewicz et al. (2013)	Evaluation of rRNA as a microbial activity marker	Critically assessed the utility of rRNA as an indicator of microbial activity, clarifying its strengths and limitations	Refined eRNA methodology and improved its application in ecological monitoring
Cristescu and Hebert (2018)	Integrated use of eDNA and eRNA for biodiversity monitoring	Investigated the combined use of eRNA and eDNA for invasive species detection and biodiversity monitoring	Promoted the integration of genetic approaches for comprehensive environmental monitoring

(continued)

Table 7.1 (continued)

Reference	Innovation	Key contribution	Scientific impact
Littlefair et al. (2022)	Environmental nucleic acids: A field-based comparison for monitoring freshwater habitats using eDNA and eRNA	Conducted field-based comparison of eDNA and eRNA for monitoring freshwater habitats; found that eRNA provides a snapshot of active biota while eDNA provides historical presence. Demonstrated the complementary nature of eDNA and eRNA in ecosystem monitoring	Highlights the importance of using both eDNA and eRNA to get comprehensive data on community composition and ecosystem health, emphasizing eRNA for real-time activity and eDNA for past biotic presence
	Development of nucleic acid probes for microbial systematics	Developed rRNA-targeted probes for identifying and quantifying specific microbial taxa in environmental samples	Enhanced eRNA methodologies for targeted organism detection in environmental microbiology
	Use of rRNA to detect lateral gene transfer among extremophiles	Applied rRNA to study horizontal gene transfer among extremophiles, illustrating eRNA's potential in understanding microbial evolution	Highlighted eRNA's role in studying microbial interactions and evolution in extreme environments
	eRNA to identify archaeal dominance in oceanic ecosystems	Used eRNA to identify archaeal dominance in the mesopelagic zone of the Pacific Ocean	Established eRNA as a tool for studying microbial community structure in marine ecosystems
DeLong and Karl (2005)	Combined rRNA with genomic approaches for microbial oceanography	Integrated rRNA analysis with genomic methods to provide insights into microbial oceanography and activity	Provided a framework for using eRNA in conjunction with genomic approaches in marine environmental studies
Blazewicz et al. (2013)	Critical evaluation of rRNA as a microbial activity marker	Provided insights into rRNA as a microbial activity marker, emphasizing its use and limitations	Clarified the strengths and weaknesses of eRNA in microbial activity assessments
Tsuri et al. (2021)	mRNA typing of eRNA	Demonstrated messenger RNA typing in eRNA from aquatic environments, specifically zebrafish	Opened new avenues for functional analysis of aquatic vertebrate communities through mRNA profiling

(continued)

Table 7.1 (continued)

Reference	Innovation	Key contribution	Scientific impact
Marshall et al. (2021)	eRNA for age prediction of eDNA	Introduced eRNA as a tool to estimate the age of eDNA signals	Improved reliability in biodiversity monitoring by enhancing temporal resolution of detection
Yates et al. (2023)	Integration of eDNA and eRNA	Leveraged environmental DNA (eDNA) and environmental RNA (eRNA) to expand beyond species detection to ecological functions	Enabled a broader application of environmental nucleic acids for understanding ecological processes
Li et al. (2024)	eRNA for biodiversity monitoring in marine environments	Validated eRNA technique in assessing fish biodiversity in offshore waters	Provided a new method for comprehensive biodiversity assessments in marine ecosystems

7.2 Understanding eRNA Dynamics in Environmental Systems

Environmental RNA (eRNA) presents distinct advantages over eDNA due to its transient nature, reflecting the active transcription processes in living organisms (Wood et al. 2019; Marshall et al. 2021). Unlike eDNA, eRNA degrades rapidly, offering a temporal and spatial snapshot of living organisms (Pochon et al. 2017; Cristescu 2019). This property allows researchers to potentially differentiate between conspecifics based on their physiological states (Tsuri et al. 2021; Yates et al. 2023), marking a significant advancement in ecological research methodologies. Jo (2023) contribute foundational insights by investigating the degradation rates of Zebrafish (*Danio rerio*) mitochondrial and nuclear eDNA and eRNA under varying environmental conditions. Their findings underscore that while high temperatures accelerate degradation of both eDNA and eRNA, alkaline conditions selectively enhance eRNA degradation. They further highlight that eRNA remains detectable over time, suggesting its potential utility in quantifying "living" biotic communities.

In a complementary study, Kagzi et al. (2023) employ metabarcoding to assess Arthropoda community diversity using eDNA and eRNA from nuclear and mitochondrial markers in seminatural mesocosms. While their results do not consistently show differences in degradation between eDNA and eRNA markers across genomic sources, they emphasize the context-specific variability in degradation rates. This underscores the necessity for continued comparative research to refine the application of eRNA in ecological assessments. Scrivener et al. (2023) reinforce the potential of eRNA to detect active biological assemblages while cautioning about the multifaceted factors influencing eRNA degradation. Their insights emphasize the complexity of interpreting eRNA data, considering environmental conditions, genomic sources, taxonomic variations, and physiological states of source organisms.

7.3 Harnessing eRNA for Enhanced Organism Detection and Ecological Monitoring

Environmental RNA (eRNA) holds immense potential for revolutionizing the detection of organisms and advancing biological monitoring in aquatic ecosystems. While the study of environmental DNA (eDNA) has become commonplace in ecological research, the investigation of eRNA remains in its early stages (Cristescu and Hebert 2018; Cristescu 2019). Unlike eDNA, which is highly stable, RNA has traditionally been regarded as too fragile to persist in environmental media, leading to the assumption that it would degrade rapidly, rendering it undetectable in meaningful quantities. However, recent breakthroughs have challenged this view, showing that environmental mRNA can indeed be readily extracted and detected under laboratory conditions (Tsuri et al. 2021; Wood et al. 2019; Marshall et al. 2019). Moreover, decay rates for eRNA are variable but can be surprisingly comparable to those of eDNA (Wood et al. 2019), with mRNA detectable in water samples for up to 13–72 hours after production (Marshall et al. 2021). Interestingly, in some cases, eRNA concentrations can even surpass those of corresponding eDNA markers (Marshall et al. 2021).

In natural environments, eRNA has been detected at high concentrations, particularly in proximity to target organisms (Von Ammon et al. 2019). The relatively fast turnover rate of eRNA has sparked renewed interest in its use for real-time organism detection, potentially offering a higher degree of spatial and temporal resolution compared to eDNA (Cristescu 2019). This ability to reflect the recent presence of organisms provides a more dynamic snapshot of biological activity. While eDNA is often shed in large quantities by deceased organisms, eRNA is typically associated with metabolically active organisms, making it a more precise indicator of living assemblages (Cristescu and Hebert 2018; Pochon et al. 2017; Laroche et al. 2018). As such, eRNA-based methods may help mitigate the issue of false positives often encountered in eDNA surveys, which can result from the persistence of DNA from dead or decaying organisms.

However, focusing solely on detecting living organisms or recently occupied habitats may limit the full potential of eRNA. Its advantages go far beyond the scope of traditional eDNA surveys, particularly because eRNA reflects the transcriptomic profile of the organisms that produce it. This distinction is critical, as organisms with nearly identical genetic profiles may exhibit significant transcriptomic differences based on factors such as developmental stage, physiological state, genetic background, or phenotype (Ciani et al. 2020). Given that RNA plays a pivotal role in gene expression, either as non-coding RNA (ncRNA) with regulatory functions or as mRNA encoding proteins (Rinn and Chang 2012), eRNA can provide insights into active biological processes that eDNA cannot. The transcriptional activity of genes varies depending on an organism's ecological niche, physiological condition, life-history stage, or phenotype (Prado-lima and Val 2015), and this variability can be captured in the eRNA released into the environment. Since RNA is produced only for genes actively transcribed within tissues, the eRNA collected from the environment may reflect the transcriptional signature of an organism at a given time, allowing for the detection of its metabolic state, life stage, or even environmental adaptations (Wood et al. 2019; Marshall et al. 2021). This represents an underutilized but potentially transformative application of eRNA in ecological studies.

By capturing such transcriptomic signatures, eRNA can push monitoring capabilities beyond species-level detection to encompass more specific biological insights, including detecting different forms within a species, such as life-history stages, sexes, or phenotypes. Moreover, it can provide a window into the physiological states of organisms in their natural habitats. For instance, eRNA could reveal upregulated genes associated with stress responses, reproduction, or developmental changes, offering real-time insights into the health and behavior of wildlife populations. At a broader scale, environmental metatranscriptomics, the study of the complete set of RNA transcripts in environmental samples could revolutionize community-level monitoring by tracking the physiological responses of entire ecosystems. By analyzing the expression of genes related to metabolic pathways, nutrient cycling, or stress responses, researchers could assess the impact of environmental changes, such as pollution or climate shifts, on ecological health. This community-level approach could provide a more comprehensive understanding of ecosystem dynamics, complementing traditional biodiversity assessments.

Despite the promise of eRNA, several technical challenges remain, particularly regarding its collection, preservation, and degradation. RNA is inherently less stable than DNA, and environmental factors such as temperature, pH, and microbial activity can accelerate its degradation. Therefore, additional research is needed to optimize methods for stabilizing eRNA in environmental samples, ensuring accurate detection and quantification. Techniques such as immediate sample filtration, the use of chemical stabilizers, and rapid freezing have shown promise in preserving eRNA (Wood et al. 2019; Von Ammon et al. 2019), but further refinement is essential for widespread application. In conclusion, eRNA represents a powerful and largely untapped tool for ecological monitoring, offering insights into organism detection, physiological states, and ecosystem-level processes. Its ability to reflect real-time biological activity and transcriptomic profiles makes it an invaluable asset for advancing environmental science. By continuing to explore its production, degradation, and preservation, researchers can fully harness the potential of eRNA for both species-level detection and broader ecological assessments (Marshall et al. 2021).

7.4 High-Specificity eRNA Assays: Advancing Biological Monitoring Precision

The advent of environmental DNA (eDNA) has revolutionized species detection and biodiversity monitoring, offering a non-invasive means to assess ecosystems with remarkable sensitivity. However, eDNA techniques are inherently limited to species-level identification, constraining their ability to resolve intraspecies variation, such as life-history stages or phenotypic differences among conspecifics (Sigsgaard et al. 2020). This limitation arises from the genetic uniformity of individuals within a species, as DNA sequences remain largely unchanged across developmental transitions. Consequently, eDNA cannot distinguish between juvenile and adult life stages or phenotype-specific variations. To transcend these constraints, environmental RNA (eRNA) has emerged as a promising alternative, leveraging the dynamic nature of gene expression to reveal biologically relevant variations within populations (Ciani et al. 2020). Unlike DNA, RNA profiles fluctuate in response to developmental stage, environmental stressors, and physiological state, offering an opportunity to distinguish between different conspecific forms based on their transcriptomic signatures. High-specificity eRNA assays targeting unique transcript variants termed environmental transcript types can enable the identification of distinct life-history stages or phenotype-associated expression patterns in environmental samples. However, challenges remain, including the scarcity of comprehensive transcriptomic datasets and the inherent instability of eRNA in natural environments (Tsuri et al. 2021). Despite these hurdles, eRNA holds immense potential to enhance biological monitoring, extending detection capabilities beyond species identification and providing real-time insights into population dynamics, environmental responses, and phenotypic plasticity in diverse ecosystems. As transcriptomic technologies advance, eRNA-based monitoring could become a cornerstone of ecological research, conservation biology, and environmental science (Tsuri et al. 2021; Ciani et al. 2020).

7.5 eRNA for Assessing Organismal Physiological State

Environmental changes induce distinct physiological responses in organisms, making them valuable indicators of ecological stress. For instance, heat-shock proteins are upregulated during thermal stress, a phenomenon well documented in aquatic organisms (Jeffries 2012; Chen et al. 2018). These stress-induced molecular responses are crucial biomarkers, offering insights into organismal health and environmental perturbations. A growing body of research demonstrates that exposure to stressors ranging from temperature fluctuations and pH shifts to hypoxia, pollution, and pathogens—leads to significant transcriptomic alterations. Such changes have been observed in diverse taxa, including corals, macroinvertebrates, zooplankton, molluscs (Jeffries et al. 2014), and vertebrates (Oomen and Hutchings 2017; Akbarzadeh et al. 2018). Crucially, these physiological shifts are reflected in the environmental RNA (eRNA) released into surrounding habitats, presenting a novel, noninvasive avenue for biological monitoring. Unlike conventional methods requiring tissue biopsies, eRNA-based assessments offer real-time insights into organismal health without disrupting ecosystems (Jeffries et al. 2014; Miller et al. 2017; Cristescu 2019). Unlike eDNA, which primarily detects species presence, eRNA captures dynamic biological activity, revealing stress responses and metabolic states (Tsuri et al. 2021). Recent studies indicate that eRNA detection is reliable in species-rich environments, particularly when organismal densities exceed detection thresholds (Von Ammon et al. 2019; Doi et al. 2019; Yates et al. 2023). As this emerging field advances, eRNA stands poised to revolutionize ecological monitoring, offering a high-resolution tool for assessing biodiversity, ecosystem health, and conservation priorities (Cristescu 2019; Von Ammon et al. 2019).

7.6 Methodology for eRNAs

7.6.1 Molecular Approaches to Detect and Quantify Environmental RNA

The advancement of molecular technologies has ushered in new avenues for detecting and quantifying environmental RNA (eRNA), providing critical insights into the health and presence of organisms within ecosystems. A high-specificity approach is crucial for the accurate detection of eRNA, and several methods have proven effective in achieving this goal. One of the most widely used techniques is quantitative polymerase chain reaction (qPCR), which is renowned for its sensitivity in detecting or quantifying specific DNA sequences. In environmental DNA (eDNA) studies, qPCR is commonly used due to its ability to provide absolute quantification of target sequences, even at low concentrations (Klymus et al. 2015). The inclusion of hydrolysis probes, such as TaqMan™, further enhances the specificity of the assay. When this technique is adapted for RNA detection through reverse transcription qPCR (RT-qPCR) it becomes highly applicable for eRNA analysis (Wood et al. 2019). However, care must be taken to avoid co-amplification of similar RNA sequences, either from closely related species that coexist in the same environment or from homologous genes within a single species (Tsuri et al. 2021). Given the limited availability of comprehensive transcriptomic libraries for most organisms, it is critical to test assay specificity on related species to ensure accuracy.

An alternative and highly sensitive method is digital droplet PCR (ddPCR), which has emerged as particularly effective for detecting low-abundance RNA sequences in environmental samples. Compared to RT-qPCR, ddPCR offers superior precision, robustness against inhibitors, and improved performance when the target RNA sequence is scarce (Taylor et al. 2017; Capo et al. 2020). Despite these advantages, the infrastructure and cost requirements for ddPCR are significantly higher, which may limit its accessibility for widespread use in environmental monitoring.

For broader assessments of biodiversity, metabarcoding approaches have become increasingly popular. This method utilizes universal primers to amplify eDNA or eRNA sequences from a wide range of taxa in an environmental sample, which are then compared to standardized species-specific sequence databases (Taberlet et al. 2012). Metabarcoding is particularly useful for capturing community-level biodiversity, offering a broad scope of resolution across taxa (Valentini et al. 2016). In eRNA studies, this method allows for the detection of RNA transcripts from genes that are conserved across multiple species, making it valuable for inventory-based applications. However, metabarcoding often lacks the sensitivity of species-specific approaches, cannot provide absolute quantification of copy numbers, and relies on extensive gene libraries for accurate identification (Harper et al. 2018). Despite these limitations, its utility in assessing community diversity renders it a powerful tool for broad ecological investigations. Metatranscriptomics represents cutting-edge eRNA detection technologies, leveraging next-generation sequencing (NGS) platforms such as RNA-seq to capture and sequence all RNA present in environmental samples. Originally developed to study microbial communities, metatranscriptomics enables the quantification of RNA from metabolically active microorganisms, which are often abundant in environmental samples (Taberlet et al. 2012). The application of metatranscriptomics to macro-organismal (eukaryotic) RNA remains an area of active research, as macro-organismal eRNA exists outside the organism and is likely to degrade rapidly, making its detection more challenging. Nevertheless, the potential for metatranscriptomics to expand our understanding of eRNA dynamics in natural environments is significant, particularly as technologies improve and methodologies evolve to better capture the transient nature of macro-organismal eRNA.

In summary, the molecular approaches used to detect and quantify eRNA, whether through high-specificity targeted methods like qPCR and ddPCR or broader, community-level approaches like metabarcoding and metatranscriptomics, offer invaluable insights into the health and diversity of organisms in their natural environments. Each method presents distinct advantages and limitations, from the precision and sensitivity of species-specific assays to the comprehensive biodiversity assessments enabled by universal primers and sequencing technologies. As eRNA research continues to develop, these molecular tools will become increasingly refined, allowing for more accurate and noninvasive monitoring of ecosystems, contributing to conservation efforts and our understanding of environmental changes (Klymus et al. 2015; Taberlet et al. 2012; Capo et al. 2020).

7.6.2 Normalizing Upregulated eRNA Biomarkers Using Housekeeping Genes or eDNA

The potential of environmental RNA (eRNA) biomarkers for assessing ecosystem health lies in their ability to provide high-resolution insights into specific environmental conditions. Ideally, eRNA biomarkers would be exclusively expressed under certain environmental stresses or by specific transcript types, making them perfect indicators of ecological changes. However, many potential biomarker genes are not strictly exclusive and exhibit upregulated expression in response to particular environmental triggers. These genes may still be expressed at low levels even in the absence of a stressor, or they could be expressed during different developmental stages of organisms, complicating their interpretation as biomarkers (Akbarzadeh et al. 2018; Miller et al. 2017). Additionally, RNA expression can vary significantly among individuals and can be affected by the quantity of starting material in a sample (Debes et al. 2012; Bustin et al. 2009). To overcome these challenges, gene expression studies often employ normalization strategies, one of the most common being the use of "housekeeping" genes. These genes are assumed to be expressed at stable levels across different environmental conditions, providing a baseline against which to measure the expression of other genes (Bustin et al. 2009). When applied to eRNA assays, upregulated genes can be normalized against housekeeping genes, assuming the latter can be detected reliably in eRNA in consistent quantities. However, identifying a stable "housekeeping" gene within the environmental RNA context is challenging, as little research has explored the behavior of different RNA forms in eRNA or their degradation rates in nature (Marshall et al. 2021). This gap in knowledge calls for further investigation to determine the most suitable housekeeping genes for normalizing eRNA biomarkers. Another critical issue in the use of upregulated eRNA biomarkers is the potential influence of organism abundance. Changes in the number of organisms in an ecosystem can affect the concentration of eRNA detected, complicating the use of these biomarkers to infer environmental stress or conditions. For instance, organisms with low gene expression levels might still contribute to eRNA signals simply by being more abundant, obscuring the actual biological or environmental drivers of upregulated eRNA (Yates et al. 2023). Therefore, to use upregulated eRNA biomarkers effectively, they must also be normalized against an index of organism abundance.

One promising method for this normalization is to use environmental DNA (eDNA), which has been shown to correlate strongly with species abundance in natural environments. This relationship may extend to eRNA as well, but given the inherent variability in individual transcription levels and the relatively unstable nature of RNA, correlations between eRNA and organism abundance are expected to be positive but potentially weaker than those observed with eDNA. Even for "housekeeping" genes, variability in eRNA levels could pose challenges to reliable normalization. An alternative approach to normalize upregulated eRNA biomarkers is to compare the concentration of eRNA to eDNA markers from the same organism. This method offers the advantage of incorporating both molecular signals into the analysis, thus providing a more accurate representation of organism presence and gene expression. However, environmental conditions, such as temperature, can impact the relationship between eDNA production and organism abundance, particularly in poikilothermic species, where eDNA production rates increase with rising temperatures (Jo 2023). Therefore, carefully considering these factors is essential when interpreting the relationship between eRNA, eDNA, and organism abundance.

The successful application of eRNA biomarkers requires thoughtful normalization strategies using housekeeping genes or eDNA as reference points. Spatial sampling across ecosystems is also necessary to account for environmental heterogeneity and interindividual variation. This would ensure that eRNA-based population assessments accurately reflect both local and broader ecological dynamics, paving the way for more comprehensive insights into ecosystem health and resilience (Akbarzadeh et al. 2018; Yates et al. 2023; Marshall et al. 2021).

7.6.3 Understanding eRNA Production, Degradation, and Improving Sample Preservation

Environmental RNA (eRNA) offers remarkable potential for ecological monitoring by providing real-time molecular snapshots of gene expression in diverse ecosystems. However, the mechanisms of eRNA production, its degradation in natural environments, and best practices for preserving eRNA samples remain underexplored areas of study. Advancing our understanding of these factors is crucial to unlocking the full potential of eRNA as a biomonitoring tool.

7.6.3.1 eRNA Production and Its Tissue Origins

One of the primary challenges in eRNA research is understanding how transcription rates in various tissues contribute to the composition of eRNA. While transcriptional activities vary significantly across different tissues within an organism (Ciani et al. 2020), it is unclear whether the eRNA detected in environmental samples represents a homogenized mixture of RNA from multiple cell types or whether it is biased toward specific organs or tissue systems. For example, the major sources of environmental DNA (eDNA) are cells shed from the organism's surface and the excretion of metabolic wastes and reproductive fluids (Stewart and Taylor 2019). This pattern of eDNA production suggests that eRNA could similarly reflect gene expression from tissues directly interfacing with the external environment, such as the skin, respiratory organs, and digestive systems, as well as from reproductive organs in species that employ external fertilization (Tsuri et al. 2021). Unravelling whether eRNA primarily reflects certain tissue-specific transcripts or whether it captures a broader, systemic gene expression profile is a critical research frontier.

7.6.3.2 Quantifying eRNA Degradation Rates

The degradation of eRNA in natural environments poses another significant obstacle for its application as a reliable biomarker. While RNA molecules are known to have variable intracellular stability, with half-lives ranging from less than 2 hours to over 16 hours depending on their structure and function (Clark et al. 2012), the degradation rates of different RNA forms in environmental contexts remain poorly understood. Initial research has shown that nuclear eRNA degrades more quickly than mitochondrial eRNA, while ribosomal RNA (rRNA) and messenger RNA (mRNA) exhibit variable degradation rates (Marshall et al. 2021). Several factors contribute to the stability of RNA, including secondary structure, the presence of RNA-binding proteins, and active degradation mechanisms within cells (Clark et al. 2012; Geisler and Coller 2013). For eRNA-based biomonitoring applications, identifying RNA species that are relatively stable in environmental conditions would be ideal, though these species need not be limited to mRNA. Long non-coding RNAs (lncRNAs), for example, have been implicated in a range of regulatory processes, including sex- and tissue-specific gene regulation, making them promising candidates for eRNA assays (Rinn and Chang 2012; Geisler and Coller 2013). Therefore, future studies should prioritize the identification of eRNA molecules with high environmental stability, particularly those that can serve as reliable indicators of ecosystem health.

7.6.3.3 Enhancing eRNA Sample Preservation Techniques

Preserving the integrity of eRNA in environmental samples is paramount for accurate downstream analysis. The inherent instability of RNA, which is highly susceptible to degradation by environmental factors such as temperature, pH, and microbial activity, necessitates the development of robust preservation techniques. Several promising strategies have been explored to improve the collection and stabilization of eRNA. Immediate filtration of environmental samples is one approach that can minimize RNA degradation by reducing the time RNA molecules remain exposed to degrading factors (Hinlo et al. 2017). The use of chemical reducing agents in filter storage buffers has also been shown to stabilize RNA by preventing oxidative damage (van der Poel-van de Luytgaarde et al. 2013). Additionally, freezing samples as soon as they are collected has emerged as an effective method for preserving both the quantity and quality of RNA in environmental samples (Wood et al. 2019; Von Ammon et al. 2019). Despite the challenges associated with RNA degradation, there is compelling evidence that eRNA exists in environmental mediums and can be successfully detected using appropriate collection and preservation methods (Cristescu 2019; Von Ammon et al. 2019). As research continues to refine these methodologies, the utility of eRNA as a biomonitoring tool will likely expand, offering new insights into the health and dynamics of ecosystems.

In summary, improving our understanding of eRNA production and degradation, alongside advancements in sample preservation techniques, will be critical for the future application of eRNA in ecological monitoring. By addressing these challenges, researchers can harness the full potential of eRNA to provide nuanced insights into environmental change and organismal responses (Clark et al. 2012).

7.6.4 Insights into Developing High-Specificity eRNA Assays

Recent advancements in environmental RNA (eRNA) technology have paved the way for highly specific biomarker detection using approaches like quantitative PCR (qPCR) and droplet digital PCR (ddPCR). These targeted methods are proving to be reliable for the detection and quantification of specific eRNA genes in environmental samples, as demonstrated by pioneering studies (Wood et al. 2019; Marshall et al. 2021). However, the identification and development of high-specificity eRNA assays require a well-planned approach, and we propose two complementary strategies, each offering unique advantages and challenges: the "library-first" approach and the "eRNA-first" approach.

7.6.4.1 The Library-First Approach: Potential Biomarker Libraries from Tissue Transcriptomes

The "library-first" approach focuses on constructing potential eRNA biomarker libraries from tissue-specific or whole-organism transcriptomes. This approach relies on quantifying RNA expression profiles under controlled experimental conditions. For example, exposing organisms to a gradient of environmental stressors can help identify differentially expressed genes that are either upregulated or downregulated in specific tissues or treatments. These transcriptomic libraries would serve as a valuable resource for developing high-specificity assays targeting key biomarkers, which can later be validated using eRNA samples from natural environments or experimental setups (Tsuri et al. 2021). The advantage of this approach lies in its controlled nature, where both tissue and eRNA samples can be simultaneously collected, facilitating a direct comparison of gene expression patterns between tissue-specific RNA and eRNA. Additionally, it allows researchers to establish causal linkages between environmental stressors and transcriptional responses, enabling a more comprehensive understanding of organismal stress response mechanisms.

There is already a wealth of transcriptomic data available for certain species that document gene expression in response to environmental stressors, which could be leveraged for eRNA biomarker development. For instance, species like Daphnia and *Xenopus laevis* have well-characterized transcriptomic responses to various stressors and life-history stages, such as juvenile-to-adult metamorphosis (Spanier et al. 2017). Moreover, socioeconomically significant species, including various fish species, have been extensively studied for transcriptional biomarkers in response to immune challenges, temperature changes, pollutants, maturation, and metamorphosis (Ciani et al. 2020). Among these, Pacific salmonids hold particular promise for eRNA-based studies due to their high densities during spawning, which increases the likelihood of eRNA recovery. Additionally, these species experience a range of environmental challenges, such as temperature fluctuations, oxygen deprivation, and immune stress, making them ideal candidates for developing eRNA assays. However, the library-first approach carries the inherent risk that tissue-specific RNA sequences identified in laboratory conditions may not be detectable in eRNA samples from natural environments. The eRNA signature of an organism may differ significantly from tissue transcriptomes, which could result in considerable effort being invested in developing biomarkers that may not be useful for environmental monitoring (Tsuri et al. 2021). Despite this uncertainty, targeting tissue types that are more likely to produce detectable eRNA such as digestive, respiratory, epidermal, and reproductive tissues can mitigate some of these risks, offering a promising direction for biomarker discovery.

7.6.4.2 The eRNA-First Approach: Directly Profiling Environmental RNA

In contrast to the library-first method, the "eRNA-first" approach aims to directly profile the eRNA produced by organisms in their environment and then develop targeted assays based on the genes detected. This approach circumvents the uncertainty regarding the representativeness of tissue transcriptomes in eRNA samples, as it focuses on the RNA sequences that are actually released into the environment. To implement the eRNA-first approach, researchers would need to collect high-volume eRNA samples from experimental replicates where organisms are exposed to a gradient of environmental stressors. These samples would then be subjected to high-throughput sequencing techniques, such as RNA-seq, to capture a comprehensive snapshot of the eRNA present in the environment (Wang et al. 2009). However, environmental samples typically contain a mixture of RNA from various sources, including microorganisms like bacteria, algae, and microeukaryotes that coexist with the target organisms. Therefore, bioinformatics tools must be employed to map eRNA sequences to the reference genomes or transcriptomes of the target species, allowing researchers to identify organism-specific transcripts amidst the complexity of environmental RNA (Culviner et al. 2020; Broman et al. 2021).

Despite its potential, the eRNA-first approach faces significant technical challenges. Environmental samples often contain low concentrations of nucleic acids, and the quality of the RNA can be compromised due to degradation. Moreover, the presence of RNA from prokaryotic organisms, which are abundant in environmental samples, may overshadow the recovery of eukaryotic eRNA (Stat et al. 2017). Similar to the challenges faced with environmental DNA (eDNA), the dominance of microbial RNA could obscure the detection of macro eukaryotic eRNA in complex samples. Several techniques have been developed to improve the recovery of eukaryotic RNA from environmental samples. For example, methods that deplete bacterial rRNA or selectively capture eukaryotic mRNA through poly-A hybridization can enhance the relative concentration of eukaryotic RNA in environmental samples (Damon et al. 2012; Shakya et al. 2019). Additionally, conducting experiments in controlled environments, such as aquaria, where the focal species are present in high densities and microbial growth is minimized, can improve eukaryotic eRNA recovery by increasing its relative abundance compared to microbial RNA. Preliminary experiments in aquaria have shown promising results, with several studies reporting high concentrations of species-specific eRNA in invertebrate species (Wood et al. 2019; Marshall et al. 2021). Ultimately, both the library-first and eRNA-first approaches offer valuable pathways to developing high-specificity eRNA assays. The choice of approach depends on the research objectives and the available resources. The library-first approach provides a controlled and systematic method for biomarker discovery but carries the risk of mismatches between tissue transcriptomes and environmental RNA. In contrast, the eRNA-first approach directly addresses the complexities of environmental RNA but requires sophisticated bioinformatics and experimental techniques to overcome the challenges of RNA degradation and microbial interference. Despite these challenges, both approaches hold significant promise for advancing the field of environmental RNA research. By harnessing the power of transcriptomics, researchers can develop highly specific eRNA assays that provide insights into the functional activity of organisms in response to environmental stressors. As eRNA technologies continue to evolve, they are likely to become indispensable tools for monitoring ecosystem health and understanding the molecular responses of organisms to their changing environments.

7.6.5 Insights into Environmental Transcriptomics for Ecosystem Monitoring

An exciting frontier in environmental nucleic acids (eNAs) research is the potential to utilize bulk mRNA released by organisms into their surroundings to monitor both the taxonomic composition and the gene expression profiles of complex macro-organism communities. Traditionally, metagenomics has been employed to reconstruct the taxonomic composition of ecosystems through DNA analysis, providing insights into community structure based on species' presence (Sunagawa et al. 2015). However, DNA does not reflect the real-time responses of these communities to environmental factors. In contrast, metatranscriptomics, which relies on RNA, offers a dynamic glimpse into the genes actively expressed by organisms at the time of sampling, thereby enabling researchers to examine how ecosystems respond to specific environmental conditions (Shi et al. 2011; Moran et al. 2013). Metatranscriptomic analysis is particularly sensitive to environmental shifts, making it a powerful tool for capturing community-wide functional profiles. Although traditionally applied to microbial studies where mRNA is extracted from bulk samples containing live bacteria, only a few studies have explored its potential for macro-organism communities. By measuring the abundance of transcripts in response to environmental stimuli, this approach can reveal which metabolic pathways are being actively regulated, providing critical insights into ecosystem health. Despite being costly at present, metatranscriptomics could eventually be integrated into environmental monitoring programs to provide early warnings about ecosystem stressors or disturbances, helping to manage and conserve biodiversity more effectively.

Environmental transcriptomics has mostly been applied to microbial communities, where RNA profiles are likely dominated by organismal RNA from living microorganisms. In microbial metatranscriptomics, mRNA constitutes a small proportion of total RNA, with most reads being ribosomal RNA (rRNA) and transfer RNA (tRNA). However, extending this approach to macro-organisms could provide valuable insights by capturing extraorganismal mRNA from environmental samples, independent of the progenitor organisms themselves. This presents a noninvasive, molecular approach to studying functional activities within entire communities, without requiring prior knowledge of species composition or specific gene expression under varying environmental conditions. A key benefit of environmental transcriptomics is its ability to reveal functional processes within ecosystems. Researchers can identify active metabolic pathways by analyzing gene expression and link them to particular environmental conditions or stressors. This could be especially valuable in understanding ecosystem responses to changes such as pollution, climate shifts, or habitat alteration. Moreover, this method does not necessitate the identification of individual species, making it an ideal tool for assessing ecosystem-level conditions across multiple trophic levels (Cristescu 2019).

However, significant technical challenges remain. Eukaryotic organisms, having larger and more complex genomes than prokaryotes, require higher sequencing coverage to capture transcript abundance accurately. Total microbial RNA samples generally comprise only 1–5% mRNA, with the rest comprising rRNAs and tRNAs. Thus, macro-organismal mRNA likely represents a small fraction of total eRNA, making its recovery and analysis more difficult. Fortunately, advancements in molecular techniques, such as improved mRNA isolation methods, are steadily overcoming these challenges, enhancing the recovery of eukaryotic mRNA (Shakya et al. 2019). Another significant hurdle is the interpretation of complex data. A major limitation is the absence of a clear baseline or 'control' condition against which transcriptomic data can be compared. In the context of ecosystem monitoring, researchers often rely on baseline conditions that represent optimal environmental states. In the absence of such controls, changes in RNA expression can be challenging to interpret. One possible solution involves comparing RNA transcript levels with corresponding DNA gene abundances. Any discrepancies between the two might reflect differential gene expression or changes in ecosystem functionality.

Despite its potential, metatranscriptomics has yet to be fully validated as a biomonitoring tool, largely due to its cost and the complexity of the data generated. Nonetheless, environmental transcriptomics holds great promise for assessing community-wide responses to environmental conditions across various trophic levels. As technology advances and costs decline, this approach could become a cornerstone of ecosystem monitoring, providing unprecedented insights into how ecosystems function and respond to changing environmental conditions (Cristescu 2019). Ultimately, environmental transcriptomics could pave the way for more nuanced and accurate assessments of ecosystem health, contributing to more informed conservation and management strategies.

7.6.6 Advancing Ecological and Biological Inquiries Beyond Community Composition Using eDNA or eRNA

The application of environmental DNA (eDNA) and environmental RNA (eRNA) has expanded beyond mere species detection to address broader ecological and biological questions, though challenges and uncertainties remain. In this Special Issue, five innovative studies exemplify the cutting-edge applications and explore potential variabilities associated with these analyses. Di Muri et al. (2023) demonstrate the efficacy of eDNA metabarcoding in delineating the spatiotemporal distribution of Arctic char (*Salvelinus alpinus*) spawning activity in Windermere, UK, aligning well with conventional netting surveys. Their findings underscore the capability of eDNA metabarcoding to accurately define the temporal and spatial dynamics of fish spawning in lentic environments.

Sutherland et al. (2023) pioneered the use of eRNA to simultaneously assign taxonomies to microbes and analyze gene expressions. Conducted in a shellfish hatchery in British Columbia, Canada, their study reveals temporal shifts in microbial community composition and functional gene expression linked to abiotic factors. This dual approach enhances our ability to assess environmental conditions based on the temporal distribution of transcribed functional genes. Condachou et al. (2023) employ eDNA metabarcoding to simulate the variability in estimates of functional diversity for fish communities. Their study illustrates the potential of eDNA to estimate community functional characteristics, contingent upon accurate taxonomic assignments. Despite challenges in taxonomic granularity, their findings suggest manageable impacts on functional diversity estimates. Querejeta et al. (2023) innovate by using metabarcoding to identify plant species from pollen loads of wild bees in sunflower crops in France. Their unique approach explores the association between bee species traits and the taxonomic diversity of pollinated plants, shedding light on the functional role of wild bees in agroecosystems. Coston-Guarini et al. (2023) address the suitability of eDNA analysis for ecological impact assessments (EIA), simulating eDNA detection probabilities under varying conditions such as population dynamics and mobility. Their study underscores the importance of robust sampling strategies and environmental factors in ensuring reliable predictions of eDNA reactivity for EIA applications. These studies highlight the transformative potential of eDNA and eRNA analyses in advancing ecological and biological research beyond species detection, paving the way for nuanced insights into ecosystem dynamics, functional diversity, and ecological impacts. Continued methodological refinements and interdisciplinary collaborations will further harness the full potential of these molecular tools in environmental science and conservation management.

7.6.7 Unveiling the Potential of eRNA: Beyond Species Detection

Environmental RNA (eRNA) has transcended its initial role as a biomolecular tool for species detection, emerging as a transformative frontier in ecological and genetic research. Unlike static DNA fragments, eRNA offers a dynamic snapshot of active biological processes, providing unprecedented insight into gene expression patterns, physiological states, and real-time ecosystem responses. This paradigm shift has unlocked new dimensions in biodiversity assessment, environmental health monitoring, and ecological interactions. The rapid evolution of eRNA methodologies underscores the necessity of understanding its release, persistence, and degradation dynamics across diverse environments. While still an emerging field, advancements in high-throughput sequencing and transcriptomic profiling have begun to unveil the intricate molecular signals embedded within environmental matrices. These developments pave the way for novel applications, such as tracking metabolic shifts, detecting stress responses, and refining species-specific ecological interactions.

Despite current challenges, including RNA stability, differential degradation rates, and methodological standardization, eRNA holds immense promise for revolutionizing ecological surveillance. By bridging molecular precision with ecosystem-scale monitoring, it offers a powerful lens into the functional dynamics of biodiversity, marking the dawn of a new era in environmental science. The continued refinement of eRNA-based approaches will undoubtedly shape the future of ecological exploration and conservation.

7.6.8 Outstanding Questions and Future Directions in eRNA-Based Ecological Surveillance

Despite its transformative potential, the field of environmental RNA (eRNA) research remains in its early stages, with numerous fundamental questions yet to be answered. A critical challenge lies in the reliable quantification of eRNA, particularly in understanding whether its presence reflects a homogenous representation of RNA from various species and tissues or whether certain biological sources disproportionately contribute to its composition. Further investigation is needed to determine the extent to which eRNA signatures vary among individuals with distinct life-history stages, phenotypes, and sexes, offering a promising avenue for noninvasive ecological and evolutionary studies. One of the most compelling prospects of eRNA lies in its ability to capture transcriptional responses to environmental stressors. However, key uncertainties remain: Can eRNA accurately reflect real-time physiological states? To what degree do environmental conditions influence transcriptional profiles detected in eRNA pools? A deeper exploration of these questions will be essential for establishing eRNA as a robust biomarker for ecosystem health and resilience.

Additionally, understanding the relative concentrations of different RNA classes mRNA, noncoding RNAs (ncRNA), microRNAs (miRNA), and ribosomal RNA (rRNA) is crucial for refining detection methodologies. The turnover rates of these molecules in diverse environments, along with their degradation kinetics postcollection, must be systematically studied to optimize sample handling and improve data reliability. Advances in RNA preservation techniques, such as the use of RNase inhibitors or chemical stabilizers, will further enhance the feasibility of eRNA-based biomonitoring in field conditions. As we refine our understanding of eRNA's ecological dynamics, it has the potential to emerge as a cornerstone of environmental monitoring, offering unprecedented molecular resolution for biodiversity assessment and conservation science. The integration of high-throughput sequencing and transcriptomic profiling will pave the way for a new era of ecological surveillance, where biodiversity is not just catalogued but dynamically monitored in real time.

7.6.9 Critical Insights and Perspectives on Environmental RNA (eRNA)

7.6.9.1 Unravelling the Potential of eRNA in Ecological Surveillance

The emergence of environmental RNA (eRNA) as a novel biomonitoring tool has opened new frontiers in ecological surveillance, offering real-time insights into organismal activity and ecosystem health. Unlike environmental DNA (eDNA), which serves as a historical footprint of species presence, eRNA reflects dynamic biological processes, capturing active gene expression across trophic levels. However, several fundamental challenges must be addressed before eRNA can be fully integrated into ecological monitoring frameworks. These challenges span from its inherent instability and rapid degradation to the complexities of quantification, tissue-specific expression biases, and the interpretation of transcriptional signals in environmental contexts.

7.6.9.2 Challenges in eRNA Quantification and Interpretation

A primary question in eRNA research is whether it can be reliably quantified to represent the metabolic activity of organisms within an ecosystem. Given that RNA production varies across tissues, developmental stages, and environmental conditions, interpreting eRNA signatures requires a nuanced understanding of gene expression dynamics. Unlike genomic DNA, which remains largely stable, RNA expression levels fluctuate in response to physiological states and environmental stressors. Therefore, distinguishing between background noise and biologically meaningful signals is essential for robust ecological interpretations.

Another key consideration is whether eRNA collected from environmental samples represents a homogenized mixture of RNA sources or if it is biased toward specific tissues, phenotypes, or species. The preferential degradation of certain RNA types such as ribosomal RNA (rRNA) versus messenger RNA (mRNA) further complicates this assessment. Investigating the relative stability of different RNA species, including long non-coding RNAs (lncRNA), microRNAs (miRNA), and transfer RNAs (tRNA), will be crucial in refining eRNA-based approaches for biodiversity monitoring.

7.6.9.3 eRNA as an Early Warning System for Ecosystem Health

Perhaps the most exciting prospect of eRNA lies in its potential to serve as an early-warning system for environmental disturbances. Since RNA is the immediate product of gene expression, shifts in eRNA profiles can indicate organismal responses to stressors such as pollution, temperature fluctuations, or habitat degradation. If specific transcriptional signatures can be consistently linked to environmental stress conditions, eRNA could provide a molecular-level diagnostic tool for ecosystem health assessments. However, identifying which genes whether coding or noncoding is consistently represented in eRNA pools remains a significant challenge. Understanding the turnover rates of different RNA types across diverse habitats will be essential for optimizing eRNA assays and improving their reliability in conservation biology.

7.6.9.4 Temporal Dynamics and Degradation of eRNA

A fundamental limitation of eRNA is its rapid degradation compared to DNA, making it a highly transient biomarker of biological activity. Unlike DNA, which can persist in the environment for extended periods, RNA is subject to enzymatic breakdown and environmental decay. This short-lived nature, however, can be an advantage, as it allows researchers to capture snapshots of active metabolic states rather than merely detecting past presence. Investigating RNA degradation kinetics under varying conditions (e.g., temperature, UV exposure, microbial activity) will be vital for establishing standardized protocols for eRNA-based ecological monitoring.

7.6.9.5 Optimizing eRNA Collection, Preservation, and Extraction

The efficiency of eRNA detection is heavily influenced by sampling strategies, preservation techniques, and extraction methodologies. RNA is highly susceptible to hydrolysis and enzymatic degradation, necessitating immediate stabilization upon collection. The choice of filtration methods, sample volumes, and chemical stabilizers such as 2-mercaptoethanol or RNA later can significantly impact RNA integrity. Developing optimized protocols tailored to different ecosystems (marine, freshwater, terrestrial) will be critical for maximizing eRNA recovery and ensuring data reproducibility across studies.

7.6.9.6 Future Directions and Research Imperatives

Despite its challenges, eRNA holds immense promise as a next-generation tool for ecological surveillance. Addressing the complexities of RNA stability, quantification, and interpretation will pave the way for its broader application in biodiversity monitoring, species conservation, and environmental diagnostics. By integrating eRNA into molecular ecology, researchers can move beyond static assessments of species presence toward dynamic analyses of ecosystem function and resilience. As sequencing technologies advance and bioinformatics tools become more sophisticated, eRNA is poised to revolutionize our understanding of ecological processes, offering a window into the living, breathing complexity of the natural world.

7.7 The Molecular Landscape of eRNA: Key Concepts and Innovations

7.7.1 Environmental DNA (eDNA): The Molecular Footprint of Organisms

Environmental DNA (eDNA) refers to genetic material shed by organisms into their surroundings, such as water, soil, or air. This extraorganismal DNA exists in various forms—free-floating, encapsulated within cells, or adsorbed onto organic or mineral surfaces. Widely used in ecological research, eDNA enables species identification without direct observation, offering a noninvasive tool for biodiversity assessment and ecosystem monitoring.

7.7.2 Environmental Nucleic Acids (eNA): Expanding the Molecular Landscape

Environmental nucleic acids (eNA) encompass both DNA and RNA found outside of living organisms within environmental media. The concept of eNA extends beyond species detection, encompassing genetic signals that reveal ecological interactions, community composition, and metabolic activity. While eDNA informs on species presence, eRNA refines this understanding by identifying active organisms and their physiological states.

7.7.3 Environmental RNA (eRNA): A Real-Time Window into Ecosystem Function

Environmental RNA (eRNA) consists of RNA molecules shed into the environment through cellular processes, extracellular vesicles, or passive release. Unlike eDNA, which can persist for extended periods, eRNA degrades rapidly, making it an indicator of recent biological activity. This unique property positions eRNA as a powerful tool for tracking metabolically active species, resolving taxonomic identities at finer scales, and assessing real-time ecological dynamics.

7.7.4 Environmental (Meta)Transcriptomics: Unravelling Gene Expression Across Ecosystems

Environmental transcriptomics also referred to as metatranscriptomic when applied to microbial communities—examines RNA expression profiles within environmental samples. This approach captures gene activity across trophic levels, providing insights into ecosystem functions, species interactions, and stress responses. By integrating eRNA-based transcriptomics, researchers can monitor gene expression patterns across diverse biomes, enhancing our understanding of ecological processes at a molecular level.

7.7.5 Environmental Transcript-Type: A Molecular Signature of Life Forms

The term "environmental transcript-type" describes any biological entity that produces distinct or highly upregulated eRNA signatures. This could include specific developmental stages, sexes, phenotypes, genotypes, or even localized populations within a species. By detecting transcript-type variations, eRNA analyses enable researchers to distinguish subtle ecological and evolutionary differences that remain undetectable through traditional eDNA-based methods.

7.7.6 Gene Expression: The Blueprint of Life in Motion

Gene expression is the process by which genetic instructions encoded in DNA are converted into functional products, including proteins and regulatory RNAs. This molecular mechanism underlies all biological activity, from cellular metabolism to organismal adaptation. In the context of eRNA, gene expression data offer a direct link to the physiological state of organisms, enabling researchers to assess responses to environmental stressors, habitat conditions, and ecological disturbances.

7.7.7 Metabarcoding: Deciphering Biodiversity Through High-Throughput Sequencing

Metabarcoding is a genetic identification technique that extracts DNA or RNA from environmental or bulk samples to analyze community composition. This approach, often leveraging high-throughput sequencing, is instrumental in assessing biodiversity across microbial, meiofaunal, and macrofaunal assemblages. While traditionally applied to eDNA, emerging methodologies now integrate eRNA metabarcoding to refine species detection and track ecological changes with temporal specificity.

7.7.8 Metatranscriptomics: The Voice of Ecosystems in Molecular Form

Metatranscriptomics refers to the large-scale sequencing of RNA molecules from environmental samples to profile active gene expression. Originally developed for microbial ecology, this approach now extends to eukaryotic communities, offering a dynamic perspective on ecosystem function. By revealing which genes are being actively transcribed, metatranscriptomics provides an unparalleled depth of insight into environmental processes, from nutrient cycling to species interactions.

7.7.9 Messenger RNA (mRNA): The Molecular Messenger of Life

Messenger RNA (mRNA) serves as an intermediary between genetic blueprints stored in DNA and the proteins essential for cellular function. These transient RNA sequences carry instructions for protein synthesis, dictating organismal responses to environmental stimuli. Within eRNA research, mRNA signatures provide a snapshot of active metabolic pathways, offering precise indicators of species presence, physiological status, and ecosystem health.

7.7.10 Noncoding RNA (ncRNA): The Silent Regulators of Ecological Function

Noncoding RNA (ncRNA) represents a vast and functionally diverse class of RNA molecules transcribed from DNA regions that do not encode proteins. Unlike messenger RNA (mRNA), which serves as a template for protein synthesis, ncRNAs play critical roles in post-transcriptional regulation, cellular communication, and environmental adaptation. Among these, long noncoding RNAs (lncRNAs) modulate gene expression through chromatin remodeling and transcriptional interference, while microRNAs (miRNAs) fine-tune gene expression by targeting specific mRNA transcripts for degradation or translational repression. Ribosomal RNA (rRNA) forms the structural and catalytic core of ribosomes, driving protein synthesis, and transfer RNA (tRNA) facilitates amino acid transport during translation. The presence of ncRNAs in environmental samples provides unique insights into the regulatory dynamics of ecosystems, revealing how organisms respond at the molecular level to environmental changes. Their detection in eRNA surveys could serve as powerful indicators of physiological stress, species interactions, and ecosystem stability.

7.7.11 Quantitative PCR (qPCR): Precision Detection of Molecular Signatures

Quantitative PCR (qPCR) has revolutionized the field of molecular ecology by enabling the real-time detection and quantification of target DNA sequences. This technique relies on polymerase chain reaction (PCR) amplification, where specific DNA sequences are amplified and their fluorescence is monitored over successive cycles. Using intercalating dyes or fluorescently labeled probes, qPCR allows researchers to estimate the abundance of specific genetic markers with unparalleled sensitivity. In the context of eRNA, qPCR serves as a crucial tool for validating transcriptomic data, quantifying gene expression in response to environmental stressors, and detecting rare or transient RNA molecules in environmental samples. By integrating qPCR into eRNA-based monitoring frameworks, researchers can achieve high-resolution assessments of biodiversity and ecosystem health.

7.7.12 RNA-Seq: Unlocking the Depths of the Environmental Transcriptome

RNA sequencing (RNA-seq) represents a transformative approach to transcriptome analysis, leveraging next-generation sequencing technologies to generate comprehensive RNA sequence data. Unlike qPCR, which targets predefined genetic sequences, RNA-seq provides an unbiased, genome-wide snapshot of transcriptional activity across all expressed genes. By sequencing millions of short RNA fragments (reads) and assembling them using advanced bioinformatics tools, RNA-seq enables the identification of novel genes, alternative splicing events, and differential gene expression patterns. In ecological surveillance, RNA-seq facilitates the discovery of previously uncharacterized eRNA signatures, offering unprecedented insights into community structure, species interactions, and environmental stress responses. The integration of RNA-seq with eRNA-based methodologies holds immense promise for advancing conservation biology, allowing for the real-time assessment of ecosystem resilience at the molecular level.

7.7.13 RT-qPCR: Bridging RNA Detection with Functional Ecology

Reverse transcription quantitative PCR (RT-qPCR) extends the capabilities of qPCR by specifically targeting RNA molecules. In this technique, RNA is first converted into complementary DNA (cDNA) through reverse transcription, followed by real-time PCR amplification. This process can be performed in a two-step approach, where cDNA synthesis and qPCR occur separately, or in a one-step reaction, where both steps are combined for greater efficiency. RT-qPCR is particularly valuable in eRNA research, as it enables precise quantification of gene expression changes in response to environmental variables. By targeting stress-responsive genes, RT-qPCR can be used to assess the physiological state of organisms in their natural habitats, providing a functional perspective on biodiversity monitoring. The method's high sensitivity makes it an essential tool for detecting rare transcripts and validating RNA-seq findings in ecological studies.

7.7.14 Environmental Stressors: Molecular Triggers of Ecosystem Change

Environmental stressors represent external variables that push ecosystems beyond their natural equilibrium, triggering molecular, physiological, and ecological disruptions. These stressors can range from temperature fluctuations, pollutants, and habitat degradation to invasive species and climate-induced shifts. At the molecular level, exposure to stressors often induces plastic gene expression, allowing organisms to mount adaptive responses. The study of eRNA enables researchers to capture these dynamic transcriptional shifts, revealing how species cope with environmental perturbations in real time. Monitoring stress-responsive eRNA transcripts could serve as an early-warning system for detecting ecological imbalances before they manifest as large-scale biodiversity losses. This application of eRNA in environmental risk assessment underscores its potential as a cutting-edge tool for conservation science.

7.7.15 Transcription: The Molecular Pulse of Life

Transcription, the fundamental process by which genetic information is converted into RNA, serves as the molecular foundation of life. This intricate mechanism involves the synthesis of RNA from a DNA template, mediated by RNA polymerases and a network of transcription factors. The resulting RNA molecules including both coding and noncoding transcripts govern cellular function, organismal development, and environmental adaptation. In ecological studies, the detection of eRNA provides a direct window into transcriptional activity within ecosystems. By analyzing transcriptional signatures in environmental samples, researchers can infer species activity, metabolic states, and ecological interactions, offering new dimensions to biodiversity monitoring and ecosystem surveillance.

7.7.16 The Transcriptome: A Living Record of Ecological Dynamics

The transcriptome encompasses the complete set of RNA transcripts expressed in an organism at a given time, including both protein-coding and noncoding RNAs. Unlike the static nature of the genome, the transcriptome is highly dynamic, reflecting real-time physiological responses to environmental stimuli. By profiling eRNA from environmental samples, scientists can reconstruct the molecular landscapes of entire ecosystems, identifying active species, monitoring gene expression shifts, and detecting early signs of ecological stress. The integration of transcriptomic data into conservation strategies enhances our ability to predict environmental changes, optimize habitat management, and safeguard biodiversity with unprecedented molecular precision.

7.8 Conclusion and Future Prospective

Environmental RNA (eRNA) heralds a transformative era in ecological surveillance, redefining how biodiversity is studied and monitored at unprecedented molecular resolution. By capturing transcriptional activity from living organisms, eRNA provides a dynamic and temporally refined perspective on ecosystem health, species interactions, and environmental responses. Unlike conventional biomonitoring approaches, which often rely on static species inventories, eRNA unveils the intricate physiological and metabolic states of organisms, offering real-time insights into ecological dynamics. This molecular tool has the potential to revolutionize conservation biology, ecosystem management, and environmental policy, enabling proactive interventions in response to anthropogenic and climatic stressors. The integration of eRNA into ecological studies elevates the granularity of biomonitoring beyond species detection, facilitating the identification of life-history stages, phenotypic variations, and stress responses within populations and communities. High-resolution transcriptomic analyses derived from eRNA signatures hold the promise of unravelling the complexities of ecological adaptation, species resilience, and ecosystem stability. Furthermore, eRNA-based methodologies enable the assessment of physiological states across trophic levels, providing a window into the molecular underpinnings of community structure and function. By leveraging these molecular insights, researchers can better predict shifts in biodiversity, detect early warning signals of ecosystem perturbations, and implement evidence-based conservation strategies.

Despite its vast potential, several challenges must be addressed to fully harness the power of eRNA in ecological surveillance. The development of standardized protocols for eRNA extraction, stabilization, and sequencing remains paramount to ensure data consistency and comparability across studies. Moreover, understanding the environmental fate of eRNA, its persistence, degradation kinetics, and transport dynamics will be critical in refining its applications for environmental monitoring. Advancements in bioinformatics and molecular assays tailored for eRNA detection will further enhance its specificity, sensitivity, and scalability, enabling its widespread adoption in ecological and evolutionary research.

As technological innovations continue to refine molecular ecology, eRNA stands at the frontier of a paradigm shift in biomonitoring. Its unparalleled capacity to capture real-time biological processes at the molecular level positions it as a cornerstone of next-generation ecological assessments. By bridging molecular biology with conservation science, eRNA empowers researchers and policymakers with actionable insights to safeguard biodiversity and ecosystem integrity in an era of rapid environmental change. The future of ecological surveillance is poised to be revolutionized by eRNA, unlocking a deeper, more nuanced understanding of the living world through the lens of molecular precision.

References

Akbarzadeh A, Günther OP, Houde AL, Li S, Ming TJ, Jeffries KM et al (2018) Developing specific molecular biomarkers for thermal stress in salmonids. BMC Genomics 19:1–28

Bertucci A, Pierron F, Ye T, Gonzalez P, Couture P, Baudrimont M (2019) Identification and expression of microRNAs in European eels Anguilla Anguilla from two natural sites with different pollution levels. Environ Pollut 250:274–283

Blazewicz SJ, Barnard RL, Daly RA, Firestone MK (2013) Evaluating rRNA as an indicator of microbial activity in environmental communities: limitations and uses. ISME J 7(11):2061–2068

Broman E, Bonaglia S, Norkko A, Creer S, Nascimento FJ (2021) High throughput shotgun sequencing of eRNA reveals taxonomic and derived functional shifts across a benthic productivity gradient. Mol Ecol 30(13):3023–3039

Bustin SA, Vandesompele J, Pfaffl M (2009) Standardization of qPCR and RT-qPCR. Gen Eng Biotechnol News 2009b, 29

Capo E, Spong G, Königsson H, Byström P (2020) Effects of filtration methods and water volume on the quantification of brown trout (*Salmo trutta*) and Arctic char (Salvelinus alpinus) eDNA concentrations via droplet digital PCR. Environ DNA 2(2):152–160

Chen B, Feder ME, Kang L (2018) Evolution of heat-shock protein expression underlying adaptive responses to environmental stress. Mol Ecol 27(15):3040–3054

Coston-Guarini J, Hinz S, Mirimin L, Guarini JM (2023) A new simulation framework to evaluate thesuitability of eDNA for marine and aquatic environmental impact assessments. Environmental DNA 5(5):1116–1130

Cristescu ME, Hebert PD (2018) Uses and misuses of environmental DNA in biodiversity science and conservation. Annual Review of Ecology, Evolution, and Systematics 49(1):209–230

Ciani E, Fontaine R, Maugars G, Nourizadeh-Lillabadi R, Andersson E, Bogerd J et al (2020) Gnrh receptor gnrhr2bbα is expressed exclusively in lhb-expressing cells in Atlantic salmon male parr. Gen Comp Endocrinol 285:113293

Clark MB, Johnston RL, Inostroza-Ponta M, Fox AH, Fortini E, Moscato P et al (2012) Genome-wide analysis of long noncoding RNA stability. Genome Res 22(5):885–898

Condachou C, Milhau T, Murienne J, Brosse S, Villéger S, Valentini A et al (2023) Inferring functional diversity from environmental DNA metabarcoding. Environ DNA 5(5):934–944

Cristescu ME (2019) Can environmental RNA revolutionize biodiversity science? Trends Ecol Evol 34(8):694–697

Culviner PH, Guegler CK, Laub MT (2020) A simple, cost-effective, and robust method for rRNA depletion in RNA-sequencing studies. MBio 11(2):10–1128

Damon C, Lehembre F, Oger-Desfeux C, Luis P, Ranger J, Fraissinet-Tachet L, Marmeisse R (2012) Metatranscriptomics reveals the diversity of genes expressed by eukaryotes in forest soils. PLoS One 7(1):e28967

Debes JH, Walsh KJ, Stark C (2012) The link between planetary systems, dusty white dwarfs, and metalpolluted white dwarfs. The Astrophysical Journal 747(2):148

DeLong EF, Wickham GS, Pace NR (1989) Phylogenetic stains: ribosomal RNA-based probes for the identification of single cells. Science 243(4896):1360–1363

DeLong EF, Karl DM (2005) Genomic perspectives in microbial oceanography. Nature 437(7057):336–342

Di Muri C, Lawson Handley L, Bean CW, Benucci M, Harper LR, James B, Li J, Winfield IJ, Hänfling B (2023) Spatio-temporal monitoring of lake fish spawning activity using environmental DNA metabarcoding. Environmental DNA 5(5):849–860

Doi H, Fukaya K, Oka SI, Sato K, Kondoh M, Miya M (2019) Evaluation of detection probabilities at the water-filtering and initial PCR steps in environmental DNA metabarcoding using a multispecies site occupancy model. Sci Rep 9(1):3581

Dragun Z, Podrug M, Raspor B (2009) The assessment of natural causes of metallothionein variability in the gills of European chub *(Squalius cephalus L.)*. Comparat Biochem Physiol Part C Toxicol Pharmacol 150(2):209–217

Geisler S, Coller J (2013) RNA in unexpected places: long non-coding RNA functions in diverse cellular contexts. Nat Rev Mol Cell Biol 14(11):699–712

Giovannoni SJ, Britschgi TB, Moyer CL, Field KG (1990) Genetic diversity in Sargasso Sea bacterioplankton. Nature 345(6270):60–63

Glover CN, Veilleux HD, Misutka MD (2025) Commentary: environmental RNA and the assessment of organismal function in the field. Comp Biochem Physiol B: Biochem Mol Biol 275:111036

Goldberg CS, Turner CR, Deiner K, Klymus KE, Thomsen PF, Murphy MA et al (2016) Critical considerations for the application of environmental DNA methods to detect aquatic species. Methods Ecol Evol 7(11):1299–1307

Harper AB, Powell T, Cox PM, House J, Huntingford C, Lenton TM, Sitch S, Burke E, Chadburn SE, Collins WJ, Comyn-Platt E (2018) Land-use emissions play a critical role in land-based mitigation for Paris climate targets. Nature communications 9(1):2938

Hinlo R, Furlan E, Suitor L, Gleeson D (2017) Environmental DNA monitoring and management of invasive fish: comparison of eDNA and fyke netting. Management of Biological Invasions 8(1):89–100

Huang Y, Shen XJ, Zou Q, Wang SP, Tang SM, Zhang GZ (2011) Biological functions of microRNAs: a review. J Physiol Biochem 67:129–139

Ikert H, Craig PM (2020) Chronic exposure to venlafaxine and increased water temperature reversibly alters microRNA in zebrafish gonads (*Danio rerio*). Comparat Biochem Physiol Part D Genom Proteom 33:100634

Iwama GK, Thomas PT, Forsyth RB, Vijayan MM (1998) Heat shock protein expression in fish. Rev Fish Biol Fish 8:35–56

Jeffries KM (2012) The effects of elevated water temperature on adult Pacific Salmon gene expression, blood physiology and mortality (Doctoral dissertation). University of British Columbia

Jeffries MA, Donica M, Baker LW, Stevenson ME, Annan AC, Humphrey MB et al (2014) Genome-wide DNA methylation study identifies significant epigenomic changes in osteoarthritic cartilage. Arthritis Rheumatol 66(10):2804–2815

Jo TS (2023) Methodological considerations for aqueous environmental RNA collection, preservation, and extraction. Anal Sci 39(10):1711–1718

Joos F, Hameau A, Frölicher TL, Stephenson DB (2023) Anthropogenic attribution of the increasing seasonal amplitude in surface ocean pCO2. Geophysical Research Letters 50(13):e2023GL102857

Kagzi K, Millette KL, Littlefair JE, Pochon X, Wood SA, Fussmann GF, Cristescu ME (2023) Assessing the degradation of environmental DNA and RNA based on genomic origin in a metabarcoding context. Environ DNA 5(5):1016–1031

Kitahashi T, Sugime S, Inomata K, Nishijima M, Kato S, Yamamoto H (2020) Meiofaunal diversity at a seamount in the pacific ocean: A comprehensive study using environmental DNA and RNA. Deep-Sea Res I Oceanogr Res Pap 160:103253

Klymus KE, Richter CA, Chapman DC, Paukert C (2015) Quantification of eDNA shedding rates from invasive bighead carp *Hypophthalmichthys nobilis* and silver carp *Hypophthalmichthys molitrix*. Biol Conserv 183:77–84

Laroche O, Pochon X, Tremblay LA, Ellis JI, Lear G, Wood SA (2018) Incorporating molecular-based functional and co-occurrence network properties into benthic marine impact assessments. FEMS Microbiology Ecology 94(11):fiy167

Li W, Jia H, Zhang H (2024) Evaluating the effectiveness of the eRNA technique in monitoring fish biodiversity–A case study in the Qingdao offshore, China. Global Ecology and Conservation 51:e02888

Littlefair JE, Rennie MD, Cristescu ME (2022) Environmental nucleic acids: A field-based comparison for monitoring freshwater habitats using eDNA and eRNA. Molecular Ecology Resources 22(8):2928–2940

Marshall IP, Ren G, Jaussi M, Lomstein BA, Jørgensen BB, Røy H, Kjeldsen KU (2019) Environmental filtering determines family-level structure of sulfate-reducing microbial communities in subsurface marine sediments. ISME J 13(8):1920–1932

Marshall NT, Vanderploeg HA, Chaganti SR (2021) Environmental (e) RNA advances the reliability of eDNA by predicting its age. Sci Rep 11(1):2769

Matozzo V, Gagné F, Marin MG, Ricciardi F, Blaise C (2008) Vitellogenin as a biomarker of exposure to estrogenic compounds in aquatic invertebrates: a review. Environ Int 34(4):531–545

Miller CL, Van Laar TA, Chen T, Karna SR, Chen P, You T, Leung KP (2017) Global transcriptome responses including small RNAs during mixed-species interactions with methicillin-resistant *Staphylococcus aureus* and *Pseudomonas aeruginosa*. Micro Biologyopen 6(3):e00427

Moran DD, Lenzen M, Kanemoto K, Geschke A (2013) Does ecologically unequal exchange occur?. Ecological Economics 89:177–186

Olsen GJ, Lane DJ, Giovannoni SJ, Pace NR, Stahl DA (1986) Microbial ecology and evolution: a ribosomal RNA approach. Annu Rev Microbiol 40(1):337–365

Oomen RA, Hutchings JA (2017) Transcriptomic responses to environmental change in fishes: insights from RNA sequencing. Facets 2(2):610–641

Oris JT, Roberts AP (2007) Statistical analysis of cytochrome P4501A biomarker measurements in fish. Environ Toxicol Chem 26(8):1742–1750

Pace NR, Stahl DA, Lane DJ, Olsen GJ (1986) Advances in microbial ecology

Pochon X, Zaiko A, Fletcher LM, Laroche O, Wood SA (2017) Wanted dead or alive? Using metabarcoding of environmental DNA and RNA to distinguish living assemblages for biosecurity applications. PloS one 12(11):e0187636

Prado-Lima M, Val AL (2015) Differentially expressed genes in the pituitary of the Amazonian fish Arapaima gigas. International Journal of Fisheries and Aquaculture 7(8):132–141

Querejeta M, Marchal L, Pfeiffer P, Roncoroni M, Bretagnolle V, Gaba S, Boyer S (2023) Environmental variables and species traits as drivers of wild bee pollination in intensive agroecosystems – a metabarcoding approach. Environ DNA 5(5):1078–1091

Rinn JL, Chang HY (2012) Genome regulation by long noncoding RNAs. Annu Rev Biochem 81(1):145–166

Scrivener KL, Matschei T, Georget F, Juilland P, Mohamed AK (2023) Advances in hydration and thermodynamics of cementitious systems. Cement and Concrete Research 174:107332

Shakya M, Lo CC, Chain PS (2019) Advances and challenges in meta-transcriptomic analysis. Front Genet 10:904

Shi Y, Tyson GW, Eppley JM, DeLong EF (2011) Integrated meta-transcriptomic and metagenomic analyses of stratified microbial assemblages in the open ocean. ISME J 5(6):999–1013

Sigsgaard EE, Jensen MR, Winkelmann IE, Møller PR, Hansen MM, Thomsen PF (2020) Population-level inferences from environmental DNA current status and future perspectives. Evol Appl 13(2):245–262

Sohel MH (2016) Extracellular/circulating microRNAs: release mechanisms, functions and challenges. Achieve Life Sci 10(2):175–186

Spanier KI, Jansen M, Decaestecker E, Hulselmans G, Becker D, Colbourne JK, Orsini L, De Meester L, Aerts S (2017) Conserved transcription factors steer growth-related genomic programs in daphnia. Genome Biol Evol 9(6):1821–1842

Stat M, Huggett MJ, Bernasconi R, DiBattista JD, Berry TE, Newman SJ, Harvey ES, Bunce M (2017) Ecosystem biomonitoring with eDNA: metabarcoding across the tree of life in a tropical marine environment. Sci Rep 7(1):12240

Stewart K, Taylor SA (2019) Leveraging eDNA to detect and monitor hybrid zones (No. e27996v1). PeerJ Preprints

Sunagawa S, Coelho LP, Chaffron S, Kultima JR, Labadie K, Salazar G, Djahanschiri B, Zeller G, Mende DR, Alberti A, Cornejo-Castillo FM (2015) Structure and function of the global ocean microbiome. Science 348(6237):1261359

Sutherland BJ, Finke JF, Saunders R, Warne S, Schulze AD, Strohm JH, Chan AM, Suttle CA, Miller KM (2023) Meta-transcriptomics reveals a shift in microbial community composition and function during summer months in a coastal marine environment. Environ DNA 5(5):920–933

Taberlet P, Coissac E, Pompanon F, Brochmann C, Willerslev E (2012) Towards next-generation biodiversity assessment using DNA metabarcoding. Mol Ecol 21(8):2045–2050

Taylor SC, Laperriere G, Germain H (2017) Droplet digital PCR versus qPCR for gene expression analysis with low abundant targets: from variable nonsense to publication quality data. Sci Rep 7(1):2409

Tsuri K, Ikeda S, Hirohara T, Shimada Y, Minamoto T, Yamanaka H (2021) Messenger RNA typing of environmental RNA (eRNA): a case study on zebrafish tank water with perspectives for the future development of eRNA analysis on aquatic vertebrates. Environ DNA 3(1):14–21

Urich T, Lanzén A, Qi J, Huson DH, Schleper C, Schuster SC (2008) Simultaneous assessment of soil microbial community structure and function through analysis of the meta-transcriptome. PloS one 3(6):e2527

Valentini A, Taberlet P, Miaud C, Civade R, Herder J, Thomsen PF, Bellemain E, Besnard A, Coissac E, Boyer F, Gaboriaud C (2016) Next-generation monitoring of aquatic biodiversity using environmental DNA metabarcoding. Mol Ecol 25(4):929–942

Van Cleef KA, Kaplan LA, Crivello JF (2000) The relationship between reproductive status and metallothionein mRNA expression in the common killifish, *Fundulus heteroclitus*. Environ Biol Fishes 57:97–105 van der Poel-van de Luytgaarde SC, Geertsma-Kleinekoort WM, Goudswaard CS, Hogenbirk-Hupkes PE, van Hoven-Beijen MA, van de Werf M, Chu IW, van Kapel J, Valk PJ (2013) Addition of β-mercaptoethanol is a prerequisite for high-quality RNA isolation using QIA symphony technology as demonstrated by detection of molecular aberrations in hematologic malignancies. Genet Test Mol Biomarkers 17(6):475–480

Von Ammon U, Wood SA, Laroche O, Zaiko A, Lavery SD, Inglis GJ, Pochon X (2019) Linking environmental DNA and RNA for improved detection of the marine invasive fanworm *Sabella spallanzanii*. Front Mar Sci 6:p.483759

Wang Z, Gerstein M, Snyder M (2009) RNA-Seq: a revolutionary tool for transcriptomics. Nat Rev Genet 10(1):57–63

Wood SA, Biessy L, Latchford JL, Zaiko A, von Ammon U, Audrezet F, Cristescu ME, Pochon X (2019) Release and degradation of environmental DNA and RNA in a marine system. Sci Total Environ 704:135314–135314

Yates MC, Furlan E, Thalinger B, Yamanaka H, Bernatchez L (2023) Beyond species detection leveraging environmental DNA and environmental RNA to push beyond presence/absence applications. Environ DNA 5(5):829–835

Proteomics in Fish Health and Aquaculture

8

Basanta Kumar Das, Vikash Kumar, and Biswajit Mandal

8.1 Introduction

The aquatic environment, encompassing over 70% of the Earth's surface and with more than 95% consisting of seawater, represents a vast and dynamic ecosystem teeming with diverse biological, chemical, and physical energy resources. Among its numerous contributions to human civilization, the provision of aquatic proteins from fish, invertebrates, seaweeds, and algae stands out as a critical factor in meeting global dietary demands. The growing human population, coupled with the need for sustainable food production, has underscored the importance of aquaculture in providing a reliable source of high-quality protein. This rapidly expanding industry, however, faces a multitude of challenges, primarily disease outbreaks that threaten fish health, productivity, and economic stability (Rodrigues et al. 2016). The increasing vulnerability of farmed fish to bacterial, viral, fungal, and parasitic infections, exacerbated by intensive farming practices, climate change, and global trade, necessitates innovative approaches to disease management and pathogen control. Proteomics, an advanced omics technology, offers groundbreaking solutions by providing deep insights into host-pathogen interactions, immune responses, and biomarker discovery, ultimately enhancing fish health management and aquaculture sustainability (Jitjumnong et al., 2025). The application of proteomics in fish health has emerged as a transformative tool in diagnosing pathogens, understanding disease resistance mechanisms, and optimizing aquaculture productivity (Rodrigues et al. 2012). Proteomic methodologies, particularly mass spectrometry (MS)-based techniques, enable the high-throughput analysis of the entire protein complement of an organism, revealing critical aspects such as protein abundance, post-translational modifications, tissue distribution, and protein-protein interactions (Cox et al. 2011).

B. K. Das (✉) · V. Kumar · B. Mandal
ICAR-Central Inland Fisheries Research Institute, Barrackpore, Kolkata, India

© The Author(s), under exclusive license to Springer Nature Singapore Pte Ltd. 2026

B. K. Das et al. (eds.), *Bioinformatics Tools for Predictive Ecology and Fisheries*, Sustainable Aquatic Food Systems, https://doi.org/10.1007/978-981-95-2573-7_8

Such approaches have revolutionized aquatic disease surveillance, facilitated early detection of infectious agents and uncovered potential therapeutic targets. For instance, matrix-assisted laser desorption/ionization time-of-flight mass spectrometry (MALDI-TOF MS) has proven highly effective in identifying bacterial fish pathogens and detecting antimicrobial resistance biomarkers, thus playing a pivotal role in biosecurity programs. Proteomics also bridges gaps in epidemiological models by integrating pathogen characterization, host immune responses, and environmental data, thereby strengthening disease monitoring and control strategies (Jean Beltran et al., 2017). Beyond pathogen diagnosis, proteomics significantly contributes to understanding the physiological responses of fish to environmental stressors, pollutants, and aquaculture management practices. The emerging field of ecotoxicproteomics provides a molecular-level assessment of how fish respond to environmental contaminants, offering crucial insights into endocrine disruption, oxidative stress, and immune dysregulation caused by pollutants (Rotchell and Ostrander 2003). Additionally, proteomic studies on mobile genetic elements in aquatic microorganisms have shed light on antibiotic resistance mechanisms and microbial adaptation strategies (Smalla and Sobecky 2002). This wealth of information is invaluable for designing targeted interventions to mitigate disease risks and ensure the sustainability of aquaculture operations.

The forensic potential of proteomics in aquatic environments is another fascinating frontier. Proteins preserved in submerged bones and tissues serve as time-sensitive biomarkers in forensic investigations, with studies demonstrating the utility of the fetuin-A protein in estimating postmortem submersion times in criminal cases (Mizukami et al. 2020). Such applications underscore the interdisciplinary nature of proteomic research, extending beyond aquaculture to environmental science, public health, and forensic biology. As the aquaculture industry scales up to meet the growing demand for animal protein, there is an urgent need for integrated health management strategies that combine proteomics with genomics, transcriptomics, and metabolomics. This multiomics approach holds immense potential for improving disease resistance in farmed species through selective breeding programs, vaccine development, and nutritional interventions (Choudhary et al., 2024). Advances in analytical chemistry, optical biosensors, and molecular diagnostics further bridge the gap between aquatic veterinary medicine and high-precision disease diagnostics (Oskoueian et al. 2016). The future of fish health management hinges on leveraging these cutting-edge technologies to enhance productivity while ensuring ecological balance and food security.

Proteomics revolutionizes fish health management and aquaculture sustainability by offering cutting-edge molecular insights into pathogen identification, disease resistance, and environmental stress responses. This chapter delves into the transformative role of proteomic technologies, ranging from mass spectrometry to bioinformatics-driven protein analysis, in diagnosing bacterial, viral, and parasitic infections. It further explores their applications in pathogen surveillance, antimicrobial resistance tracking, and biosecurity strategies. Ecotoxic proteomics sheds light on pollutant-induced stress, while molecular profiling enhances our understanding of disease resilience and therapeutic targets. By integrating proteomics with

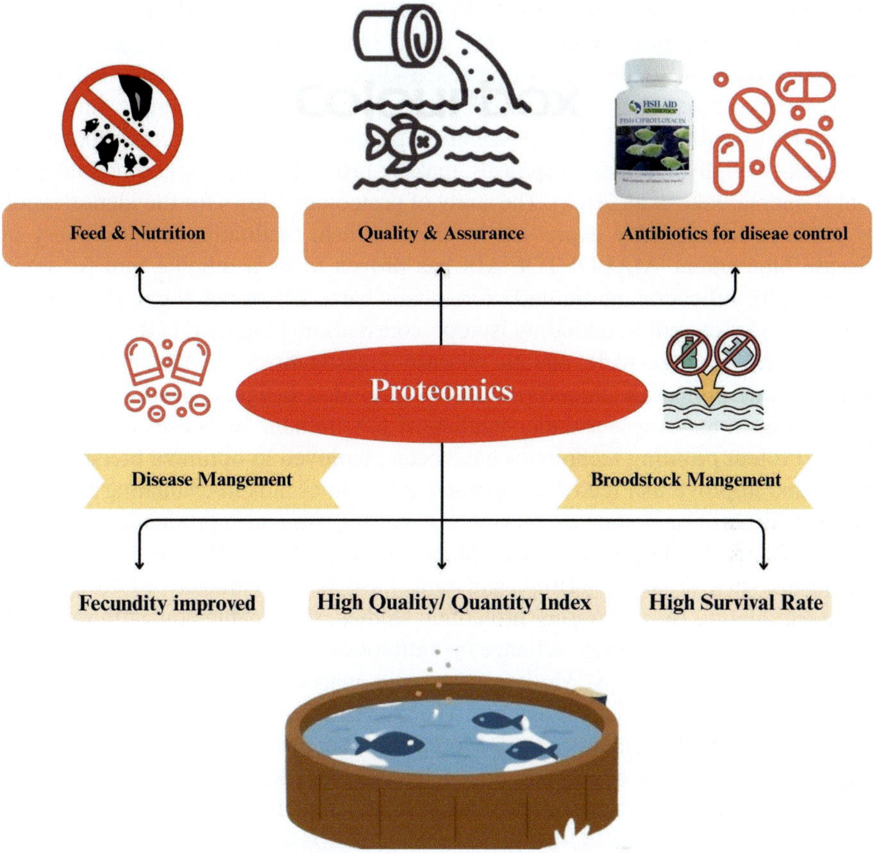

Fig. 8.1 Application of proteomics tools in aquaculture

multiomics approaches, aquaculture is advancing toward precision breeding and optimized health management (Fig. 8.1). With emerging innovations and forensic applications, proteomics stands as a cornerstone for sustainable aquaculture and global food security in an era of environmental change.

8.2 Proteomics as Biomonitoring Tools

Proteomics has emerged as a transformative tool in aquaculture and fisheries, enabling a deeper understanding of disease mechanisms, pathogen identification, and overall fish health management. This field provides valuable insights into protein expression changes in response to environmental stressors, infectious diseases, and metabolic conditions. By integrating proteomics with other omics technologies, aquaculture practices can be optimized for sustainability, disease control, and food safety. Proteomics serves as a powerful approach to monitor fish health by identifying biomarkers associated with diseases, enhancing early detection, and improving intervention strategies. The rapid identification of pathogenic agents through proteomic techniques surpasses traditional microbiological methods in both speed and accuracy (Nyman 2001). As proteins are the primary functional molecules in an organism, their post-transcriptional modifications contribute to the complexity of physiological responses, making proteomics a crucial biomonitoring tool (Ponomarenko et al. 2016). The study of proteomes allows for the identification of proteins differentially expressed in diseased fish, facilitating the discovery of disease biomarkers. MALDI-TOF MS has proven to be a reliable tool for rapid pathogen identification, particularly for aquatic bacterial species (Feng et al. 2020). Label-based proteomics, including isotope-coded affinity tags and iTRAQ, has been instrumental in large-scale aquatic disease investigations (Dieme et al. 2014). Additionally, proteomic fingerprinting has provided insights into host-pathogen interactions, improving our understanding of fish immune responses (Zhang et al. 2022). Protein profiling techniques have been employed to optimize breeding programs and improve fish feed formulations. Proteomics aids in evaluating feed efficiency and nutritional status, ensuring optimal growth and reproduction (Moreira et al. 2021; Rodrigues et al. 2012). Moreover, protein hydrolysates from marine organisms such as krill, shrimp, and tuna have been characterized to enhance aquafeed quality. By integrating proteomic studies, sustainable aquaculture strategies can be devised to reduce reliance on antibiotics and enhance disease resistance in farmed fish populations. Modern proteomic analyses can be broadly classified into gel-based and gel-free approaches. The conventional two-dimensional gel electrophoresis (2D-GE) remains a useful tool for global protein separation, despite its limitations in resolving membrane-associated proteins (Sanchez et al. 2011). On the other hand, mass spectrometry-based techniques, particularly liquid chromatography–mass spectrometry (LC-MS), have revolutionized protein identification and quantification (Niessen-Ruenzi and Ruenzi 2019). Tandem mass tagging (TMT) and label-free quantification methods have also been successfully applied in aquatic proteomic research (Betti et al. 2021). A comparative analysis suggests that integrating 2D-GE with MS-based approaches enhances protein identification and provides robust datasets for biomarker discovery. Despite its advantages, proteomics research in aquaculture faces significant hurdles. The lack of comprehensive protein databases for aquatic species limits accurate protein identification, necessitating the continuous expansion of resources such as UniProt and Pfam (https://www.uniprot.org/; http://pfam.xfam.org/). Additionally, the complexity of protein structures requires advanced bioinformatics tools for functional annotation and pathway analysis (Cui et al. 2014). Standardization of protein extraction methods remains another challenge, as no universal protocol has been established for diverse aquatic species (An and Scherer 2020). Advances in sample preparation techniques, such as the TRIzol method for gill and liver tissues, have shown promise (Wu et al. 2013), but further optimization is needed for reproducibility. The integration of proteomics with genomics, transcriptomics, and metabolomics will play a pivotal role in advancing aquaculture sustainability. With the increasing application of artificial intelligence in protein analysis, more accurate predictive models can be developed for disease prevention and treatment. Furthermore, the development of protein-based biosensors for rapid pathogen detection in aquaculture facilities holds immense potential for mitigating economic losses due to disease outbreaks (Avissar and Schreiber 2000). As global demand for seafood continues to rise, proteomics-driven innovations will be essential for ensuring food security and maintaining the ecological balance of aquatic environments.

Proteomics has revolutionized fish health management by providing insights into disease mechanisms, facilitating early pathogen detection, and improving aquaculture productivity. Advancements in mass spectrometry, protein quantification, and bioinformatics tools have expanded the capabilities of proteomics in fisheries research. Despite challenges such as data standardization and protein database limitations, the future of aquatic proteomics is promising, with potential applications in biomonitoring, food safety, and sustainable aquaculture practices. By leveraging proteomics, the aquaculture industry can transition towards more resilient and disease-resistant fish populations, contributing to long-term food security and environmental conservation.

8.3 History of Matrix-Assisted Laser Desorption/Ionization Time-of-Flight Mass Spectrometry (MALDI-TOF MS) in Fish Health and Aquaculture

Matrix-assisted laser desorption/ionization time-of-flight mass spectrometry (MALDI-TOF MS) has revolutionized the field of microbial proteomics, offering rapid, accurate, and cost-effective pathogen identification in aquaculture. Initially developed by Karas et al. (1985, 1987) and later refined through Tanaka et al.'s (1988) "soft desorption ionization" technique, MALDI-TOF MS has emerged as a gold-standard tool in clinical microbiology. The integration of MALDI with TOF mass analyzers has paved the way for its extensive application in proteomics and metabolomics (Tanaka 2003). The American Food and Drug Administration (FDA) has approved MALDI-TOF MS-based systems, such as the Clin-TOF (Bioyong Technology, China), MassARRAY System (Agena Bioscience, USA), VitekMS (bioMérieux Clinical Diagnostics, France), and MALDI Biotyper (Bruker, Germany), for microbial identification. These platforms, enriched with vast microbial mass spectrum libraries, have enhanced pathogen detection in aquaculture.

MALDI-TOF MS enables precise identification of bacterial, fungal, viral, and parasitic pathogens affecting fish health, aiding early diagnosis and improving disease management strategies (Seng et al. 2009). The technique's high-throughput capability has facilitated its adoption in fish pathology research and aquaculture productivity management. Recent applications include the identification of zoonotic pathogens in veterinary microbiology, detection of foodborne pathogens, and screening for antibiotic resistance. In aquaculture, MALDI-TOF MS has proven instrumental in diagnosing diseases caused by bacterial and fungal pathogens in fish farms (Assis et al. 2017). Furthermore, its role in environmental microbiology has advanced the study of aquatic ecosystems, allowing for the identification of waterborne pathogens and contaminants (Singhal et al. 2015). The future of MALDI-TOF MS in aquaculture lies in its integration with proteomics-based approaches for precision disease diagnosis and biomarker discovery. The ability to analyze pathogen-host interactions at the protein level will enable targeted therapeutic interventions, enhancing fish health and sustainability. As the field progresses, MALDI-TOF MS will remain pivotal in safeguarding global aquaculture productivity and food security.

8.4 General Principles and Methodology of MALDI-TOF MS in Fish Health and Aquaculture

Matrix-assisted laser desorption/ionization-time of flight mass spectrometry (MALDI-TOF MS) has emerged as a transformative tool in fish health diagnostics, revolutionizing pathogen detection with precision, speed, and efficiency (Singhal et al. 2015). This technique functions by ionizing ribosomal proteins from microbial samples, generating unique peptide mass fingerprints that serve as molecular barcodes for species identification (López-Cortés et al. 2017). The fingerprinting approach ensures rapid differentiation of bacterial, fungal, and protozoan pathogens affecting aquaculture systems, facilitating timely interventions for disease management (Fig. 8.2). The methodology of MALDI-TOF MS follows a structured three-step process. First, microbial samples are prepared either directly from infected tissues or from cultured isolates. Depending on the microbial cell wall structure and biosafety considerations, three approaches—direct sample spotting, on-target extraction, or full protein extraction—can be employed for sample preparation. Among these, full extraction is particularly effective for identifying fastidious and

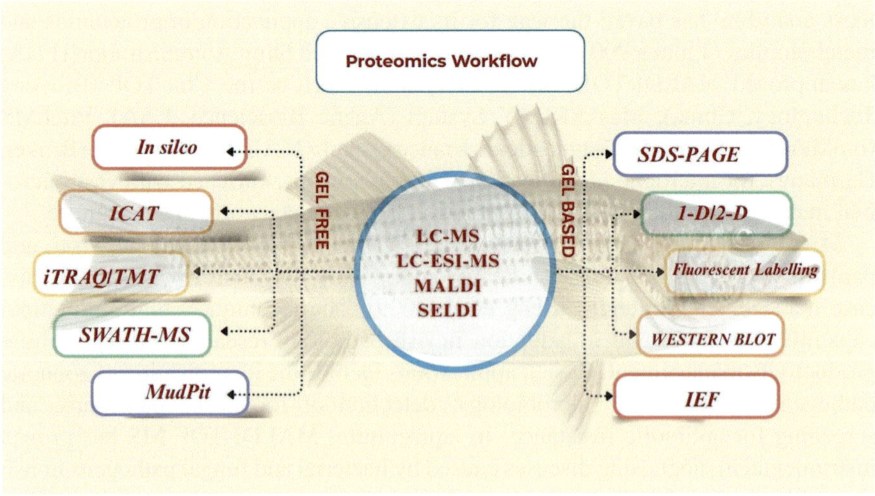

Fig. 8.2 Proteomics workflow for mass spectrometry

challenging-to-detect pathogens. The prepared sample is then mixed with a crystallizing matrix solution, commonly cyano-4-hydroxycinnamic acid (CHCA) or trifluoroacetic acid, which facilitates the desorption and ionization process (Kazazić et al. 2019). In the second step, the sample-matrix mixture is placed into the MALDI-TOF instrument and subjected to a high-energy laser beam. This induces the formation of protonated ions, which are accelerated through a vacuum chamber, allowing their time-of-flight (TOF) to be recorded based on mass-to-charge (m/z) ratios. The smallest analytes reach the detector first, followed by larger ones, generating a characteristic spectrum or peptide mass fingerprint (De Marco and Burnham, 2014). The third and final step involves data analysis, where the acquired mass spectra are matched against extensive reference libraries to identify pathogens with high confidence. Identification is based on logarithmic score values, with scores $\geq$2.00 confirming species-level identification and scores between 1.70 and 2.00 indicating genus-level classification (Singhal et al. 2015).

Proteomic studies have provided valuable insights into the molecular responses of fish to infectious diseases, identifying critical modifications in proteins associated with immune defense, metabolic regulation, and cellular stress Tables 8.1 and 8.2. By integrating MALDI-TOF MS into aquaculture disease surveillance, researchers and veterinarians can achieve unprecedented accuracy in pathogen diagnosis, ultimately improving fish health, enhancing productivity, and securing sustainable food production systems (Piamsomboon et al., 2020).

8.5 Protein Databases and Computational Tools in Fish Proteomics

The continuous expansion of protein databases has revolutionized proteomic research, enhancing protein identification and unravelling proteoform complexity. Among the most comprehensive repositories, UniProt (https://www.uniprot.org/, accessed on 27 January 2022) remains the gold standard, comprising two platforms: Swiss-Prot, containing 565,928 manually curated protein entries, and TrEMBL, which holds 225,013,025 computationally annotated sequences. Pfam (http://pfam.xfam.org/, accessed on 27 January 2022) further enriches proteomic research by categorizing 19,632 protein families with millions of associated sequences, facilitating functional and evolutionary insights.

For homology-based protein identification, BLASTP (https://blast.ncbi.nlm.nih.gov/Blast.cgi, accessed on 27 January 2022) remains an indispensable tool, allowing rapid sequence comparison and discovery of novel or unknown proteins. Structural proteomics, crucial for understanding protein function, is significantly advanced by databases such as AlphaFold (https://alphafold.ebi.ac.uk/, accessed on 2 February 2022), which predicts high-accuracy 3D protein structures, and the Protein Data Bank (https://www.rcsb.org/, accessed on 2 February 2022), a repository of experimentally determined macromolecular structures. Expasy (https://www.expasy.org/, accessed on 2 February 2022) further provides a suite of

Table 8.1 Proteomic approaches for the identification of aquatic pathogens in fish health and aquaculture

Pathogen	Disease	Proteomic methodology	Reference
Viral pathogens			
Singapore grouper iridovirus (SGIV)	Systemic disease	SDS-PAGE + MALDI-TOF-MS	Song et al. (2004)
Singapore grouper iridovirus (SGIV)	Systemic disease	1-DE + MALDI-TOF/TOF-MS/MS, LC-MALDI-TOF/TOF-MS/MS	Zhou et al. (2011)
Cyprinid herpesvirus 3 (CyHV-3)	Koi herpesvirus (KHV) disease	SDS-PAGE + LC-MS/MS	Vancsok et al. (2017)
Bacterial pathogens			
Photobacterium damselae subsp. *piscicida*	Pasteurellosis	2-DE + LC-nanoESI-Q-TOF MS/MS	Chuang et al. (2016)
Photobacterium damselae subsp. *damselae*	Vibriosis	2-DE + MALDI-TOF/TOF-MS	Puentes et al. (2017)
Photobacterium damselae	Pasteurellosis	MALDI-TOF-MS	Pérez-Sancho et al. (2016)
Edwardsiella tarda	Hemorrhagic septicemia	2-DE + MALDI-TOF MS	Tan et al. (2004); Srinivasa Rao et al. (2004)
Edwardsiella tarda	Hemorrhagic septicemia	DE + ESI MS/MS	Buján et al. (2016)
Yersinia ruckeri	Enteric redmouth disease	Nano LC-MS/MS	Kumar et al. (2017)
Yersinia ruckeri	Enteric red mouth disease	Nano LC-MS/MS	Kumar et al. (2016)
Yersinia ruckeri	Enteric redmouth disease	Gel-free 1-D + nLC-ESI-MS/MS 2-	Ormsby et al. (2019)
Flavobacterium columnare	Columnaris disease	2-D LC-ESI-MS/MS + 2-DE MALDI-TOF/TOF MS	Dumpala et al. (2010)
Flavobacterium psychrophilum	Bacterial Coldwater disease	2-DE + LC-MS/MS	LaFrentz et al. (2009)
Flavobacterium psychrophilum	Bacterial Coldwater disease	MALDI-TOF-MS	Pérez-Sancho et al. (2017)
Vibrio parahaemolyticus	Vibriosis	2-DE + LC-ESI-Q-TOF MS/MS	Shen et al. (2020)
Streptococcus agalactiae	Streptococcosis	NanoUPLC-HDMSE	Tavares et al. (2018)
Streptococcus iniae and *S. parauberis*	Streptococcosis	MALDI-TOF-MS	Yolanda et al. (2019)
Streptococcus parauberis	Streptococcosis	MALDI-TOF-MS	Kim et al. (2015)
Vagococcus salmoninarum	Cold-water streptococcosis	MALDI-TOF-MS	Torres-Corral et al. (2019)
Streptococcus iniae	Streptococcosis	2-DE + MALDI-TOF-MS	

(continued)

Table 8.1 (continued)

Pathogen	Disease	Proteomic methodology	Reference
Piscirickettsia salmonis	Salmonid rickettsial syndrome	MALDI-TOF-MS	López-Cortés et al. (2017)
Tenacibaculum spp.	Tenacibaculosis	MALDI-TOF-MS	Fernández-Álvarez et al. (2017, 2018)
Mycobacterium marinum	Mycobacteriosis	MALDI-TOF-MS	
Fungal pathogens			
Saprolegnia parasitica	Saprolegniasis	iTRAQ + SDS-PAGE + Nano-LC-MS/MS	Srivastava et al. (2018)
Parasitic pathogens			
Caligus rogercresseyi	Sea lice infestation	MALDI-TOF-MS	

Table 8.2 Summary of modified proteins identified in fish infectious diseases using proteomics

Etiological agent	Host species	Tissue	Modified protein groups	Reference
Viral pathogens				
Infectious spleen and kidney necrosis virus (ISKNV) (Megalocytivirus)	Zebrafish (*Danio rerio*)	Spleen	Cytoskeletal proteins, stress response, lipoprotein and carbohydrate metabolism, signal transduction, proteolysis, metabolic and catabolic processes	Xiong et al. (2011)
Megalocytivirus	Turbot (*Scophthalmus maximus*)	Spleen	Cytoskeletal proteins, molecular biosynthesis, cellular signal transduction, chaperone proteins	Zhang et al. (2013)
Salmonid alphavirus subtype 3 (SAV3)	Atlantic salmon (*Salmo salar*)	Serum	Humoral components of immunity	Braceland et al. (2013)
Rhabdovirus	Zebrafish (*Danio rerio*)	Fins	Glycolytic pathway proteins, cytoskeletal components	Encinas et al. (2013)
Spring viremia of carp virus	Zebrafish (*Danio rerio*)	Plasma	Vitellogenin, Gig2	Medina Gali et al. (2019)
Viral hemorrhagic septicemia virus (VHSV)	Rainbow trout (*Oncorhynchus mykiss*)	Red blood cells, head kidney	Proteins related to viral transcription	Nombela et al. (2019)
Cyprinid herpesvirus-2	Crucian carp (*Carassius carassius*)	Head kidney	Cytoskeletal, transport, immunologic, intracellular, and physiological proteins	Liu et al. (2018)

(continued)

Table 8.2 (continued)

Etiological agent	Host species	Tissue	Modified protein groups	Reference
Bacterial pathogens				
Aeromonas salmonicida	Rainbow trout (*Oncorhynchus mykiss*)	Liver	Complement system proteins, acute-phase response proteins	Mulder et al. (2007)
Aeromonas salmonicida	Rainbow trout (*Oncorhynchus mykiss*)	Spleen	Immune system proteins, signaling molecules	Long et al. (2015)
Aeromonas hydrophila	Common carp (*Cyprinus carpio*)	Intestinal mucosa	Stress and immune response proteins	Di et al. (2017)
Aeromonas hydrophila	Zebrafish (*Danio rerio*)	Gills	Stress and immune response proteins	Lu et al. (2013)
Edwardsiella ictaluri	Channel catfish (*Ictalurus punctatus*)	Head kidney	Macrophage function, cellular stress response, energy metabolism	Booth et al. (2011)
Yersinia ruckeri	Rainbow trout (*Oncorhynchus mykiss*)	Head kidney, spleen	Immune system, metabolic regulation, cellular adhesion, response to stimuli	Rajan et al. (2013)
Yersinia ruckeri	Rainbow trout (*Oncorhynchus mykiss*)	Intestine	Metabolic processes, cellular regulation	Addis et al. (2010)
Vibrio anguillarum	Atlantic cod (*Gadus morhua*)	Mucus	Proteins involved in immune defense	Rajan et al. (2013)
Streptococcus parauberis	Olive flounder (*Paralichthys olivaceus*)	Kidney	Immune response proteins, glycoprotein synthesis	Park et al. (2009)
Moraxella sp.	Gilthead seabream (*Sparus aurata*)	Kidney	Mitochondrial proteins, oxidative stress response, inflammation	Addis et al. (2010)
Parasitic pathogens				
Ichthyophthirius multifiliis	Common carp (*Cyprinus carpio*)	Mucus	Proteins involved in immune and inflammatory responses	Saleh et al. (2019)
Lepeophtheirus salmonis	Atlantic salmon (*Salmo salar*)	Mucus	Proteins involved in glycolysis, peptide synthesis, immune defense	Provan et al. (2013)
Amyloodinium ocellatum	Gilthead seabream (*Sparus aurata*)	Plasma	Acute-phase response proteins, lipid transport, inflammation, wound healing	Moreira et al. (2017)
Neoparamoeba perurans	Atlantic salmon (*Salmo salar*)	Gill, skin mucus	Proteins involved in cell signaling and inflammatory pathways	Medina Gali et al. (2019)
Neoparamoeba perurans	Atlantic salmon (*Salmo salar*)	Gill, skin mucus	Proteins regulating the cell cycle, cytoskeletal integrity, oxidative metabolism, immune defense	Lopez-Cortes et al. (2017)

bioinformatics tools for protein sequence analysis, structure prediction, and functional characterization.

In the field of aquatic proteomics, the Fish Proteomic Database (http://www.cifri.res.in/Fishprot/, accessed February 22, 2022) has been developed for biomarker discovery and disease diagnostics. With the integration of artificial intelligence and computational biology, proteomics is evolving into a powerful biomonitoring tool, unlocking novel study opportunities for aquaculture sustainability and ecosystem health.

8.5.1 Identification of Bacterial Fish Pathogens with MALDI-TOF MS

Proteomics-driven diagnostics have revolutionized fish health management, and matrix-assisted laser desorption/ionization time-of-flight mass spectrometry (MALDI-TOF MS) has emerged as an indispensable tool in aquaculture for the rapid and precise identification of bacterial fish pathogens. This state-of-the-art technique enables both direct identification of whole-cell bacterial agents (Piamsomboon et al. 2020) and the detection of antibiotic resistance biomarkers, allowing real-time surveillance of antimicrobial-resistant strains, such as those resistant to beta-lactam, carbapenem, methicillin, and vancomycin (Zhu et al. 2002; Fernández-Álvarez et al. 2017; Sandalakis et al. 2017; Popović et al. 2021; Florio et al. 2020). This rapid profiling capability not only facilitates early disease intervention but also promotes sustainable aquaculture practices by mitigating outbreaks and ensuring the health of farmed aquatic species.

Studies have demonstrated the application of MALDI-TOF MS in identifying single and co-infecting bacterial agents in both freshwater and marine aquaculture species, minimizing losses due to high-mortality bacterial diseases (Nissa et al. 2021). The economic and ecological impact of bacterial pathogens especially those from *Aeromonadaceae*, *Pseudomonadaceae*, *Mycobacteriaceae*, *Streptococcaceae*, *Vibrionaceae*, and *Enterobacteriaceae* extends beyond aquaculture, as many are zoonotic and pose significant public health concerns. The increasing adoption of MALDI-TOF MS in aquaculture diagnostics is evident from a growing number of publications, with Google Scholar data showing a rising trend in its application for pathogen identification from 2010 to 2023. Table 8.2 comprehensively lists 50 major bacterial pathogens 37 Gram-negative and 13 Gram-positive detected in diverse aquatic organisms, including eggs, larvae, juveniles, and adult fish, as well as shellfish and shrimp. By integrating MALDI-TOF MS with other molecular techniques, aquaculture health management is transitioning toward a more precise, data driven approach, ensuring food security and sustainability in global fisheries.

8.6 Applications of Proteomics in Aquatic Studies

Proteomics has emerged as a revolutionary tool in aquatic sciences, offering insights into diverse fields such as aquaculture, environmental monitoring, and aquatic natural product development. The application of proteomic methodologies enhances our ability to study protein structure, expression dynamics, and functional interactions, enabling advancements in sustainable aquaculture and seafood safety. Despite the immense biodiversity of aquatic organisms, challenges remain in protein characterization due to the incomplete genome sequences of many species, underscoring the need for further refinement in aquatic omics research.

8.6.1 Proteomics in the Aquatic Food Industry

Aquatic organisms, ranging from microalgae to fish and marine mammals, contribute significantly to global food security. Proteomics plays a pivotal role in assessing the nutritional quality, safety, and metabolic pathways of seafood, helping to optimize production and ensure sustainable aquaculture practices. By 2030, global consumption of aquatic animal-source foods is projected to rise, while terrestrial meat consumption is expected to decline due to health concerns and evolving dietary preferences (Golden et al. 2021).

8.6.1.1 Advancing Aquaculture Sustainability

Aquaculture provides essential nutrients such as eicosapentaenoic acid (EPA) and docosahexaenoic acid (DHA), crucial for human health. Genomic and proteomic research is instrumental in improving aquaculture productivity by identifying causal genes and proteins that influence fish growth, immunity, and stress responses (Abdelrahman et al. 2017). With the advent of next-generation sequencing (NGS), genome and proteome data for several aquaculture species have been established, enabling advanced functional studies. The rapid expansion of proteomics applications has addressed key aquaculture challenges, including fish welfare, nutrition, disease resistance, and seafood quality. Fish growth and metabolism are critical for optimizing aquaculture yields, and proteomic tools have been used to decipher metabolic pathways involved in growth enhancement. For instance, dietary supplementation with β-glucan in rainbow trout (*Oncorhynchus mykiss*) significantly modulated myosin isoforms, improving muscle composition and immunity (Ghaedi et al. 2016). Similarly, replacing fish oil with vegetable oil in Salmonid diets altered their intestinal proteome response, improving resilience against oxidative stress.

8.6.1.2 Proteomics in Aquatic Feed and Waste Management

Proteomic techniques are integral to developing nutritionally optimized, cost-effective feed formulations with reduced environmental impact. Excess feed materials contribute to aquatic pollution, necessitating the identification of feed waste-degrading enzymes and microbes to enhance waste management. Sustainable macroalgae farming, integrated with proteomics, is a promising approach to improving aquaculture sustainability. Notably, a novel non-animal emulsifier peptide isolated from *Eucheuma denticulatum* has shown potential for food applications (Yesiltas et al. 2021). Comparative proteomic studies on *Neoporphyra haitanensis* have identified stress-responsive proteins, reinforcing the role of proteomics in optimizing aquaculture practices (Wang et al. 2020).

8.6.1.3 Proteomics in Fish Stress and Disease Management

Environmental stressors profoundly affect fish growth, health, and biochemical composition. Proteomic analyses have been instrumental in identifying metabolic changes associated with stress, enabling the development of stress-resilient aquaculture systems (Addis et al. 2010). In Chinese shrimp (*Fenneropenaeus chinensis*), proteomic studies revealed distinct protein expression profiles in response to culture medium stress, facilitating strategies to improve farming conditions (Jian et al. 2012). Furthermore, Pacific geoduck clams exhibited remarkable resistance to acidified environments, temperature fluctuations, and dissolved oxygen variations, demonstrating the potential of proteomics in identifying stress-tolerant species for aquaculture.

8.6.1.4 Proteomics in Seafood Safety and Pathogen Detection

Ensuring seafood safety is paramount for global food security, necessitating stringent quality control measures. Proteomic applications extend to detecting nutritional variations, seafood authentication, biomarker development for freshness and contamination, and allergen identification (Silva et al. 2011). Regulatory frameworks, such as those established by the European Union, mandate accurate species labelling to prevent seafood fraud. Proteomic techniques have been employed to differentiate commercial fish species and identify sex-specific markers for seafood traceability. Proteomic studies also aid in assessing biotic and abiotic hazards in seafood products. Waste-derived pathogens, including bacteria, viruses, and parasites, pose significant health risks, while chemical contaminants such as heavy metals and microplastics can compromise food quality. A recent proteomic and metabolomic study on *Litopenaeus vannamei* highlighted the toxicological effects of microplastic ingestion, revealing alterations in metabolic pathways and gut microbiota (Duan et al. 2021). Nanoparticle pollution in aquaculture environments presents emerging challenges, with targeted proteomics playing a crucial role in evaluating its impact.

8.6.1.5 Proteomics in Antibiotic Resistance and Aquatic Disease Management

Antibiotic resistance is a growing concern in aquaculture, threatening fish health and economic sustainability. Proteomic approaches have been instrumental in characterizing resistance mechanisms and identifying novel strategies to combat bacterial infections. In *Aeromonas hydrophila*, proteomics confirmed tetracycline resistance, while biofilm formation in *Edwardsiella tarda* was linked to specific protein regulatory networks (Abdelrahman et al. 2017). Furthermore, proteomic techniques have been used to detect multidrug-resistant strains and elucidate bacterial pathogenesis in aquaculture settings.

8.6.1.6 Proteomics in Aquatic Virology and Bacteriology

Proteomics has revolutionized pathogen diagnosis in aquaculture by enabling rapid identification of bacterial and viral infections. Protein interaction mapping has facilitated the discovery of therapeutic targets for white spot syndrome virus in shrimp (Jian et al. 2012). Additionally, proteomic methods have been applied to differentiate toxic and nontoxic dinoflagellates responsible for harmful algal blooms. These advancements underscore the potential of proteomics in mitigating disease outbreaks and improving aquaculture biosecurity.

8.6.1.7 Proteomics in Seafood Allergen Detection

Seafood allergies present significant health risks, necessitating advanced detection methods. Proteomics-based biosensors offer rapid and accurate identification of allergenic proteins in fish and shellfish. Targeted proteomics has been employed to detect β-parvalbumin, a primary allergen in fish, and to characterize the complete amino acid sequence of shellfish tropomyosin, enhancing allergen traceability and consumer safety (Silva et al. 2011). As proteomics continues to evolve, its integration with metabolomics, transcriptomics, and bioinformatics will further revolutionize aquaculture research and seafood safety. Novel biomarker discovery, precision breeding programs, and sustainable aquafeed development will benefit from proteomic advancements. Moreover, emerging technologies such as mass spectrometry-based proteomics and machine learning-driven proteome analysis hold promise for optimizing aquaculture productivity and ensuring food security in a rapidly changing environment.

By leveraging proteomics, researchers can address critical challenges in aquaculture, from enhancing fish health and productivity to mitigating environmental and food safety risks. As we move forward, the continued development and application of proteomic technologies will play a crucial role in shaping the future of sustainable aquaculture and global food security.

8.6.2 Proteomics in Aquatic Environmental Pollution and Monitoring

Proteomics has become an invaluable tool in aquatic environmental monitoring, providing critical insights into how organisms respond to pollution at the molecular level. The proteome shaped by both genetic and environmental factors undergoes biochemical modifications in response to stressors, making proteomics a powerful approach for studying ecotoxicological effects in aquatic life (Chia et al. 2025). However, the incomplete genome sequences of many aquatic species remain a limitation, hindering comprehensive protein identification (Picotti et al. 2012). Despite this challenge, proteomic studies have successfully detected environmental stress responses, particularly from pharmaceutical pollutants that alter phytoplankton protein profiles, offering novel insights into toxicity assessment. Ecotoxicoproteomics, an emerging discipline over the past two decades, leverages advanced methodologies such as two-dimensional polyacrylamide gel electrophoresis (2D-PAGE) and selected reaction monitoring (SRM) to quantify targeted proteins across diverse environmental samples (Picotti et al. 2012). A significant breakthrough in this field includes the identification of engineered nanomaterials (ENMs) as aquatic contaminants. These materials penetrate cellular membranes, induce intracellular disturbances, and cause DNA damage in aquatic organisms. The full spectrum of ENM-induced damage, however, remains underexplored. Recent proteomic studies on *Gammarus fossarum* have tested SRM-based biomarker panels for multibiomarker development, paving the way for precision monitoring of aquatic pollutants.

Proteomics plays a crucial role in assessing marine and freshwater ecosystem health, identifying naturally occurring biomarkers for environmental stress. Nevertheless, vast oceanic regions remain unexplored, limiting the discovery of novel stress-related proteins. Proteoforms structurally distinct protein variants are an emerging focus for tracking aquatic stress responses at the enzymatic and transcriptomic levels. Advanced proteomic tools now facilitate the detection of protein isoforms and post-translational modifications linked to chemical pollution, contributing to a refined understanding of anthropogenic impacts on aquatic ecosystems. Studies indicate that silver nanoparticles significantly alter protein folding and transmembrane transport, underscoring the need for expanded research on nanoparticle toxicity in aquatic environments (Casabianca et al. 2021). The role of environmental proteins (eProteins) in pollution monitoring is another growing research area. A study on genetically engineered Bt maize revealed that Cry proteins from plant residues persist in nearby water bodies, suggesting that eProteins might function as pseudo-persistent contaminants. This finding highlights the potential of proteomic technologies to trace pollution sources and assess their ecological impacts.

Plastic pollution, particularly microplastics and nano-plastics, presents an escalating threat to aquatic environments. These contaminants disrupt essential biological processes, impair photosynthetic efficiency in phytoplankton, and trigger heteroaggregate formation, severely affecting aquatic ecosystems (Liu et al. 2020). Despite their evident toxicity, large-scale investigations into their molecular mechanisms remain limited. Proteomics offers a promising avenue for exploring how microplastics alter metabolic pathways, biodistribution, and bioaccumulation patterns in aquatic organisms. Recent studies on the edible shrimp *Litopenaeus vannamei* using integrated proteomics have revealed significant changes in hemolymph protein profiles and immune-related proteins upon microplastic exposure. These findings underscore the value of proteomic approaches in assessing the ecological and potential human health risks microplastics pose. Researchers can unravel the intricate molecular mechanisms underlying pollution-induced stress in aquatic organisms by integrating next-generation proteomics with biomonitoring strategies. This knowledge enhances our ability to mitigate environmental hazards and supports the development of sustainable aquaculture and fisheries management practices, ensuring long-term food security and ecological balance.

8.6.3 Natural Aquatic Proteins and Marine-Derived Medicine

The exploration of natural aquatic proteins has revolutionized marine-derived medicine, linking scientific advancements with economic and technological innovations. Marine bioresources provide a rich repository of biomaterials, including polysaccharides (chitin, alginates, fucoidans, carrageenans, ulvans, agar), structural proteins (spongin, collagen, gelatin, keratin, conchiolin, conchixes), and biominerals (corals, shells), all of which hold immense potential in biomedical and industrial applications (Khrunyk et al. 2020). Proteomic approaches are central to identifying and characterizing these proteins, enabling the sustainable development of marine biomaterial-based industries.

A comprehensive understanding of protein structures and functional domain specificity is crucial for biotechnological advancements. Chitin, a structurally robust biopolymer derived from crustacean exoskeletons, is widely utilized in agriculture, medicine, food, and cosmetics. Its thermostability (260–360 °C) enhances its industrial applicability, but its physicochemical properties vary across species due to differences in chitinolytic enzyme activity (Khrunyk et al. 2020). Recent studies have identified 11 chitinolytic enzymes in *Pseudoalteromonas flavipulchra*, providing insights into their domain architecture and enzymatic secretory mechanisms, which may enhance chitin bioconversion strategies (Ren et al. 2022). The field of marine-derived medicine is rapidly advancing, with proteomic strategies playing a pivotal role in the discovery of novel bioactive compounds for human health applications. Marine-based gold nanoparticles, such as chitosan, have demonstrated promise in drug delivery, particularly in oncology, where they are being explored for targeted cancer therapy (Banerjee et al. 2022). However, further proteomic investigations are needed to fully elucidate the molecular interactions of chitosan-based drug carriers, potentially leading to diagnostic kit development and more effective anticancer treatments.

Collagen, another marine-derived biomaterial, is gaining attention in regenerative medicine and tissue engineering due to its biocompatibility and sustainability (Liu et al. 2022). However, to harness its full potential as a biomedical scaffold, integrative proteomic and omics-based studies are essential to optimize its functional properties. Additionally, metabolic pathway analyses of bioactive marine compounds have been a key focus of marine medicine, driving the discovery of 68 unique and patented natural marine products between 2018 and 2021, with most showing significant biomedical promise (Papon et al. 2022). The ongoing expansion of aquatic proteomics is poised to transform these industries, fostering sustainable innovations and unlocking new frontiers in marine biotechnology and aquaculture.

8.7 Challenges and Recommendations in Aquatic Proteomic Applications

Proteomics has revolutionized aquatic sciences, yet its applications face substantial technical and analytical challenges. One primary limitation is the efficiency of protein extraction, which is highly dependent on protocol selection. Given the diversity of aquatic organisms, a universal, cost-effective, and standardized extraction protocol remains a pressing need. While this presents inherent difficulties, family-specific extraction methods tailored to distinct aquatic taxa could mitigate these constraints. Similarly, discrepancies in analytical methodologies hinder comparative assessments of marine algae-based protein compositions, further underscoring the necessity of a standardized protocol.

Another major hurdle lies in quantifying complex protein hydrolases in aquatic species. While gel- and mass spectrometry (MS)-based techniques have demonstrated efficacy in terrestrial organisms, their adaptation to aquatic protein hydrolases is constrained by inherent biochemical complexities. Many aquatic proteins are contaminated with nonprotein substances such as polysaccharides, complicating their quantification. Therefore, results obtained using conventional techniques require validation through complementary methods, such as amino acid analysis (Knight et al. 2003). Two-dimensional differential in-gel electrophoresis (2D-DIGE) coupled with isobaric tagging for relative and absolute quantitation (iTRAQ) has shown promise for aquatic protein characterization (Martyniuk et al. 2012). Additionally, liquid chromatography–mass spectrometry (LC–MS)-based quantification has been instrumental in proteome profiling, as demonstrated in *Daphnia pulex* and *Daphnia longicephala*, where 531 and 317 proteins were identified, respectively, providing novel insights into the Daphnia genome. Furthermore, the tandem mass tag (TMT) labeling approach offers high-throughput, multiplexed proteomic applications with improved automation (Pappireddi et al. 2019). A critical limitation of proteomic applications in aquatic research is the lack of complete genome data, restricting accurate protein identification. Integrating proteomics with genomics, transcriptomics, metabolomics, and metagenomics could provide a more holistic understanding of environmental stress responses and biological interactions. The development of a dedicated aquatic protein database is crucial, as existing platforms lack comprehensive coverage of aquatic taxa. The emergence of AlphaFold as a powerful tool for protein structure prediction highlights the potential for structural characterization, yet a tailored database for aquatic organisms is needed to unlock the full potential of proteomics in fish health and aquaculture.

To address these challenges, we propose sector-specific recommendations. In the food industry, proteomics can enhance seafood safety by enabling allergen testing, improving feed quality through alternative protein sources, and identifying biomarkers for disease diagnostics (Sandalakis et al. 2017). Environmental monitoring can benefit from species-specific proteomic databases for biomarker species, pollution impact assessments, and antibiotic resistance tracking (Fernández-Álvarez et al. 2017). In natural product research, proteomics facilitates the discovery of bioactive molecules, sustainable harvesting of marine resources, and non-animal protein identification using high-tech underwater sampling and nuclear magnetic resonance (NMR) spectroscopy (Kwon et al. 2020). Proteomics has revolutionized aquatic sciences, offering unprecedented insights into fish health, aquaculture sustainability, and environmental monitoring. However, significant challenges persist across various sectors, necessitating strategic solutions to enhance proteomic applications in aquatic research. Table 8.3 presents a comprehensive overview of key

Table 8.3 Common challenges and recommended solutions in aquatic proteomics

Sector	Challenges	Recommendations
Food industry	Seafood safety and human health risks Sustainable aquaculture development through modified feed, disease management, and pollution control Quality of the aquatic product Pathogens and biotoxins in aquatic food Animal welfare and ethics	Development of advanced seafood allergy testing and ELISA assays for quality control Optimization of aquatic feed quality with alternative protein sources Integration of proteomic biomarkers for early disease detection in aquaculture farms Application of macroalgae-based proteomics to identify novel aquaculture species Proteomic-driven sustainable aquaculture and conservation strategies for edible aquatic species High-throughput proteomic techniques for disease management and welfare monitoring
Environmental monitoring	Monitoring pollutants and assessing their impact Conservation and management of aquatic biodiversity Increasing antibiotic resistance in aquatic ecosystems Aquatic biodiversity assessment Environmental toxicity due to anthropogenic stressors	Multibiomarker application for aquatic toxicology studies Development of species-specific proteomic databases for precise biomarker identification Establishment of global protein baseline datasets to monitor proteomic responses to environmental stressors Discovery of novel aquatic species through proteomic profiling Utilization of eProteins for biomonitoring and ecological impact assessments
Natural products	Isolation and identification of novel bioactive compounds Evaluation of biological activities for pharmaceutical applications Drug discovery from marine and freshwater organisms Predicting protein domain architecture for functional studies Sustainable harnessing of biodiversity for bioresources Non-animal protein identification for alternative food sources	Comprehensive proteomic analysis to assess biodiversity and species interactions before sampling Sustainable extraction methodologies to minimize ecological disturbance Implementation of bio- or chemo-synthesis techniques for novel compound production Development of high-tech underwater sampling equipment for minimal environmental impact Advanced nuclear magnetic resonance (NMR) spectroscopy for precise molecular structure elucidation Application of seaweed proteomics for novel protein and peptide discovery

challenges and innovative recommendations to bridge existing knowledge gaps and advance proteomic methodologies for improved aquatic health, sustainability, and biodiversity conservation. A synergistic, multiomics-driven approach is necessary to propel aquatic proteomics forward. Bridging knowledge gaps in aquatic proteoforms, developing standardized methodologies, and leveraging computational advancements will transform proteomics into an indispensable tool for fish health management, aquaculture sustainability, and food security.

8.8 Conclusion and Future Prospects

The advent of proteomics has revolutionized fish health management and aquaculture sustainability, offering unprecedented molecular insights into pathogen detection, disease resistance, and environmental stress responses. This chapter underscores the transformative potential of proteomics, particularly mass spectrometry-based approaches such as MALDI-TOF MS, in identifying bacterial, viral, and parasitic infections with remarkable precision. Beyond diagnostics, proteomic technologies have illuminated host-pathogen interactions, facilitated biomarker discovery, and enhanced our understanding of antimicrobial resistance-critical aspects for strengthening biosecurity and disease management in aquaculture. Integrating proteomics with multiomics frameworks, including genomics, transcriptomics, and metabolomics, is paving the way for precision breeding, nutritional optimization, and novel therapeutic interventions. Ecotoxicoproteomics, a rapidly emerging field, has further extended the applications of proteomics into environmental monitoring, shedding light on pollutant-induced stress and endocrine disruption in aquatic ecosystems. As aquaculture scales up to meet global protein demands, proteomic-driven innovations will be instrumental in mitigating disease outbreaks, optimizing fish growth, and enhancing seafood safety. Despite its vast potential, proteomics in aquaculture faces challenges, including the need for extensive protein databases, standardization of methodologies, and accessibility of high-throughput analytical platforms. However, ongoing advancements in computational proteomics, artificial intelligence, and optical biosensors are bridging these gaps, fostering a new era of precision aquaculture. Future research should emphasize interdisciplinary collaboration, expanding proteomic applications to address food security concerns and environmental sustainability. As we navigate the complexities of global aquaculture, proteomics emerges as a cornerstone technology, ensuring the health and productivity of farmed fish while safeguarding aquatic biodiversity. By harnessing its full potential, the aquaculture industry can transition toward a more resilient and sustainable future, securing high-quality protein sources for future generations.

References

Abdelrahman M, El-Sayed M, Sato S, Hirakawa H, Ito SI, Tanaka K, Mine Y, Sugiyama N, Suzuki M, Yamauchi N, Shigyo M (2017) RNA-sequencing-based transcriptome and biochemical analyses of steroidal saponin pathway in a complete set of Allium fistulosum—A. cepa monosomic addition lines. PLoS One 12(8):e0181784

Addis MF, Cappuccinelli R, Tedde V, Pagnozzi D, Porcu MC, Bonaglini E, Roggio T, Uzzau S (2010) Proteomic analysis of muscle tissue from gilthead sea bream (Sparus aurata, L.) farmed in offshore floating cages. Aquaculture 309(1–4):245–252

An YA, Scherer PE (2020) Mouse adipose tissue protein extraction. Bio-protocol 10(11):e3631-e3631

Assis J, Berecibar E, Claro B, Alberto F, Reed D, Raimondi P, Serrão EA (2017) Major shifts at the range edge of marine forests: the combined effects of climate changes and limited dispersal. Sci Rep 7(1):44348

Avissar S, Schreiber G (2000) G protein measurements: an empirical tool for the diagnosis of affective disorders. Drug Dev Res 50(3–4):316–323

Banerjee A, Ray A, Barpanda A, Dash A, Gupta I, Nissa MU, Zhu H, Shah A, Duttagupta SP, Goel A, Srivastava S (2022) Evaluation of autoantibody signatures in pituitary adenoma patients using human proteome arrays. PROTEOMICS–Clinical Applications 16(6):2100111

Betti M, Ciacci C, Abramovich S, Frontalini F (2021) Protein extractions from Amphisteginalessonii: protocol development and optimization. Life 11(5):418

Booth JG, Eilertson KE, Olinares PDB, Yu H (2011) A bayesian mixture model for comparative spectral count data in shotgun proteomics. Molecular & Cellular Proteomics 10(8)

Braceland M, McLaughlin M, Eckersall PD, Bhide MR (2013) Rapid protein production of Atlantic Salmon, Salmo salar, serum amyloid A (SAA) in an inducible Leishmania tarentolae expression system. In Farm animal proteomics (pp. 59-62). Wageningen Academic.

Bujan J, Yanoviak SP, Kaspari M (2016) Desiccation resistance in tropical insects: causes and mechanisms underlying variability in a Panama ant community. Ecology and evolution 6(17):6282–6291

Casabianca S, Bellingeri A, Capellacci S, Sbrana A, Russo T, Corsi I, Penna A (2021) Ecological implications beyond the ecotoxicity of plastic debris on marine phytoplankton assemblage structure and functioning. Environ Pollut 290:118101

Chia R, Moaddel R, Kwan JY, Rasheed M, Ruffo P, Landeck N, Reho P, Vasta R, Calvo A, Moglia C, Canosa A (2025) A plasma proteomics-based candidate biomarker panel predictive of amyotrophic lateral sclerosis. Nature Medicine1-11

Choudhary RK, Kumar BVS, Sekhar Mukhopadhyay C, Kashyap N, Sharma V, Singh N, Salajegheh Tazerji S, Kalantari R, Hajipour P, Singh Malik Y (2024) Animal wellness: the power of multiomics and integrative strategies: multiomics in improving animal health. Veterinary Medicine International 2024(1):4125118

Chuang MY, Tsai WC, Kuo TY, Chen HM, Chen WJ (2016) Comparative proteome analysis reveals proteins involved in salt adaptation in *Photobacterium damselae* subsp. *piscicida*. J Basic Microbiol 56(12):1234–1243

Cox J, Neuhauser N, Michalski A, Scheltema RA, Olsen JV, Mann M (2011) Andromeda: a peptide search engine integrated into the MaxQuant environment. Journal of proteome research 10(4):1794–1805

Cui X, Li SC, He L, Li M (2014) Fingerprinting protein structures effectively and efficiently. Bioinformatics 30(7):949–955

DeMarco ML, Burnham CAD (2014) Diafiltration MALDI-TOF mass spectrometry method for cultureindependent detection and identification of pathogens directly from urine specimens. American Journal of Clinical Pathology 141(2):204–212

Dieme C, Yssouf A, Vega-Rúa A, Berenger JM, Failloux AB, Raoult D, Parola P, Almeras L (2014) Accurate identification of Culicidae at aquatic developmental stages by MALDI-TOF MS profiling. Parasites & vectors 7(1):544

Di G, Li H, Zhang C, Zhao Y, Zhou C, Naeem S, Li L, Kong X (2017) Label-free proteomic analysis of intestinal mucosa proteins in common carp (Cyprinus carpio) infected with Aeromonas hydrophila. Fish & Shellfish Immunology 66:11–25
Duan Y, Li J, Wang F, Wei J, Yang Z, Sun M, Liu J, Wen M, Huang W, Chen Z, Lu Z (2021) Protein modifications throughout the lung cancer proteome unravel the cancer-specific regulation of glycolysis. Cell Reports 37(12)
Dumpala PR, Gülsoy N, Lawrence ML, Karsi A (2010) Proteomic analysis of the fish pathogen *Flavobacterium columnare*. Proteome Sci 8(1):26
Encinas P, Garcia-Valtanen P, Chinchilla B, Gomez-Casado E, Estepa A, Coll J (2013) Identification of multipath genes differentially expressed in pathway-targeted microarrays in zebrafish infected and surviving spring viremia carp virus (SVCV) suggest preventive drug candidates. PLoS One 8(9):e73553
Feng J, Zhang L, Xia X, Hu W, Zhou P (2020) Effect of geographic variation on the proteome of sea cucumber (Stichopus japonicus). Food Res Int 136:109498
Fernández-Álvarez C, Torres-Corral Y, Saltos-Rosero N, Santos Y (2017) MALDI-TOF mass spectrometry for rapid differentiation of Tenacibaculum species pathogenic for fish. Applied Microbiology and Biotechnology 101(13):5377–5390
Fernández-Álvarez C, Torres-Corral Y, Santos Y (2018) Comparison of serological and molecular typing methods for epidemiological investigation of Tenacibaculum species pathogenic for fish. Applied Microbiology and Biotechnology 102(6):2779–2789
Florio W, Baldeschi L, Rizzato C, Tavanti A, Ghelardi E, Lupetti A (2020) Detection of antibioticresistance by MALDI-TOF mass spectrometry: an expanding area. Frontiers in cellular and infection microbiology 10:572909
Ghaedi G, Keyvanshokooh S, Azarm HM, Akhlaghi M (2016) Proteomic analysis of muscle tissue from rainbow trout (Oncorhynchus mykiss) fed dietary β-glucan. Iran J Vet Res 17(3):184
Golden CD, Koehn JZ, Shepon A, Passarelli S, Free CM, Viana DF, Matthey H, Eurich JG, Gephart JA, Fluet-Chouinard E, Nyboer EA (2021) Aquatic foods to nourish nations. Nature 598(7880):315–320
Jean Beltran PM, Federspiel JD, Sheng X, Cristea IM (2017) Proteomics and integrative omic approaches for understanding host–pathogen interactions and infectious diseases. Molecular systems biology 13(3):922
Jian FENG, ZHANG J, Lei ZHU (2012) Proteome analysis and tissue array for profiling protein markers associated with type B thymoma subclassification. Chinese Medical Journal 125(16):2811–2818
Jitjumnong J, Taweechaipaisankul A, Lin JC, Wongchanla S, Chuwatthanakhajorn S, Lin CJ, Khang LTP, Linh NV, Sangsawad P, Dinh-Hung N, Tang PC (2025) An Overview of Advancements in Proteomic Approaches to Enhance Livestock Production and Aquaculture. Animals 15(13):1946
Karas M, Bachmann D, Hillenkamp F (1985) Influence of the wavelength in high-irradiance ultraviolet laser desorption mass spectrometry of organic molecules. Analytical chemistry 57(14):2935–2939
Karas M, Bachmann D, Bahr U, Hillenkamp F (1987) Matrix-assisted ultraviolet laser desorption of non-volatile compounds. Int J Mass Spectrom Ion Process 78:53–68
Kazazić SP, TopićPopović N, Strunjak-Perović I, Babić S, Florio D, Fioravanti M, Bojanić K, Čož-Rakovac R (2019) Matrix-assisted laser desorption/ionization time of flight mass spectrometry identification of Vibrio (Listonella) anguillarum isolated from sea bass and sea bream. PLoS One 14(11):e0225343
Khrunyk Y, Lach S, Petrenko I, Ehrlich H (2020) Progress in modern marine biomaterials research. Mar Drugs 18(12):589
Kim EY, Kim WK, Oh KJ, Han BS, Lee SC, Bae KH (2015) Recent advances in proteomic studies of adipose tissues and adipocytes. International journal of molecular sciences 16(3):4581–4599
Knight ZA, Schilling B, Row RH, Kenski DM, Gibson BW, Shokat KM (2003) Phosphospecific proteolysis for mapping sites of protein phosphorylation. Nature biotechnology 21(9):1047–1054

Kumar G, Hummel K, Ahrens M, Menanteau-Ledouble S, Welch TJ, Eisenacher M, Razzazi-Fazeli E, El-Matbouli M (2016) Shotgun proteomic analysis of Yersinia ruckeri strains under normal and iron-limited conditions. Vet Res 47(1):100

Kumar G, Hummel K, Welch TJ, Razzazi-Fazeli E, El-Matbouli M (2017) Global proteomic profiling of Yersinia ruckeri strains. Vet Res 48(1):55

Kwon D, Park JM, Duong VA, Hong SJ, Cho BK, Lee CG, Choi HK, Kim DM, Lee H (2020) Comparative proteomic profiling of marine and freshwater Synechocystis strains using liquid chromatography-tandem mass spectrometry. J Mar Sci Eng 8(10):790

LaFrentz BR, LaPatra SE, Call DR, Wiens GD, Cain KD (2009) Proteomic analysis of Flavobacterium psychrophilum cultured in vivo and in iron-limited media. Dis Aquat Org 87(2):171–182

Lu S, Zhang J, Lian X, Sun L, Meng K, Chen Y, Sun Z, Yin X, Li Y, Zhao J, Wang T (2019) A hidden human proteome encoded by 'non-coding' genes. Nucleic acids research 47(15):8111–8125

Liu B, Jiang S, Li M, Xiong X, Zhu M, Li D, Zhao L, Qian L, Zhai L, Li J, Lu H (2018) Proteomewide analysis of USP14 substrates revealed its role in hepatosteatosis via stabilization of FASN. Nature communications 9(1):4770

Liu W, Zhao Y, Shi Z, Li Z, Liang X (2020) Ecotoxicoproteomic assessment of microplastics and plastic additives in aquatic organisms: a review. Comparative Biochemistry and Physiology Part D: Genomics and Proteomics 36:100713

Liu S, Lau CS, Liang K, Wen F, Teoh SH (2022) Marine collagen scaffolds in tissue engineering. Curr Opin Biotechnol 74:92–103

Long M, Zhao J, Li T, Tafalla C, Zhang Q, Wang X, Gong X, Shen Z, Li A (2015) Transcriptomic and proteomic analyses of splenic immune mechanisms of rainbow trout (Oncorhynchus mykiss) infected by Aeromonas salmonicida subsp. salmonicida. Journal of proteomics 122:41–54

López-Cortés XA, Nachtigall FM, Olate VR, Araya M, Oyanedel S, Diaz V, Jakob E, Ríos-Momberg M, Santos LS (2017) Fast detection of pathogens in salmon farming industry. Aquaculture 470:17–24

Lopez-Lopez V, Perez-Sanz F, de Torre-Minguela C, Marco-Abenza J, Robles-Campos R, Sanchez-Bueno F, Pons JA, Ramirez P, Baroja-Mazo A (2021) Proteomics in liver transplantation: a systematic review. Frontiers in immunology 12:672829

Martyniuk CJ, Alvarez S, Denslow ND (2012) DIGE and iTRAQ as biomarker discovery tools in aquatic toxicology. Ecotoxicol Environ Saf 76:3–10

Medina-Gali R, Bello-Perez M, Ciordia S, Mena MC, Coll J, Novoa B, del Mar Ortega-Villaizan M, Perez L (2019) Plasma proteomic analysis of zebrafish following spring viremia of carp virus infection. Fish & Shellfish Immunology 86:892–899

Mizukami H, Hathway B, Procopio N (2020) Aquatic decomposition of mammalian corpses: a forensic proteomic approach. J Proteome Res 19(5):2122–2135

Mohd I, Mushtaq ST (2025) Aquaculture for Improving Productivity, Income, and Nutrition. Aquaculture: Enhancing Food Security and Nutrition. p.373.

Moreira LM, Soares MR, Facincani AP, Ferreira CB, Ferreira RM, Ferro MI, Gozzo FC, Felestrino ÉB, Assis RA, Garcia CCM, Setubal JC (2017) Proteomics-based identification of differentially abundant proteins reveals adaptation mechanisms of Xanthomonas citri subsp. citri during Citrus sinensis infection. BMC microbiology 17(1):155

Moreira M, Schrama D, Farinha AP, Cerqueira M, Raposo de Magalhães C, Carrilho R, Rodrigues P (2021) Fish pathology research and diagnosis in aquaculture of farmed fish: a proteomics perspective. Animals 11(1):125

Mulder IE, Wadsworth S, Secombes CJ (2007) Cytokine expression in the intestine of rainbow trout (Oncorhynchus mykiss) during infection with *Aeromonas salmonicida*. Fish Shellfish Immunol 23(4):747–759

Nissa MU, Pinto N, Mukherjee A, Reddy PJ, Ghosh B, Sun Z, Ghantasala S, Chetanya C, Shenoy SV, Moritz RL, Goswami M (2021) Organ-based proteome and post-translational modification profiling of a widely cultivated tropical water fish, Labeo rohita. Journal of proteome research, 21(2):420–437

Niessen-Ruenzi A, Ruenzi S (2019) Sex matters: gender bias in the mutual fund industry. Manag Sci 65(7):3001–3025

Nombela I, Lopez-Lorigados M, Salvador-Mira ME, Puente-Marin S, Chico V, Ciordia S, Mena MC, Mercado L, Coll J, Perez L, Ortega-Villaizan MDM (2019) Integrated transcriptomic and proteomic analysis of red blood cells from rainbow trout challenged with VHSV point towards novel immunomodulant targets. Vaccines 7(3):63

Nyman TA (2001) The role of mass spectrometry in proteome studies. Biomol Eng 18:221–227

Ormsby MJ, Grahame E, Burchmore R, Davies RL (2019) Comparative bioinformatic and proteomic approaches to evaluate the outer membrane proteome of the fish pathogen *Yersinia ruckeri*. J Proteome Res 199:135–147

Oskoueian E, Eckersall PD, Bencurova E, Dandekar T (2016) Application of proteomic biomarkers in livestock disease management. In Agricultural Proteomics Volume 2: Environmental Stresses (pp. 299–310). Cham: Springer International Publishing.

Papon N, Copp BR, Courdavault V (2022) Marine drugs: biology, pipelines, current and future prospects for production. Biotechnol Adv 54:107871

Pappireddi N, Martin L, Wühr M (2019) A review on quantitative multiplexed proteomics. Chembiochem 20(10):1210–1224

Park YK, Nho SW, Shin GW, Park SB, Jang HB, Cha IS, Ha MA, Kim YR, Dalvi RS, Kang BJ, Jung TS (2009) Antibiotic susceptibility and resistance of *Streptococcus iniae* and *streptococcus parauberis* isolated from olive flounder (*Paralichthysolivaceus*). Vet Microbiol 136(1–2):76–81

Pérez-Sancho M, Vela Ana I, Awad M, Kostrzewa M, Domínguez L, Fernández-Garayzábal JF (2016) Differentiation of *Photobacterium damselae* subspecies using matrix-assisted laser-desorption/ionization time-of-flight mass spectrometry (MALDI-TOF MS) in fish isolates. Aquaculture 464:159–164

Pérez-Sancho M, Vela AI, Wiklund T, Kostrzewa M, Domínguez L, Fernández-Garayzábal JF (2017) Differentiation of Flavobacterium psychrophilum from Flavobacterium psychrophilum-like species by MALDI-TOF mass spectrometry. Res Vet Sci 115:345–352

Piamsomboon P, Thanasaksiri K, Murakami A, Fukuda K, Takano R, Jantrakajorn S, Wongtavatchai J (2020) Streptococcosis in freshwater farmed seabass Lates calcarifer and its virulence in Nile tilapia Oreochromis niloticus. Aquaculture 523:735189

Picotti P, Aebersold R (2012) Selected reaction monitoring–based proteomics: workflows, potential, pitfalls and future directions. Nature methods 9(6):555–566

Ponomarenko EA, Poverennaya EV, Ilgisonis EV, Pyatnitskiy MA, Kopylov AT, Zgoda VG, Lisitsa AV, Archakov AI (2016) The size of the human proteome: the width and depth. Int J Anal Chem 2016:7436849

Popovic R, Celardo I, Yu Y, Costa AC, Loh SH, Martins LM (2021) Combined transcriptomic and proteomic analysis of perk toxicity pathways. International Journal of Molecular Sciences 22(9):4598

Provan F, Jensen LB, Uleberg KE, Larssen E, Rajalahti T, Mullins J, Obach A (2013) Proteomic analysis of epidermal mucus from sea lice–infected A tlantic salmon, S almo salar L. Journal of fish diseases 36(3):311–321

Puentes JP (2017) Etnobotánica urbana: El conocimiento botánico local sobre las plantas alimenticias y medicinales, y sus usos, en la conurbación Buenos Aires-La Plata (Argentina)

Rajan B, Lokesh J, Kiron V, Brinchmann MF (2013) Differentially expressed proteins in the skin mucus of Atlantic cod (*Gadus morhua*) upon natural infection with Vibrio anguillarum. BMC Vet Res 9:1–11

Ren XB, Dang YR, Liu SS, Huang KX, Qin QL, Chen XL, Zhang YZ, Wang YJ, Li PY (2022) Identification and characterization of three chitinases with potential in direct conversion of crystalline chitin into N, N′-diacetylchitobiose. Mar Drugs 20(3):165

Rodrigues PM, Silva TS, Dias J, Jessen F (2012) Proteomics in aquaculture: applications and trends. J Proteome 75(14):4325–4345

Rodrigues T, Reker D, Schneider P, Schneider G (2016) Counting on natural products for drug design. Nat Chem 8(6):531–541

Rotchell J, Ostrander G (2003) Molecular markers of endocrine disruption in aquatic organisms. J Toxic Environ Health, Part B 6(5):453–496

Saleh S, Staes A, Deborggraeve S, Gevaert K (2019) Targeted proteomics for studying pathogenic bacteria. Proteomics 19(16):1800435

Sanchez BC, Ralston-Hooper K, Sepúlveda MS (2011) Review of recent proteomic applications in aquatic toxicology. Environ Toxicol Chem 30(2):274–282

Sandalakis V, Goniotakis I, Vranakis I, Chochlakis D, Psaroulaki A (2017) Use of MALDI-TOF mass spectrometry in the battle against bacterial infectious diseases: recent achievements and future perspectives. Expert Rev Proteomics 14(3):253–267

Seng P, Drancourt M, Gouriet F, La Scola B, Fournier PE, Rolain JM, Raoult D (2009) Ongoing revolution in bacteriology: routine identification of bacteria by matrix-assisted laser desorption ionization time-offlight mass spectrometry. Clinical infectious diseases 49(4):543–551.

Shen B, Yi X, Sun Y, Bi X, Du J, Zhang C, Quan S, Zhang F, Sun R, Qian L, Ge W (2020) Proteomic and metabolomic characterization of COVID-19 patient sera. Cell 182(1):59–72

Silva TS, Cordeiro O, Richard N, Conceição LE, Rodrigues PM (2011) Changes in the soluble bone proteome of reared white seabream (*Diplodussargus*) with skeletal deformities. Comparat Biochem Physiol Part D Genom Proteom 6(1):82–91

Singhal N, Kumar M, Kanaujia PK, Virdi JS (2015) MALDI-TOF mass spectrometry: an emerging technology for microbial identification and diagnosis. Frontiers in microbiology 6:791

Smalla K, Sobecky PA (2002) The prevalence and diversity of mobile genetic elements in bacterial communities of different environmental habitats: insights gained from different methodological approaches. FEMS Microbiol Ecol 42(2):165–175

Srinivasa Rao PS, Yamada Y, Tan YP, Leung KY (2004) Use of proteomics to identify novel virulence determinants that are required for Edwardsiella tarda pathogenesis. Mol Microbiol 53(2):573–586

Srivastava V, Rezinciuc S, Bulone V (2018) Quantitative proteomic analysis of four developmental stages of Saprolegnia parasitica. Frontiers in microbiology 8:2658

Song WJ, Qin QW, Qiu J, Huang CH, Wang F, Hew CL (2004) Functional genomics analysis of Singapore grouper iridovirus: complete sequence determination and proteomic analysis. J Virol 78(22):12576–12590

Tan W, Wang , Drake TJ (2004) Molecular beacons. Current Opin Chem Biol 8(5):547–553

Tanaka K (2003) The origin of macromolecule ionization by laser irradiation. AngewandteChemie-International Edition 42:3860–3870

Tanaka K, Waki H, Ido Y, Akita S, Yoshida Y, Yoshida T, Matsuo TJ (1988) Protein and polymer analyses up to m/z 100 000 by laser ionization time-of-flight mass spectrometry. Rapid Commun Mass Spectrom 2(8):151–153

Tavares GC, Carvalho AF, Pereira FL, Rezende CP, Azevedo VA, Leal CA, Figueiredo HC (2018) Transcriptome and proteome of fish-pathogenic Streptococcus agalactiae are modulated by temperature. Frontiers in Microbiology 9:2639

Torres-Corral Y, Santos Y (2019) Identification and typing of Vagococcus salmoninarum using genomic and proteomic techniques. Journal of Fish Diseases 42(4):597–612

Vancsok C, Peñaranda MMD, Raj VS, Leroy B, Jazowiecka-Rakus J, Boutier M et al (2017) Proteomic and functional analyses of the virion transmembrane proteome of cyprinid herpesvirus 3. J Virol 91(21):10–1128

Wang D, You W, Chen N, Cao M, Tang X, Guan X et al (2020) Comparative quantitative proteomics reveals the desiccation stress responses of the intertidal seaweed *NEOPORPHYRA haitanensis*. J Phycol 56(6):1664–1675

Wu H, Ji C, Wei L, Zhao J (2013) Evaluation of protein extraction protocols for 2 DE in marine ecotoxicoproteomics. Proteomics 13(21):3205–3210

Xiong XP, Dong CF, Xu X, Weng SP, Liu ZY He JG (2011) Proteomic analysis of zebrafish (Danio rerio) infected with infectious spleen and kidney necrosis virus. Developmental & Comparative Immunology 35(4):431–440

Yesiltas B, Gregersen S, Lægsgaard L, Brinch ML, Olsen TH, Marcatili P et al (2021) Quantitative proteomics and bioinformatics identified emulsifier peptides derived from seaweed, methanotrophic bacteria, and potato proteins. Food Chem 362:130217

Yolanda TC, Clara FÁ, Ysabel S (2019) Proteomic and molecular fingerprinting for identification and tracking of fish pathogenic Streptococcus. Aquaculture 498:322–334

Zhang Y, Fonslow BR, Shan B, Baek MC, Yates III JR (2013) Protein analysis by shotgun/bottom-up proteomics. Chemical reviews 113(4):2343–2394

Zhang H, Zong R, He H, Huang T (2022) Effects of hydrogen peroxide on Scenedesmus obliquus: cell growth, antioxidant enzyme activity and intracellular protein fingerprinting. Chemosphere 287:132185

Zhu H, Pan S, Gu S, Bradbury EM, Chen X (2002) Amino acid residue specific stable isotope labeling for quantitative proteomics. Rapid communications in mass spectrometry 16(22):2115–2123

Zhou S, Wan Q, Huang Y, Huang X, Cao J, Ye L et al (2011) Proteomic analysis of Singapore grouper iridovirus envelope proteins and characterization of a novel envelope protein VP088. Proteomics 11(11):2236–2248

Application of Metagenomic Tools in Aquatic Virology

9

Amrita Mohanty, Anand Kumar, and Payoja Mohanty

9.1 Introduction

The practice of farming fish and other aquatic animals dates back at least 4000 years to pre-feudal China. Historical references to fish ponds appear in the Old Testament and in Egyptian hieroglyphics from the Middle Kingdom (2050–1652 BC). During Roman times, fish farms were widespread in Europe, and a recent study of land forms in the Bolivian Amazon has uncovered a complex network of fish weirs that predates the Hispanic era (Suttle 2013). Despite its ancient roots, aquaculture mainly functioned as a subsistence farming activity until the mid-twentieth century, when pioneering husbandry methods for salmon, trout, and various tropical fish and shrimp species emerged and gained traction (Palermo et al. 2019). Today, aquaculture is a significant global industry, with total annual production surpassing 50 million tonnes and an estimated value of nearly US$80 billion. Between 1970 and 2007, it grew at an average annual rate of 6.9%, making it the fastest-growing sector for animal food production, and it is expected to soon surpass capture fisheries as the primary source of seafood (Fuhrman 1999; Edwards and Rohwer 2005).

For centuries, various diseases in aquatic animals have been reported, with most being either non-infectious (such as tumors) or attributed to common endemic pathogens, primarily parasites and bacteria, which were familiar to observers and

A. Mohanty (✉)
ICAR-Central Institute of Fisheries Education, Mumbai, India
e-mail: amritha.fbtpb101@cife.edu.in

A. Kumar
College of Fisheries, Kishanganj, Bihar, India

ICAR-Central Inland Fisheries Research Institute, Barrackpore, Kolkata, India

P. Mohanty
ICAR-Central Inland Fisheries Research Institute, Barrackpore, Kolkata, India

© The Author(s), under exclusive license to Springer Nature Singapore Pte Ltd. 2026

B. K. Das et al. (eds.), *Bioinformatics Tools for Predictive Ecology and Fisheries*, Sustainable Aquatic Food Systems, https://doi.org/10.1007/978-981-95-2573-7_9

practitioners in traditional aquaculture. However, over the past century, the advent of novel intensive aquaculture practices, increased global movement of aquatic animals and their products, and various anthropogenic stresses on aquatic ecosystems have contributed to the emergence of many new diseases in fish and shrimp (Jacquet et al. 2010; Costa and Holmes 2024). In this chapter, we categorize these emerging diseases as: (i) new or previously unidentified diseases; (ii) known diseases that have first appeared in a new species (expanding host range); (iii) known diseases that have emerged for the first time in a new location (expanding geographic range); and (iv) recognized diseases exhibiting a new presentation (different signs) or higher virulence due to changes in the causative agent (Priya and Kappalli 2022; Zhang et al. 2022).

Emerging disease outbreaks frequently lead to significant, often severe, losses in fish and shrimp populations, resulting in considerable economic impacts on commercial aquaculture and posing threats to valuable wild aquatic species (Hurwitz and Sullivan 2013). However, the spread of these diseases and their effects are often worsened by additional challenges, including the following: (i) delays in developing diagnostic tools for confirming diseases or identifying pathogens, allowing infected animals to remain unnoticed; (ii) limited understanding of current or potential host ranges; (iii) insufficient knowledge of existing geographic distributions; (iv) lack of insight into key epidemiological factors (such as replication cycles, transmission modes, reservoirs, vectors, and stability); and (v) inadequate comprehension of strain differences and their relationships to known pathogens (Weynberg 2018). Efforts to curb pathogen spread are frequently hampered by limited capacity in some developing countries to implement effective quarantine and biosecurity measures, as well as the illegal or poorly regulated international movement of live aquatic animals (Winter et al. 2014).

Significant emerging viral pathogens in fish belong to numerous families of well-established vertebrate viruses, some of which also affect humans or livestock. Nevertheless, the ecology of viral diseases in fish significantly differs from that in humans and other land vertebrates. Key distinctions include the following: (i) a limited number of fish viruses are vectored by arthropods; (ii) while water acts as a stabilizing medium, currents are less effective than aerosols for long-distance virus transmission; (iii) wild reservoir species typically exist at very low densities, with exceptions noted in schooling and spawning aggregates; (iv) fish are poikilotherms, making temperature crucial in influencing the disease process by modulating both the virus's replication rate and the host's immune response, along with other physiological resistance factors; (v) sexual transmission of fish viruses between adults is uncommon, though some viruses can be found in high concentrations within spawning fluids, and a few can be passed vertically from parents to offspring, either intra-ovum or on the egg surface. Similar to avian diseases, migratory fish can act as carriers for the long-range spread of viral pathogens.

The global expansion of finfish aquaculture, along with notable advancements in fish health surveillance, has resulted in the identification of several viruses that are unprecedented in scientific literature. Many of these viruses are endemic within native populations and may opportunistically spill over to infect fish within aquaculture facilities. Additionally, other well-documented fish viruses (such as the channel catfish virus and *Onchorhynchus masou* virus) can inflict considerable losses in aquaculture; however, there appears to be no significant increase in their host or geographic range. In the subsequent sections, we will examine the principal emerging fish virus diseases that lead to substantial losses in aquaculture and are proliferating in their host or geographic extent. Due to the potential for transmission through commercial trade in finfish, numerous diseases are classified as notifiable by the World Organization for Animal Health (OIE).

9.2 Viral Diseases

Outbreaks of viral diseases in cultured fish and shrimp have been reported with increasing frequency over the past two decades. The significance of this issue cannot be underestimated, as these occurrences have resulted in severe mortalities. This chapter aims to provide fundamental information regarding viral disease challenges affecting economically important fish species, including catfish (*Clarias* spp. and *Ictalurus* spp.), snakeheads (*Ophicephalus striatus*), carp (*Cyprinus* spp.), tilapia, milkfish (*Chanos chanos*), grouper (*Epinephelus* spp.), rabbitfish (*Siganus* spp.), sea bass (*Lates calcarifer*), mullet (*Mugil cephalus*), and penaeid shrimp (Rowley et al. 2024).

Viruses are extremely small organisms measuring between 10 and 300 nanometers (nm). To observe them, an electron microscope is essential. Due to their minuscule size, viruses can easily pass through filters with a pore size of 0.5 microns. The fundamental structure of a virion includes a capsid that encloses a nucleic acid genome (Fig. 9.1). The capsid comprises identical protein subunits known as capsomeres, while the genome can either be ribonucleic acid (RNA) or deoxyribonucleic acid (DNA). Collectively, these viral components form the nucleocapsid, which may be surrounded by a lipid envelope. Viruses lacking this envelope are referred to as naked viruses (Blanco-Rodriguez and Di Nunzio 2021; Lostroh 2024).

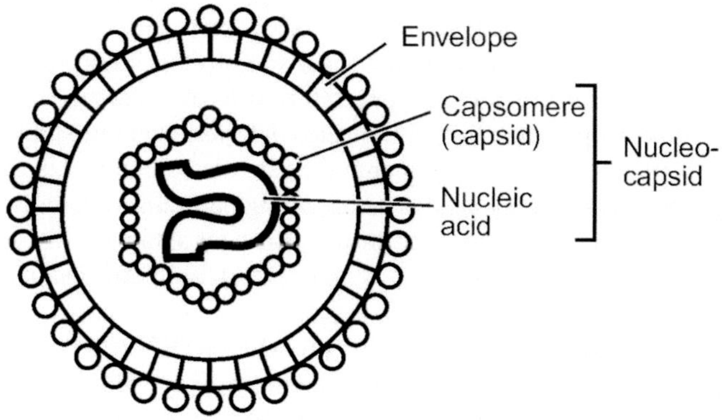

Fig. 9.1 Structure of an enveloped virus

Viruses exhibit cubical, polyhedral, helical, or complex morphology and symmetry. These microbes are obligate parasites, multiplying exclusively within live hosts. In animal cells, the virus initially attaches to specific cell surface components known as receptors. Subsequently, the entire virion penetrates the cell, utilizing the metabolic machinery and pathways of the living cell to replicate its nucleic acid and synthesize protein subunits. Thereafter, these fundamental structural components are assembled to package the new viral particles. The newly formed virions are typically released from the cell through lysis. Cells infected by certain viruses develop inclusion or occlusion bodies within the infected cell as the replication cycle progresses. These bodies are formed due to the accumulation of virions or viral components, although some inclusion bodies may not contain virions. They may be singular, multiple, intracytoplasmic, or intranuclear and can be detected through histopathology using a compound microscope. Aquatic viruses are transmitted from fish or shrimp to other fish or shrimp, from water to fish or shrimp, or from reservoirs to fish or shrimp via horizontal transmission. Disease transmission may also occur from brooders to eggs or fry through vertical transmission. Known reservoirs of viral pathogens include farmed fish and crustaceans, imported fish and crustaceans, wild fish and crustaceans, other aquatic animals and plants, as well as survivors of viral epizootics.

The diagnosis of viral infections can be established through a combination of various methodologies, including clinical signs of disease, detection of inclusion or occlusion bodies, electron microscopy (EM), and infection enhancement bioassays. Tissue culture techniques employ established fish cell lines, namely, BB (Brown Bullhead), BF2 (Bluegill Fin), CCO (Channel Catfish Ovary), CFS (Catfish Spleen), CHSE (Chinook Salmon Embryo), EPC (Epithelioma Papulosum Cyprini), FHM (Fathead Minnow), GCK-84 (Grass Carp Kidney), GCG (Grass Carp Gonad), and GCF (Grass Carp Fin), as well as RTG-2 (Rainbow Trout Gonad), SBK-2 (Sea Bass Kidney), SHS (Snakehead Spleen), and SSN-1 (Striped Snakehead Fry), are utilized. To date, however, no continuous cell line derived from shrimp has been established. Filtrates prepared from homogenized tissues of virus-infected fish, when inoculated onto monolayers of susceptible cells, result in cell lysis or destruction known as the cytopathic effect (CPE). Serial dilution of the filtrate furnishes information regarding the estimated titer of the virus, expressed as the tissue culture infection dose (TCID). Alternatively, the focal site of CPE induced by each virion leads to the development of a clear zone referred to as a plaque. In such plaque assays, the number of plaques formed indicates the estimated viral particles present in a given sample. Other serological techniques are also employed in the diagnosis of viral infections, with widely used tests including neutralization index (NI) determination, western blot, enzyme-linked immunosorbent assay (ELISA), fluorescent antibody technique (FAT), and indirect fluorescent antibody technique (IFAT). Recent advancements in molecular biology, such as polymerase chain reaction (PCR), reverse transcriptase-polymerase chain reaction (RT-PCR), and DNA probe methodologies, have been developed and are currently applied extensively for the diagnosis of viral infections in shrimp (Bootland and Leong 2011).

9.3 Classification of Virome

Aquatic viruses can be classified into various groups on the basis of host (animal, plant, microorganism, and planktonic state), molecular biological features (structure and composition of the genome and envelope, type of infection and replication), ultrastructure morphology and size (helical, polyhedral, or complex viruses range from 5 to 400 nm or more; the Pithoviridae giant virus family even has 1.5 μm in length), and location (freshwater, ocean, or extreme aquatic environments), but they are also interrelated in a complicated network of ecological roles and relationships. Aquatic viruses have the main groups of aquatic animal viruses (emerging and reemerging; e.g., Iridovirus, Rhabdovirus, virioplanktons (e.g., cyanophages, phycoviruses, and archaea viruses; Cyanomyophage, Ampullavirus), giant viruses (Megavirales; e.g., Pithovirus, Tupanvirus, Mimivirus), etc.)

The potential for transmission of certain aquaculture animals (including fish and crustaceans) viruses has been investigated and evaluated (Oidtmann et al. 2018). Emerging and reemerging viral diseases are on the increase. For instance, the rapid spread worldwide of fish herpesvirus, amphibian ranaviruses (Sriwanayos et al. 2020), shrimp and crayfish nimaviruses (Ke et al. 2021), and Tilapia lake virus (TiLV) (Thawornwattana et al. 2021) highlights the constant threat of emerging and reemerging aquatic animal viral diseases.

New aquatic animal pathogenic viruses are constantly arising. Freshwater bodies are estimated to be populated with over 10,000 fish species, which form about 40% of the world's fish communities and a quarter of highly varied vertebrate populations on a global level (Lundberg et al. 2000). Freshwater aquaculture plays a significant role in aquatic food supplies and nutrition security (Belton et al. 2020; Naylor et al. 2021; Gui 2024), especially in China, which accounts for the world's aquaculture production (Gui and Zhu 2012; Wang et al. 2019). At the same time, viral diseases in aquaculture are also on the rise (Kibenge 2019; Mugimba et al. 2021). In aquaculture, wild and farmed aquatic animals belonging to the same column of water suffer from identical viral issues. On this note, research, avoidance, and intervention of aquaculture viral disease have important effects on aquaculture animals, the entire aquatic ecosystem, fishers' livelihoods, and overall national economic stability. Plenty of new, recurrent, and interspecies transmissible aquatic animal viruses emerged during the past several decades. Some representative species have been isolated and characterized, such as iridoviruses, herpesviruses, rhabdoviruses, reoviruses, nodaviruses, nimaviruses, and orthomyxoviruses (e.g., infectious salmon anemia virus, ISAV) (Zhang and Gui 2015; Kibenge 2019; Zhang et al. 2020; Abbadi et al. 2021). Most of these viruses are capable of infecting various farmed and wild aquatic animals, such as fish, shrimp, amphibians, reptiles, and aquatic mammals. Viral diseases may lead to major losses in aquaculture (Kibenge (2019)).

Most aquatic animal viruses are roughly spherical, but there are still different shapes, such as bullet-shaped, rod-shaped, brick-shaped, wheel-shaped, filamentous-shaped, and pleomorphic or irregular-shaped (Lodish et al. 2000; Gui and Zhang 2018). A crucian carp herpesvirus (*Carassius auratus* herpesvirus, CaHV) was isolated from diseased crucian carp that had acute gill hemorrhages and high mortality.

The virus particles have icosahedral capsids with a diameter of about 97 nm, exhibiting typical herpesvirus morphology. The typical symptom of crucian carp caused by CaHV is significantly different from other known fish herpesviruses (CyHV-1, CyHV-2, CyHV-3, or KHV), which are characterized by acute gill hemorrhages and high mortality. The pathological changes in the gills and head-kidney increased with time after CaHV infection. There were a large number of virus particles in the head-kidney cells, showing that the head-kidney is the primary target organ for CaHV invasion and replication (Fang et al. 2016). For example, Koi herpesvirus (KHV) disease is a devastating disease of ornamental koi and common carp varieties and such infectious disease contributed to the decline of iconic populations of salmon in the Northeast Pacific, metatranscriptomic sequencing and analysis of dead and moribund cultured salmon revealed that novel arenavirus, reovirus, and nidovirus are impactful. These previously unrecognized viruses are also threats to wild salmon (Mordecai et al. 2019).

9.4 Processing of Viromes

Recent studies have rigorously explored the structure of viral communities and virome populations within different freshwater ecosystems, informing them regarding watershed attributes. Through conducting a thorough virome study, mostly focusing on viral pathogens and fecal viral markers, these works afford insightful knowledge about how the community structures of viromes are influenced by particular land uses (Fig. 9.2). These results are critical for evaluating the risks of waterborne illness and for supporting macroscopic environmental management efforts intended to protect public health. The research of interest sought to investigate the following: (1) virome distribution and diversity in tropical freshwater environments, such as reservoirs and tributaries; (2) virome community differences among different land use types; and (3) the presence of human-specific pathogenic viruses, with a focus on novel zoonotic and previously unknown viral risks in surface waters. These findings not only advance our knowledge of aquatic viral ecology but also highlight the need to incorporate virome surveillance into risk assessment strategies. The capacity to analyze and interpret aquatic virome data is an innovative strategy for quantifying viral hazards and guiding evidence-based watershed and public health management strategies (Table 9.1).

9.5 Metagenomics: A Breakthrough into the Viral Diversity

The term metagenomics was first coined by Handelsman in 1998 and is referred to as the study of the genetic material (DNA and RNA) from all organisms within a specific environment, including those that cannot be cultured in a lab. Researchers can use metagenomic methods to study viruses in oceans, on land, and in living organisms. The first viral metagenomics study was done about 10 years ago. Viral metagenomics took off when Breitbart et al. put out a paper on viral spread in

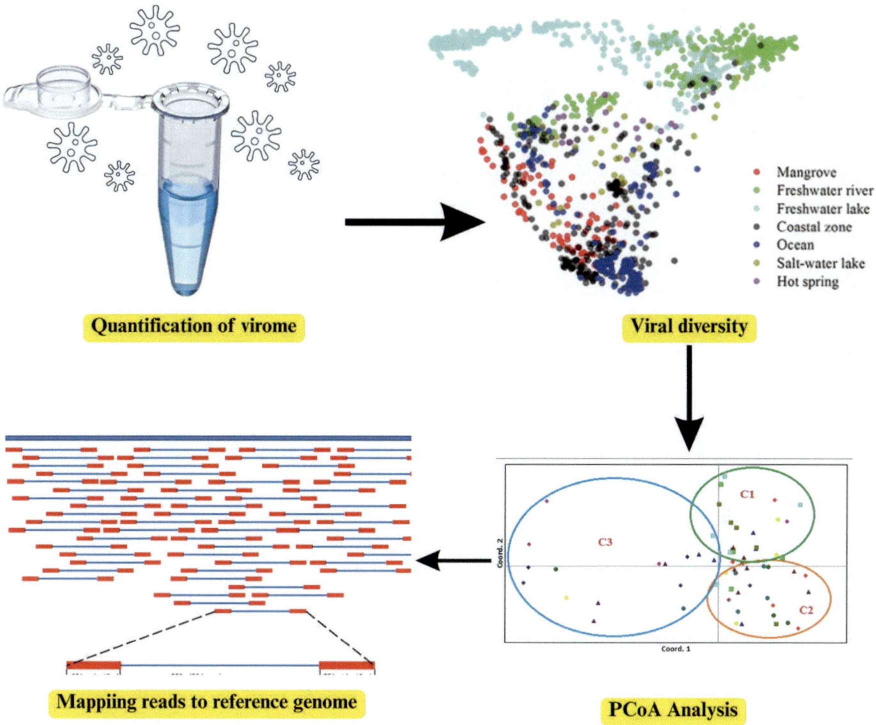

Fig. 9.2 Representation of virome processing from freshwater habitat

seawater. They used about 1000 sequences and reported that there were 300–7000 new viral types in the ocean. Of these, 35% were phages. The metagenomic sequences included bacteria, archaea, and eukarya. Most were major families of dsDNA phages, yet over 65% showed no clear match to known viruses. Earlier, diagnostic approaches like PCR, immunology tests, and wide-range arrays were not the best techniques to find new viruses. But metagenomics has brought a revolutionary impact on new virus groups and shown how they transform over time (Macedo et al. 2024). This method looks at virus genes from aquatic ecosystems using tools like nanopore and Illumina. It helps analyze unknown viruses (Alavandi and Poornima 2012). Software like Megan, PathSeq, and CAMERA analyzes sequences (Seshadri et al. 2007; Kostic et al. 2011), while MetaVir, CheckV, and VirSorter help find and sort virus data. Oceans have helped grow metagenomics databases, shown by MarCat, MarDB, and MarRef (Klemetsen et al. 2018). Tools like DeepVirFinder, ppr-meta, Vibrant, and VirSorter2 show great results, with high true rates and low false rates (Wu et al. 2024). Metagenomics can improve the understanding of viral diversity and emerging illnesses in aquaculture environments, as well as enhance biosecurity in aquatic ecosystems (Fig. 9.3; Table 9.2).

Table 9.1 Classification of fish viruses by habitat, genomic types, viral taxonomy, and host susceptibility (Walker et al. 2022, Munang'Andu et al., 2017, Zhang et al. 2022, OIE 2009)

Virus species	Genus	Family	Genome type	Susceptible fish families	Ecosystem
Anguillid herpesvirus	Cyprinivirus	Alloherpesviridae	dsDNA	Anguillidae	Freshwater
Cyprinid herpesvirus				Cyprinidae	
Cyprinid herpesvirus 3					
Acipenserid herpesvirus 2	Ictalurivirus	Alloherpesviridae	dsDNA	Acipenseridae	Both
Ictalurid herpesvirus					
Salmonid herpesvirus	Salmonivirus	Alloherpesviridae	dsDNA	Salmonidae	Both
Tilapia lake virus	Tilapinevirus	Amnoonviridae	+ssRNA	Cichlidae, Cyprinidae, Osphronemidae	Freshwater
Infectious pancreatic necrosis virus (IPNV)	Aquabirnavirus	Birnaviridae	dsRNA	Salmonidae, Scophthalmidae, Anguillidae	Both
Lates calcarifer birnavirus	Blosnavirus	Birnaviridae	dsRNA	Latidae	Marine
Atlantic salmon calicivirus	Salovirus	Caliciviridae	+ssRNA	Salmonidae	Both
Epizootic hematopoietic necrosis virus (EHNV)	Ranavirus	Iridoviridae	dsDNA	Percidae	Freshwater
European catfish virus	Ranavirus	Iridoviridae	dsDNA	Siluridae	Freshwater
Infectious spleen and kidney necrosis virus (ISKNV)	Megalocytivirus	Iridoviridae	dsDNA	Sparidae, Carangidae, Scombridae, Scaridae, etc.	Marine
Barfin flounder nervous necrosis virus (BFNNV)	Betanodavirus	Nodaviridae	+ssRNA	Pleuronectidae, Gadidae,	Marine
Infectious salmon anemia virus	Isavirus	Orthomyxoviridae	−ssRNA	Salmonidae	Both
Atlantic salmon Aquaparamyxovirus	Aquaparamyxovirus	Paramyxoviridae	−ssRNA	Salmonidae	Both
Bluegill picornavirus	Limnipivirus	Picornaviridae	+ssRNA	Centrarchidae	Freshwater
Atlantic salmon swim bladder sarcoma virus (SSSV)	Epsilontretrovirus	Retroviridae	RT	Salmonidae	Both
Viral hemorrhagic septicemia virus (VHSV)	Novirhabdovirus	Rhabdoviridae	−ssRNA	Numerous (50+ marine and freshwater families)	Both (widely distributed)

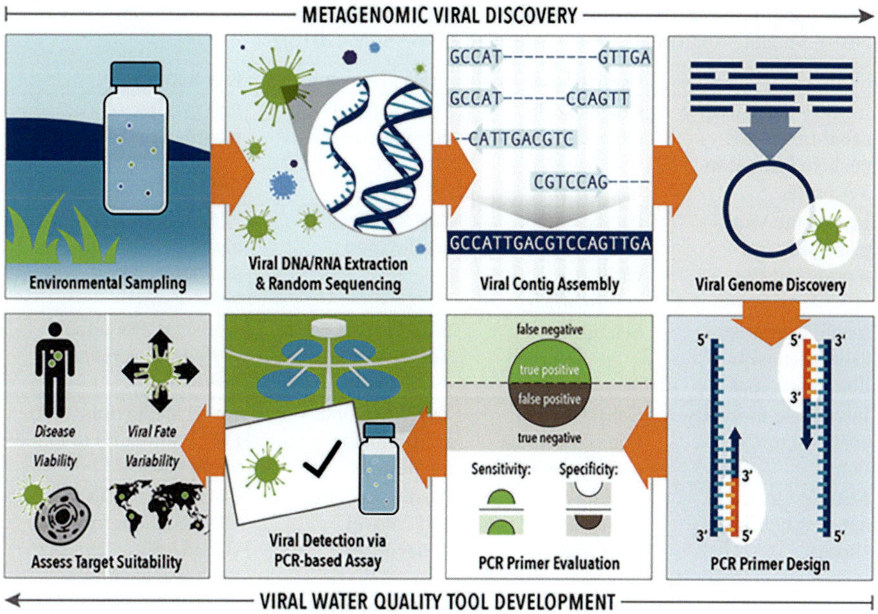

Fig. 9.3 Illustration of metagenomic viral DNA discovery

9.6 Significance of Aquatic Viromes in Fisheries

Viruses such as Nodavirus, Iridovirus, and Koi herpesvirus cause significant financial losses by infecting seafood such as shellfish and fish (Volpe et al. 2023). Metagenomic research aids the detection of new viral hazards by enabling quick and timely responses (Macedo et al. 2024). Viral infections have an indirect impact on the well-being of fish health by altering the host-associated microbiota (Munn 2006). Viral predation upon harmful microorganisms may contribute to preventing outbreaks of infectious diseases. However, specific viral strains may upset the microbiological balance and increase vulnerability to diseases (Elbehery and Deng 2022) (Table 9.3).

9.7 Implications of Virome Research for Fisheries Management

Metagenomics facilitates the fast detection of many viral infections, enabling early warning in the aquaculture sector (Alavandi and Poornima 2012). These approaches are essential for reducing large-scale losses in aquaculture operations. Hence, bacteriophages enable a more sustainable option toward antibiotics to combat

Table 9.2 Discovery of viruses in aquatic species using metagenomic tools (Palermo et al., (2019), Munang'Andu et al., (2017), Alavandi & Poornima (2012))

Study	Host species	Virus discovered	Key findings
Identify the causative agent from CPE-positive filtrates in Vero cells	Ringed seals (*Phoca hispida*)	SePV-1 (Seal picornavirus 1)	Full genome of a novel picornavirus identified
Investigate viral pathogens in a mortality event	Pacific harbor seals (*Phoca vitulina richardsii*)	SealAV (Seal anellovirus)	Novel anellovirus clustered with sea lion and domestic cat anelloviruses
Identify viruses associated with fibropapillomas	Sea turtles	STTV1 (Sea Turtle Tumor Virus 1)	Novel ssDNA virus (~1800 nt), circular genome
Investigate mortality event in captive sea lions	California sea lions (*Zalophus californianus*)	ZcAV (Zalophus californianus anellovirus)	Novel anellovirus, 35% ORF1 similarity to feline anelloviruses
Identify viruses in apparently healthy shrimp	Shrimp (*Penaeus duorarum*)	Novel circovirus and 2 novel nodaviruses	Circovirus: ssDNA, 1955 nt; Noda viruses shared <60% identity to known nodaviruses
Compare viral communities in healthy vs diseased corals	Coral (*Diploria strigosa*)	Herpes-like viruses, cyanophages, vibriophages	Viral communities varied with health status; presence of algae- and plant-like viruses
Characterize marine viral ecology across ecosystems	Multiple marine animals, environments	Diverse viromes with overexpressed sulfur metabolism genes	Viromes differ by habitat; fish-associated viromesare enriched in sulfur metabolism genes
Characterize uncultured viral communities offshore	Unspecified marine hosts	dsDNA-tailed phages, algal viruses	>65% of sequences had no significant match to known viruses
Analyze marine sediment viromes	Marine sediment ecosystems	dsDNA lysogenic phages	Dominated by lysogenic dsDNA phages
Analyze viral diversity in four oceanic regions	Various marine hosts	Cyanophages, ssDNA phages (new clade), prophages	Regional variation with unique viral groups in the Sargasso Sea and Arctic
Characterize marine RNA viruses	Unknown marine organisms	Unspecified marine RNA viruses	Metagenomic characterization of RNA virus diversity in ocean waters

infections caused by bacteria such as *Vibrio* outbreaks, and phage-based therapies tend to be ecologically friendly as well as effective in reducing antibiotic resistance (Fiedler et al. 2023). Viromes act as biomarkers of viral diversity change, indicate environmental stresses like pollution and changing climates, and may detect water quality changes, letting fishermen and managers implement remedial action (Fig. 9.4) (Bibby et al. 2019).

Table 9.3 Illustration of major aquatic viromes affecting wild and cultured species in aquatic environments (Costa & Holmes (2024), Speare (2006), Kim & Leong (2004))

Virus name	Affected species	Impact on fisheries	Geographical distribution	Transmission	Symptoms/ disease
White spot syndrome virus (WSSV)	Shrimp (penaeids) and crabs	High mortality in shrimp aquaculture	Global (tropical and subtropical regions)	Waterborne consumption of infected tissues	White spots on the exoskeleton, high mortality
Tilapia lake virus (TiLV)	Tilapia	Economic losses in tilapia aquaculture	Asia, Africa, Latin America	Waterborne, direct contact	Lethargy, skin lesions, and organ
Koi herpesvirus (KHV)	Common carp, koi	High mortality rates in carp farming	Asia, Europe, North America	Direct contact, waterborne	Gill necrosis, lethargy, and skin lesions
Infectious hematopoietic necrosis virus (IHNV)	Salmonids (e.g., trout and salmon)	Mortality in juveniles, economic losses	North America, Europe, Asia	Horizontal (waterborne), vertical	Necrosis of hematopoietic tissues, and high mortality
Viral hemorrhagic septicemia virus (VHSV)	Salmonids, cod, herring, and pike	Mass fish kills, trade restrictions	North Atlantic, Pacific, and Europe	Waterborne	Hemorrhages in internal organs, skin, and muscles
Yellow head virus (YHV)	Shrimp (*Penaeus monodon*)	High mortality in shrimp aquaculture	Southeast Asia	Waterborne, direct contact	Yellowing of the cephalothorax, high mortality
Betanodavirus (viral nervous necrosis—VNN)	Groupers, seabass, and cod	High mortality in marine finfish aquaculture	Asia, the Mediterranean, and Europe	Waterborne, vertical	Neurological symptoms, abnormal swimming, and mortality
Iridoviruses (Megalocyto viruses)	Grouper, tilapia and ornamental fishes	Severe mortality and trade losses	Asia, Australia, and South America	Waterborne	Enlarged cells (megalocytosis) and high mortality

9.8 Challenges

Considering its immense potential for virome studies in aquaculture, various challenges pose obstacles to its broad applications. One of the major issues involves the complex nature and broad spectrum of aquatic viromes due to the unavailability of any universal genetic markers, which renders it difficult to link viruses to a specific host. Another key impediment lies in the limits of metagenomic approaches, such as biases in the technology for sequencing, difficulties in assembling fragmented viral

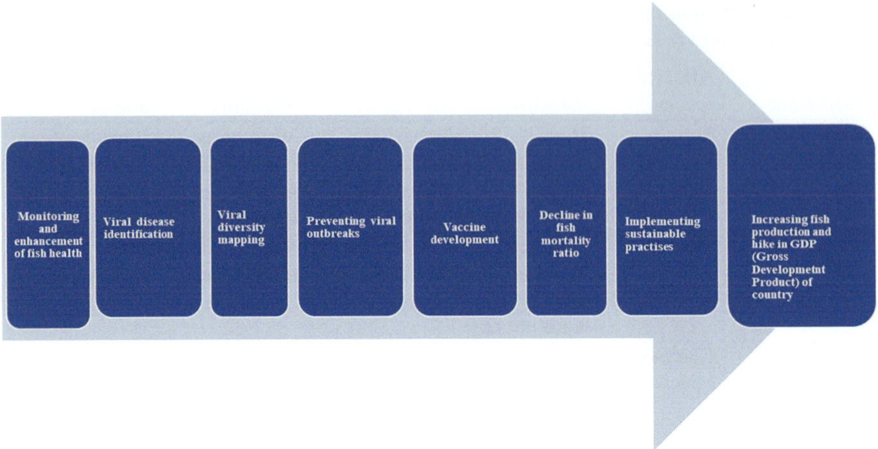

Fig. 9.4 A flow diagram describing the impact of virome research in fishery management and the resulting improvement in the ecosystem's biosecurity

genomes, and a lack of complete viral reference databases. Even sequencing and computational biology are highly expensive and demand specialized expertise. Lastly, various legal and ethical norms make it difficult to apply virome research for disease management. Unfortunately, a significant proportion of viral genomes are still unidentified, which restricts our knowledge about their ecological significance and effects on the fishing industry. Metagenomics offers powerful insights into fish virology, but it also encounters several challenges. High-throughput sequencing requires advanced, costly computational tools such as *DIAMOND*, *UBLAST*, and *Kaiju*. Genome assembly is complicated by the risk of generating chimeric sequences, even with modern assemblers like *MetAMOS*, *MetaVelvet*, and *MetaSPAdes*, often requiring manual correction. Inaccurate or poorly curated reference databases can mislead analysis, and larger databases slow down processing. Interpretation is further limited by surveillance bias, as studies often focus only on clinical samples, overlooking asymptomatic or environmental reservoirs. This restricts our understanding of viral diversity and hampers early outbreak prediction.

9.9 Conclusion and Future Prospects

Aquatic viromes are crucial elements of the aquatic environment's stability, and studying viral diversity and disease management. Metagenomic research has transformed our knowledge concerning these viromes, providing insights into their involvement with the ecology of bacteria, the functioning of ecosystems, and disease dynamics. In addition, environmental surveillance implementing viromics could prove essential for monitoring water quality, foreseeing disease outbreaks, and studying the impacts of climate change on the aquatic virus population. Considering current obstacles, which include sequencing limits, data alignment, regulatory limits, ongoing research alliances, better viral databases, and technical breakthroughs, we will lead the way toward better applications of virome research in aquaculture. These findings have the potential to transform the fishing industry by promoting sustainable management and innovations. Ongoing research, along with cross-disciplinary collaboration, shall prove necessary for realizing the full potential of aquatic viromes for fisheries. The upcoming decade of virome research in aquaculture seems promising, especially with emerging technology and increasing emphasis on sustainable aquaculture approaches. Long-read sequencing (Nanopore, PacBio) and metagenomic deep sequencing are intended for more extensive viral genome assembly and novel aquatic virus identification. These advancements, when integrated with artificial intelligence (AI) and machine learning, will boost viral analysis, thereby enabling real-time monitoring of virome dynamics and earlier disease diagnosis. Establishing complete viral genomics databases exclusively for aquaculture and fisheries will boost detection and functional annotations. Furthermore, phage therapy and viral biocontrol technologies have recently emerged as substitutes for antibiotics, enabling tailored remedies to infections by bacteria while decreasing antimicrobial resistance in the aquaculture sector. Aquatic viruses have been employed for numerous applications: vaccine development, bioengineering, and enzyme synthesis. Integrating virome research studies in fisheries management strategies will support sustainable aquaculture, increase farm output, prevent disease transmission, and, as a result, promote conservation of the environment.

References

Abbadi, M., Rosani, U., et al. (2021) Tracing the invertebrate herpesviruses in the global sequence datasets. Frontiers in Marine Science. https://doi.org/10.3389/fmars.2023.115974

Alavandi SV, Poornima M (2012) Viral metagenomics: a tool for virus discovery and diversity in aquaculture. Indian J Virol 23:88–98

Belton, B., Little, D. C., Zhang, W., Edwards, P., Skladany, M., & Thilsted, S. H. (2020) Farming fish in the sea will not nourish the world. Nat Commun 11(1):5804

Bibby K, Crank K, Greaves J, Li X, Wu Z, Hamza IA, Stachler E (2019) Metagenomics and the development of viral water quality tools. NPJ Clean Water 2(1):9

Blanco-Rodriguez G and Di Nunzio F (2021) The viral capsid: a master key to access the host nucleus. Viruses 13(6):1178

Bootland LM, Leong JAC (2011) Infectious haematopoietic necrosis virus. In Fish diseases and disorders. Volume 3: viral, bacterial and fungal infections (pp. 66-109). Wallingford UK: CABI

Costa VA, Holmes EC (2024) Diversity, evolution, and emergence of fish viruses. J Virol 98(6):e00118–e00124

Edwards RA, Rohwer F (2005) Viral metagenomics. Nat Rev Microbiol 3: 504–510. *View Article*

Elbehery AH, Deng L (2022) Insights into the global freshwater virome. Front Microbiol 13:953500

Fang J, Deng Y, Wang J, Tao LI, Chen Z, Zhou L, Zhang Q (2016) Pathological changes in acute viral crucian gill hemorrhagic disease. Chin J Fish Sci 23(2):336–343

Fiedler AW, Gundersen MS, Vo TP, Almaas E, Vadstein O, Bakke I (2023) Phage therapy minimally affects the water microbiota in an Atlantic salmon (Salmo salar) rearing system while still preventing infection. Sci Rep 13(1):19145

Fuhrman JA (1999) Marine viruses and their biogeochemical and ecological effects. Nature 399(6736):541–548

Gui JF (2024) Chinese wisdom and modern innovation of aquaculture. Water Biol Secur 3(3):100271
Gui J, Zhu Z (2012) Molecular basis and genetic improvement of economically important traits in aquaculture animals. Chin Sci Bull 57(15):1751–1760
Gui JF, Zhang QY (2018) Viruses in marine ecosystems: from open waters to coral reefs. In: Weynberg KD, Leong PSH, Evans MN (eds) Viruses in marine ecosystem (pp. 13–35). Academic Press
Hurwitz BL, Sullivan MB (2013) The Pacific Ocean Virome (POV): a marine viral metagenomic dataset and associated protein clusters for quantitative viral ecology. PLoS One 8(2):e57355
Jacquet S, Miki T, Noble R, Peduzzi P, Wilhelm S (2010) Viruses in aquatic ecosystems: important advancements of the last 20 years and prospects for the future in the field of microbial oceanography and limnology. Adv Oceanogr Limnol 1(1):97–141
Ke X, Kim J, Santos G (2021) Viromes of human-managed aquatic systems: compositional structure and ecological implications. Front Microbiol 12:671925. https://doi.org/10.3389/fmicb.2021.671925
Kibenge FS (2019) Emerging viruses in aquaculture. Curr Opin Virol 34:9–103
Kim CH, Leong JA (2004) Fish viruses. Ency Virol 558
Klemetsen T, Raknes IA, Fu J, Agafonov A, Balasundaram SV, Tartari G et al (2018) The MAR databases: development and implementation of databases specific for marine metagenomics. Nucleic Acids Res 46(D1):D692–D699
Kostic AD, Ojesina AI, Pedamallu CS, Jung J, Verhaak RG, Getz G, Meyerson M (2011) PathSeq: software to identify or discover microbes by deep sequencing of human tissue. Nat Biotechnol 29(5):393–396
Lodish HF, Berk A, Zipursky SL, Matsudaira P, Baltimore D, Darnell J (2000) Photosynthetic stages and light-absorbing pigments. Molecular Cell Biology, 4th edn. W. H. Freeman, New York
Lostroh P (2024) Molecular and cellular biology of viruses. CRC Press
Lundberg JG, Kottelat M, Smith GR, Stiassny ML, Gill AC (2000) So many fishes, so little time: an overview of recent ichthyological discovery in continental waters. Ann Missouri Bot Gard 2–62
Macedo DB, dos Anjos TMC, De Los Santos EFF, Rodrigues MDN, Alegria OVC, Ramos RTJ (2024) New perspectives on metagenomic analysis for pathogen monitoring in sustainable freshwater aquaculture production: a systematic review. Front Freshw Sci 2:1459233
Mordecai GJ, Miller KM, Di Cicco E, Schulze AD, Kaukinen KH, Ming TJ, ... Suttle CA (2019) Endangered wild salmon infected by newly discovered viruses. Elife 8:e47615
Mugimba KK, Byarugaba DK, Mutoloki S, Evensen Ø, Munang'andu HM (2021) Challenges and solutions to viral diseases of finfish in marine aquaculture. Pathogens 10(6):673
Munang'Andu HM, Mugimba KK, Byarugaba DK, Mutoloki S, Evensen Ø (2017) Current advances on virus discovery and diagnostic role of viral metagenomics in aquatic organisms. Front Microbiol 8:406
Munn CB (2006) Viruses as pathogens of marine organisms—from bacteria to whales. J Mar Biol Assoc U K 86(3):453–467
Naylor RL, Hardy RW, Buschmann AH, Bush SR, Cao L, Klinger DH, ... Troell M (2021) A 20-year retrospective review of global aquaculture. Nature 591(7851):551–563
Oidtmann B, Dixon P, Way K, Joiner C, Bayley AE (2018) Risk of waterborne virus spread–review of survival of relevant fish and crustacean viruses in the aquatic environment and implications for control measures. Rev Aquacult 10(3):641–669
OIE (2009) Chapter 2.4.6: Bovine tuberculosis. Terrestrial Manual. World Organization for Animal Health, Paris
Palermo CN, Fulthorpe RR, Saati R, Short SM (2019) Metagenomic analysis of virus diversity and relative abundance in a eutrophic freshwater harbour. Viruses 11(9):792
Priya TJ, Kappalli S (2022) Modern biotechnological strategies for vaccine development in aquaculture–prospects and challenges. Vaccine 40(41):5873–5881

Rowley AF, Baker-Austin C, Boerlage AS, Caillon C, Davies CE, Duperret L, ... Coates CJ (2024) Diseases of marine fish and shellfish in an age of rapid climate change. Iscience 27(9)
Seshadri R, Kravitz SA, Smarr L, Gilna P, Frazier M (2007) CAMERA: a community resource for metagenomics. PLoS Biol 5(3):e75
Speare DJ (2006) Aquatic animal health code. Can Vet J 47(12):1217. https://doi.org/10.3389/fmars.2023.1159754
Sriwanayos P, Subramaniam K, Stilwell NK, Imnoi K, Popov VL, Kanchanakhan S, ... Waltzek TB (2020) Phylogenomic characterization of ranaviruses isolated from cultured fish and amphibians in Thailand. Facets 5(1):963–979
Suttle CA (2013) Viruses: unlocking the greatest biodiversity on Earth. Genome 56(10):542–544
Thawornwattana Y, Dong HT, Phiwsaiya K, Sangsuriya P, Senapin S, Aiewsakun P (2021) Tilapia lake virus (TiLV): genomic epidemiology and its early origin. Transbound Emerg Dis 68(2):435–444
Volpe E, Errani F, Mandrioli L, Ciulli S (2023) Advances in viral aquatic animal disease knowledge: the molecular methods' contribution. Biology 12(3):466
Walker PJ, Bigarré L, Kurath G, Dacheux L, Pallandre L. (2022). Revised taxonomy of rhabdoviruses infecting fish and marine mammals. Animals (Basel) 12:1363. https://doi.org/10.3390/ani12111363
Wang J, Beusen AH, Liu X, Bouwman AF (2019) Aquaculture production is a large, spatially concentrated source of nutrients in Chinese freshwater and coastal seas. Environ Sci Technol 54(3):1464–1474
Weynberg KD (2018) Viruses in marine ecosystems: from open waters to coral reefs. Adv Virus Res 101:1–38
Winter C, Garcia JA, Weinbauer MG, DuBow MS, Herndl GJ (2014) Comparison of deep-water viromes from the Atlantic Ocean and the Mediterranean Sea. PLoS One 9(6):e100600
Wu LY, Wijesekara Y, Piedade GJ, Pappas N, Brussaard CP, Dutilh BE (2024) Benchmarking bioinformatic virus identification tools using real-world metagenomic data across biomes. Genome Biol 25(1):97
Zhang Q, Gui JF (2015) Virus genomes and virus-host interactions in aquaculture animals. Sci China Life Sci 58(2):156–169
Zhang W, Duan C, Zhang H, Weng S, He J, Dong C. (2020) Widespread outbreaks of the emerging mandarin fish ranavirus (MRV) both in natural and ISKNV-FKC vaccinated mandarin fish Siniperca chuatsi in Guangdong, South China, 2017. Aquaculture 520:734989. https://doi.org/10.1016/j.aquaculture.2020.734989
Zhang QY, Ke F, Gui L, Zhao Z (2022) Recent insights into aquatic viruses: emerging and reemerging pathogens, molecular features, biological effects, and novel investigative approaches. Water Biol Secur 1(4):100062

Bioinformatics Tools for Aquaculture Disease Management

10

Arpita Roy, Kampan Bisai, Pratyasha Bhowal,
Rajat Banerjee, and Basanta Kumar Das

10.1 Introduction

Bioinformatics, an interdisciplinary science that combines biology, computer science, mathematics, and information technology, has evolved into an essential tool of contemporary life sciences. Its growth has been propelled by monumental advances in high-throughput sequencing, genome editing, molecular modeling, and computational power. Today, bioinformatics provides the critical infrastructure required to decode complex biological systems from molecular to ecological scales by enabling the storage, analysis, and interpretation of vast and diverse biological datasets. Bioinformatics has evolved from a supporting tool to a crucial framework for precision biology, systems immunology, and ecosystem-level diagnostics in aquatic life (Vinithkumar 2006).

Aqua-informatics, an emerging field that adapts bioinformatics techniques and methodologies for use in aquaculture, is one of the most promising by-products of this revolution. Fish health is the primary issue, and aqua-informatics is a game-changer as aquaculture becomes the fastest-growing food production sector in the world. Through cloud-based platforms and real-time monitoring systems,

A. Roy
ICAR-Central Inland Fisheries Research Institute, Barrackpore, Kolkata, India

Department of Biotechnology and Dr. B. C. Guha Centre for Genetic Engineering and Biotechnology, University of Calcutta, Kolkata, West Bengal, India

K. Bisai · B. K. Das (✉)
ICAR-Central Inland Fisheries Research Institute, Barrackpore, Kolkata, India

P. Bhowal · R. Banerjee
Department of Biotechnology and Dr. B. C. Guha Centre for Genetic Engineering and Biotechnology, University of Calcutta, Kolkata, West Bengal, India

© The Author(s), under exclusive license to Springer Nature Singapore Pte Ltd. 2026

B. K. Das et al. (eds.), *Bioinformatics Tools for Predictive Ecology and Fisheries*, Sustainable Aquatic Food Systems, https://doi.org/10.1007/978-981-95-2573-7_10

aqua-informatics integrates biological, environmental, and production data to enable a comprehensive understanding of aquatic health dynamics. More importantly, it facilitates the development of evidence-based strategies that can minimize stress-induced vulnerabilities, particularly during early developmental stages such as hatching and larval transitions, when immune competency is limited (Castro et al. 2016).

Global trends in environmental sustainability and nutrition highlight how urgent it is to improve fish health. As the human population continues to grow at an accelerated rate, there is an increasing demand for high-quality, high-protein foods, especially fish. However, the aquaculture environment poses unique challenges for disease prevention and control. Aquatic systems are natural reservoirs for a multitude of pathogens, as seawater contains over a million bacteria and millions of viruses per milliliter. Fish are constantly exposed to these microbes through direct contact with water, increasing the risk of disease, particularly during periods of physiological stress or environmental fluctuation such as salinity shifts, temperature variation, or high stocking densities (Rodger 2016; Jeffries et al. 2014).

Epidemiological data indicate that bacterial infections are responsible for over half (54.9%) of disease outbreaks in aquaculture, followed by viral (22.6%), parasitic (19.4%), and fungal (3.1%) diseases (Martin and Król 2017). Species such as groupers are particularly vulnerable due to their high commercial value and sensitivity to environmental stressors (Shapawi et al. 2014). Although fish possess innate immune defenses such as mucosal barriers, antimicrobial peptides, and cellular responses these mechanisms are often insufficient against chronic or virulent pathogens (Low et al. 2014; Kordon et al. 2018). Thus, for long-term disease control, it becomes crucial to comprehend and improve adaptive immunity (Smith et al. 2019; Semple and Dixon 2020).

Aquatic health science has seen a revolution because of omics technologies that include proteomics, transcriptomics, metabolomics, and genomics. These multi-omics techniques offer a thorough understanding of immune system modulation, host-pathogen interactions, and the genetic markers of disease resistance and vulnerability. Researchers can find biomarkers for early diagnosis, customize vaccines to target particular disease strains, and track immune responses with previously unheard-of precision by incorporating multi-omics data into bioinformatics pipelines. The secret to transforming fish health surveillance lies in the intersection of aqua-informatics and omics. Aquaculture is moving toward a future where disease outbreaks are not only treated but also expected and prevented in advance thanks to machine learning, digital health platforms, and predictive modeling. Precision aquaculture is a paradigm change that improves productivity, lessens the need for antibiotics, promotes conservation of the environment, and insures a sustainable food source for future generations.

10.2 Bioinformatics Tools and Databases for Aquaculture Health

Bioinformatics tools are specialized software applications designed to collect, manage, retrieve, and analyze vast volumes of biological data, particularly from high-throughput multi-omics platforms such as genomics, transcriptomics, proteomics, metabolomics, and epigenomics. These tools transform unstructured raw datasets into logically organized frameworks, enabling researchers to decipher complex biological phenomena at multiple levels (Khalil and Hill 2005). The foundation of contemporary precision fish health is formed by these technologies, which are essential for comprehending fish physiology, immunological responses, disease development, and environmental adaptation (Fig. 10.1).

10.2.1 Genomics and Genetic Insights in Fish Health

Aquaculture and fisheries have been transformed by genomics, which offers comprehensive insights into the genetic makeup of fish species. Comprehensive assessments of genetic diversity, population structure, and adaptive aspects in both wild and cultivated fish stocks are made possible by high-throughput sequencing methods when combined with strong bioinformatics platforms. Finding genetic markers associated with environmental tolerance, growth efficiency, and disease resistance requires the use of genomic analyses. In aquaculture, genomics facilitates marker-assisted and genomic selection for features including enhanced feed conversion, faster growth, and resistance to pathogens, hence supporting selective breeding efforts. Genome-wide association studies (GWAS), for instance, are useful in identifying loci linked to resistance to bacterial or viral infections. Additionally, by identifying decreases in genetic variability brought on by overfishing, habitat degradation, or climate change, genomic monitoring of wild populations promotes conservation and helps create marine protected zones and restocking plans.

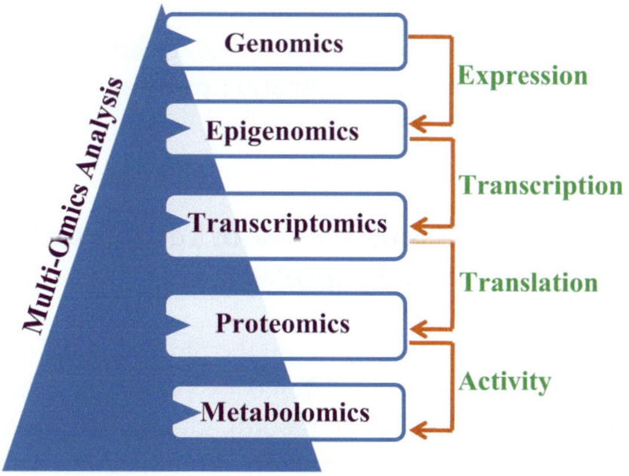

Fig. 10.1 Bioinformatics tools for aquaculture research

10.2.2 Epigenomics: Non-genetic Influences on Fish Health

The study of heritable, reversible changes in how genes are expressed that do not require modifications to the DNA sequence is known as epigenomics. Fish development, immunological responses, stress tolerance, and metabolic adaptation are among the vital physiological processes that are regulated by these alterations, which include DNA methylation, histone modification, and chromatin remodeling. Environmental elements such as temperature, salinity, hypoxia, and pathogen exposure can cause epigenetic modifications in aquaculture that affect the health and efficiency of fish. Early exposure to adverse conditions, for instance, might epigenetically trigger the immune system or stress response pathways, resulting in phenotypic impacts that endure a lifetime. These epigenetic alterations in response to environmental stressors are mapped using sophisticated techniques like whole-genome bisulfite sequencing and ChIP-seq (Chromatin Immunoprecipitation Sequencing). In addition to genetic data, epigenomic markers are becoming important predictive indicators in selective breeding. Furthermore, non-genetic therapies, such as probiotics, functional diets, or hatchery environmental modification, may be developed to epigenetically improve development and immunity in aquaculture stocks because many epigenetic modifications are reversible.

10.2.3 Transcriptomics: Gene Expression in Disease and Immunity

The comprehensive analysis of the RNA transcriptome, or the entire collection of RNA molecules expressed in a specific organ or cell type, is known as transcriptomics. Transcriptomic profiling is used in fish health research to show how gene expression changes in response to stress, infection, or immunization. Differentially expressed genes (DEGs) linked to tissue healing, inflammatory signaling, and immunological activation can be found using methods like RNA-Seq. This aids in the knowledge of host-pathogen interactions, immune status monitoring, and the identification of new biomarkers for disease diagnosis. For instance, after a viral infection, fish may exhibit an increase in interferon-stimulated genes (ISGs), which may be a sign of innate antiviral defensive mechanisms. By identifying antigen targets and immune-modulating pathways, transcriptomic data also aid in the creation of vaccines.

10.2.4 Proteomics: Protein Level Insights in Fish Disease Response

A quantitative perspective on the physiological condition of fish during both normal and diseased states is offered by proteomics, the extensive study of proteins and their roles. Proteomics records changes in protein abundance, post-translational modifications, and interactions in immediate detail, as opposed to transcriptomics and genomics, which predict possible responses. Protein signatures linked to immunological pathways, stress indicators, and metabolic changes during infection have been identified by proteomic analysis, especially those based on mass spectrometry. For example, variations in the quantity of complement proteins, heat shock proteins, or lysozymes can be used as indicators of environmental stress or immunological activity. However, for a thorough understanding, proteomic data might not be enough on their own. Proteomics allows for multi-dimensional analysis of host responses when combined with transcriptomics and metabolomics, providing a greater understanding of resilience mechanisms and fish-pathogen interactions.

10.2.5 Metabolomics

Like the other omics, metabolomics has grown quickly and is seen to be a key tool for identifying and explicating complicated phenotypes in a variety of biological systems. Metabolomics is a complementary technology that is commonly used in conjunction with transcriptomics and proteomics, as it is a downstream consequence of gene and protein expression. Metabolomics uses instruments to screen a range of samples for chemical hazards in a high-throughput fashion, connecting the intricate biological processes and explaining the results of any deviation through combining in vitro systems with in vivo animal biology (Mohanty et al. 2019). For more significant effects, the tools must be user-friendly and available online to the global scientific research community. However, according to Danish et al. (2017), these tools can be loosely divided into four groups:

10.2.5.1 Homology and Related Tools

These techniques determine structure, function, and evolutionary lineage by comparing query sequences such as an unknown gene or protein from a fish with known sequences in global databases (including GenBank, UniProt, and PDB).

By comparing unknown genes to databases of model fish or mammals, we can find novel immune genes or antimicrobial peptides in species such as salmon or tilapia. Also, by comparing bacterial or viral sequences to recognized toxins or immune evasion proteins, pathogen virulence factors can be identified. It also discovers the evolutionary conservation of immunological receptors like TLRs and MHC in fish species.

10.2.5.2 Protein Function Analysis

In order to predict a protein's function in physiological processes, including immunity, metabolism, and cell signaling, protein function tools examine motifs, domains, signal peptides, and structural features in protein sequences, for example, Phobius, SignalP, TMHMM, and InterProScan (Pfam, PROSITE, and SMART). These tools sort or separate unidentified proteins that are expressed during infection into chemokines, cytokines, or inflammatory response-related enzymes, and also by locating conserved domains in viral surface proteins and vaccination antigens. These tools examine secreted proteins that support barrier immunity in serum or mucus.

10.2.5.3 Structural Analysis

Structural bioinformatics techniques show the mechanism of action, binding locations, and interaction partners of proteins by predicting or comparing their three-dimensional conformation. Function at the molecular level is informed by structure, especially in cases when sequencing data is unclear. For instance, Chimera, PyMOL, I-TASSER, Phyre2, SWISS-MODEL, and others. These tools inform vaccine development and model fish antibody fragments (such as IgM and IgT) or T-cell receptors. These tools help by using pathogen-host protein docking simulation to investigate how bacterial or viral effectors take over host machinery, and also help in finding sites for structure-based drug design, aquaculture pathogens.

10.2.5.4 Sequence Analysis

Sequence analysis tools offer an extensive examination of DNA or protein sequences to reveal regulatory elements, functional sections, evolutionary history, and the effects of mutations. For instance, ProtScale (ExPASy), Clustal Omega, MAFFT, CpGPlot/CpGReport, MEGA (Molecular Evolutionary Genetics Analysis), help in phylogenetic analysis of immune-related genes in wild and farmed fish populations to track introgression or genetic drift. These also Locate CpG islands, which are areas rich in cytosine-guanine dinucleotides and are used to investigate how immune or stress response genes are epigenetically regulated. Find harmful mutations in genes related to metabolic adaptation or illness resistance, anticipate membrane-spanning domains, which are essential for receptor and transporter functions, and analyze hydropathy plots.

10.3 Some Commonly Used Bioinformatics Tools in Fisheries and Aquaculture

10.3.1 BLAST (The Basic Local Alignment Search Tool)

A fundamental method for comparing fish nucleotide or protein sequences to publicly accessible reference libraries is called BLAST. It assists in annotating unknown genes or proteins and identifies areas of similarity. These tools help in immune-related gene sequencing, fish genomes, such as interleukins, Toll-like receptors, or antimicrobial peptides, and also to identify genes relevant to aquaculture pathogens such as *Vibrio* spp. or *Aeromonas hydrophila,* bacterial toxins, or viral capsid proteins.

- PSI-BLAST (Position-Specific Iterated): Used to find immunological protein homologs that are far away.
- PHI-BLAST: For identifying proteins in immune signaling pathways that have conserved motifs.
- BLAST 2 Sequences: Used to compare gene variations between fish with disease and those in good condition.

10.3.2 FASTA (FAST-All)

An algorithm for quick sequence matching compares nucleotide or peptide query sequences with a sequence database.

Uses in the Health of Fish:

- To identify immune genes that are differently expressed, compare transcriptome data from healthy and infected tissues.
- Determine which species have functional analogs of stress-related proteins or cytokines (e.g., from zebrafish to catfish or tilapia).
- Use marker genes or DNA barcodes to quickly scan ambient or gut microbiome samples for pathogens.

10.3.3 EMBOSS (The European Molecular Biology Open Software Suite)

A new, open-access software analysis program includes approximately 100 programs (applications) for tasks such as codon usage analysis for small genomes, protein motif identification and domain analysis, nucleotide sequence pattern analysis, sequence alignment, database searching with sequence patterns, and others.

Uses in the Health of Fish:

- Codon usage analysis: Take into account the target fish species' codon bias when optimizing synthetic gene sequences
- Motif identification: Look for conserved domains in MHC molecules or antimicrobial peptides.
- Pattern searches: Look for transcription factor binding sites in immune gene regulatory areas.
- Phylogenetic tree generation: Gain insight into how genes that fight disease have evolved among aquaculture species.

10.3.4 ClustalW

A common purpose multiple sequence alignment program for DNA or proteins determines the best match for the selected sequences and compares and contrasts them to visualize their identity in order to provide physiologically important multiple sequence alignments of divergent sequences.

Applications in Fish Health Include:

- Tracking mutation hotspots influencing virulence or vaccination efficacy by aligning bacterial or viral strains.
- To assess genetic diversity, compare immune genes (such as interferons and immunoglobulins) among species or groups.
- To track the origins of outbreaks or evolutionary links, support phylogenetic investigations of fish diseases.

10.3.5 RasMol and Protein Explorer (Derivative of RasMol)

Commonly used to display proteins, DNA, and smaller molecules structurally. Some other bioinformatics tools used in visualization process such as Cn3D, Swiss PDB viewer, Hex, Vega, Bioeditor, Bioviewer, and Chime.

10.4 Bioinformatics Tools Used in Database Management

Some bioinformatics tools are commonly used to maintain various database, including genomics or transcriptomics, proteomics, and high-throughput analysis which are summarized in Table 10.1

10.5 Utilization of Bioinformatics in Aquaculture Health

Modern aquaculture has been transformed by bioinformatics, which offers strong computational frameworks for managing, analyzing, and interpreting the constantly increasing amount of biological data produced by omics technologies, high-throughput sequencing, and diagnostic tools. In addition to organizing and curating vast amounts of biological data, its main goal is to create specialized databases, software, and algorithms that bring forward biological innovations that are pertinent to aquatic creatures. Bioinformatics is becoming more and more important in the context of aquaculture health for examining immune system regulation, host-pathogen interactions, environmental stress responses, and disease resistance mechanisms (Fig. 10.2). By enabling aquaculture researchers to evaluate genomic, transcriptomic, proteomic, and metabolomic data in biologically meaningful ways, these computational tools promote evidence-based decision-making in the management of fish health (Table 10.2).

10.5.1 Disease Diagnosis and Aquatic Health Management

Molecular and DNA-based diagnostic techniques combined with bioinformatics tools have made it possible to diagnose fish diseases early, before clinical signs manifest. By focusing on certain genetic sequences, sequencing the entire genome of infectious agents, or utilizing high-throughput technologies like nucleic acid microarrays to examine DNA and gene expression patterns, these sophisticated methods make it possible to identify infections (Altinok and Kurt 2003). A number of high-impact pathogens in aquaculture, such as *Renibacterium salmoninarum* (the cause of bacterial kidney disease), infectious hematopoietic necrosis virus (IHNV),

Table 10.1 Bioinformatics tools used in "Omics" database management (Mohanty et al. 2019)

Frequently used bioinformatics tools	Databases
Basic Local Alignment Search Tool (BLAST): Analyzes databases with a large number of preserved sequences against nucleotide or protein sequences. Primer-BLAST: Utilizes Primer 3 to create PCR primers that match an amino acid template. Splign: Utilized to generate alignments between cDNA and genomic sequences. GenBank: BankIt/Sequin: Both are online tools for submitting sequences to the GenBank database. Sequence Read Archive (SRA) Submission: Contains sequencing data from next-generation sequencing platforms, such as the Life Technologies AB SOLiD System, Roche 454 GS System, and Illumina.	Genomics/transcriptomics
Translate: Converts a sequence of nucleotides into a series of proteins. FindMod: Estimate possible post-translational changes in proteins. Mascot: MS/MS ion search and sequence query from London-based Matrix Science Ltd. ProtParam: The physical-chemical characteristics of a protein sequence, such as its atomic and amino acid compositions, extinction coefficient, and isoelectric point. ScanProsite: Checks an arrangement against the UniProt Knowledgebase or the target sequence against PROSITE. Protein Data Bank (PDB) Archive: Data regarding the three-dimensional forms of complex assemblies, proteins, and nucleic acids. MS-Fit: Identifying a protein in a sequence database that best matches the mass spectrometry data.	Proteomics
Kyoto Encyclopedia of Genes and Genomes (KEGG) Pathway: A group of databases used for translational research, systems biology modeling and simulation, metagenomics, metabolomics, and other thematic investigations. Biocarta.	High-throughput analysis

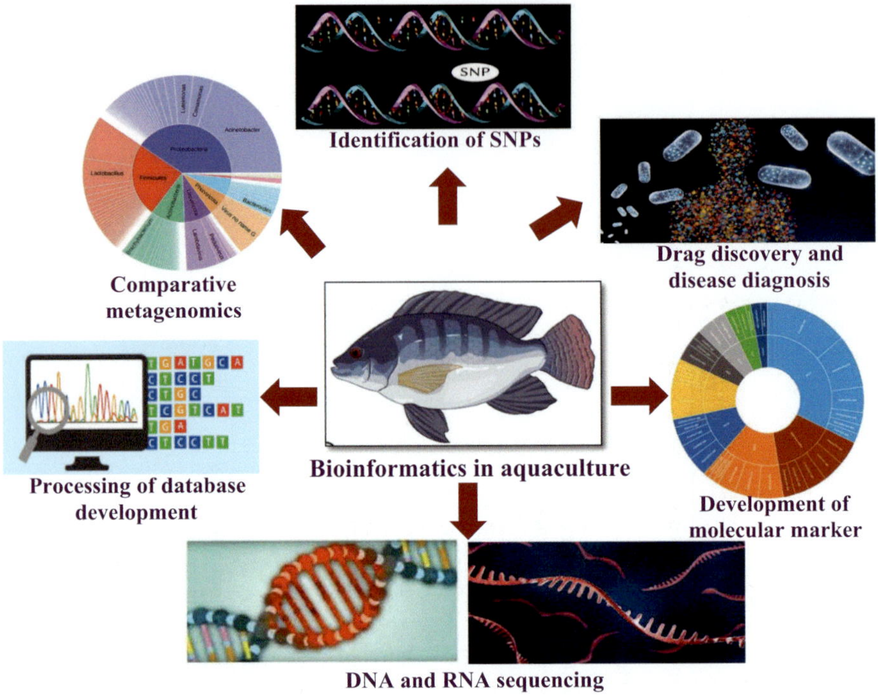

Fig. 10.2 Application of bioinformatics in fisheries and aquaculture

Table 10.2 Sources and utilization of bioinformatics data in different subject areas (Kamble and Khairkar 2016)

Data Sources	Research Areas
Genomes	1. Phylogenetic analysis 2. Analyzing the connections between particular genes and disease 3. Description of the metabolic processes involving protein content 4. Structure-based gene assignments
Raw DNA sequence	1. Determination of exons and introns 2. Keeping code and non-coding areas apart 3. The study of forensics 4. Predicting gene products
Protein sequence	1. Different techniques for aligning sequences 2. Algorithms for sequence comparison 3. Finding the preserved sequence regions
Gene expression	1. Connecting transcriptional data to biochemical, structural, and sequence data 2. Connecting patterns of expression
Macromolecular structure	1. Characterization of protein geometry 2. Secondary, tertiary structure, and 3D structural alignment algorithms 3. The interactions between molecules

viral hemorrhagic septicemia virus (VHSV), and viral nervous necrosis virus (VNNV), have currently been successfully detected using bioinformatics-assisted DNA detection techniques.

- *Nucleic acid vaccine development*, where computational tools are used to identify antigenic regions of pathogen genomes, boosts host immunity through targeted immunization strategies.
- *Identification of stress-resilient proteins*, which support enhanced physiological responses in fish exposed to environmental fluctuations or disease challenges.
- In silico *drug discovery*, where molecular modeling and docking tools facilitate the design of therapeutics that can inhibit specific pathogen proteins or enhance fish immune functions (Viarengo et al. 2007; Banerjee et al. 2017)

10.5.2 Fish Nutritional Aspects

Nutrigenomics, the study of the relationship between nutrition and gene expression, has emerged as an important branch in aquaculture science due to the increased focus on sustainable aquaculture and the growing demand for safe, wholesome, and premium seafood. Nutrigenomics, which is supported by bioinformatics tools and multi-omics platforms, enables researchers to investigate the effects of various feed compositions on aquatic creatures' growth performance, immunological responses, metabolic pathways, and even food safety profiles.

Nutrigenomics provides critical insights into how fish respond at the molecular level to various dietary components. Bioinformatics platforms such as RNA-Seq pipelines, gene ontology (GO) tools, and pathway enrichment software (e.g., KEGG and DAVID) help interpret large-scale transcriptomic data, linking nutritional inputs to health-related outcomes (Ngoh et al. 2015; Ganguly et al. 2018). The integration of nutrigenomics with bioinformatics accelerates the development of species-specific, functional feeds tailored to:

- Enhance growth performance and feed conversion ratios
- Boost your immune system readiness against diseases
- Improve reproductive output and fry quality
- Reduce metabolic disorders linked to unbalanced or high-fat diets

Beyond growth and immunity, bioinformatics is increasingly used to evaluate the nutritional and biochemical quality of edible fish tissue, through identifying nutrient biomarkers (e.g., essential fatty acids, taurine, selenium that influence taste, texture, and health benefits), detecting bioaccumulation of contaminants or metabolic by-products due to poor nutrition, and predicting alterations in muscle composition under varying feeding regimes or stress conditions.

An emerging area of concern is the potential for fish-derived food allergens, particularly those that trigger IgE-mediated hypersensitivity in consumers. Bioinformatics platforms enable the prediction of allergenic peptides using databases like AllergenOnline or AllerTOP, Sequence comparison with known allergens to assess cross-reactivity risks, and also analysis of how feed additives (e.g., soy-based proteins, marine oils) influence the expression of allergenic proteins in fish muscle (Mohanty et al. 2019). This supports regulatory compliance, consumer safety labeling, and the design of hypoallergenic aquaculture products.

Bioinformatics-enhanced nutrigenomics also intersects with environmental sustainability and fish welfare, reducing the reliance on fishmeal and fish oil through nutritionally equivalent alternatives (e.g., insect meal and microalgae), supporting gut health monitoring via meta-transcriptomics of microbiota, and tracking dietary impacts on oxidative stress, inflammation, and gut barrier function.

10.6 Challenges of Bioinformatics Associated Aquaculture Health

Bioinformatics has emerged as the cornerstone for advancements in fish health management, disease diagnosis, vaccine development, and selective breeding in the age of genome-based biotechnology and precision aquaculture. Addressing the complex issues facing contemporary aquaculture requires the capacity to handle, mine, and comprehend large biological datasets, such as those produced by transcriptomics, proteomics, metabolomics, and high-throughput sequencing. However, there are still a number of significant obstacles that prevent the practical application of bioinformatics in aquaculture health, despite its crucial importance (Rana et al. 2020). The absence of bioinformatics knowledge among aquaculture scientists is one of the main drawbacks. The majority of aquaculture professionals, veterinarians, and fish biologists have extensive training in wet-lab methods, but they frequently lack the computational abilities needed to manage complicated datasets. On the other hand, software engineers and informatics specialists might be very skilled in computation, but they might not have the biological knowledge required to analyze molecular data in relation to fish physiology, immunology, or aquaculture systems. The efficient conversion of genetic discoveries into useful health interventions, including the identification of illness biomarkers, the discovery of immunogens, or the modeling of environmental stress in farmed fish populations, is hampered by this interdisciplinary gap.

The need for professionals with a combined expertise in biology and bioinformatics, or "training-scientists," is growing steadily but is still unfulfilled in the majority of countries. Aquaculture organizations and research facilities find it challenging to include bioinformatics workflows in their health management plans due to a lack of cross-trained staff. Finding a balance between learning computational biology and conducting experimental aquaculture research is still a major educational problem for biology students. Consequently, important analyses like metagenomic surveillance of pathogens, gene expression profiling during infections, or multi-omics integration for disease modeling are frequently neglected or outsourced.

10.7 Current Advancements in Infectious Fish Disease Management

One of the most important factors in aquaculture for lowering mortality and financial losses is probably effective disease management. However, because pathogenic threats, especially those brought on by viral agents, are complex and constantly changing, controlling fish infections continues to be a chronic challenge. The increasing mortality rates in intensive fish farming systems around the world have been mostly caused by these viral diseases. To address a variety of pathogenic diseases, aquaculture operations have traditionally depended on the extensive use of antibiotics, chemotherapeutic drugs, and pesticides. Even though these treatments provide temporary respite and have shown some promise in treating bacterial and parasite illnesses, their long-term effects such as antibiotic resistance, environmental contamination, and microbial ecosystem disruption frequently outweigh their benefits. Despite being used constantly, these techniques have not shown long-lasting gains in fish health. It is encouraging to observe that a number of immunologically based disease prevention techniques have emerged in recent years with the goal of improving innate and adaptive immune responses in fish raised for food. Among the two most promising strategies are the following:

- *Vaccination*: Targeted vaccination techniques employing live-attenuated, recombinant, inactivated, or DNA-based vaccines have demonstrated effectiveness in preventing bacterial and viral illnesses in fish like carp, tilapia, and salmon.
- *Natural Immunostimulants*: Substances that come from plants, algae, or microorganisms, like probiotics, essential oils, and β-glucans, have shown immune-boosting qualities that increase resistance to disease, lessen stress, and aid in healing.

10.7.1 Active and Passive Immunization

All vertebrates, including aquatic organisms, use several kinds of innate and adaptive immune responses that protect themselves against infections that are invasive. Both passive and active forms of vaccination are a part of these defense mechanisms and are essential to aquaculture disease prevention efforts.

Passive immunity refers to the temporary immune protection conferred to fish through the external introduction of antibodies or immune factors. This process does not require the host to mount an immune response; instead, it relies on *administered antibodies* to neutralize pathogens already present in the organism. In fish, this mechanism becomes particularly important when individuals are already infected and require rapid immune support to enhance survival. Studies suggest that *tolerance mechanisms* such as the generation of superfluous antibodies can help infected fish manage viral load and inflammation without immediate pathogen elimination (Bailey et al. 2019). However, passive immunity offers *short-term protection* and lacks the immunological memory required for long-lasting disease resistance.

In contrast, *active immunization* involves the stimulation of the host's own immune system through exposure to antigens—usually via *vaccines*—to produce a specific and long-lasting defense against future infections. Vaccination is currently the *most effective strategy* for controlling infectious diseases in aquaculture and for reducing dependency on antimicrobial drugs.

- *First-generation vaccines*, often formulated using inactivated or live-attenuated pathogens, are widely used in aquaculture to combat viral and bacterial diseases.
- *Live and attenuated vaccines* are particularly effective in inducing both *humoral and cellular immune responses*, providing comprehensive protection in species such as salmon, trout, tilapia, and carp.
- Despite early successes, *vaccine efficacy can be inconsistent*, especially against viral pathogens with *high antigenic variability*. The *evolution of pathogen strains*, *production costs*, and *cold-chain maintenance* continue to challenge vaccine development and large-scale implementation in aquaculture systems (Mohd-Aris et al. 2019).

Antibiotics remain commonly used in aquaculture to *treat bacterial infections* and prevent disease outbreaks, especially when vaccination is not feasible. However, their overuse poses significant risks to fish health, environmental integrity, and public safety:

- The *development of antimicrobial resistance (AMR)* in aquatic pathogens threatens treatment efficacy and may compromise human health through zoonotic transmission.
- Residual antibiotics can *accumulate in fish tissues*, raising concerns about *food safety and consumer exposure*.
- Antibiotics can disrupt *microbial biodiversity* in aquatic environments, altering ecosystem function and facilitating the horizontal transfer of resistance genes among microbial communities. The use of multiple-omics technologies for immunization of fish during illness is depicted in Fig. 10.3.

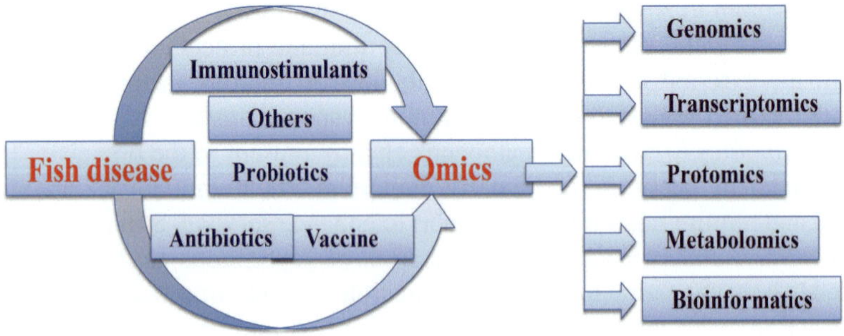

Fig. 10.3 Aquatic health management through omics technologies

10.7.2 Immunostimulants

In aquatic organisms, particularly aquaculture species like *Oreochromis niloticus* (Nile tilapia), immunostimulants, whether synthetic or natural, are essential for boosting both innate and adaptive immune responses. These materials could be chemical molecules, medicinal compounds, or biologically active substances that have been isolated from microbes, fungi, plants, and animals. Their main purpose is to strengthen the body's defenses against infections, which lessens the need for chemical therapies and antibiotics, which may have long-term effects on the environment and public health. Levamisole is a well-known synthetic immunostimulant that has been shown to be useful in aquatic species. It is a deworming medication that has historically been utilized in terrestrial animals. It works by altering immunological processes, such as raising antibody production and macrophage activity. On the other hand, the immunomodulatory properties of β-glucans, which are naturally occurring polysaccharides that are isolated from the cell walls of yeast (*Saccharomyces* spp.), fungi, and cereals, are well known. By stimulating immune cells like neutrophils and macrophages, these chemicals improve the body's ability to fight off bacterial infections like those brought on by *Aeromonas hydrophila*, a prevalent pathogen that causes hemorrhagic septicemia and high mortality rates in freshwater fish.

The activation and amplification of different innate immune system components, which operate as the first line of defense against invasive pathogens, are the main methods by which immunostimulants improve fish immunity. Since these innate reactions are non-specific, they can be used instantly against a wide range of microorganisms without any prior exposure. The use of immunostimulants improves the following important immunological pathways:

- *Lysozyme Activity*
  Peptidoglycan, a key structural element of bacterial cell walls, is broken down by the hydrolytic enzyme lysozyme, which is essential to the fish's innate immunological system. Although it also has some activity against Gram-negative bacteria, it is more efficient against Gram-positive ones. Immunostimulants that raise lysozyme levels improve the fish's capacity to immediately break down and lyse bacterial cells, which helps eliminate pathogens and stop infections from starting.
- *Complement System Activation*
  The complex cascade of plasma proteins known as the complement system identifies, opsonizes, and lyses invading microbes. Complement proteins like C3 and C5 which are activated enable the following:
  - *Opsonization*, marking pathogens for destruction by phagocytes.
  - *Chemotaxis*, attracting immune cells to the site of infection.
  - *Membrane attack complex (MAC) formation*, which punctures the membrane of microbial invaders, causing cell lysis. Immunostimulants enhance the pro-

duction and activation of complement proteins, increasing the efficiency and speed of microbial elimination.

- *Phagocytic Activity*
  Phagocytosis is a crucial cellular mechanism where specialized immune cells like macrophages and neutrophils engulf and digest invading pathogens. Enhanced phagocytic activity means these cells:
  - More actively seek out pathogens.
  - Internalize them into phagosomes.
  - Destroy them using enzymatic digestion and oxidative burst.
- *Bactericidal Functions*
  Beyond engulfing pathogens, immune cells possess direct bactericidal mechanisms. These include the release of antimicrobial peptides (AMPs), enzymes (e.g., proteases), and other cytotoxic compounds into the phagosome or extracellular environment. Immunostimulants stimulate these mechanisms, increasing the overall killing capacity of immune cells and more effectively reducing the pathogen load.
- *Respiratory Burst Activity*
  Respiratory burst, also known as oxidative burst, is a rapid release of reactive oxygen species (ROS) such as superoxide anions, hydrogen peroxide, and hydroxyl radicals by phagocytes upon pathogen encounter. These ROS are highly toxic to microbes and play a vital role in the following:
  - Killing engulfed pathogens inside phagosomes
  - Inhibiting pathogen replication
  - Modulating inflammatory responses

There are three primary methods used to provide these immunostimulants to fish:

1. Oral administration (via feed)
2. Intraperitoneal injection
3. Immersion (bath treatment)

Oral delivery is the method most commonly employed in commercial aquaculture settings. Large fish populations can be treated without putting them under unnecessary stress thanks to the use of medicated diets. However, the immunostimulants' bioavailability through this route may be limited due to either inadequate intestinal absorption or their degradation in the digestive tract. Although intraperitoneal injection is more effective in ensuring precise dosage and rapid systemic absorption, it is time-consuming, traumatizing for fish, and unfeasible for large-scale farming operations. It is mostly used in experimental settings or high-value species when it makes financial sense to handle each individual. Despite being non-invasive, immersion is less effective in consistently delivering dosages and is primarily utilized for smaller fish or larval stages.

Immunostimulants have limitations despite their demonstrated advantages in improving disease resistance and lowering mortality. The possibility of chemical residues is one of the primary worries, particularly when it comes to synthetic formulations that can contain pharmacological or hormone chemicals. If left unchecked, these contaminants can linger in fish tissues and endanger consumer food safety. Furthermore, immunostimulant abuse or improper application might eventually cause immune system fatigue or decreased efficacy, underscoring the necessity of carefully monitored, evidence-based dosage guidelines.

10.7.3 Other Strategies

According to Cheng et al. (2015), selecting parental broodstock based on species traits rather than genotypes is no longer an option because traditional breeding inevitably leads to inbreeding and loss of fitness.

Translational and genomic advances have made genomic parental brood selection feasible. Advances in genomes, for example, have enabled the creation of hybrid species. A stronger innate immune system should make hybrid groupers more resistant to infection (Zhu et al. 2013). E. fuscoguttatus (female) x Epinephelus lanceolatus (male) is a successful hybrid. Using insecticides is another way to prevent and control parasite infestations, according to another study. Farmers often use excessive doses of insecticides to prevent argulosis from animals like fish lice or *Argulus* sp. An infectious parasite called argulosis causes severe bodily damage to saltwater fish before killing the organisms.

Additionally, probiotics are utilized to reduce disease in aquaculture. Probiotics are members of the beneficial microbiota of the host. According to Dawood et al. (2019), a probiotic should benefit the host in addition to being safe. The non-invasive, non-pathogenic qualities of potential probiotics can be modified for industrial processes to allow for commercial manufacturing. There are several ways that probiotics may help prevent bacterial infections. This involves competing for essential nutrients and adhesion sites, producing compounds that make the environment inhospitable for infections, and then modifying the immune systems of the hosts. A variety of contemporary omics technologies can provide a thorough understanding of the problem being studied, which can lead to various suggestions for addressing the infectious disease outbreak among cultured fish. The development of computer methods and the dissemination of enormous amounts of quantitative data are the primary forces behind the advent of systems biology.

10.8 Application of Multi-omics for Identification of Biomarker

The use of omics techniques in fish disease research has greatly improved our knowledge of disease mechanisms and the discovery of biomarkers associated with host defense responses and pathogen virulence. When combined with bioinformatics, these high-throughput technologies provide enormous volumes of data, which speeds up the search for possible biomarkers for improvements in diagnosis and treatment. Much research on fish disease has revealed a variety of molecular indicators, including genes, mRNAs, proteins, and metabolites. These indicators aid in the differentiation of several elements impacting an organism's reaction to infection. They also aid in the early diagnosis of infections and offer information on the intricate relationships between other influencing factors and disease-related systems. In clinical trials, biomarkers help with patient stratification, toxicity assessment, and drug efficacy evaluation. Biomarkers help with patient stratification in clinical trials, toxicity assessment, and therapeutic efficacy evaluation. In environmental research, biomarkers aid in determining exposure to pollutants and toxins (Quezada et al. 2017).

10.9 Conclusion and Future Prospective

Bioinformatics has the ability to drastically change aquaculture research by establishing a single, comprehensive information base about aquaculture. In addition to saving aquaculture scientists a tremendous amount of time and effort, such information interchange would reduce haphazard financing for relevant research. Therefore, bioinformatics-based research is crucial to boosting aquaculture's sustainability and economic feasibility in the country. However, financial and logistical support from the public and private sectors is required for aqua-informatics to succeed.

The development of vaccinations and the investigation of pathogen-host interactions and processes have both heavily relied on omics approaches. The omics technique is also being used to improve the diet of fish by adding the supplements that are required to sustain and stimulate immune response. Despite a few limitations, these techniques are widely employed and generally reliable for many disease-related studies outside of fish diseases. Combining these omics could improve fish health and address fish sickness issues. Applications of proteomics, metabolomics, and transcriptomics in particular are stabilizing the fish export industry and affecting the increasing mortality rate. It is expected that the integration of numerous omics would enhance our comprehension of fish immunological mechanisms during disease infection, as several layers of omics data can be used to reflect an organism's properties at different bimolecular levels. Because of the reliable database and easy outreach, the emerging field of integrative data analysis utilizing a bioinformatics technique would produce a reliable and accurate diagnosis concerning fish disease biomarkers. Although all of the main fish species used for import and export are now being studied, much more can be done because there are still a lot more illnesses and fish species to cover. These studies are yielding more and more fascinating findings that are motivating a number of additional research projects. In conclusion, this study looks at the most common approaches and emphasizes the various efforts to use multi-omics technologies to combat fish sickness. Furthermore, multi-omics, a relatively new technology that is now working toward the discovery of biomarkers to assist in managing disease in aquaculture, is effectively explained to readers in this study.

References

Altinok İ, Kurt İ (2003) Molecular diagnosis of fish diseases: a review. Turk J Fish Aquat Sci 3(2):131–138

Bailey C, Strepparava N, Wahli T, Segner H (2019) Exploring the immune response, tolerance and resistance in proliferative kidney disease of salmonids. Dev Comp Immunol 90:165–175

Banerjee S, Mahanty A, Mohanty S, Mazumder DG, Cash P, Mohanty BP (2017) Identification of potential biomarkers of hepatotoxicity by plasma proteome analysis of arsenic-exposed carp *Labeo rohita*. J Hazard Mater 336:71–80

Castro R, Jouneau L, Tacchi L, MacQueen DJ, Alzaid A, Secombes CJ, Martin SAM, Boudinot P (2016) Disparate developmental patterns of immune responses to bacterial and viral infections in fish. Fish Shellfish Immunol 53:92

Cheng SS, Senoo S, Siddiquee S, Rodrigues KF (2015) Genetic variation in the mitochondrial genome of the giant grouper *Epinephelus lanceolatus* (Bloch, 1790) and its application for the identification of broodstock. Aquac Rep 2:139–143

Danish M, Trivedi RN, Kanyal P, Khati A, Agrawal R (2017) Importance of biotechnology in fish farming system: an overview. Prog Res J 12(1):7–14

Dawood MA, Koshio S, Abdel-Daim MM, Van Doan H (2019) Probiotic application for sustainable aquaculture. Rev Aquac 11(3):907–924

Ganguly S, Mahanty A, Mitra T, Mohanty S, Das BK, Mohanty BP (2018) Nutrigenomic studies on hilsa to evaluate flesh quality attributes and genes associated with fatty acid metabolism from the rivers Hooghly and Padma. Food Res Int 103:21–29

Jeffries KM, Hinch SG, Sierocinski T, Pavlidis P, Miller KM (2014) Transcriptomic responses to high water temperature in two species of Pacific salmon. Evol Appl 7(2):286–300

Kamble A, Khairkar R (2016) Basics of bioinformatics in biological research. Int J Appl Sci Biotech 4(4):425–429

Khalil IG, Hill C (2005) Systems biology for cancer. Curr Opin Oncol 17(1):44–48

Kordon AO, Karsi A, Pinchuk L (2018) Innate immune responses in fish: antigen presenting cells and professional phagocytes. Turk J Fish Aquat Sci 18(9):1123–1139

Low CF, Shamsudin MN, Chee HY, Aliyu-Paiko M, Idrus ES (2014) Putative apolipoprotein A-I, natural killer cell enhancement factor and lysozyme g are involved in the early immune response of brown-marbled grouper, Epinephelusfuscoguttatus, Forskal, to Vibrio alginolyticus. J Fish Dis 37(8):693–701

Martin SA, Król E (2017) Nutrigenomics and immune function in fish: new insights from omics technologies. Dev Comp Immunol 75:86–98

Mohanty BP, Mohanty S, Mitra T, Mahanty A, Ganguly S, Singh S (2019) Omics technology in fisheries and aquaculture. Adv Fish Res 7:1–30

Mohd-Aris A, Muhamad-Sofie MHN, Zamri-Saad M, Daud HM, Ina-Salwany MY (2019) Live vaccines against bacterial fish diseases: a review. Vet World 12(11):1806

Ngoh SY, Tan D, Shen X, Kathiresan P, Jiang J, Liew WC, Thevasagayam NM, Kwan HY, Saju JM, Prakki SR, Goh CH (2015) Nutrigenomic and nutritional analyses reveal the effects of pelleted feeds on Asian seabass (*Lates calcarifer*). PLoS One 10(12):e0145456

Quezada H, Guzmán-Ortiz AL, Díaz-Sánchez H, Valle-Rios R, Aguirre-Hernández J (2017) Omics-based biomarkers: current status and potential use in the clinic. Bol Méd Hosp Infant México (English Ed) 74(3):219–226

Rana KMS, Ahammad K, Salam MA (2020) Bioinformatics: scope and challenges in aquaculture research of Bangladesh-a review. Int J Agric Res Innov Technol 10(2):137–145

Rodger HD (2016) Fish disease causing economic impact in global aquaculture. In: Fish vaccines, pp 1–34. https://doi.org/10.1007/978-3-0348-0980-1_1

Semple SL, Dixon B (2020) Salmonid antibacterial immunity: an aquaculture perspective. Biology 9(10):331

Shapawi R, Ebi I, Yong ASK, Ng WK (2014) Optimizing the growth performance of brown-marbled grouper, *Epinephelus fuscoguttatus*, by varying the proportion of dietary protein and lipid levels. Anim Feed Sci Technol 191:98–105
Smith NC, Rise ML, Christian SL (2019) A comparison of the innate and adaptive immune systems in cartilaginous fish, ray-finned fish, and lobe-finned fish. Front Immunol 10:2292
Viarengo A, Lowe D, Bolognesi C, Fabbri E, Koehler A (2007) The use of biomarkers in biomonitoring: a 2-tier approach assessing the level of pollutant-induced stress syndrome in sentinel organisms. Comp Biochem Physiol C Toxicol Pharmacol 146(3):281–300
Vinithkumar NV (2006) Marine bioinformatics. Adv Bio Tech 1:1–12
Zhu LY, Nie L, Zhu G, Xiang LX, Shao JZ (2013) Advances in research of fish immune-relevant genes: a comparative overview of innate and adaptive immunity in teleosts. Dev Comp Immunol 39(1–2):39–62

Integrative Approaches to Aquatic Toxicology Using Bioinformatics

11

Ankita Pati, Ayushman Gadnayak, Dibyajyoti Uttameswar Behera, Ananya Nayak, and Mahesweta Sahoo

11.1 Introduction

Aquatic ecosystems are essential for the global environment as they perform ecological functions by recycling nutrients, reducing floods, sustaining stream-flow, purifying water, restoring groundwater and producing habitat for wildlife and humans (Nimma et al. 2025). Often ecosystems undergo threatening remarks from various pollutants including pharmaceuticals, garbage disposal, heavy metals, microplastics, habitat loss, industrial chemicals, deforestation and pesticides. To overcome this, aquatic toxicology is beneficial for protecting natural resources, safeguarding human health, and understanding the major impact of chemical contaminants and harmful substances on marine ecosystems and organisms. Based on this, Lin et al. (2025) stated that aquatic toxicology denotes a complete knowledge of how the dangerous and unfavourable substances affect the aquatic species in the environment by identifying their toxicity level and predicting "metabolite disease associations" covering the bioinformatics field. In this regard, bioinformatics is utilised to analyse the hazardous reactions caused by chemical substances and environmental pollution. Therefore, bioinformatics investigates genetic information and designs strategies to manage and conserve them properly. This chapter provides a clear and detailed idea about how bioinformatics improves the knowledge of aquatic toxicology according to ecological and molecular perspectives.

A. Pati · D. U. Behera · A. Nayak · M. Sahoo
Centre for Biotechnology, Siksha 'O' Anusandhan Deemed to be University, Bhubaneswar, India

A. Gadnayak (✉)
ICAR-Central Inland Fisheries Research Institute, Barrackpore, Kolkata, India

© The Author(s), under exclusive license to Springer Nature Singapore Pte Ltd. 2026

B. K. Das et al. (eds.), *Bioinformatics Tools for Predictive Ecology and Fisheries*, Sustainable Aquatic Food Systems, https://doi.org/10.1007/978-981-95-2573-7_11

11.2 Role of Bioinformatics in Aquatic Toxicology

Bioinformatics is utilised in aquatic to assess and evaluate biological information and analyse the possible risks of chemical substances in water. This is used by perfecting the toxicity and its adversities, analysing bioconcentration, annotating, and manipulating genome data. The major areas of bioinformatics usually include the thorough analysis of omics technologies, the use of authentic resources and databases and the integration of omic datasets.

Omic technologies evaluate the functional and structural biological system and determine the adverse impact and toxicity of molecular organisms caused by chemical substances, affecting aquatic ecosystems' genes, proteins and metabolite levels (Fig. 11.1). Omics technologies usually consist of proteomics determining proteins, genomics determining genes, metabolomics identifying metabolites and transcriptomics detecting mRNA. While exposed to toxicants, the biomarkers are being identified to detect the adverse toxicological results affecting the aquatic species or organisms. Therefore, the technologies are utilised to decrease the overall sequencing costs, develop clinical interpretations, identify the major functions of miRNAs and improve both the modelling and simulation procedures while examining toxicity in aquatic ecosystems.

To gain sufficient knowledge and data about toxicological studies, publicly reliable, available and authentic databasessuch as Kyoto Encyclopaedia of Genes and Genomes (KEGG), AOP-Wiki, ENCODE and ToxCast are used (Fig. 11.2) (Jiang et al. 2024). These notable databases provide detailed knowledge about the

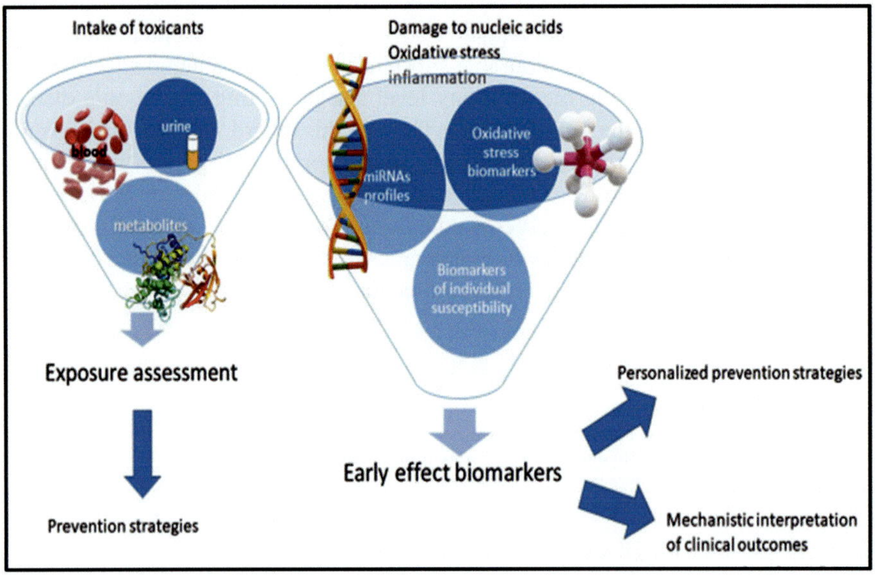

Fig. 11.1 The process showing how the biomarkers assess the risks when toxicants are exposed (Source: Capone et al. 2024)

Source	Version	URL
CTD	v2021-10	http://ctdbase.org/, accessed on 13 October 2021
AOP-Wiki	v2022-12	https://aopwiki.org/, accessed on 10 December 2022.
KEGG	v0.7.2	https://www.kegg.jp/, accessed on 5 October 2020.
DrugBank	v2020-12-15	https://go.drugbank.com/, accessed on 15 December 2020.
DisGeNet	v7.0	https://www.disgenet.org/, accessed on 15 October 2022.
Disease Ontology	v2021-10-11	https://disease-ontology.org/, accessed on 11 October 2022.
Human Phenotype Ontology	v2021-10-10	https://hpo.jax.org/app/, accessed on 10 October 2022.
PhosphoSitePlus	v6.6	https://www.phosphosite.org/
UbiBrowser	v2.0	http://ubibrowser.ncpsb.org.cn, accessed on 18 October 2022.
ENCODE	v120	https://www.encodeproject.org/, accessed on 8 October 2022.
STRINGdb	v11.0	https://string-db.org/, accessed on 18 December 2020.

Fig. 11.2 Publicly reliable databases regarding toxicological information

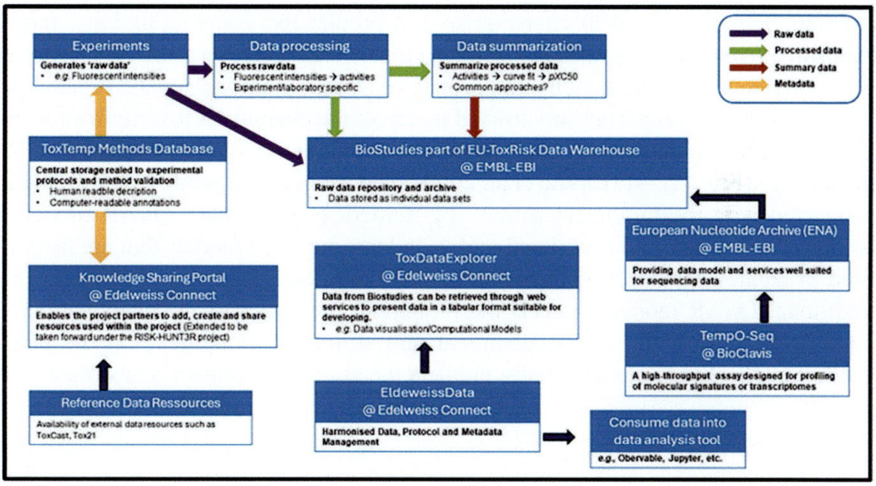

Fig. 11.3 Integration of omic datasets to understand toxicant effects

toxicogenomic pathways, diseases, mechanisms and chemicals. The databases are essential in providing varied information about the toxicological mechanisms, developing innovative therapeutic targets and analysing their chemical side effects.

The collected genomic and molecular information is combined with omic datasets to recognise the molecular changes, final predictions, and marginal correlations and detect complex phenotypes across biological states in aquatic ecosystems. Hardy et al. (2024) mentioned that the "EU-ToxRisk project" is implicated in assessing the toxicological mechanisms, analysing the chemical outcomes, and reviewing both processing protocols and data generation. Information is archived from sustainable databases and resources (Fig. 11.3). Omics data sets are combined by generating and processing raw data, summarising through experimental protocols, and archiving and sequencing them to profile molecular transcriptomes and signatures. With the help of bioinformatics, the researcher determines the biomarkers that recognise the aquatic toxicological mechanisms and actively detect the environmental risks.

11.3 Computational Approaches for Aquatic Toxicology

The computational approaches are essential for aquatic toxicology as these incorporate statistics, computer science, mathematics and modelling to understand the major impact of chemicals on aquatic species. "Toxicokinetic and Toxicodynamic (TK-TD)", usually assess how chemical substances enter the body and react with the toxicant exposures. The "Physiologically based pharmacokinetic (PBPK)" model is utilised to activate the toxicants of the "absorption, distribution, metabolism, and excretion (ADME)" among aquatic species (Fig. 11.4) (Ozbek et al. 2024). Nanoparticles (NP) are applied to enhance the distribution of tissue and pharmacokinetic properties along with the small molecules related to systemic circulation. This is utilised to characterise and predict the systemic toxicity, target exposure, disposition, and efficacy of the nanoparticles. Therefore, the PBTK model is initiated to predict the toxicity and growth of toxicants among aquatic organisms.

The molecular properties are utilised to detect the chemical's toxicity by a well-known approach known as "quantitative structure-activity relationship (QSAR) modelling" (Fig. 11.5) (Tropsha et al. 2024). The chemical structure of the model is connected with the biological activity by numerical descriptors. QSAR model is being differentiated into traditional and deep learning (DL) models that are usually used to creates cheminformatics tools to plan and manage organic synthesis. The traditional QSAR model connects the biological properties of the chemical compounds along with the physicochemical and structural properties. On the other hand, the DL-QSAR model usually predicts the significant impact of chemicals on

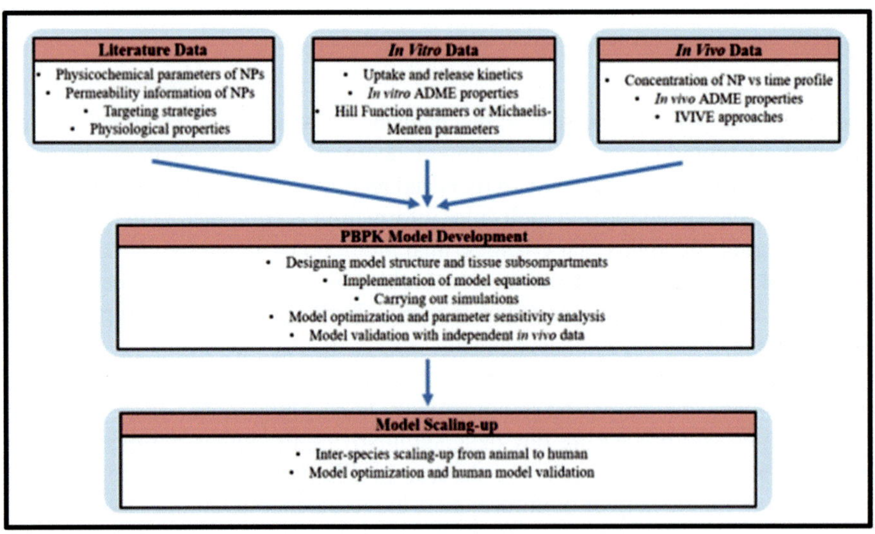

Fig. 11.4 Use of the PBPK model in the development of nanoparticle

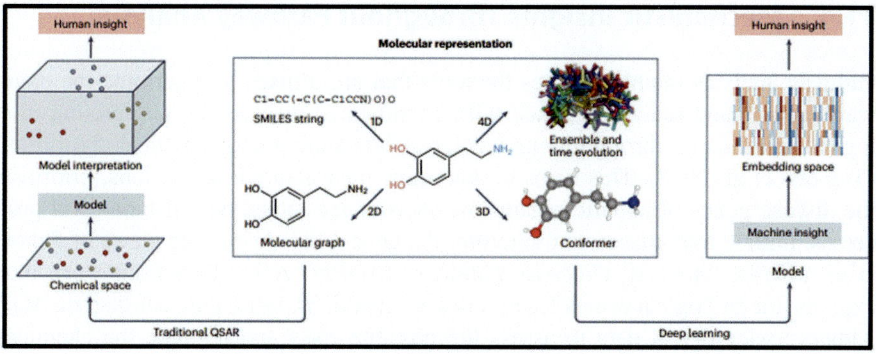

Fig. 11.5 Differentiation between deep and traditional QSAR models

$$R^2 = 1 - \frac{\sum_{i=1}^{n}(Y_i - \hat{Y}_i)^2}{\sum_{i=1}^{n}(Y_i - \overline{Y}_i)^2}$$

Fig. 11.6 Mathematical formula in machine learning

the organisms. Quantum computing accelerates the QSAR model with the help of democratised and open-source resources while designing a "computer-aided drug" (Tropsha et al. 2024). Therefore, the atoms present in the molecular structure of the atom are arranged to identify the toxicity and infectiousness of the complicated biological processes in the aquatic ecosystem.

According to Godasiaei (2025), "*machine learning (ML) algorithms*" usually predict the unseen as well as new information in a large data set or learning variables (Fig. 11.6). The algorithm determines the in vitro, in vivo, and clinical endpoints that decrease experimental time and cost and overcome the ethical risks. Therefore, the below figure shows the statistical metric that is used to analyse the accuracy, efficacy, and precision of the regression models by assessing each mathematical criterion. Here, R^2 refers to the "coefficient of determination" that mainly evaluates the variability of the learning data, and highlights and captures the latest data pattern across the dataset.

"Multi-Omics Integration" usually analyses the complicated biological system by determining targets and biomarkers to overcome the disease and keep the aquatic organisms safe from chemical hazards. Multi-omic integration exposes biological insights and helps the researcher gain complete knowledge about this, detect and optimise the treatments, and design and determine signalling pathways and the responses of the toxicants (Jendoubi 2021). This helps scientists understand the major challenges affecting the aquatic ecosystem by linking the biochemical mechanisms to the phenotypical and toxicological outcomes.

11.4 Mechanistic Insights Throughout Pathway Analysis

Pathway analysis mainly denotes the tools that are utilised to determine the pathways, routes and functions linked with complicated diseases by interpreting biological information through biomolecular understanding and statistical algorithms (Trapotsi et al. 2022). This helps to determine unstructured interactions, prioritise the disease genes of aquatic organisms, uncover the interaction of biological processes, initiate hypotheses, and interpret the genes' list. As per the views of Sahoo et al. (2024), "*Adverse Outcome Pathways (AOPs)*" refers to a framework that explains the biological events leading to poor health impact on the ecosystem. AOP gathers toxicological data to assess the possible risks and regulate the chemical substances by integrating heterogeneous information and "AOP-Wiki" from reliable resources. This framework helps to organise the biological events into modular forms known as "key events (KE)", linked with others through "key event relationships (KERs). Therefore, the toxicity pathways are surely understood through AOP and the biological information is integrated through the "Comparative Toxicogenomic Database (CTD)". AOPs benefit aquatic ecosystems by detecting biological results, evaluating chemical risks, decreasing species testing, developing casual routes or pathways and designing the latest testing techniques. Biological instruments and omic technologies are used to analyse the effect of contaminants on organisms and decrease uncertainties while assessing the risks in aquatic ecosystems.

"Pathway Enrichment Analysis (PEA)" is another method that determines the biological pathways and explores the functional disease that raises the risk rate (Jiang et al. 2022). This method helps in developing connections between phenotypes and genotypes, considering the biological procedures and interpreting the omics information. Based on this regard, certain tools analyse the biological pathway and determine how they affect the overall biological procedure in aquatic ecosystems. Among them, "Gene Ontology" is one of the most common analyses that detects the cellular components and molecular and biological functions, determines "differentially expressed genes" and interprets the regulated gene products (Jiang et al. 2022).

11.5 Case Studies

Bioinformatics is applied in aquatic ecosystems to profile, annotate genomes, modelling and initiate biological networks. Bioinformatics is applicable in a diverse range within the aquatic ecosystem. While predicting the toxicity of heavy metals, RNA-seq is used to elevate the level of "stress response genes" among aquatic fishes through exposure to cadmium. Cadmium (Cd) is a dangerous heavy metal that creates a major problem for the aquatic ecosystem, affecting aquatic organisms such as "green fin horse-faced filefish", exposed to nearly 5.0 mg/L Cd for 7 days and influencing its severity harmfully (Zhang et al. 2024). Therefore, fishes prioritise the regulation of ions on exposure to heavy metals by decreasing the absorption of metals from the water, reducing the accumulation of metals in gills and increasing ion transporters that sustain osmotic balance. Based on this, Fu et al. (2020) mentioned that the role of microRNAs (miRNAs) is essential to managing the expression of heavy metals expression where both miRNA and mRNA respond to cadmium stress. The enrichment pathway and "Cd responding genes" sequence with transcriptome and are associated with ribosome biogenesis.

Another case study observed where QSAR models on aquatic invertebrates detect the effect of pesticides. It has been noted that pesticides usually contaminate the aquatic environment, increasing nitrate, ammonium, sulphate and nitrite in aquatic systems and warning the ecological environment. The acute toxicity of harmful chemicals and pesticides is assessed by QSAR and "interspecies quantitative structure toxicity–toxicity relationship (iQSTTR)" models and is associated with statistical parameters (Li et al. 2024). The pesticides' toxicity is detected from a comprehensive database and is utilised to assess "ecotoxicological risks", fill data gaps and design innovative "green pesticide" products.

The other case study reveals that the oil spill affects the metabolic profiling of the aquatic organisms and the identified biomarkers assess the possible ecological risks. Based on this, Monteiro et al. (2024) stated that "Polycyclic Aromatic Hydrocarbons (PAHs)" are a common pollutant that affects the environment leading to teratogenic effects and oxygenic stress. Therefore, "environmental quality guidelines' are drawn and the biomarkers always try to assess the risks by gaining a deep knowledge regarding "PAH interaction mechanisms". It has been recognised that highly concentrated oil spills hamper microbial growth and affect aquatic organisms such as Atlantic killifish, top smelt, water fleas, and zebrafish. This is improved through a bioremediation process that eliminates and breaks down oil spills from the aquatic ecosystem.

11.6 Advancements in Molecular Docking and Simulation

Molecular docking, as well as simulation, is regarded as a computational method that is utilised to study and get knowledge about the chemicals' toxicity within the aquatic ecosystem. Molecular docking identifies how the minute molecules interact with the binding site of the targeted proteins by understanding the biological procedures at an atomic level (Fig. 11.7) (Aliko et al. 2024). Docking simulations are managed by utilising essential enzymes including "superoxide dismutase (SOD1)" and "zebrafish hatching enzyme1 (ZHE1)", gaining wide knowledge about the molecular mechanism, that underlies the "zinc oxide nanoparticles (ZnO-NPs)" embryotoxic impact on zebrafish embryos. Therefore, simulation and molecular docking are applied to examine the ZnO NPs' intermolecular interactions against the embryonic proteins.

Molecular dynamics and docking usually detect the toxic potentials and the binding affinities of the harmful chemicals, affecting the aquatic organisms. Along with this, the molecular targets of the aquatic toxicants are determined by elucidating the unknown protein structure. Molecular docking simulations understand the

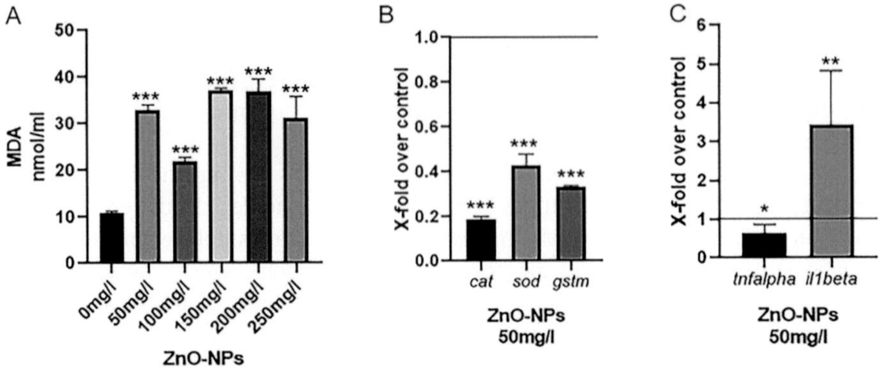

Fig. 11.7 Evaluation of embryotoxic impact on zebrafish, underlying ZnO-NPs

		Times (hours)				
Indices	Groups (μg L^{-1})	12	24	48	72	96
Gill	Control	175.89 ± 15.15	183.74 ± 14.12	181.46 ± 17.11	167.04 ± 15.61	179.13 ± 11.01
	55	156.47 ± 11.48	165.34 ± 10.99	102.38 ± 11.73^a	117.39 ± 12.67^a	108.29 ± 12.06^a
	110	100.9 ± 9.54^a	94.25 ± 8.14^a	85.90 ± 8.77^a	69.99 ± 7.94^b	51.68 ± 5.09^b
Kidney	Control	206.53 ± 16.08	195.27 ± 11.01	214.63 ± 10.78	204.13 ± 10.48	205.92 ± 16.14
	55	189.6 ± 11.67	177.18 ± 10.95	150.5 ± 8.84^a	109.1 ± 6.92^a	101.46 ± 11.38^a
	110	92.73 ± 9.15^a	101.73 ± 8.47^a	79.18 ± 5.42^b	91.64 ± 7.14^a	60.11 ± 5.17^b
Liver	Control	197.72 ± 10.49	203.46 ± 19.19	193.99 ± 17.11	191.09 ± 15.09	211.94 ± 21.27
	55	184.6 ± 8.96	133.79 ± 15.07^a	111.3 ± 12.07^a	105.6 ± 11.54^a	96.71 ± 11.74^a
	110	105.7 ± 7.28^a	121.91 ± 10.15^a	89.37 ± 9.17^a	71.36 ± 9.12^b	50.68 ± 4.83^b

Fig. 11.8 Effect of SOD on the liver, kidney and gills of aquatic organisms

biological procedures, evaluate the toxicity mechanism, model the molecular dynamics and detect the chemical-physical characteristics of the interactions and compounds. The above figure (Fig. 11.7) shows that Superoxide dismutase (SOD) protects the liver tissue, kidney and gills from oxidative destruction and balances the level of enzyme activity, forming superoxide radicals, in an aquatic ecosystem (Fig. 11.8) (Taysi et al. 2024). It has been observed that molecular docking is essential in aquatic toxicology to understand the associations between biological targets and ecological pollutants. This usually predicts the major impact of pharmaceutical compounds affecting the aquatic organisms, chooses the ecotoxicity test series correctly, understands the drugs' actions and determines the susceptibility of the species due to the impact of the chemicals.

Molecular docking studies showed that the herbicides impede aquatic enzymes often leading to environmental imbalance. "Aquatic toxicity control programmer" is outlined to enhance the optimum combination of microplastics and plasticisers that are associated with the eco-friendly "phthalic acid ester (PAE) by-products (Chen et al. 2021). This surely reduces the level of aquatic toxicity by decreasing the translocation and uptake of herbicides, enhancing sequestration, metabolism and adjusting the sequences of amino acids.

11.7 Epigenetic and Environmental Toxicology

Histone modifications and methylation of DNA are considered epigenetic modifications that play an essential role in responding to aquatic toxicants. DNA methylation is essential as it is often utilised as a biomarker to track the health and recovery of the aquatic ecosystem, get adjusted to ecological stress, manage gene expression and maintain genetic sequence. As per this, Kim et al. (2024) opined that the "DNA methylase (DNMT)" enzyme controls the DNA methylation along with the translocation genes and the methylation rates differ among the aquatic organisms, permitting transcription, eliminating transcriptional noise and initiating gene expression. Along with this, the changes in the histone modification acknowledge the ecological factors and their potential for chromatin dynamics as well as gene expression. Therefore, histone modification explicitly shows how gene expression is affected by the ecological factor in aquatic ecosystems. The comprehensive knowledge regarding epigenetic patterns and mechanisms is expanded among the aquatic organisms, which are exposed to ecological stressors.

It has been observed that bioinformatics tools mainly determine the epigenetic biomarkers to predict aquatic toxicity and relate to the possible epigenetic changes into phenotypic results. Bioinformatics tools usually detect and validate the biomarkers, sequence the technologies, and improve the genomics and polymerase chain reaction and termination (Fig. 11.9) (Clark and Lillard Jr 2024). DNA sequencing is essential in bioinformatics as it helps to get complete knowledge about the genomic data, become associated with genetic diseases and variations, and enable diagnostics. It has been observed that bioinformatics is essential for "epigenomic profiling", to evaluate both genomic and genetic information to determine aquatic species, and track the major changes caused due to the degradation of the aquatic ecosystem.

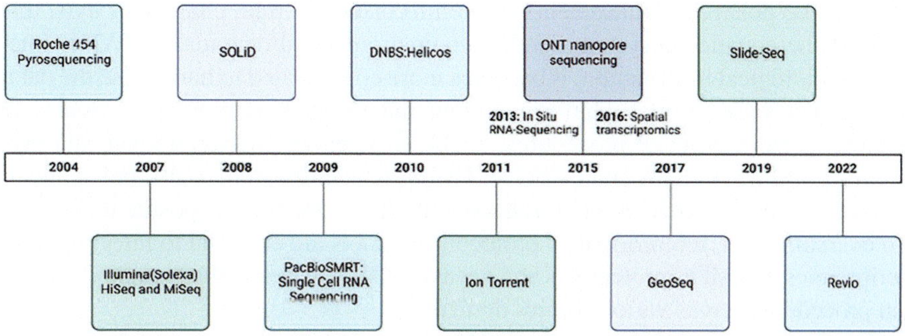

Fig. 11.9 Sequencing technologies improving genomics (Reproduced from: Clark et al. 2024)

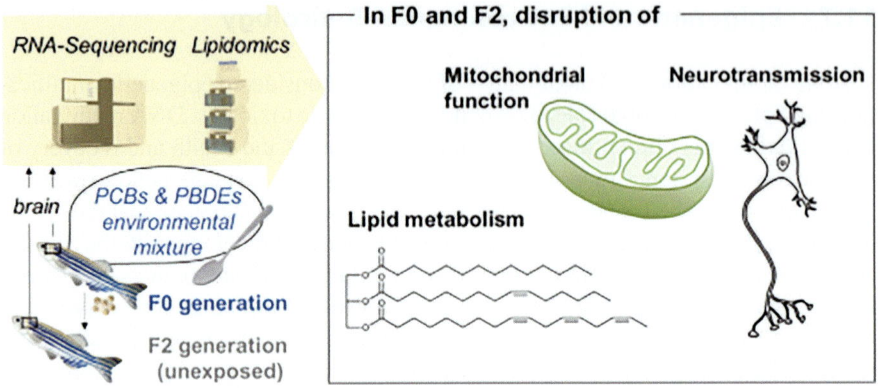

Fig. 11.10 Graphical representation of the impact of PCB on aquatic organisms (Reproduced from : Blanc et al 2021)

Recent studies revealed that "polychlorinated biphenyls (PCB)", a common pollutant, induce the "heritable epigenetic changes" among the aquatic species and are required for ecological epigenomics. PCB exists in aquatic environments where the fish usually absorb it from the sediments and water (Blanc et al. 2021). This determines the molecular changes, displays the regulation of lipid metabolism and mitochondrial function and explicit "energy homeostasis". Lipids are essential for aquatic organisms by acting as a "'energy storing" signalling and structuring molecule (Fig. 11.10). This can be improved by eliminating the fatty areas from the aquatic species, incinerating, and activating carbon filtration.

11.8 Challenges

Besides the positive advancements of bioinformatics, certain challenges exist that hamper the aquatic ecosystem, leading to environmental degradation. While integrating biological information, it becomes more complicated to harmonise the dataset from a varied omics platform. Often, the researchers experience issues in evaluating the biological procedures, unable to integrate transcriptomics and proteomic profiling (Belalov and Li 2025). Therefore, the researchers also fail to understand how the chemicals are interrelated with the stressors on exposure to toxicity. To overcome this, bioinformatics protocols and tools are essential to integrate transcriptomics as well as proteomics and get a deep knowledge of the complex biological procedures across various omics domains.

On the other hand, Rosner et al. (2024) stated that a lack of proper standardised and uniform protocols often leads to the limitation of reproducibility. Therefore, the researcher often fails to predict correct information regarding how chemical toxicity affects the aquatic ecosystem, leading to poor documentation, control and

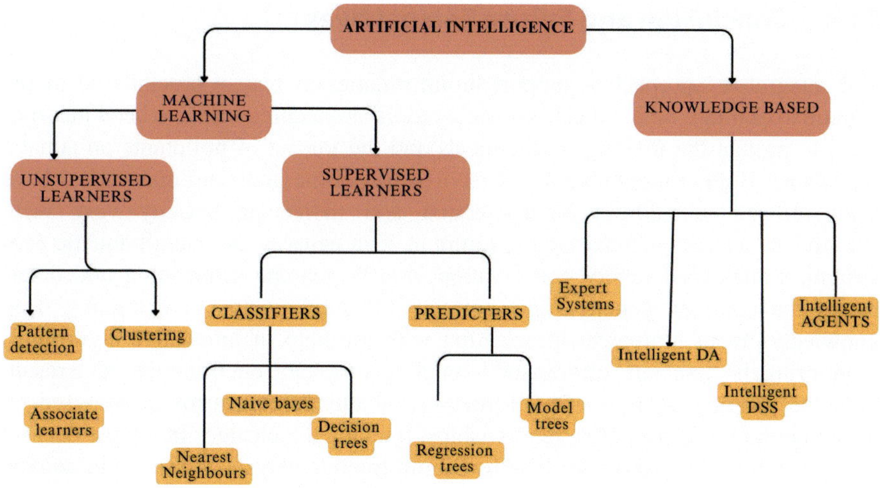

Fig. 11.11 AI and machine learning in toxicology

interoperability. Due to a lack of consistency, toxicology guidelines are not established properly and are unable to apply standardised methods for data interpretation, analysis, evaluation and collection. To overcome this, transparency is crucial to develop experimental designs establish more consistent protocols to predict the sensitivity level of different chemicals, become more proficient in laboratory maintenance and be able to measure physiological parameters.

Another common issue is the lack of sufficient resources to handle and deal with large datasets, leading to poor computational power. This makes it difficult for the researchers to manage the infrastructure and store the biological information properly. Due to poor computational power, the researcher faces risks while gathering sensitive data from groups and is unable to protect them from breaches, leading to ethical issues (Fig. 11.11) (Banaee et al. 2024). While handling large datasets, there exists the possibility of unauthorised access, cyber-attacks, security risks and leakage of data. To mitigate this, the researcher needs to be more aware that the collected information does not create inaccuracies and biases and proper training is essential to handle the datasets.

On the other hand, Ditria et al. (2022) opined that an automated approach is being facilitated that helps the researchers supervise the bottlenecks, incorporate automation, identify the roadblocks and make the environment safer. Along with this, cloud computing also plays a crucial role in tracking water quality, accessing, processing and integrating biological data, retrieving and storing real-time information and minimising energy consumption. This surely helps to analyse and regulate both the protein and genome expression in an aquatic ecosystem.

11.9 Conclusion and Future Perspectives

The integrative approaches support bioinformatics to play a crucial role in the aquatic toxicology field, where innovative tools and techniques have been incorporated to predict the toxicity of chemicals and the impact of pollutants on aquatic organisms. Both computational and environmental methods and approaches have been utilised to address environmental and molecular toxicity very well. Bioinformatics is beneficial for the future as it safeguards the overall aquatic ecosystem, values biodiversity conservation, assists genome sequencing and determines the complex genetic traits correctly. It surely helps in extracting deep knowledge about biological information with the help of innovative algorithms, monitoring the toxicants' effects and inhabiting an aquatic environment. As a result, both the genomic and genetic information of aquatic organisms is analysed by understanding transformative relationships, tracking the changes in a large data set and evaluating the ecosystem health. Bioinformatics plays a vital role in aquatic toxicology and based on future perspectives, cloud computing, machine learning, artificial intelligence (AI) and global collaborations will quickly overcome the issues in the toxicology field. AI technology is used to detect the chemicals' toxicity, handle large datasets, initiate chemical screening and extract the correct information from varied omics datasets (Kleinstreuer and Hartung 2024). This surely helps in maintaining the evaluation of chemical safety, determining toxicity endpoints, enhancing trust and raising transparency in aquatic ecosystems. Along with this, global collaborations can also be essential to explore the knowledge expansion regarding aquatic toxicology, enhance the characterisation related to chemical contaminants, detect pollutants and address ecological issues.

References

Aliko V, Vasjari L, Istifli ES, Gjonaj G, Impellitteri F, Faggio C et al (2024) Molecular docking analysis and in vivo assessment of zinc oxide nanoparticle toxicity in zebrafish larvae. Aquat Toxicol 276:107112

Banaee M, Zeidi A, Faggio C (2024) Application of artificial intelligence and machine learning in computational toxicology in aquatic toxicology. Pollution 10(1):210–235

Belalov I, Li X (2025) Methods and tools for OMICS data integration. In: Molecular pathway analysis using high-throughput OMICS molecular data. Academic Press, Cambridge, pp 271–297

Blanc M, Alfonso S, Bégout ML, Barrachina C, Hyötyläinen T, Keiter SH, Cousin X (2021) An environmentally relevant mixture of polychlorinated biphenyls (PCBs) and polybrominated diphenylethers (PBDEs) disrupts mitochondrial function, lipid metabolism and neurotransmission in the brain of exposed zebrafish and their unexposed F2 offspring. Sci Total Environ 754:142097

Blanc M, Morin B, Pilote M, Bouchard M, Fournier M (2021) Polychlorinated biphenyls (PCBs) immunotoxicity in aquatic organisms: A review. Environ Res, 195:110822. https://doi.org/10.1016/j.envres.2021.110822

Capone P, Chiarella A, Sisto R (2024) Advanced technologies in genomic toxicology: Current trends and future directions. Computational Toxicol, 370:000444. https://doi.org/10.1016/j.comtox.2023.100444

Chen X, Li X, Li Y (2021) Toxicity inhibition strategy of microplastics to aquatic organisms through molecular docking, molecular dynamics simulation and molecular modification. Ecotoxicol Environ Saf 226:1128

Clark AJ, Lillard JW Jr (2024) A comprehensive review of bioinformatics tools for genomic biomarker discovery driving precision oncology. Genes 15(8):1036

Clark DP, Lillard SJ (2024) Next-generation sequencing technologies: The next wave of genomics. Trends Biotechnol, 42(4):450–464. https://doi.org/10.1016/j.tibtech.2023.09.004

Ditria EM, Buelow CA, Gonzalez-Rivero M, Connolly RM (2022) Artificial intelligence and automated monitoring for assisting conservation of marine ecosystems: a perspective. Front Mar Sci 9:918104

Fu J, Wang M, Chaudhry MT, Tian Y, Liu C (2020) Combined RNA-Seq with small RNA revealed ribosome biogenesis and oxidative stress associated with cadmium response in carp (*Cyprinus carpio Ls.*) Hepato-pancreas. Aquaculture 518:734817

Godasiaei SH (2025) Predictive modeling of microplastic adsorption in aquatic environments using advanced machine learning models. Sci Total Environ 958:178015

Hardy B, Mohoric T, Exner T, Dokler J, Brajnik M, Bachler D, Athar A (2024) Knowledge infrastructure for integrated data management and analysis supporting new approach methods in predictive toxicology and risk assessment. Toxicol In Vitro 100:105903

Jendoubi T (2021) Approaches to integrating metabolomics and multi-omics data: a primer. Meta 11(3):184

Jiang X, Zhang H, Wang X, Zhang X, Ding K (2022) Comprehensive analysis of the association between human diseases and water pollutants. Int J Environ Res Public Health 19(24):16475

Jiang P, Zhang Z, Yu Q, Wang Z, Diao L, Li D (2024) ToxDAR: a workflow software for analyzing toxicologically relevant proteomic and transcriptomic data, from data preparation to toxicological mechanism elucidation. Int J Mol Sci 25(17):9544

Kim MS, Kim DH, Lee JS (2024) A review of environmental epigenetics in aquatic invertebrates. Mar Pollut Bull 208:117011

Kleinstreuer N, Hartung T (2024) Artificial intelligence (AI)it's the end of the tox as we know it (and I feel fine). Arch Toxicol 98(3):735–754

Li Y, Fan T, Ren T, Zhang N, Zhao L, Zhong R, Sun G (2024) Ecotoxicological risk assessment of pesticides against different aquatic and terrestrial species: using mechanistic QSTR and iQSTTR modelling approaches to fill the toxicity data gap. Green Chem 26(2):839–856

Lin Y, Yang X, Zhang M, Cheng J, Lin H, Zhao Q (2025) CLSSATP: contrastive learning and self-supervised learning model for aquatic toxicity prediction. Aquat Toxicol 279:107244

Monteiro V, da Silva DD, Martins M, de Pinho PG, Pinto J (2024) Metabolomics perspectives of the ecotoxicological risks of polycyclic aromatic hydrocarbons: a scoping review. Environ Res 249:118394

Nimma D, Devi OR, Laishram B, Ramesh JVN, Boddupalli S, Ayyasamy R, Arabil A (2025) Implications of climate change on freshwater ecosystems and their biodiversity. Desalin Water Treat 321:100889

Ozbek O, Genc DE, Ulgen KO (2024) Advances in physiologically based pharmacokinetic (PBPK) modeling of nanomaterials. ACS Pharmacol Transl Sci 7(8):2251–2279

Rosner A, Ballarin L, Barnay-Verdier S, Borisenko I, Drago L, Drobne D et al (2024) A broad-taxa approach as an important concept in ecotoxicological studies and pollution monitoring. Biol Rev 99(1):131–176

Sahoo AK, Chivukula N, Ramesh K, Singha J, Marigoudar SR, Sharma KV, Samal A (2024) An integrative data-centric approach to derivation and characterization of an adverse outcome pathway network for cadmium-induced toxicity. Sci Total Environ 920:170968

Taysi MR, Kirici M, Kirici M, Tuzun B, Poustforoosh A (2024) Antioxidant enzyme activities, molecular docking studies, MM-GBSA, and molecular dynamic of chlorpyrifos in freshwater fish *Capoetaumbla*. J Biomol Struct Dyn 42(1):163–176

Trapotsi MA, Hosseini-Gerami L, Bender A (2022) Computational analyses of mechanism of action (MoA): data, methods and integration. RSC Chem Biol 3(2):170–200

Tropsha A, Isayev O, Varnek A, Schneider G, Cherkasov A (2024) Integrating QSAR modelling and deep learning in drug discovery: the emergence of deep QSAR. Nat Rev Drug Discov 23(2):141–155

Zhang X, Zhang W, Zhao L, Zheng L, Wang B, Song C, Liu S (2024) Mechanisms of gills response to cadmium exposure in greenfin horse-faced filefish (*Thamnaconus septentrionalis*): oxidative stress, immune response, and energy metabolism. Animals 14(4):561

Exploring Marine Microorganisms for Enhancing Aquaculture Sustainability

12

Debasmita Mohanty, Punam Kumari, Asit Kumar Bera, Barsha Baisakhi, and Basanta Kumar Das

12.1 Introduction

With the ongoing growth of the global population, there has been a significant increase in the demand for food that is produced sustainably, highlighting the crucial importance of aquatic ecosystems in fulfilling these requirements. The planet's water bodies, which encompass more than 70% of its surface, yield a variety of protein-rich seafood, such as fish, invertebrates, and algae, drawn from marine, brackish, and freshwater ecosystems. The consumption of seafood, sourced from both capture and aquaculture, has experienced an annual growth rate of 3.1% from 1961 to 2017, surpassing the growth rates of other livestock and animal production sectors (FAO 2020). As wild fisheries have plateaued, aquaculture has become an essential player, demonstrating a mean annual growth rate of 5.3% from 2001 to 2018 and representing 52% of seafood produced for human consumption by 2018 (FAO 2020). Nonetheless, the ongoing growth of aquaculture encounters obstacles, including outbreaks of infectious diseases, environmental issues, and the effects of climate change, highlighting the need for innovative and sustainable approaches to guarantee its future sustainability. Health management in aquaculture has always depended on antibiotics and chemotherapeutics; however, recent advancements

D. Mohanty · B. Baisakhi
ICAR-Central Inland Fisheries Research Institute, Barrackpore, Kolkata, West Bengal, India

Department of Bioscience and Biotechnology, Fakir Mohan University, Balasore, Odisha, India

P. Kumari
Department of Bioscience and Biotechnology, Fakir Mohan University, Balasore, Odisha, India

A. K. Bera · B. K. Das (✉)
ICAR-Central Inland Fisheries Research Institute, Barrackpore, Kolkata, West Bengal, India

© The Author(s), under exclusive license to Springer Nature Singapore Pte Ltd. 2026

B. K. Das et al. (eds.), *Bioinformatics Tools for Predictive Ecology and Fisheries*, Sustainable Aquatic Food Systems, https://doi.org/10.1007/978-981-95-2573-7_12

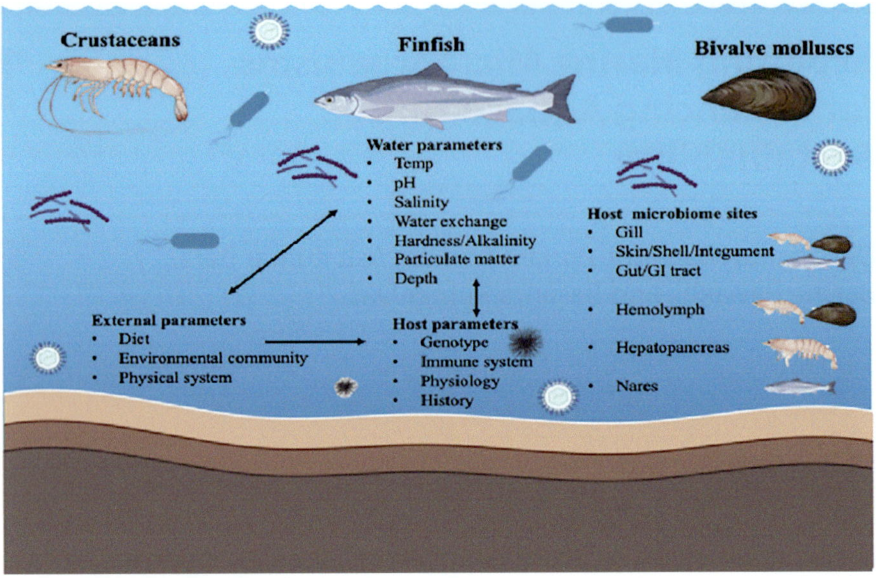

Fig. 12.1 Schematic representation of the microbiome and its associated environment, facilitating host-microbe interactions

highlight the importance of environmentally sustainable alternatives, including targeted vaccines and selective breeding for disease-resistant lines. An integrated approach necessitates the examination of both the cultured animals and their environment, acknowledging the essential contribution of microbial communities to growth, disease resistance, and overall animal welfare. The dynamic function of these microbial communities across a range of cultured animal groups, including invertebrates like molluscs and crustaceans as well as vertebrates such as finfish, is increasingly recognized as vital for the sustainability of aquaculture. Utilizing these microbial ecosystems presents a valuable opportunity for creating robust and efficient aquaculture systems (Kumar et al. 2021; 2023; Mohanty et al. 2025) (Abdelsalam et al. 2023) (Fig. 12.1).

12.2 What Is a Microbiome?

Many definitions exist for a microbiome, which is a population of commensal, symbiotic, and pathogenic microorganisms in a bodily area or environment. The microbiome is a community of microorganisms, including bacteria, viruses, fungi, parasites, and Archaea, as well as their downstream products and functions. Aquatic animals possess microbial communities in several body parts, including the gut, skin, nasal and oral cavities, gills, haemolymph, shells, integument, and hepatopancreas. Some microorganisms are found in all cultured animal phyla, such as gills. The microbiome composition is complex, involving host factors like genotype and immune system, external environmental influences like diet, culture system parameters, microbes, and water physiochemical parameters like temperature, salinity, and dissolved oxygen. Microbiome composition in host niches is influenced by these characteristics, and commensal communities impact host health. Determinants of host microbiome composition in aquatic animals under culture conditions are influenced by both environmental and host-associated variables. These factors together determine the structure and function of the host microbiome. Environmental conditions and the host, especially the water microbial population, might interact. Aquaculture can assess and adjust environmental conditions and host characteristics like genetics. Microbiomes are found in all cultivated species, including finfish, bivalve molluscs, and crustaceans. Mucosal tissues including stomach, gills, and skin include common microbial habitats, but some are exclusive to certain taxa, such as invertebrate haemolymph. Although host microbiome composition impacts host health, the underlying functional and mechanistic processes are little known in aquaculture (Berg et al. 2020).

12.3 Origins of Commensal Microbes in Aquaculture: Establishing the Host Microbiome

Aquatic organisms, unlike their terrestrial counterparts, are continuously exposed to water teeming with microbes, which can offer adaptive benefits but also pose potential risks. In aquaculture, environmental conditions are highly variable, often influenced by factors such as weather, temperature fluctuations, and farming practices. Many aquaculture species reproduce externally, releasing eggs that are immediately subjected to microbial colonization from their surroundings. Following hatching, the external mucosal surfaces of these organisms are rapidly colonized by environmental microbes, with the hatchery environment significantly influencing the diversity and richness of these initial microbial communities. Conversely, the internal gut mucosa remains free of microbes until the organism begins feeding, introducing a distinct microbial population from the environment and diet. Notably, sterile or sanitized conditions can reduce the variety of microbes available for colonization, affecting the establishment of stable microbial communities in aquaculture species (Minich et al. 2021).

12.4 Navigating Microbial Colonization: Host Strategies for Selective Association

The initial colonization of mucosal surfaces by commensal microbes involves intense interactions with the host immune system. Both invertebrates and vertebrates share common processes during this early phase, where the acceptance or exclusion of microbial taxa is influenced by a combination of environmental factors, microbial characteristics, and the host's innate immune system. Germline-encoded cell receptors, when present, play a critical role in shaping the microbial community, enabling the immune system to tolerate beneficial commensals while effectively preventing colonization by opportunistic or harmful microbes. These cell receptors are expected to vary across different mucosal surfaces, presenting intriguing mechanistic questions. However, the selection of symbionts is not exclusively driven by the host immune system. Environmental microbes may also establish colonization based on their ability to thrive in the nutrient-rich biochemical environment of mucosal surfaces, regardless of the host's influence. Unlike crustaceans and molluscs, which lack lymphocytes and immunoglobulins (IgGs) in their immune systems, fish possess an adaptive immune system capable of immunological memory that begins to develop at first feeding. Gut colonization in fish occurs between the opening of the mouth and the onset of feeding, exposing them to a distinct set of potential microbial colonizers during this time. Consequently, the mechanisms governing gut colonization differ between invertebrates and vertebrates, with fish relying on separate immune pathways. In fish and other vertebrates, IGs play a central role in mucosal immunity by coating resident microbes, thereby contributing to microbial selectivity (Vonaesch et al. 2018).

12.5 Host Genetic Factors Driving Microbiome Composition

Host genetics can influence the establishment of microbial associations through mechanisms such as microbial recognition, immune-mediated selection, and the determination of suitable biochemical niches. This highlights the potential for environmental conditioning in aquaculture to enhance host–microbe interactions. Understanding the influence of aquaculture systems on both host and environmental microbes is crucial to implementing such strategies. Fish of different species raised in similar environments often exhibit distinct microbiome compositions, suggesting the involvement of genetic factors. Within a species, genetic differences can also play a role, as evidenced by variations in intestinal microbiota among genetically distinct populations. These observations indicate that host genetics may contribute significantly to the selection and establishment of microbial communities.

12.6 Eco-Friendly Approaches in Aquaculture: The Role of Marine Microbiome

Marine bacteria are essential in the promotion of sustainable fisheries, as they provide eco-friendly alternatives to conventional aquaculture practices. This is particularly true in the context of combating bacterial pathogens and improving the health of cultured species. These microorganisms generate a diverse selection of bioactive compounds, such as antibiofilm agents and antimicrobial metabolites, which are highly effective against drug-resistant pathogens that are frequently encountered in aquaculture environments. Marine bacteria aid in the prevention of bacterial infections by inhibiting quorum sensing and biofilm formation. This reduces the need for conventional antibiotics and mitigates associated risks, including environmental

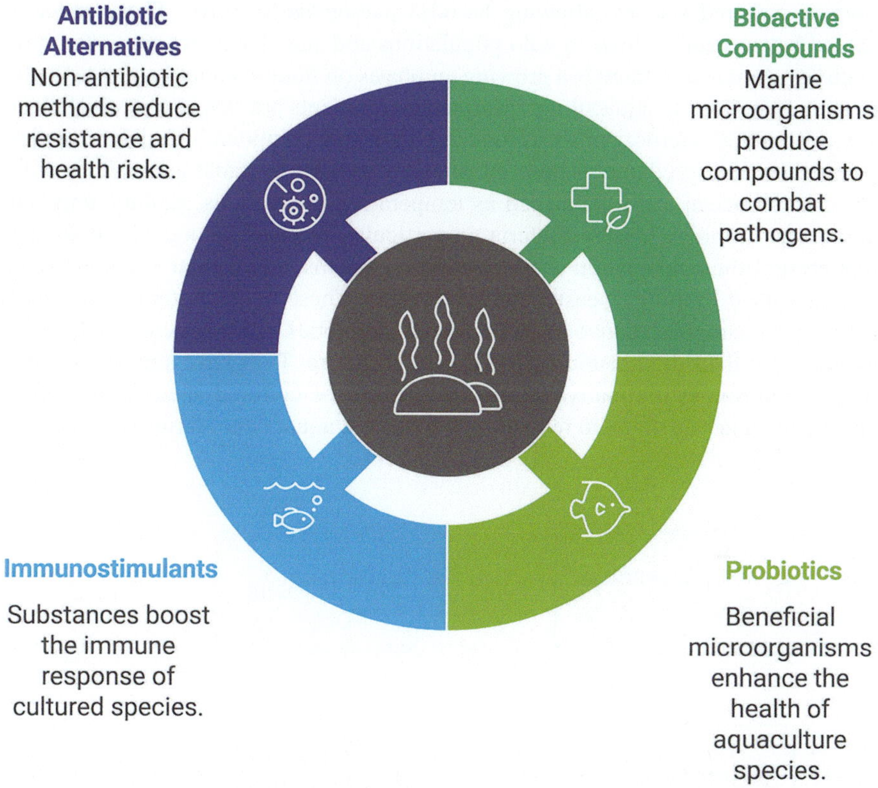

Fig. 12.2 Diagram illustrating the issues encountered by the aquaculture industry in relation to antibiotic use and how marine microorganisms provide solutions to address these challenges

contamination and an increase in antibiotic resistance. Furthermore, the overall health, growth, and resilience of aquaculture species are further improved by the use of marine microorganisms as probiotics, feed additives, and immunostimulants. In addition to promoting responsible management of aquatic resources and minimizing the environmental footprint, their integration into contemporary aquaculture practices also supports disease control and aligns with goals for sustainability (Fig. 12.2).

12.7 Role of Marine Microbiome in Controlling Aquaculture Disease Outbreaks

The long-term use of marine microorganisms to control aquaculture disease outbreaks can lead to microbial resistance, as certain species develop resistance to their effects. The presence of antibiotic residues and resistant bacteria in seafood poses significant health risks to consumers, and antibiotics disrupt the normal microbial flora of cultured species, allowing harmful pathogens to thrive. These concerns extend beyond aquaculture to wild populations and can also affect ecosystem productivity. As a result, there is a growing emphasis on finding natural alternatives for treating infections in aquaculture. Marine environments are rich in microbial diversity, with many microorganisms producing bioactive compounds to defend against competitors and predators. These metabolites are highly stable and able to withstand harsh ocean conditions such as temperature fluctuations, salinity, and contaminants. Epibiotic marine bacteria, in particular, produce secondary metabolites that prevent the colonization of other microorganisms, and similar compounds are also produced by fungi, yeasts, and microalgae. These novel, natural compounds offer a promising alternative to conventional antibiotics and are increasingly being explored for their therapeutic potential in aquaculture. They have shown effectiveness in improving immunity, health, and growth of cultured species, as well as enhancing water quality and reducing infections in aquaculture farms (Fig. 12.3).

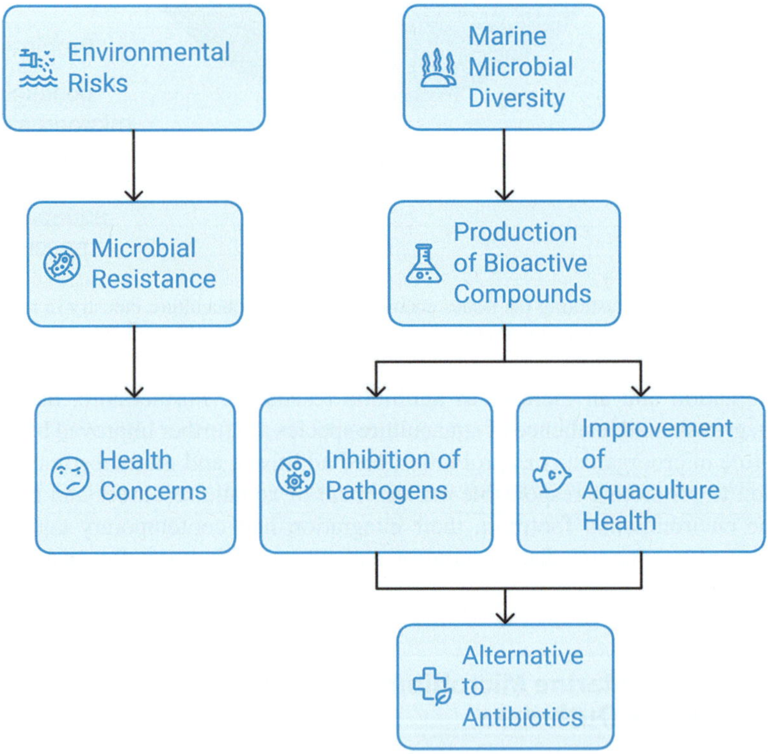

Fig. 12.3 The diagram highlights the key issues faced by the aquaculture industry due to antibiotic use, including antibiotic resistance, disruption of microbial balance, and biofilm formation by pathogens. It also depicts innovative solutions offered by marine microorganisms, such as the production of antibiofilm agents, quorum-sensing inhibitors, and natural antimicrobial compounds, which contribute to sustainable disease management practices in aquaculture

12.8 Marine Microbiome as Natural Antibiofilm Agents for Aquaculture

The efficiency of several antibiofilm compounds made from marine bacteria in fighting aquaculture infections is highlighted in this section. Biofilms are collections of bacteria that adhere to both living and non-living surfaces and are covered in extracellular polymeric substances (EPS). These structures serve as pathogen reservoirs, increasing the likelihood of recurring infections and encouraging the growth of bacterial populations resistant to antibiotics. Biofilm development is one of the survival strategies used by several diseases linked with aquaculture. Scanning electron microscopy has shown that bacteria like *Vibrio harveyi*, *V. alginolyticus*, *Pseudomonas aeruginosa*, and *P. gessardii* may form biofilms on surfaces like glass. Microorganisms from phyla such as Proteobacteria, Bacteroidota, Fusobacteriota, Patescibacteria, and Verrucomicrobiota are frequently prevalent in the early phases of biofilm development on biofilters in aquaculture systems (Qu et al. 2021). Since pathogenic strains like Pseudomonas species exhibit virulence factors, quorum-sensing abilities, and antibiotic resistance in the tissues of farmed fish, such as Nile tilapia, biofilms produced by *Vibrio*, *Pseudomonas*, *Flavobacterium*, *Aeromonas*, and *Streptococcus* present serious challenges in aquaculture. Because biofilm-forming organisms are naturally resistant, conventional therapies like peroxides and antibiotics frequently don't work against them. As a result, novel antibiofilm compounds derived from marine microorganisms that include traits including quorum-sensing inhibition, surfactant activity, bactericidal effects, and anti-adhesion capabilities provide encouraging answers to biofilm-related problems in aquaculture (Fig. 12.4).

12.9 Microbicidal Agents

The bactericidal effects of bioactive substances derived from both Gram-positive and Gram-negative bacteria have been demonstrated to successfully suppress the production of biofilms by pathogens in aquaculture. Bacillus species are among the Gram-positive bacteria that are regularly researched for possible uses in aquaculture. The *lantibiotic subtilomycin*, for instance, is produced by *Bacillus subtilis* and has been shown to effectively suppress the growth of pathogens associated with aquaculture, such as *Aeromonas hydrophila*, *V. anguillarum*, and *Alteromonas* species. Similar to this, *B. licheniformis* produces BLDZ1, an antibiotic protein that has the ability to break up and eradicate mature biofilms of pathogens such as *Pseudomonas aeruginosa*, *V. harveyi*, *V. alginolyticus*, and *P. gessardii*. Targeting and dissolving the bacterial cell membrane is how this protein functions. Furthermore, it has been demonstrated that the cell-free supernatants of *Bacillus* species possess anti-adhesive qualities, which inhibit the first bacterial attachment during biofilm development. Beyond *Bacillus*, *Salinibacterium*, a Gram-positive bacterium from the Arctic, has also demonstrated effectiveness in controlling the fish pathogen *Photobacterium damselae*.

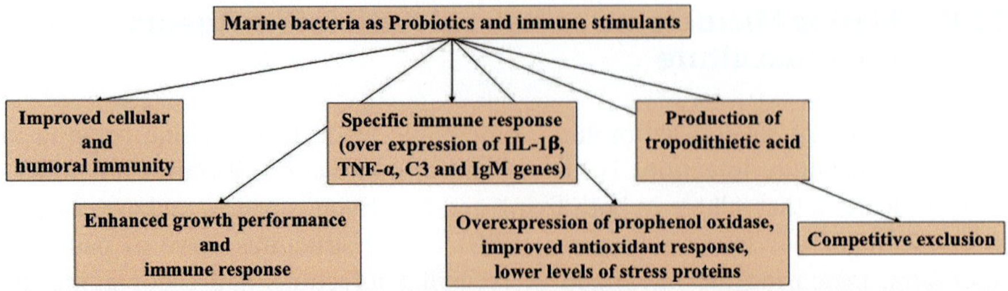

Improved cellular and humoral immunity	Enhanced growth performance and immune response	Specific immune response (over expression of IIL-1β, TNF-α, C3 and IgM genes)	Overexpression of prophenol oxidase, improved antioxidant response, lower levels of stress proteins	Production of tropodithietic acid	Competitive exclusion
Flavobacterium sp. *Shewanella* sp. *Aeromonas* sp. *Vibrio* sp.	*Providencia* sp. *Lactobacillus* sp. *Enterococcus* sp. *Alcaligenes* sp. *Klebsiella* sp. *Proteus* sp.	*Vibrio* sp.	*Bacillus* sp.	*Phaeobacter* sp. *Ruegeria* sp.	*Staphylococcus* sp. *Streptomyces* sp. *Lactococcus* sp.

Fig. 12.4 Various mechanisms by which marine microorganisms disrupt biofilms of aquaculture pathogens

Certain Gram-negative bacteria also produce compounds that fight aquaculture pathogens. For example, *V. coralliilyticus* produces andrimid, a peptide antibiotic, while *P. halotolerans* produces holomycin, both of which are effective against *V. anguillarum*. Additionally, supernatants from *Pseudoalteromonas* sp. 3J6 contain proteinaceous exoproducts that inhibit biofilm formation and disrupt established biofilms of *V. tapetis*, a pathogen causing brown ring disease in Manila clams. These bioactive compounds, including subtilomycin, andrimid, holomycin, and the antimicrobial protein BLDZ1, are key players in preventing biofilm formation in aquaculture settings. Other substances from *Salinibacterium* and *Pseudoalteromonas* also exhibit biofilm inhibition (Table 12.1). By scaling up production processes and cloning the genes responsible for antimicrobial protein synthesis, these marine-derived compounds could become valuable tools for managing infections in aquaculture. While resistance may develop against some antibiotics, other substances like exopolysaccharides and proteins may remain effective. Therefore, marine-derived natural products continue to offer promising alternatives for combating drug-resistant bacterial infections (Liu MiaoMiao et al. 2019).

12.10 Non-adhesive Substances

Certain bioactive compounds derived from marine bacteria exhibit non-adhesive properties, playing a significant role in preventing biofilm-associated infections in aquaculture environments. Biosurfactants, a diverse group of secondary metabolites produced by microorganisms, are known for their hydrophilic and hydrophobic

Table 12.1 Summary of marine bacteria with antibiofilm potential through bactericidal, surfactant, and anti-adhesive activities

Marine bacteria	Effective against pathogens	Tested models	Bioactive compound
Bacillus subtilis	*Aeromonas hydrophila, V. anguillarum, and Alteromonas* sp.	–	Subtilomycin
Bacillus licheniformis	*Vibrio harveyi, V. alginolyticus, P. gessardii, and P. aeruginosa*	*Artemia salina*	Protein BLDZ1
Salinibacterium	*Photobacterium damselae*	–	Not identified
Vibrio coralliilyticus	*Vibrio anguillarum*	–	Andrimid
Photobacterium halotolerans	*Vibrio anguillarum*	–	Holomycin
Pseudoalteromonas sp.	*Vibrio tapetis*	–	Proteinaceous exoproducts
Biosurfactant and anti-adhesive compounds of Staphylococcus lentus	*Vibrio harveyi, P. aeruginosa*	*Artemia salina*	Glycolipid
Bacillus indicus, B. pumilus	*Vibrio harveyi*	–	Non-proteinaceous
Brevibacterium casei	*Vibrio vulnificus, V. fischeri, V. parahaemolyticus, V. alginolyticus, and V. harveyi*	–	Polyhydroxy butyrate
Pseudomonas	*Flavobacterium* sp.	–	Polysaccharides and proteins
Pseudomonas stutzeri	*Pseudomonas aeruginosa PAO1*	–	Exopolysaccharide

components. These compounds are considered more environmentally friendly than their chemical alternatives due to their lower toxicity, effectiveness in varying conditions, and biodegradability. Different types of biosurfactants, such as glycolipids, phospholipids, and lipopolysaccharides, have been identified for their role in reducing microbial adhesion and infection rates. Surfaces coated with biosurfactants can effectively lower microbial attachment, thereby mitigating the prevalence of infections in aquaculture systems.

Several studies highlight the effectiveness of surfactant molecules in combating biofilm-related infections. For instance, glycolipid biosurfactants produced by *Streptomyces lentus* were found to prevent biofilm formation by aquaculture pathogens like *V. harveyi* and *P. aeruginosa* by blocking pathogen adhesion. This strain of bacteria also protected brine shrimp larvae from infections. Furthermore, other compounds have been identified that inhibit bacterial attachment during the early stages of biofilm formation. For example, supernatants from *Bacillus* species were able to prevent the initial attachment and biofilm formation of *V. harveyi*, while also removing mature biofilms without causing bacterial death. Additionally, compounds like polyhydroxy butyrate (PHB), produced by *B. casei*, prevented the adhesion of various aquaculture pathogens.

In conclusion, bioactive compounds such as biosurfactants, proteins, and polysaccharides from marine bacteria show promise in preventing biofilm formation in aquaculture systems. These compounds not only reduce bacterial attachment but also minimize the risk of developing antibiotic-resistant pathogens, making them valuable for sustainable aquaculture practices.

12.11 Quorum-Sensing Regulation Disruption

In a process known as quorum sensing (QS), which permits cell-to-cell communication, bacterial gene expression is frequently controlled by chemical signalling molecules. The expression of virulence factors, such as poisons, siderophores, lytic enzymes, and adhesion molecules, as well as biofilm formation, bioluminescence, conjugation, antibiotic synthesis, swarming, nodulation, and sporulation, is all regulated by QS systems. Vibrio species are among the many aquatic pathogens that use QS to control the expression of virulence genes. Therefore, it is thought that one of the best ways to lower bacterial infections in aquaculture is to interfere with or inhibit these processes. When it comes to reducing the virulence and pathogenicity of aquaculture infections, quorum-sensing inhibitors (QSIs) that disrupt cell-to-cell communication are particularly useful. Numerous QSIs produced by marine microorganisms have been thoroughly investigated. By (i) breaking down autoinducer QS molecules, including N-acyl homoserine lactones (AHLs) made by Gram-negative bacteria, and (ii) generating QSIs themselves, these bacteria primarily interfere with QS-related processes.

12.12 Exploring AHL-Degrading Bacteria for Quorum-Sensing Disruption

Aquaculture farms use certain marine bacteria that can break down acyl homoserine lactones (AHLs) to control diseases. For instance, it was discovered that a strain of *Pseudomonas* sp. (PsDAHP1) from the intestines of Indian white prawns (*Fenneropenaeus indicus*) affected the hydrophobicity of *V. parahaemolyticus* cells and prevented the formation of biofilms. By reducing pathogen colonization in the intestines and gills, oral treatment of this strain to zebrafish decreased mortality. Quorum quenching, achieved through AHL degradation, occurs via enzymes such as AHL acylases and AHL lactonases. *Shewanella* sp., isolated from the intestine of Ayu fish, produces AHL acylases that degrade AHLs responsible for exoprotease production in *Aeromonas* species. Additionally, *B. licheniformis*, isolated from shrimp intestines, produces AHL lactonases that break the lactone ring of AHLs, reducing biofilm formation in *V. parahaemolyticus*. These bacteria, capable of modulating quorum sensing (QS), have potential as biocontrol agents to combat infections in aquaculture, offering an alternative approach to traditional methods of infection management.

12.13 Quorum-Sensing Inhibitors from Marine Microorganisms

Quorum-sensing inhibitors (QSIs), which interfere with the quorum-sensing (QS) process, are known to be produced by a number of marine microorganisms. These QSIs work by either disrupting the signalling pathways, creating molecules that resemble AHLs, or competing for receptors. For instance, *Pseudomonas* sp. FF16 and *Raoultella planticola* R5B1, both obtained from Chilean salmonid farms, were reported to efficiently suppress QS in a research that screened over 80 isolates. These bacteria generated homoserine lactones (HSLs), which prevented fish pathogens like *V. anguillarum*, *A. hydrophila*, and *Flavobacterium psychrophilum* from forming biofilms. It is possible that these isolates could be used as probiotics in the salmon farming sector. In a similar vein, *Synechococcus* sp.'s hexadecanoic acid decreased the generation of proteases in *V. vulnificus*, bioluminescence in *V. harveyi*, and biofilm development in several *Vibrio* species, including *V. vulnificus*, *V. alginolyticus*, *V. harveyi*, and *V. parahaemolyticus*. QS-related genes were downregulated in *V. harveyi*, according to a transcriptomic study, and in vivo investigations verified that it effectively shielded *A. salina* from *Vibrio* infections by lowering intestine colonization.

Marine Gram-positive bacteria have also shown QS inhibition. For instance, the marine actinomycete Streptomyces albus demonstrated QSI activity and was able to disrupt biofilms of *Vibrio* species. Recent studies have also highlighted that polysaccharides extracted from various marine actinomycetes were effective against the biofilms of aquaculture pathogens like *V. harveyi* through QSI. Additionally, a sterol derivative from *Nocardiopsis alba* inhibited bioluminescence in *V. harveyi* without toxicity. Certain members of the Bacillaceae family, such as *Halobacillus salinus* associated with seagrass, also produced phenethylamide metabolites that showed QSI activity by competing with AHLs for receptor binding, inhibiting bioluminescence in *V. harveyi*. Bacillus species were also effective in preventing biofilm formation in *P. aeruginosa*, reducing the production of virulence factors like protease, pyocyanin, and pyoverdin in a QS-dependent manner.

12.14 Other Marine Bacteria-Derived Agents for Biofilm Management

Apart from the previously mentioned substances, it has been discovered that additional biomolecules or extracts from marine microorganisms prevent the formation of biofilms. The processes behind these impacts are yet unclear, though. For example, a marine strain of Pseudomonas stutzeri 273 developed an exopolysaccharide that was demonstrated to both disrupt pre-existing biofilms and stop Pseudomonas aeruginosa PAO1 from forming new ones. Comprising glucosamine, rhamnose, glucose, and mannose, this exopolysaccharide decreased the synthesis of virulence components such as rhamnolipids, exoprotease, and pyocyanin. Additionally, it reduced the synthesis of extracellular DNA and hydrogen peroxide (H_2O_2), both of

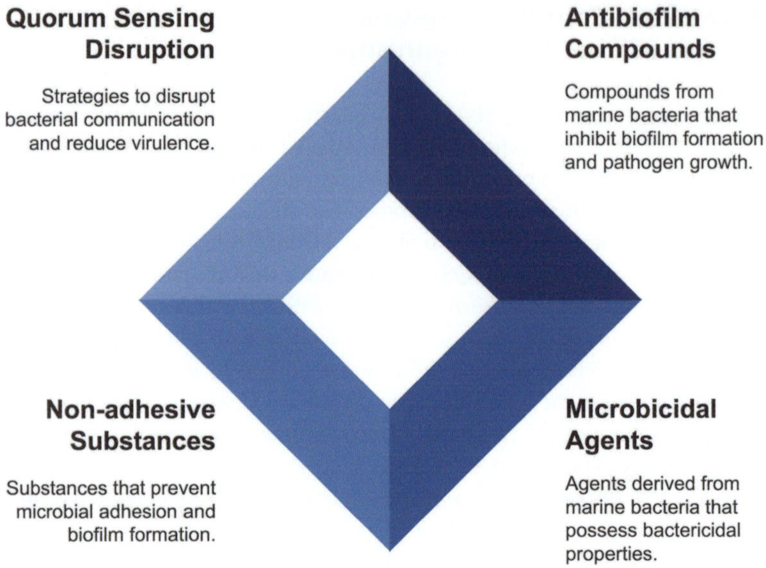

Fig. 12.5 Mechanistic insights into the disruption of aquaculture pathogen biofilms by marine microorganisms

which are necessary for the formation of biofilms. In a different instance, the *P. piscicida* strain DE2-B secreted proteolytic enzymes that suppressed fish infections, such as *Vibrio* species. Several genes encoding proteases with possible antibacterial qualities were found in the bacterium's genome (Wibowo et al. 2023) (Fig. 12.5).

12.15 Utilizing Marine Microalgae for Improved Aquaculture Practices

Marine microalgae are highly valued in aquaculture for their rich content of lipids, proteins, polysaccharides, pigments, vitamins, and minerals. They also serve as live feed for zooplankton used in aquaculture. These microalgae contribute to improving the quality of farmed products, enhancing colour and overall health. Various marine microalgae have been explored for use in aquaculture, primarily as feed supplements and probiotics.

Notable examples include *Tetraselmis* sp., which has been used to reduce microbial load, particularly *Vibrio* species, in Artemia cultures. Microalgae like *Tetraselmis chui* and *Chlorella minutissima* have also been effective in improving water quality, reducing *Vibrio* concentrations, and enhancing fish immunity when incorporated into diets. Studies have shown that including microalgae in fish diets can reduce the need for fish meal without compromising fish growth, while boosting immune responses.

Additionally, marine microalgae such as *Isochrysis* sp. and *Schizochytrium* sp. have replaced fish meal and oil in diets for rainbow trout, contributing to more sustainable feed formulations. Pavlova sp. has been used in diets for Atlantic salmon, improving growth and fatty acid profiles. Furthermore, certain species of the genus Haslea produce a blue pigment called marennine, which has antioxidant properties and protects oysters from *Vibrio* species while improving their appearance and market value (Wei et al. 2022).

Microalgal communities also play a role in nutrient removal, bioremediation, and wastewater purification in aquaculture systems. For instance, the association of *Rhodobacter azotofulmans* bacteria with *Chlorella vulgaris* microalgae has been shown to reduce nitrogen levels and improve fish production in recirculating ponds. Overall, marine microalgae contribute to better growth conditions, enhanced health, and sustainable practices in aquaculture.

12.16 Marine Fungi as Sources of Antimicrobial Agents for Aquaculture Applications

Over the past few decades, marine fungi have become a significant source of secondary metabolites with diverse applications in industries, pharmacology, and the environment. Several of these bioactive compounds have shown promise in combating biofilm-related infections caused by common pathogens. However, only a few studies have focused on their potential in controlling pathogens associated with aquaculture. Fungal genera like Aspergillus and Penicillium are particularly well-explored for their antimicrobial properties against aquaculture pathogens. For example, marine fungal strains, particularly from Aspergillus species, have yielded compounds like polyketides, phenolic prenylated derivatives, and terpenoids with activity against pathogens such as *A. hydrophila*, *V. anguillarum*, and *V. harveyi*. Some of these compounds exhibit strong, broad-spectrum antimicrobial properties. Similarly, Penicillium species have also produced antimicrobial compounds effective against pathogens in aquaculture, including *Vibrio* species and *A. hydrophila*. These compounds include bisabolane sesquiterpenes, bisthiodiketopiperazines, and pyranonigrin derivatives, among others (Table 12.2).

While various marine fungal species show potential in tackling aquaculture-associated infections, many of the compounds have not been thoroughly tested for their toxicity or effectiveness in real-world aquaculture conditions. There is also a need for further research to evaluate their ability to combat biofilm growth, a common challenge in aquaculture settings. The development of antifungal agents derived from marine fungi could provide alternative treatments to antibiotics traditionally used in aquaculture, reducing the risk of resistance development.

Table 12.2 Summary of bioactive compounds from marine fungi effective against aquaculture pathogens

Marine fungi	Effective against aquaculture pathogens	Bioactive compound	MIC (μg ml^{-1})
Aspergillus iizukae	*Vibrio* sp.	Not identified	–
Aspergillus ochraceus	*Aeromonas hydrophila, V. anguillarum, and V. harveyi*	Asperochrin A	8–16
Aspergillus terreus	*Aeromonas hydrophila, E. tarda, Micrococcus luteus, P. aeruginosa, V. harveyi, V. vulnificus, and V. parahaemolyticus*	Terreprenphenols A, B, C, and derivatives	2–64
Aspergillus versicolor	*Aeromonas hydrophila, E. tarda, P. aeruginosa, V. anguillarum, V. harveyi, and*	ent-Aspergoterpenin C,7-O-methyl hydroxysydonic acid, Pestalotiolactones C, D	4–8
Aspergillus wentii	*Aeromonas hydrophila, E. tarda, M. luteus P. aeruginosa, V. anguillarum, V. harveyi, and V. parahaemolyticus*	Aspewentin A, D, E, F, G, and H	0.5–32
Aspergillus wentii	*Edwardsiella tarda, V. harveyi, and V. parahaemolyticus*	Aspewentin I, J	8–32
Aspergillus sp.	*Vibrio harveyi*	Aspergillsteroid A	16
Aspergillus sp.	*Vibrio harveyi*	3-((1-Hydroxy-3-(2-methylbut-3-en-2-yl)-2- oxoindolin-3-yl) methyl)-1-methyl-3,4-dihydrobenzo[e][1,4]diazepine-2,5-dione, Austalide R	0.1–1
Aspergillus tritici	*Vibrio vulnificus, V. rotiferianus, and V. campbellii*	4-Methyl-candidusin A, aspetritone A, B	15–31
Aspergillus sp.	*Vibrio parahemolyticus, V. anguillarum, and V. alginolyticus*	4-Aspergixanthone I	1.56–3.12 μM
Aspergillus clavatus	*Aeromonas hydrophila, V. harveyi, P. aeruginosa, M. luteus, V. anguillarum, V. parahaemolyticus*	Seco-clavatustide B, 5- acetyl-2,4-dihydroxy-3-methylbenzoic acid	8–>100 μM

(continued)

Table 12.2 (continued)

Marine fungi	Effective against aquaculture pathogens	Bioactive compound	MIC (μg ml^{-1})
Aspergillus fumigatus	*Vibrio harveyi*	Trypacidin	31.25
Penicillium aculeatum	*Aeromonas hydrophilia, E. tarda, M. luteus, P. aeruginosa, V. anguillarum, V. harveyi, V. parahaemolyticus, and V. alginolyticus*	1-hydroxyboivinianin A, (7S,11S)-(+)-12-hydroxysydonic acid, 1-hydroxyboivinianin A	1–32
Penicillium adametzioides	*Aeromonas hydrophila, V. harveyi, and V. parahaemolyticus*	Adametizine A	8–32
Penicillium brocae	*Vibrio harveyi, V. parahaemolyticus*	Pyranonigrin F and A	0.5
Penicillium chermesinum	*Micrococcus luteus*	Chermesins A and B	8
Penicillium chermesinum	*Vibrio anguillarum, V. parahaemolyticus, and M. luteus*	Chermesiterpenoid B Chermesiterpenoid C	0.5–64
Penicillium citrinum	*Vibrio alginolyticus, V. parahaemolyticus*	4-Chloro-1-hydroxy-3-methoxy-6-methyl- 8-methoxycarbonyl-xanthen-9-one (1) and 2r-acetoxy-7-chlorocitreorosein	>50 μM
Penicillium citrinum	*Vibrio vulnificus, V. campbellii, and V. rotiferianus*	Penicitol D, 1-epi-citrinin H1	30–32 μM
Penicillium cyclopium	*Edwardsiella tarda, E. ictalurid, M. luteus, V. harveyi, and V. anguillarum*	9-Dehydroxysargassopenilline A, 1,2-didehydropeaurantiogriseol E	4–8
Penicillium simplicissimum	*Aeromonas hydrophila, M. luteus, P. aeruginosa, V. alginolyticus, V. harveyi, and V. parahaemolyticus*	Penicisimpins A, B, and C	4 > 64
Penicillium sp.	*Vibrio harveyi*	Penicillilactone A	8
Penicillium atrovenetum	*Lactococcus garvieae*	Not identified	–
Cladosporium cladosporioides	*Micrococcus luteus, V. harveyi*	Cladosporols F, G, H, and I	4–128
cladosporioides	*aeruginosa*	lide, ent-cladospolide F, cladospolide G, cladospolide H	
Cladosporium cladosporioides	*Edwardsiella tarda, E. ictarda*	Thiocladospolides A, B, C, and D	1–8

(continued)

Table 12.2 (continued)

Marine fungi	Effective against aquaculture pathogens	Bioactive compound	MIC (μg ml^{-1})
Cladosporium	*Edwardsiella tarda, P. aeruginosa, and V. anguillarum*	Cladocladosin A, thiocladospolides F, G	0.5–4
Acremonium sp.	*Vibrio scophthalmi, V. shilonii, and Vibrio brasiliensis*	Acremocholone and three analogues	8–32
Chaetomium sp.	*Vibrio rotiferianus, V. vulnificus, and V.campbellii*	Chaephilone C, Chaetoviridides A, B, and C	7–32
Engyodontium album	*Vibrio vulnificus*	Ethyl 3,5-dimethoxy-2-propionyl benzoate, ethyl 3,5-dihydroxy-2-propionylbenzoate	15.6
Eutypella sp.	*Vibrio vulnificus, V. alginolyticus, and A. hydrophila*	Libertellenones M, N	16–64
Fusarium solani	*Pseudomonas aeruginosa, V. parahaemolyticus*	Fusaricates H, I, J, and K, Fusolanones A, B, fusaric acid	6.25–>200
Stachybotrys sp.	*Micrococcus luteus*	Stachybomycins E	8
Trichoderma	*Micrococcus luteus Pseudoalteromonas citrea*	trichophenol A (10E)-isocyclonerotriol, (10Z)-isocyclonerotriol, trichophenol A	16
Trichoderma reesei	*Vibrio harveyi*	Not identified	–

12.17 Marine Microalgae as Natural Defenders Against Aquaculture Pathogens

Marine microalgal cultures have occasionally been successful in controlling the growth of pathogens in aquaculture. Several studies have explored coculturing microalgae with aquaculture pathogens to suppress pathogen growth. For instance, coculturing species like *Tetraselmis suecica* and *Chaetoceros calcitrans* has shown significant inhibition of *V. parahaemolyticus* (a cause of acute hepatopancreatic necrosis disease). Similarly, seawater extracts of *C. calcitrans* also exhibited inhibitory effects on the same pathogen. Other studies have demonstrated that axenic cultures of *Tetraselmis chuii* alongside *Chlorella minutissima*, *Isochrysis* sp., and *Nannochloropsis* sp. were effective in inhibiting the growth of various *Vibrio* species. Additional research found that *Chlorella* sp., *C. calcitrans*, and *Nitzschia* sp. were capable of suppressing the growth of *V. harveyi*. Other species such as *Nannochloropsis oceanica*, *C. gracilis*, *Isochrysis* sp., and *Thalassiosira weissflogii* have also been found to inhibit *V. harveyi* growth (Table 12.3). However, many of these studies have not identified the specific active compounds responsible for

Table 12.3 Overview of the roles of marine microalgae in aquaculture systems as biocontrol agents

Marine algae	Effective against pathogen/ mode of action	Tested models	Mode of supply
Tetraselmis suecica, Chaetoceros calcitrans	*Vibrio parahaemolyticus*	–	Coculture
Tetraselmis chui, C. minutissima, Isochrysis sp., *and Nannochloropsis* sp.	*Vibrio* sp.	–	Coculture
Chlorella sp.	*Vibrio harveyi*	–	Coculture
Chaetoceros calcitrans, Nitzschia sp.	*Vibrio harveyi*	–	Coculture
Nannochloropsis oceanica, C. gracilis, Isochrysis sp., *and Thalassiosira weissflogii*	*Vibrio harveyi*	–	Coculture
Marine algal endophytes, as feed supplements	*Photobacterium damselae, S. iniae, and A. salmonicida*	–	Plate assays
Dunaliella tertiolecta	*Vibrio campbellii, V. proteolyticus*	*Artemia* sp.	Feed
Tetraselmis chuii, C. minutissima	*Vibrio* sp.	Artemia	Enrichment
Tetraselmis sp., *Nannochloropsis* sp., *and Chlorella* sp.	High protein content	–	Feed
Tetraselmis chui	*Vibrio* sp. *increased immunity, improvement of water quality, and utrient recovery*	*Solea senegalensis*	Feed
Tetraselmis suecica, Tisochrysis lutea	Enhanced non-specific immune response	*European sea bass*	Feed
Isochrysis galbana	*Edwardsiella tarda better survival, growth rate, and enhanced innate immunity*	*Danio rerio*	Feed
Isochrysis sp., *Schizochytrium* sp.	Sustainable feed replacing fishmeal/fish oil	*Oncorhynchus mykiss*	Feed
Schizochytrium sp.	Sustainable feed replacing fishmeal/fish oil	*Oncorhynchus mykiss*	Feed
Pavlova sp.	Sustainable feed replacing fishmeal/fish oil, enhanced growth	*Salmo salar*	Feed
Haslea ostrearia	Extracellular marennine pigment greening of oysters	Oysters	Extracted pigment
Haslea karadagensis, H. silbo, and *H. provincialis*	*Vibrio* sp. *greening of oysters*	Oysters	Feed

pathogen inhibition. Furthermore, marine algal endophytes, including bacterial and fungal strains, have also been shown to combat aquaculture pathogens and may offer a potential solution for controlling diseases in aquaculture settings. Overall, microalgae contribute to pathogen management in aquaculture both by directly

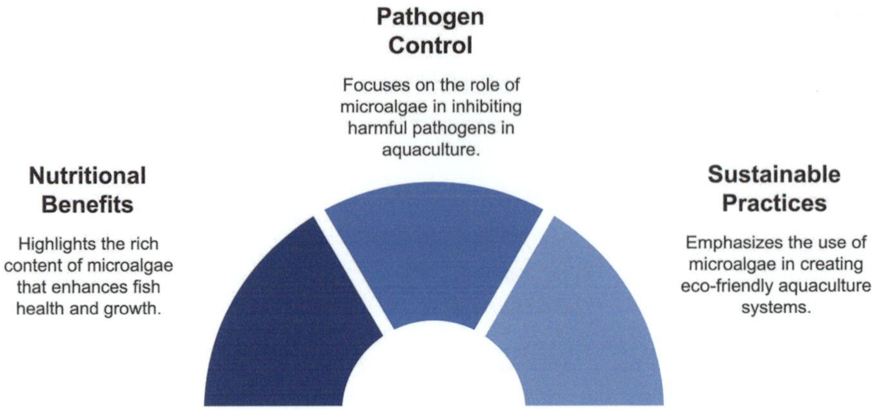

Fig. 12.6 Overview of microalgae as biocontrol agents against aquaculture pathogens

inhibiting pathogen growth and by producing bioactive compounds, making them beneficial for improving aquaculture conditions and yields (Fig. 12.6).

12.18 Marine Microorganisms as Natural Immunostimulants and Probiotic Agents for Aquatic Health

Probiotics are live microorganisms that serve as dietary supplements, benefiting host organisms by improving gut health and balance. These probiotics help prevent the growth of harmful pathogens by competing for available surfaces and nutrients and producing various inhibitory substances such as antibiotics, bacteriolytic enzymes, and bacteriocins. Additionally, they support digestion, enhance the host's innate immunity, and contribute to better water quality. The subsequent sections describe how Gram-positive and Gram-negative bacteria, yeasts, and microalgae are used to enhance aquaculture productivity.

12.19 The Role of Gram-Positive Bacteria as Probiotics and Immune Enhancers in Aquaculture

Bacillus species are among the Gram-positive bacteria whose probiotic and immunostimulatory qualities have been the subject of much research. For instance, when given just prior to exposure to *V. campbellii*, *Bacillus* species improved the survival of brine shrimp larvae. Increased prophenol oxidase mRNA expression and decreased levels of stress-related proteins including transglutaminase and heat shock protein 70 were indicators of a greater immunological response brought on by this treatment. It has also been demonstrated that *B. coagulans* enhances common carp growth, survival, immunological response, and meat quality. Likewise,

B. licheniformis enhanced the growth, immunological, and antioxidant responses of *V. parahaemolyticus*-infected striped catfish.

By producing lactic acid and biosurfactants, *Lactobacillus pentosus* H16 showed protection against *V. alginolyticus* infections in *A. franciscana nauplii*. *Litopenaeus vannamei* grew and responded better to an additional strain of *L. pentosus* that was isolated from the digestive tract of *Acanthogobius hastali*. Furthermore, it was discovered that *Streptomyces rubrolavendulae* M56, a marine actinomycete, might prevent vibriosis in aquaculture settings by competitively rejecting other *Vibrio* species, such as V. harveyi, and preventing infection of Penaeus monodon post-larvae (Augustine et al. 2016).

Staphylococcus lentus, *Lactococcus lactis*, and *Enterococcus faecium* are among the other Gram-positive bacteria that provide defence against aquaculture diseases. For example, *Enterococcus fecium* isolated from *A. hastatus* enhanced *L. vannamei* growth and immunological response. Olive flounder was shielded from *S. parauberis* infections by *Lactococcus lactis*, which is present in the digestive tracts of wild marine fish and has shown inhibitory efficacy against a variety of aquaculture pathogens. In addition, *S. lentus* effectively prevented infections in *A. salina* larvae and excluded *V. harveyi*. These results demonstrate that probiotic formulations containing *Bacillus*, *Lactobacillus*, *Streptomyces*, *Enterococcus*, *Lactococcus*, and *Staphylococcus* species can be used to improve the growth and survival of cultured aquatic organisms (Table 12.4).

12.20 Immune Modulation and Probiotic Benefits of Gram-Negative Bacteria in Aquatic Environments

In aquaculture, *Shewanella* species have demonstrated efficacy as probiotics. For example, feeding abalone (*Haliotis discus*), *S. colwelliana* and *S. olleyana* increased their immunity and survival against *V. harveyi* infections. The cellular and humoral immunity of gilt-head bream (*Sparus aurata*) was also improved by feeding them *S. putrefaciens*, especially in stressful situations like high stocking density. When added to an oil-dressed diet, *Enterobacter* species also protected rainbow trout (*Oncorhynchus mykiss*) from infection by *F. psychrophilum*.

Apart from these species, bacteria belonging to the Roseobacter group, like *Phaeobacter inhibens*, have demonstrated the ability to manage aquaculture diseases. *V. anguillarum*'s growth in Artemia culture was successfully inhibited by *P. inhibens*, the bacteria that generate the antibiotic tropodithietic acid (TDA). Likewise, Artemia's growth was aided by *Rhodobacter mobilis*, another bacterium that produces TDA. The health of cultivated species has also been enhanced by the addition of other Gram-negative marine bacteria, such as *Proteus*, *Providencia*, *Klebsiella*, and *Alcaligenes*. Furthermore, it has been noted that giving *Cyprinus carpio A. veronii*, *V. lentus*, and *F. sasangense* enhances their immunological responses (Table 12.4).

Pacific red snapper (*Lutjanus peru*) injected with lysates and ToxA produced from *V. parahaemolyticus* demonstrated improved immunological parameters,

Table 12.4 Summary of marine microorganisms (bacteria and yeasts) used as probiotics and immunostimulants in aquaculture

Marine microorganism	Effective against pathogen/ Mode of activity	Tested models
Bacillus sp.	*Vibrio* sp. and other general pathogens	*Artemia* species, *Cyprinus carpio, and Pangasius hypophthalmus*
Lactobacillus pentosus	*Vibrio alginolyticus*	*Artemia* species, *Litopenaeus vannamei*
Enterococcus faecium	Improved immune response	*Litopenaeus vannamei*
Streptomyces rubrolavendulae	*Vibrio* sp.	*Penaeus monodon*
Lactococcus lactis	*Streptococcus parauberis*	*Paralichthys olivaceus*
Staphylococcus lentus	*Vibrio harveyi*	*Artemia* species
Shewanella colwelliana, S. olleyana, and S. putrefaciens	*Vibrio harveyi*	*Haliotis discus, Sparus aurata*
Enterobacter sp.	*Flavobacterium psychrophilum*	*Oncorhynchus mykiss*
Phaeobacter inhibens, Ruegeria mobilis, Roseobacter sp., *Proteus* sp., *Providencia* sp., *Klebsiella* sp., *and Alcaligenes* sp.	*Vibrio anguillarum*, improvement in general health conditions	*Artemia* sp., *P. olivaceus*
Aeromonas veronii, V. lentus, and F. sasangense	Improved immune response	*Cyprinus carpio*
Vibrio parahaemolyticus	Improved immune response	*Lutjanus peru*
Rhodosporidium paludigenum		*Litopenaeus vannamei*
Debaryomyces hansenii, Candida tropicalis, C. aquaetextoris, C. sake, and C. utilis	White spot syndrome virus	*Penaeus monodon, F. indicus*
Hanseniaspora opuntiae	*Vibrio splendidus*	*Holothuroidea* sp.
Yarrowia lipolytica	*Vibrio parahaemolyticus, V. harveyi*	*Lutjanus peru, Artemia* sp.

increased expression of immune-related genes like IL-1b, TNF-a, C3, and IgM, and improved survival in a different study. Overall, it has been demonstrated that using these Gram-negative bacteria and chemicals from *V. parahaemolyticus* enhances the growth rates, immunological responses, and general health of aquaculture species.

12.21 Exploring the Potential of Marine Yeasts as Probiotics and Immune Enhancers

Marine yeasts play a significant role in combating aquaculture-associated infections, with certain strains showing immunostimulatory effects when included in diets. For example, the Basidiomycetous yeast *R. paludigenum* has been found to improve growth and antioxidant levels in *L. vannamei*, though most studies focus on Ascomycetous genera. Various investigations highlight the immunostimulatory potential of marine yeasts like *C. sake* and *D. hansenii*. Adding *C. sake* to the diet of *F. indicus* improved its immune function. Glucans extracted from *D. hansenii* and *C. tropicalis* enhanced immunity in *P. monodon* and provided protection against White Spot Syndrome Virus (WSSV). Yeast cells from these species also proved more effective than common baker's yeast in boosting immune responses.

Other yeasts like *C. aquetextoris* offered protection against WSSV in P. monodon, with optimal immune responses observed at a 10% diet inclusion. *Hanseniaspora opuntiae* also enhanced immunity and disease resistance in juvenile sea cucumbers, especially against *V. splendidus*. Marine yeasts such as *Y. lipolytica* have shown promise in improving immunity in species like *L. peru* and *L. vannamei* after exposure to pathogens. These yeasts regulate genes associated with immunity and oxidative stress, leading to better overall immune function and survival. *Y. lipolytica* strains, when enriched with selenium nanoparticles, have also been effective in boosting resistance to infections like *V. harveyi* in *A. salina*.

In summary, marine yeasts, such as Candida, *D. hansenii*, *Y. lipolytica*, and others, offer enhanced protection and growth benefits for cultured species, outperforming traditional feeds like *S. cerevisiae*. Their immunostimulatory properties could reduce the need for antibiotics in aquaculture, contributing to more sustainable farming practices.

12.22 Conclusion and Future Perspectives

The future of sustainable fisheries heavily relies on advancing marine microbiome research to combat the challenges faced by aquaculture systems, particularly those related to disease management, immunity enhancement, and biofilm formation. With increasing concerns about antibiotic resistance and the ineffectiveness of conventional treatments, marine microorganisms such as bacteria, yeasts, and microalgae present a promising alternative for improving the health and growth of cultured species. Aquaculture systems can reduce antibiotic dependency and enhance sustainability by harnessing the therapeutic potential of these marine microbes and their bioactive compounds. Future research should focus on understanding these microorganisms' molecular mechanisms, efficacy, and safe application in aquaculture settings, including their role in immune modulation and disease prevention. Additionally, integrating marine microbiomes into feed formulations and farm management practices can optimize productivity, improve water quality, and ensure ecological balance, paving the way for more sustainable and resilient aquaculture practices in the future.

References

Abdelsalam M, Elgendy MY, Elfadadny MR, Ali SS, Sherif AH, Abolghait SK (2023) A review of molecular diagnoses of bacterial fish diseases. Aquac Int 31(1):417–434

Augustine D, Jacob JC, Philip R (2016) Exclusion of V ibrio spp. by an antagonistic marine actinomycete *S treptomyces rubrolavendulae* M56. Aquac Res 47(9):2951–2960

Berg G, Rybakova D, Fischer D, Cernava T, Vergès MCC, Charles T, Chen X, Cocolin L, Eversole K, Corral GH, Kazou M (2020) Microbiome definition re-visited: old concepts and new challenges. Microbiome 8:1–22

FAO (2020) The state of world fisheries and aquaculture. Rome: Sustainability in action

Kumar V, Roy S, Behera BK, Swain HS and Das BK (2021) Biofloc microbiome with bioremediation and health benefits. Front Microbiol 12:741164

Kumar V, Bera T, Roy S, Vuong P, Jana C, Sarkar DJ, Devi MS, Jana AK, Rout AK, Kaur P and Das BK (2023) Investigating bio-remediation capabilities of a constructed wetland through spatial successional study of the sediment microbiome. NPJ Clean Water 6(1):8

Liu MiaoMiao LM, El-Hossary EM, Oelschlaeger TA, Donia MS, Quinn RJ, Abdelmohsen UR (2019) Potential of marine natural products against drug-resistant bacterial infections. Lancet Infect Dis 19(7):e237–e245

Minich JJ, Nowak B, Elizur A, Knight R, Fielder S, Allen EE (2021) Impacts of the marine hatchery-built environment, water and feed on mucosal microbiome colonization across ontogeny in Yellowtail Kingfish, Seriola lalandi. Front Mar Sci 8:676731

Mohanty D, Kumari M, Kumar V, Kumari P, Behera B and Das BK (2025) Genome sequence of prodigiosin-producing and tetracycline-resistant Serratia marcescens (IFSMLMEK1), identified from a freshwater bivalve Lamellidens marginalis. Microbiol Resour Announc 14(4):e01115-24

Qu J, Yang H, Liu Y, Qi H, Wang Y, Zhang Q (2021) The study of natural biofilm formation and microbial community structure for recirculating aquaculture system. In: IOP conference series: earth and environmental science, vol 742, no 1. IOP Publishing, p 012018

Vonaesch P, Anderson M, Sansonetti PJ (2018) Pathogens, microbiome and the host: emergence of the ecological Koch's postulates. FEMS Microbiol Rev 42(3):273–292

Wei M, Parrish CC, Guerra NI, Tibbetts SM, Colombo SM (2022) Dietary inclusion of a marine microalgae meal for Atlantic salmon (*Salmo salar*): impact of Pavlova sp. 459 on growth performance and tissue lipid composition. Aquaculture 553:738084

Wibowo JT, Bayu A, Aryati WD, Fernandes C, Yanuar A, Kijjoa A, Putra MY (2023) Secondary metabolites from marine-derived bacteria with antibiotic and antibiofilm activities against drug-resistant pathogens. Mar Drugs 21(1):50

Genomic Tools for Fish Breeding Programs

13

Ayushman Gadnayak, Smruti Priyambada Pradhan, and Subhasmita Behera

13.1 Introduction

Fish breeding programs have been a cornerstone of aquaculture for decades, aiming to enhance the productivity, health, and sustainability of fish populations. With the advent of genomic tools, these programs have undergone a significant transformation, allowing for more precise and efficient selection of desirable traits. This chapter delves into the importance of genomic tools in modern aquaculture, exploring their impact on fish breeding programs and the future directions of this rapidly evolving field. Fish breeding programs are designed to improve the genetic quality of fish stocks, focusing on traits such as growth rate, disease resistance, and reproductive performance. Traditional breeding methods rely on phenotypic selection, where breeders choose individuals based on observable characteristics. However, these methods are often time-consuming and may not always accurately reflect the genetic potential of the fish (Chavanne et al. 2016). Genomic tools have revolutionized fish breeding programs by providing a deeper understanding of the genetic makeup of fish populations. These tools include techniques such as genome sequencing, genotyping, and genome-wide association studies (GWAS), which allow breeders to identify and select for specific genetic markers associated with desirable traits (Abdelrahman et al. 2017). By leveraging these tools, breeders can accelerate the selection process, reduce the time required to achieve genetic gains, and improve the overall efficiency of breeding programs (Yáñez et al. 2015; Vallejo et al. 2024). Genomic selection in salmonids has led to discoveries and future perspectives, highlighting the potential for significant genetic progress (D'Agaro et al. 2021). Furthermore, the implementation of breeding programs using genomic tools

A. Gadnayak (✉) · S. P. Pradhan · S. Behera
ICAR-Central Inland Fisheries Research Institute, Barrackpore, Kolkata, India

© The Author(s), under exclusive license to Springer Nature Singapore Pte Ltd. 2026

B. K. Das et al. (eds.), *Bioinformatics Tools for Predictive Ecology and Fisheries*, Sustainable Aquatic Food Systems, https://doi.org/10.1007/978-981-95-2573-7_13

has been documented to speed up the breeding process and enhance the genetic quality of fish populations (Wells et al. 2020).

The primary objective of this chapter is to provide a comprehensive overview of the genomic tools used in fish breeding programs and their impact on modern aquaculture. The chapter will cover the basics of fish genomics, the implementation of genomic selection and marker-assisted selection (MAS), and the role of functional genomics and epigenomics in understanding fish traits. Additionally, the chapter will discuss the challenges and future directions of genomic tools in fish breeding programs, highlighting the potential for further advancements in this field (D'Agaro et al. 2021; Abdelrahman et al. 2017). The insights gained from genomics in aquaculture have enabled researchers to understand species biology better and accelerate genetic progress, contributing to the overall success of fish breeding programs (Yáñez et al. 2015; Yáñez et al. 2015).

Genomics is the study of genomes, the complete set of DNAs within an organism (Ahmad et al. 2021). In fish, genomics helps us understand the genetic basis of traits such as growth, disease resistance, and adaptation to environmental changes (Ramya and Behera 2023). By studying fish genomes, scientists can identify genes responsible for important traits, which can be used to improve aquaculture practices and conserve endangered species (Ahmad et al. 2021; Ramya and Behera 2023). Genetic markers are specific DNA sequences that can identify individuals or species and track genetic traits (Yang et al. 2020). Common types of genetic markers in fish include microsatellites, single-nucleotide polymorphisms (SNPs), and amplified fragment length polymorphisms (AFLPs) (Yang et al. 2020). Genetic mapping involves locating these markers on chromosomes to study the inheritance patterns of traits and identify genes associated with specific phenotypes (Yang et al. 2020).

Next-generation sequencing (NGS) technologies allow for rapid and cost-effective sequencing of entire fish genomes (Khatei et al. 2022; Gheyas et al. 2009). This has revolutionized our understanding of fish genetics and enabled comparative genomics studies (Ahmad et al. 2021). In this regard, techniques such as SNP arrays and high-throughput sequencing are used to identify genetic variations within and between fish populations (Wang et al. 2024). This information is crucial for breeding programs and conservation efforts (Wang et al. 2024). The computational tools and databases are essential for analyzing large genomic datasets and identifying genes of interest (Nguyen et al. 2022). Databases like FishDB provide comprehensive resources for fish genomics research (Nguyen et al. 2022). In this context, technologies like CRISPR-Cas9 allow scientists to make precise changes to the DNA of fish, enabling the study of gene function and the development of genetically improved fish strains (Soon and Zheng 2020). A list of the most important genetic tools used in fish breeding programs is shown in Table 13.1.

Table 13.1 The applications for each key genomic technology used in fish breeding programs and their specific uses

Sl. no.	Technology	Description	Application names and uses
1.	Sequencing	Next-generation sequencing (NGS) technologies allow for rapid and cost-effective sequencing of entire fish genomes	Illumina NovaSeq, PacBio Sequel, Oxford Nanopore: Used for understanding genetic diversity, identifying genetic markers, comparative genomics, detecting mutations, and discovering new genes
2.	Genotyping	Techniques such as SNP arrays and high-throughput sequencing are used to identify genetic variations within and between fish populations	Affymetrix Axiom, Illumina Infinium: Used for marker-assisted selection, population genetics, tracking inheritance of traits, and conservation genetics
3.	Bioinformatic	Computational tools and databases are essential for analyzing large genomic datasets and identifying genes of interest	BLAST, Genome Browser, FishDB: Used for data analysis, genome annotation, functional genomics, evolutionary studies, and managing genetic information
4.	Genome editing	Technologies like CRISPR-Cas9 that allow precise changes to the DNA of fish	CRISPR-Cas9, TALENs, ZFNs: Used for functional studies, developing genetically improved strains, gene knockouts, gene insertion, and studying gene function
5.	GWAS	Genome-wide association studies involve scanning the entire genome to find genetic associations with observable traits	PLINK, GCTA: Used for identifying genetic markers linked to traits like growth rate, disease resistance, and reproductive performance
6.	QTL mapping	Quantitative trait loci (QTL) mapping involves identifying genome regions associated with quantitative traits	MapQTL, R/qtl: Used to understand the genetic basis of complex traits, improve selection accuracy, and identify candidate genes
7.	Transcriptomics	Study of RNA transcripts to understand gene expression patterns	RNA-Seq, Cufflinks, DESeq2: Used for identifying active genes in different tissues, understanding regulatory mechanisms, and studying responses to environmental changes
8.	Proteomics	Study of the proteome, the entire set of proteins expressed by a genome, cell, tissue, or organism	Mass Spectrometry, Proteome Discoverer: Used for understanding protein functions, interactions, and pathways and studying post-translational modifications

(continued)

Table 13.1 (continued)

Sl. no.	Technology	Description	Application names and uses
9.	Epigenomics	Study of heritable changes in gene expression that do not involve changes to the underlying DNA sequence, such as DNA methylation and histone modification	Bisulfite Sequencing, ChIP-Seq, ATAC-Seq: Used for understanding gene regulation, studying the impact of environmental factors on gene expression, and identifying epigenetic markers for selection
10.	Metabolomics	Study of the complete set of metabolites (small molecules) within a biological sample	LC-MS, GC-MS, NMR Spectroscopy: Used for understanding metabolic pathways, studying responses to environmental changes, and identifying metabolic markers linked to health and disease

13.2 Genomic Selection in Fish Breeding

Genomic selection (GS) is a transformative breeding method that utilizes genome-wide molecular markers to predict the genetic potential of fish for complex traits. This approach significantly enhances selection accuracy compared to traditional methods, enabling faster genetic gains, cost-effective breeding, and improved disease resistance. By leveraging SNP arrays and next-generation sequencing technologies, GS captures genetic variance across numerous loci, leading to precise predictions and better breeding outcomes. Implementing GS in fish breeding involves establishing a reference population that is thoroughly genotyped and phenotyped. Statistical models are then developed to predict genomic breeding values (GBVs). These models help select candidates with the highest predicted genetic merit for breeding. The efficiency of GS lies in its ability to provide detailed genetic information, reducing reliance on extensive phenotyping and accelerating genetic improvement. Key steps include genotyping, phenotyping, model development, and strategic breeding decisions. Success stories in fish breeding illustrate the impact of GS. For instance, in Atlantic salmon breeding programs, GS has significantly improved growth rates and disease resistance. Rainbow trout programs have also seen enhanced fillet quality and stress tolerance through GS. Similarly, marine shrimp breeding has benefited from GS with faster genetic gains and improved disease resistance, demonstrating the broad applicability and benefits of genomic selection across different aquaculture species (Karlsson et al. 2020; Ozaki et al. 2001; Eibeck et al. 2024).

13.3 Genome-Wide Association Studies (GWAS)

Ever wondered how scientists pinpoint the genes that make us who we are? That is where Genome-Wide Association Studies (GWAS) come in (Sharma and Kumar 2022). These studies are like genetic treasure hunts, scouring the DNA of many individuals to find variations linked to specific traits or diseases. By comparing the genomes of people (or animals, in our case) with different traits, researchers can identify which genetic variants are more common in those with a particular characteristic (Beck et al. 2012). It is a powerful way to uncover the genetic secrets behind complex traits and conditions (Beck et al. 2012). GWAS has revolutionized this field by helping breeders identify genetic markers associated with traits like growth rate, disease resistance, and stress tolerance (Marees et al. 2018). Imagine being able to select fish that grow faster, resist disease better, or handle stress without breaking a sweat, GWAS makes that possible (Marees et al. 2018). These genetic insights are then used in selective breeding programs, making aquaculture more efficient and sustainable. It is like giving fish breeders a high-tech toolkit to craft the best possible fish for farming (Marees et al. 2018). Researchers have made some exciting discoveries using GWAS in aquaculture. For tilapia, they have pinpointed genetic variants that lead to faster growth rates. This means fish farmers can now selectively breed tilapia that grow quicker, increasing productivity (Xu et al. 2015). In salmon, genetic markers have been found that offer resistance to diseases like sea lice and infectious salmon anemia (Xu et al. 2015). This helps breed more resilient fish that can withstand these common issues. Additionally, in catfish, GWAS has identified genetic variants that enhance stress tolerance. This is crucial for maintaining healthy and well-adjusted fish in farming environments (Zenger et al. 2019). All these findings combined make fish breeding more efficient and sustainable, paving the way for a brighter future in aquaculture. The visual representing the key points of genome-wide association studies (GWAS) in aquaculture is shown in Fig. 13.1.

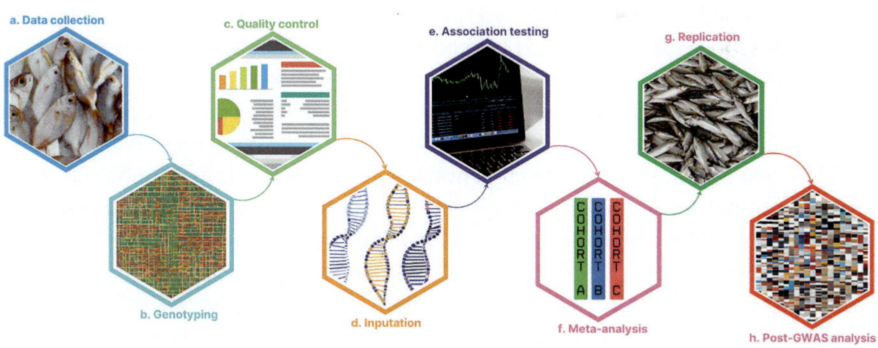

Fig. 13.1 (**a**) Collect data from cohorts or biobanks, considering confounders. (**b**) Use microarrays or sequencing for genotypic data. (**c**) Conduct quality control on genotypes. (**d**) Phase and impute genotypic data using reference populations. (**e**) Run genetic association tests, adjust for confounders, and control for multiple testing. (**f**) Combine results from multiple cohorts. (**g**) Replicate results using independent cohorts. (**h**) Conduct in silico analysis and test hypotheses experimentally

13.4 Marker-Assisted Selection (MAS)

Marker-Assisted Selection (MAS) was a technique used in selective breeding to improve desirable traits in fish by using molecular markers linked to Quantitative Trait Loci (QTL) (Eze 2019). This method allowed breeders to select individuals with favorable genetic traits more accurately and efficiently than traditional phenotype-based selection. MAS was particularly beneficial for traits that were difficult to measure, had low heritability, or were influenced by recessive genes (Eze 2019). In the world of Marker-Assisted Selection (MAS), researchers embarked on a fascinating journey. They first identified genetic markers, like Single Nucleotide Polymorphisms (SNPs), which were linked to desirable traits (Haldar 2018). Once these markers were pinpointed, fish were genotyped to check for the presence of these beneficial markers (Haldar 2018). The fish that possessed the right genetic markers were then selected for breeding (Eze 2019). These chosen fish were bred to produce offspring that would inherit the targeted desirable traits, improving the overall quality of the stock. While MAS honed in on specific markers linked to desirable traits, genomic selection (GS) took it a step further. GS used information from the entire genome to predict the breeding value of each fish (Martinez et al. 2007). This meant GS could be more effective for improving complex traits controlled by multiple genes. However, MAS was more cost-effective for traits with well-characterized markers, making it a valuable tool in the fish breeder's toolkit (Unêda-Trevisoli et al. 2017).

13.5 Functional Genomics

In the past, functional genomics played a crucial role in uncovering the complexities of fish traits. Researchers utilized advanced technologies such as transcriptomics, proteomics, and metabolomics to explore gene expression, protein functions, and metabolic pathways. This helped them gain a deep understanding of how fish adapted to their environments and responded to various stressors (Ahmad et al. 2021). When diving into functional genomics, researchers explored three fascinating areas: transcriptomics, proteomics, and metabolomics. By studying RNA transcripts, they could see which genes were active and how their expression changed in different environments, providing valuable insights into gene function (Natnan et al. 2021). Proteomics took it a step further by analyzing all the proteins fish produced, helping scientists understand protein functions and interactions. Meanwhile, metabolomics focused on the tiny molecules involved in metabolism, shedding light on how these processes influenced traits like growth and health. These combined approaches gave a comprehensive view of the inner workings of fish at the molecular level, making it a powerful tool in fish breeding and aquaculture. When it came to practical applications in fish breeding, functional genomics truly made a difference. Imagine breeders having a roadmap to pinpoint fish with the best genetic traits, such as enhanced disease resistance or faster growth rates. Genomic selection allowed them to do just that, making breeding programs much more efficient. This method ensured that only the best traits were passed down, supporting more sustainable practices in aquaculture. It was a game-changer, helping to produce healthier and more resilient fish for future generations (Zenger et al. 2019).

13.6 Epigenomics and Gene Editing

Epigenetic modifications, such as DNA methylation, histone modification, chromatin remodelling, and the action of noncoding RNAs (ncRNAs), have attracted significant interest for their practical potential in fish reproduction (Sumon et al. 2022). These modifications can influence gene expression without altering the DNA sequence, thereby affecting traits such as growth, reproduction, and disease resistance (Yu et al. 2022). For instance, DNA methyltransferase has been shown to improve reproductive gene expression and feminize zebrafish (*Danio rerio*) (Sumon et al. 2022).

CRISPR/Cas9 technology has revolutionized the field of genetic engineering by allowing precise and targeted modifications to the genome (Liu et al. 2022). In aquaculture, CRISPR/Cas9 has been used to enhance traits such as disease resistance, growth rates, and stress tolerance in various fish species, including tilapia (*Oreochromis niloticus*) and Atlantic salmon (*Salmo salar*) (Okoli et al. 2022). Despite its potential, challenges such as off-target effects and regulatory hurdles remain (Okoli et al. 2022). The application of gene-editing technologies in fish breeding raises several ethical and regulatory concerns (Okoli et al. 2022). Ethical considerations include the welfare of genetically modified organisms, potential ecological impacts, and the long-term effects on biodiversity (Ediriweera and Manjula 2021). Regulatory frameworks vary across countries, with some having stringent guidelines for the use of genetically modified organisms in aquaculture. Public acceptance and transparent risk assessments are crucial for the sustainable application of these technologies (Ediriweera and Manjula 2021).

13.7 Bioinformatics and Data Analysis

Bioinformatics has revolutionized genomic studies by providing computational tools to manage, analyze, and interpret vast amounts of biological data (Gnimpieba et al. 2024). It integrates biology, computer science, and statistics to decode the genetic information encoded in DNA sequences. This interdisciplinary field has become essential for understanding genetic variations, identifying disease causing genes, and developing personalized treatments (Gnimpieba et al. 2024; Gadnayak et al. 2022).

Fish genomics benefited significantly from various bioinformatics tools and software designed to analyze and visualize genomic data. Notable tools included FishDB (Yang et al. 2020), an integrated functional genomics database housing genomes, transcriptomes, and other omics data; Evol2Circos (Pandey et al. 2020), a web-based tool for genome synteny and collinearity analysis, which facilitated the visualization of genomic data; and Next-Generation Sequencing (NGS) technologies such as Illumina, 454 GS FLX, and PacBio, which were extensively used for high-throughput sequencing of fish genomes (Khan and Ali 2024). Effective data management, analysis, and interpretation were crucial for extracting meaningful insights from genomic data. The process involved several key steps: gathering raw genomic data and performing quality control to ensure accuracy; matching DNA or RNA sequences to a reference genome to identify variations; detecting genetic variations such as single nucleotide polymorphisms (SNPs) and insertions/deletions (indels); assigning biological significance to genomic elements through functional annotation; and using visualization tools like Circos and R to interpret the data, providing valuable insights into biological processes and evolutionary patterns (Sandhya et al. 2023).

13.8 Challenges and Future Directions

The application of genomic tools in fish breeding programs has faced several challenges. One of the primary issues is the high cost and complexity of genomic sequencing and analysis. Additionally, there is a lack of comprehensive genomic data for many fish species, which hinders the development of effective breeding strategies. Another significant challenge is the difficulty in accurately predicting complex traits such as disease resistance and growth rate due to the complex genetic architecture of these traits (Zenger et al. 2019). Despite these challenges, there are promising solutions and advancements on the horizon. Advances in sequencing technologies have significantly reduced the cost and time required for genomic analysis (Zenger et al. 2019; Gadnayak et al. 2022). The development of high-throughput sequencing techniques and bioinformatics tools has made it easier to analyze large genomic datasets. Additionally, genomic selection and marker-assisted selection are becoming more widely used, allowing breeders to select for desirable traits more efficiently. The integration of omics technologies such as transcriptomics, proteomics, and metabolomics is also enhancing our understanding of the molecular mechanisms underlying important traits (Gadnayak et al. 2022). The future of fish breeding programs looks promising with the continued integration of genomic tools (Zenger et al. 2019). Genomic selection is expected to play a crucial role in improving the efficiency and sustainability of fish breeding programs. The use of CRISPR-Cas9 gene editing and other genome editing technologies may also revolutionize fish breeding by allowing precise modifications of genetic traits (Okoli et al. 2022). Furthermore, the development of environmentally friendly breeding practices and the use of alternative feed sources are expected to enhance the sustainability of aquaculture.

References

Abdelrahman H, ElHady M, Alcivar-Warren A, Allen S, Al-Tobasei R et al (2017) Aquaculture genomics, genetics and breeding in the United States: current status, challenges, and priorities for future research. BMC Genomics 18:1–23

Ahmad SF, Jehangir M, Srikulnath K, Martins C (2021) Fish genomics and its impact on fundamental and applied research of vertebrate biology. Rev Fish Biol Fish 32:357–385

Beck T, Free RC, Thorisson GA, Brookes AJ (2012) Semantically enabling a genome-wide association study database. J Biomed Semantics 3:1–16

Chavanne H, Janssen K, Hofherr J, Contini F, Haffray P, Aquatrace Consortium et al (2016) A comprehensive survey on selective breeding programs and seed market in the European aquaculture fish industry. Aquacult Int 24:1287–1307

D'Agaro E, Favaro A, Matiussi S, Gibertoni PP, Esposito S (2021) Genomic selection in salmonids: new discoveries and future perspectives. Aquac Int 29:2259–2289

Ediriweera T, Manjula P (2021) A brief review on aquaculture genetics, machine learning, and their convergence. J Anim Breed Genomics 5(3):107–112

Eibeck A, Shaocong Z, Mei Qi L, Kraft M (2024) Research data supporting "A simple and efficient approach to unsupervised instance matching and its application to linked data of power plants". J Web Semant 80:100815

Eze F (2019) Marker-assisted selection in fish: a review. Asian J Fish Aquat Res 3(4):1–11

Gadnayak A, Sahoo S, Nayak A, Sahoo M, Dave S, Das J (2022) Analysis of emerging microbial contaminants through next-generation sequencing (NGS). In: Omics for environmental engineering and microbiology systems. CRC Press, pp 71–83

Gheyas AA, Woolliams JA, Taggart JB, Sattar MA, Das TK, McAndrew BJ, Penman DJ (2009) Heritability estimation of silver carp (*Hypophthalmichthys molitrix*) harvest traits using microsatellite-based parentage assignment. Aquaculture 294(3–4):187–193

Gnimpieba EZ, Hartman TW, Do T, Zylla J, Aryal S, Haas SJ et al (2024) Biofilm marker discovery with cloud-based dockerized metagenomics analysis of microbial communities. Brief Bioinform 25(Supplement_1):bbae429

Haldar C (2018) Importance of marker assisted selection in fish breeding in India: challenges and opportunities. Ann Aquac Res 5(1):1047

Karlsson S, Bolstad GH, Hansen H, Jansen PA, Moen T, Noble LR (2020) The potential for evolution of resistance to Gyrodactylus salaris in Norwegian Atlantic salmon. NINA Rapport

Khan A, Ali R (2024) Computational tools for cancer nanomedicine. In: Personalized and precision nanomedicine for cancer treatment. Springer Nature Singapore, Singapore, pp 277–312

Khatei A, Tripathy PS, Parhi J (2022) Molecular markers in aquaculture. In: Advances in fisheries biotechnology. Springer Nature Singapore, Singapore, pp 165–174

Liu Z, Zhou T, Gao D (2022) Genetic and epigenetic regulation of growth, reproduction, disease resistance and stress responses in aquaculture. Front Genet 13:994471

Marees AT, De Kluiver H, Stringer S, Vorspan F, Curis E, Marie-Claire C, Derks EM (2018) A tutorial on conducting genome-wide association studies: quality control and statistical analysis. Int J Methods Psychiatr Res 27(2):e1608

Martinez V, Guimarães E, Ruane J, Scherf B, Sonnino A, Dargie J (2007) Marker-assisted selection in fish and shellfish breeding schemes. In: Marker-assisted selection-current status and future perspectives in crops, livestock, forestry and fish. Food and Agriculture Organization of the United Nations (FAO), pp 329–362

Natnan ME, Low CF, Chong CM, Bunawan H, Baharum SN (2021) Integration of omics tools for understanding the fish immune response due to microbial challenge. Front Mar Sci 8:668771

Nguyen NH, Sonesson AK, Houston RD, Moghadam H (2022) Applications of modern genetics and genomic technologies to enhance aquaculture breeding. Front Genet 13:898857

Okoli AS, Blix T, Myhr AI, Xu W, Xu X (2022) Sustainable use of CRISPR/Cas in fish aquaculture: the biosafety perspective. Transgenic Res 31(1):1–21

Ozaki A, Sakamoto T, Khoo S, Nakamura K, Coimbra MR, Akutsu T, Okamoto N (2001) Quantitative trait loci (QTLs) associated with resistance/susceptibility to infectious pancreatic necrosis virus (IPNV) in rainbow trout (*Oncorhynchus mykiss*). Mol Gen Genomics 265(1):23–31

Pandey M, Kushwaha B, Kumar R, Srivastava P, Saroj S, Singh M (2020) Evol2circos: a web-based tool for genome synteny and collinearity analysis and its visualization in fishes. J Hered 111(5):486–490

Ramya VL, Behera BK (2023) Molecular markers and their application in fisheries and aquaculture. In: Biotechnological tools in fisheries and aquatic health management. Springer Nature Singapore, Singapore, pp 115–150

Sandhya MP, Raju GPA, Iyappan M (2023) Bioinformatics: exploring the past, Present and future. Multidisciplinary 127:149

Sharma A, Kumar A (2022) Genomics, transcriptomics and proteomics for computational biology. In: Genomic, proteomics, and biotechnology. CRC Press, pp 1–11

Soon TK, Zheng H (2020) Climate change and bivalve mass mortality in temperate regions. Rev Environ Contam Toxicol 251:109–129

Sumon MAA, Molla MHR, Hakeem IJ, Ahammad F, Amran RH, Jamal MT et al (2022) Epigenetics and probiotics application toward the modulation of fish reproductive performance. Fishes 7(4):189

Unêda-Trevisoli SH, da Silva FM, Di Mauro AO (2017) Marker-assisted selection and genomic selection. In: Soybean breeding. Springer International Publishing, Cham, pp 275–291

Vallejo RL, Pietrak MR, Milligan MM, Gao G, Tsuruta S, Fragomeni BO et al (2024) Genetic architecture and accuracy of predicted genomic breeding values for sea lice resistance in the St John River aquaculture strain of North American Atlantic salmon. Aquaculture 586:740819

Wang Y, Ni P, Sturrock M, Zeng Q, Wang B, Bao Z, Hu J (2024) Deep learning for genomic selection of aquatic animals. Mar Life Sci Technol:1–20

Wells JV, Dawson N, Culver N, Reid FA, Morgan Siegers S (2020) The state of conservation in North America's boreal forest: issues and opportunities. Front For Global Change 3:90

Xu K, Duan W, Xiao J, Tao M, Zhang C, Liu Y, Liu S (2015) Development and application of biological technologies in fish genetic breeding. Sci China Life Sci 58:187–201

Yáñez JM, Newman S, Houston RD (2015) Genomics in aquaculture to better understand species biology and accelerate genetic progress. Front Genet 6:128

Yang L, Xu Z, Zeng H, Sun N, Wu B, Wang C et al (2020) FishDB: an integrated functional genomics database for fishes. BMC Genomics 21:1–5

Yu Y, Hilsdorf AWS, Zhou L, Lin Q, Gao ZX (2022) Genetics and molecular breeding in aquaculture animals. Front Genet 13:1071303

Zenger KR, Khatkar MS, Jones DB, Khalilisamani N, Jerry DR, Raadsma HW (2019) Genomic selection in aquaculture: application, limitations and opportunities with special reference to marine shrimp and pearl oysters. Front Genet 9:693

Next-Generation Sequencing Data Analysis in Fisheries Research

14

Pratyasha Bhowal and Arpita Roy

14.1 Introduction

The significance of fisheries extends to global nutrition, economic stability, and ecosystem conservation. Approximately three billion individuals rely on seafood as their primary protein source, with fisheries and aquaculture sectors employing approximately 58.5 million people worldwide (FAO Report: Global Fisheries and Aquaculture Production Reaches a New Record High n.d.). Global fish production, encompassing both capture fisheries and aquaculture, exceeded 223 million metric tons in 2020, with aquaculture contributing to more than half of this total. However, the sustainability of fishing faces increasing threats from overexploitation, habitat loss, pollution, and climate change. The Food and Agriculture Organization (FAO) indicates that over 35% of global fish stocks are harvested at unsustainable levels. Aquaculture production, less than four million tons in 1970, has grown exponentially since the early 1980s, reaching 80 million tons in 2016. It now accounts for more than half of the aquatic production, specifically for human consumption, with an estimated value of $230 billion. Given the increased demand associated with population growth, a 37% increase in aquaculture production is anticipated by 2030. Consequently, fisheries research has become crucial in addressing the sustainable

P. Bhowal (✉)
Department of Biotechnology and Dr. B. C. Guha Centre for Genetic Engineering and Biotechnology, University of Calcutta, Kolkata, West Bengal, India
e-mail: pbbiotech_rs@caluniv.ac.in

A. Roy
Department of Biotechnology and Dr. B. C. Guha Centre for Genetic Engineering and Biotechnology, University of Calcutta, Kolkata, West Bengal, India

Biotechnology Laboratory, Indian Council of Agricultural Research (ICAR)-Central Inland Fisheries Research Institute, Kolkata, West Bengal, India

© The Author(s), under exclusive license to Springer Nature Singapore Pte Ltd. 2026

B. K. Das et al. (eds.), *Bioinformatics Tools for Predictive Ecology and Fisheries*, Sustainable Aquatic Food Systems, https://doi.org/10.1007/978-981-95-2573-7_14

management of aquatic resources. This discipline constitutes a wide array of scientific enquiries, from the study of fish population biology and ecology to the socio-economic aspects of fishing communities (Rather et al. 2023). Fisheries research focuses on examining fish populations, including their biological characteristics, behavioural patterns, and environmental interactions. Scientists amass data on various fish life-cycle parameters, including growth rates, reproductive patterns, migratory behaviours, and mortality statistics. This comprehensive information forms the foundation for developing precise models to forecast fish population dynamics. These models are instrumental in evaluating the consequences of fishing activities and environmental alterations on fish stocks, facilitating more informed management strategies. Concurrently, it has been observed that the rapid expansion of aquaculture, coupled with increased global transportation of live aquatic organisms, has significantly contributed to the emergence of numerous viral diseases (Nkili-Meyong et al. 2016; Walker and Winton 2010). These diseases represent a substantial threat to the sustainability of the burgeoning global industry. The extent of disease proliferation and its impact is largely determined by the capacity to swiftly identify causative agents and comprehend crucial epidemiological factors, such as transmission modes and virulence mechanisms (Bergh et al. 1989).

Fish have been used as model organisms in biological research using a variety of molecular biology techniques before the development of genome sequencing (Powers 1989). Fish are especially popular model vertebrates among molecular genetics and genomics researchers. Other eukaryotic model species, including the yeast *Saccharomyces cerevisiae*, the common fruit fly *Drosophila melanogaster*, and the roundworm *Caenorhabditis elegans*, have been sequenced concurrently with the human genome (Adams et al. 2000; Goffeau et al. 1996). Fugu fish have been proposed as a model genome (Brenner et al. 1993) because of its modest genome size (approximately 400 Mb), which is the smallest among vertebrates and appropriate for sequencing technology during the early sequencing era. Numerous fascinating results in comparative sequence analysis between fish and humans have been obtained from the successful sequencing of the fugu genome (Aparicio et al. 2002), the first vertebrate genome to be sequenced after humans. These findings include the discovery of multiple putative novel human genes and the exposure of the genomic locus configurations of fugu and humans with varying quantities of repetitive DNA (Aparicio et al. 2002). Following this, as NGS technologies have advanced, the number of fish whole-genome sequencing projects has increased annually. As of July 2025, the genomes of 2878 fish species, 2750 of which are teleost had been sequenced, mapped, and assembled according to the National Centre for Biotechnology Information (NCBI) database (Fig. 14.1), which marked a surge in the application of genomic data in research on fish physiology, genetics, evolution, immunology, and endocrinology utilizing non-model species.

Previously, genomic data were primarily obtained using Sanger sequencing techniques, which employed fluorescent dye-labelled dideoxy-nucleotide triphosphates to terminate DNA chains (Maxam and Gilbert 1977). However, this method has limitations owing to its high cost and labour-intensive nature. Next-generation sequencing (NGS) technologies have revolutionized the field, enabling the

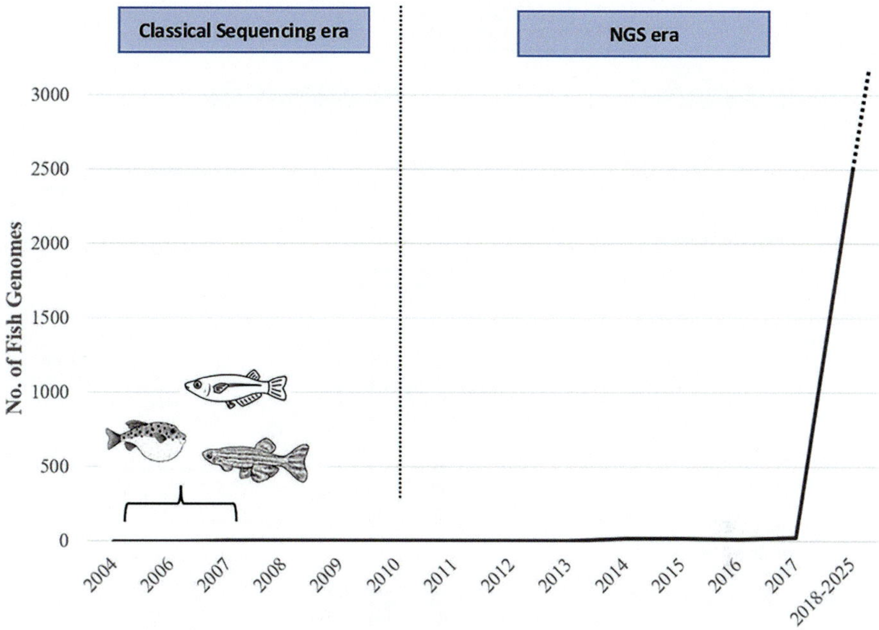

Fig. 14.1 The sequencing timeline demonstrates the recent increase in fish genome sequencing, which is rapidly increasing in the NGS era

simultaneous acquisition of vast amounts of nucleotide information from hundreds of thousands to millions of DNA molecules with sequence read lengths ranging from 30 to 1500 nucleotides (Fig. 14.2). Concurrently, significant advancements have been made in bioinformatics tools to enhance their capacity to analyse these extensive datasets and identify unique gene sequences. The emergence of NGS has had a profound impact on fisheries science by offering novel research possibilities. As a powerful tool for providing high-throughput genomic data, NGS has become instrumental in efforts related to the conservation, management, and sustainability of aquatic resources. The advent of NGS has transformed traditional genetic research, offering the ability to sequence complete genomes, transcriptomes, and environmental DNA (eDNA) with remarkable precision and speed (Cruz-Cano et al. 2024). Despite its groundbreaking potential, the implementation of NGS in fisheries studies poses several obstacles. These include the intricacies of bioinformatic analysis, substantial computational power requirements, and ethical issues surrounding the sharing of genomic data. However, continuous improvements in sequencing technologies, data processing pipelines, and cloud-based bioinformatics platforms have steadily increased the accessibility and efficacy of NGS-based investigations (Franco et al. 2021). This chapter explores the wide-ranging applications of NGS in fisheries research, outlines the analytical processes involved in handling and interpreting NGS data, and considers the challenges and prospects of this rapidly developing field. By harnessing the capabilities of NGS, scientists and

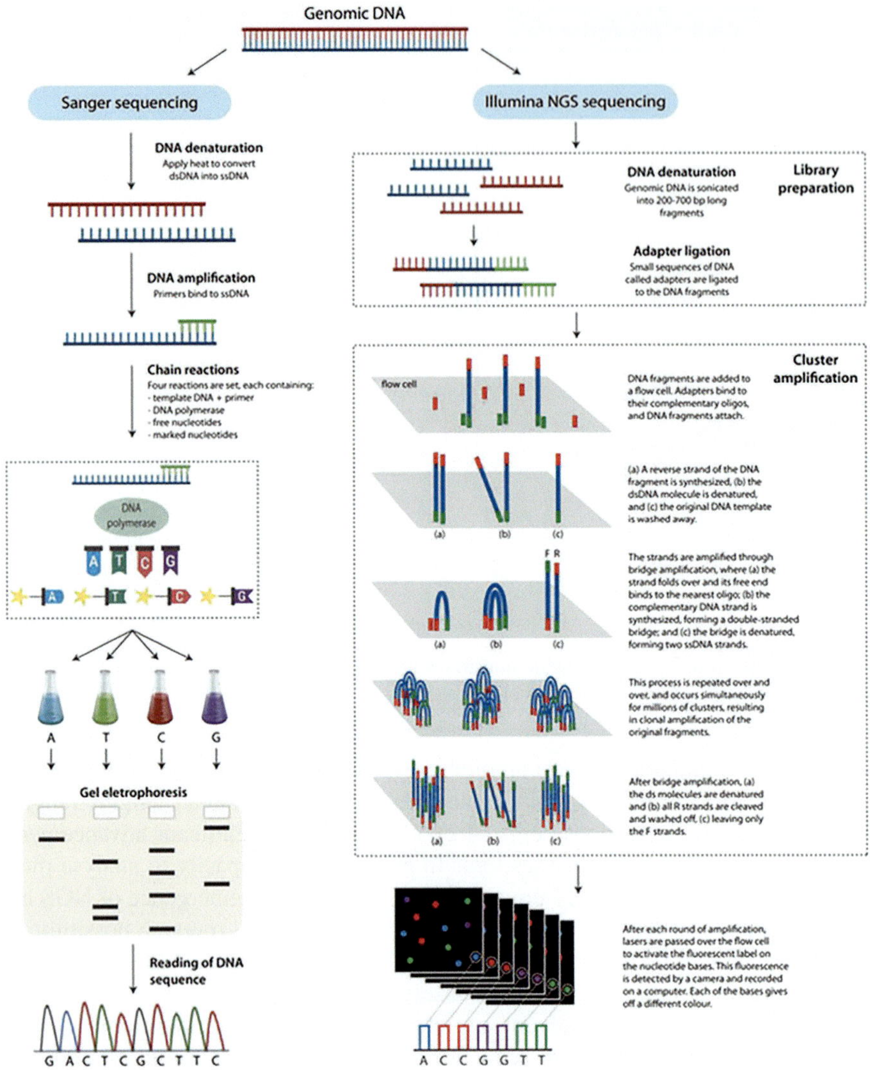

Fig. 14.2 Sanger sequencing was compared with next-generation sequencing (NGS)

policymakers can make better-informed decisions to safeguard sustainable management and preservation of aquatic biodiversity.

14.2 NGS Technologies Relevant to Fisheries Research

Research on fisheries has been transformed by NGS, making it possible to conduct extensive genomic, transcriptomic, and eDNA studies. Several NGS platforms and methodologies have offered important insights into species diversity, population genetics, evolutionary biology, aquaculture enhancement, and ecosystem monitoring. The main NGS technologies utilized in fisheries research are addressed in this section, along with their methods, benefits, drawbacks, and uses. Several NGS platforms are widely used in fisheries research, each offering unique advantages for different genomic analyses.

14.2.1 Genome-Sequencer FLX

The 454 GS FLX was one of the first commercially available and most commonly used NGS platforms for fish genomics research, developed by Roche. It uses pyrosequencing technology, which detects nucleotide incorporation through light emission (Liu et al. 2012). This platform produces longer reads (up to ~700 bp) than the early Illumina platforms, making it useful for de novo genome assembly, metagenomics, and transcriptomics (Rothberg and Leamon 2008). However, it has drawbacks, such as high error rates in homopolymeric regions and higher sequencing costs. Due to sequencing technology advancements, Roche discontinued the 454 GS FLX in 2016, and long-read platforms such as PacBio and Oxford Nanopore Technology have since taken their place in applications requiring long-read sequencing.

14.2.2 Illumina Sequencing

Illumina sequencing combines short-read (50–300 bp) high-throughput sequencing with sequencing-by-synthesis (SBS). This platform has a low error rate (~0.1%) and good precision. It is perfect for metagenomics, RNA-Seq, and population genomics owing to its high sequencing depth. It is also reasonably priced per base sequence. Bridge amplification is a method used by Illumina technology to amplify collections of DNA fragments bonded to a glass slide (Quail et al. 2012). The parallel sequencing approach inserts dye-labelled nucleotides or one fluorophore per base instead of sequential insertion. Illumina sequencing is primarily used for whole-genome sequencing (WGS) of fish species, genotyping-by-sequencing (GBS), and RAD-Seq for population genetics, SNP identification for population genetics and stock assessment, and eDNA metabarcoding for biodiversity monitoring. The low read length of this technology can make genome assembly challenging, particularly for repetitive or complicated genomes (Metzker 2010; Nakamura et al. 2011).

14.2.3 Pacific Biosciences (PacBio) SMRT Sequencing

The PacBio is a single-molecule sequencing technique designed to cut down on the time needed to collect these genes or transcriptome sequences. It is considered a 'third-generation' sequencing platform. It uses fluorescently labelled nucleotides to directly monitor polymerase activity during DNA synthesis, enabling real-time imaging of freshly synthesized DNA strands (Quail et al. 2012). This instrument can sequence single molecules (10,000–50,000 bp). Long reads facilitate high-quality de novo genome assembly by detecting structural variations, alternative splicing, and epigenetic modifications. When combined with circular consensus sequencing, this produces highly accurate 'HiFi' reads. Although this platform has a higher cost per base, it produces high-quality genome assemblies for non-model fish species. It is favourable for structural variant detection in evolutionary adaptation and epigenomics studies to investigate the environmental influences on fish development (Rhoads and Au 2015).

14.2.4 Oxford Nanopore Technologies (ONT) Sequencing

Since Oxford Nanopore Technologies (ONT) released its first nanopore sequencer, MinION, in 2014, nanopore sequencing technology and its use in fundamental and applied research have experienced significant expansion. The device is based on a protein pore, or 'nanopore', which is placed in an electrically resistant polymer membrane and functions as a biosensor (Wang et al. 2010). This platform uses nanopore-based sequencing to generate ultra-long reads (>100,000 bp) and is portable (MinION) and scalable (PromethION). This platform is suitable for field-based applications in fisheries monitoring with direct RNA sequencing capability. Compared with Illumina and PacBio, ONT sequencing has a higher error rate (~5–10%) and requires advanced bioinformatics for error correction. ONT enables rapid species identification in the field as well as eDNA monitoring for invasive species detection.

14.2.5 Ion Torrent Sequencing

The Ion Torrent NGS platform is a reliable, affordable, and rapid technology that has endured over time. It operates based on detecting the hydrogen ions generated when a new nucleotide is incorporated into the expanding DNA template (Quail et al. 2012). With a maximum read length of 400 bp and a data output ranging from 0.3 to 15 GB, the system allows for flexibility in chip sizes. It is particularly helpful for transcriptome sequencing, microbial and viral whole-genome sequencing, shallow metagenomics, and mitogenomics. Amplicon-based 16 s rRNA gene-focused metagenomics has enabled the identification of previously unculturable and unknown bacteria from their natural habitats (Rothberg et al. 2011). This platform is much faster compared to Illumina with lower upfront equipment costs. Nevertheless, this device also generates short readings with higher error rates, particularly in homopolymer regions. This platform is particularly used in target amplicon sequencing and small-scale genome sequencing studies (Mehinto et al. 2012).

14.2.6 10× Genomics Linked-Read Sequencing

An integrated microfluidics-based platform for creating linked reads and a specialized software for analysing them are offered by 10× Genomics (Mostovoy et al. 2016). Contigs can be extended, or the spaces between adjacent contigs can be filled using 10× Genomics linked-read sequencing. This platform improves and provides high-resolution genome assembly for species with complex genomes, including those with repetitive areas or significant heterozygosity, making it useful for conservation genomics, aquaculture breeding programs, and fisheries research (Liang et al. 2020). One drawback of this platform is the requirement for high-molecular-weight DNA, which can sometimes be difficult to extract from certain species.

For fisheries research, choosing the right sequencing platform is critical because different technologies provide different benefits, depending on the study's objectives. Illumina sequencing is widely used in eDNA, transcriptomics, and population genomics research owing to its high precision, scalability, and affordability. PacBio and ONT offer long-read sequencing for genome assembly and discovery of major structural variations. This enables a more detailed investigation of complex genetic locations.

14.3 Workflow for NGS Data Analysis in Fisheries Science

NGS offers high-throughput sequencing capabilities, and researchers may examine the genetic diversity, population structure, and adaptation strategies of aquatic species. Fish tissues, surrounding water samples, and microbial communities are collected at the start of the process. To obtain precise and accurate sequencing results, high-quality DNA or RNA is extracted using specific techniques. Subsequently, library preparation is performed for sequencing. Several NGS systems, including Oxford Nanopore, PacBio, and Illumina, are chosen according to the goals of the study, accuracy requirements, and read-length specifications. To forecast trends and improve fisheries management, NGS analysis increasingly incorporates machine learning and artificial intelligence (AI). Visualization tools, such as R-based plotting frameworks and IGV, aid in interpreting the sequencing results (Fig. 14.3). Sustainable aquaculture development, selective breeding initiatives, and conservation methods are all influenced by insights obtained from NGS data. NGS is still revolutionizing fisheries science by providing unmatched accuracy and effectiveness in comprehending aquatic environments.

14.3.1 Sample Collection and DNA/RNA Extraction

High-quality nucleic acids (DNA or RNA) must be extracted before sequencing. Depending on the study, different types of samples may be used, such as whole fish tissues (liver, muscle, fin clips, etc.), which are primarily used for genome assembly, species identification, and population genetics; eDNA samples (water and

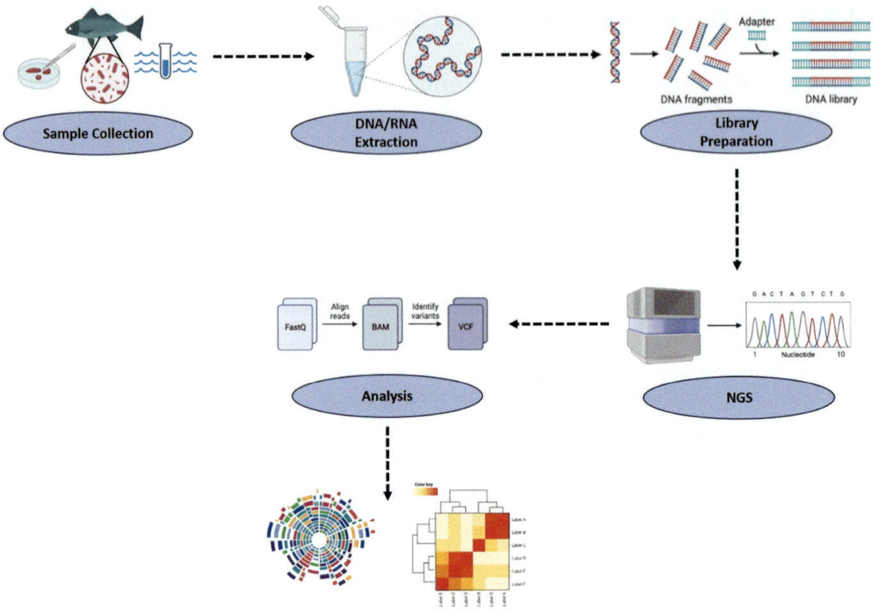

Fig. 14.3 Graphical representation of the NGS workflow, starting from sample collection and analysis

sediment), which are used for ecosystem studies and biodiversity monitoring; gut microbiome samples, which are used for nutrition, disease monitoring, and fish health; and RNA samples (from gills, liver, or muscle) for transcriptomic studies (gene expression). Agarose gel electrophoresis (for integrity check), Qubit Fluorometer (for concentration measurement), and Nanodrop (for purity evaluation) are tools used for proper DNA/RNA quality control (QC).

14.3.2 Library Preparation

NGS library preparation is a crucial phase that ensures that DNA or RNA samples are appropriately processed for compatibility with the sequencing platforms. This process involves fragmentation of nucleic acids, attachment of sequencing adapters, and amplification of the material to produce high-quality libraries. Various sequencing methods require specific library preparation techniques, each designed to suit the aims of the study and the biological materials under examination. WGS allows for a comprehensive examination of an organism's entire genome, offering insights into genetic diversity, evolutionary relationships, and adaptive characteristics in fish species. The primary stages in WGS library preparation encompass DNA fragmentation, where extracted genomic DNA is broken into smaller segments either mechanically or enzymatically to facilitate sequencing; adapter ligation, which involves attaching short synthetic DNA sequences to both ends of the fragmented DNA to enable binding to the sequencing platform; and PCR amplification, where libraries are replicated using polymerase chain reaction to increase DNA quantity, ensuring sufficient material for sequencing. RNA-Seq is extensively utilized to examine gene expression patterns, identify regulatory elements, and investigate transcriptomic responses to environmental changes in fish species. The key steps include RNA extraction, mRNA enrichment or rRNA depletion, cDNA synthesis, and library construction, which, like WGS, involve adapter ligation and size selection before sequencing. Metagenomics provides access to entire microbial communities in aquatic environments and offers insights into biodiversity, ecosystem health, and pathogen monitoring in fisheries. This approach involves analysing the total DNA from environmental samples, such as water, sediment, or fish gut microbiota. The primary steps are DNA extraction, fragmentation, adapter ligation, and shotgun sequencing or 16S rRNA sequencing. Shotgun metagenomic sequencing analyses all microbial DNA, while 16S rRNA sequencing targets the ribosomal RNA gene to identify bacterial species. The final step involves data processing.

14.3.3 NGS Platform-Based Sequencing

Following library preparation, sequencing is performed using a suitable platform, according to the study aims (Table 14.1). Read length, affordability, accuracy, and coverage depth play a role in platform selection.

14.3.4 Raw Data Processing (Quality Control and Preprocessing)

Following sequencing, the initial data were collected in the FASTQ format. Many measures are required before proceeding to ensure data quality. Quality assessment methods include FastQC, which checks per-base quality scores, GC content, and adapter contamination, and MultiQC, which combines numerous FastQC reports for batch analysis. Trimmomatic or Cutadapt is used to filter and trim quality data, remove low-quality bases, and sequence adapters. PRINSEQ was used to refine reads depending on their length, GC content, and complexity. After filtering, high-quality readings are ready for alignment or assembly processes.

Table 14.1 The top four NGS systems have transformed aquaculture genomics

NGS platform	Application in fisheries science
Illumina (HiSeq, NovaSeq, MiSeq)	Population genomics, transcriptomics, eDNA, and amplicon sequencing
PacBio (SMRT Sequencing)	Whole genome sequencing and long-read assemblies
Oxford Nanopore (ONT)	Real-time sequencing, portable field applications
Ion Torrent	Small-scale sequencing in aquaculture genetics

14.3.5 Read Alignment and Genome Assembly

Two main methodologies were used, depending on the study type. If a reference genome is available, reads are aligned using tools such as Burrows-Wheeler Aligner (BWA) for short-read mapping, Bowtie2, which is efficient for small genomes and amplicon sequencing, and STAR for RNA-Seq mapping. The output was in the SAM/BAM format and contained mapped reads. In the absence of a reference genome, the genome must be constructed from the reads. Tools for this purpose include SPAdes (for bacterial and metagenomic assemblies), MEGAHIT (for large-scale metagenomic data), and Canu and Flye (for long-read sequencing data assembly). After assembly, the genome was assessed by BUSCO (for completeness based on conserved genes) and QUAST (for genome quality evaluation).

14.4 Variant Calling and Annotation

14.4.1 SNP and Indel Discovery

Variant calling is carried out for population genetics and stock evaluation using GATK (Genome Analysis Toolkit), which finds SNPs and minor indels or FreeBayes, a Bayesian method for variant calling. ANNOVAR or SnpEff, which forecasts the functional implications of variations, is used to further annotate variant effects.

14.4.2 Structural Variant Detection

Manta, which detects structural variants such as insertions, deletions, and duplications, or Delly, which detects translocations and massive insertions, is utilized for big genome variations.

14.5 Functional Analysis and Interpretation

After variant identification, functional interpretation is done by using multi-omics approaches.

14.5.1 Gene Expression Analysis (RNA-Seq)

While DESeq2 or EdgeR identifies differentially expressed genes (DEGs), Feature Counts or HTSeq quantifies gene expression levels for transcriptomics investigations.

14.5.2 Metagenomic Analysis (eDNA & Microbiome Studies)

Kraken2 or MetaPhlAn classifies microbial species from sequencing data, whereas QIIME2 or Mothur analyses the structure of microbial communities for environmental and gut microbiome studies.

14.5.3 Pathway and Functional Enrichment Analysis

The enriched Gene Ontology (GO) terms are identified using GOseq in order to understand biological relevance. PFAM, eggNOG, and KEGG are some databases that can be utilized for potential functional prediction. Babelomics or KEGG Mapper examines metabolic pathways.

14.6 Visualization

Visualization tools are used to present findings efficiently. Mapping reads and variations may be visualized using IGV (Integrative Genomics Viewer). In R, ggplot2 is used to plot statistical findings and expression data, whereas Cytoscape is mostly used to visualize pathways.

14.7 Revolutionizing Fisheries with Next-Generation Sequencing

In the fields of eukaryotic microorganisms (Nowrousian 2010), plants (Bräutigam and Gowik 2010), animals, and human genomes (Pareek et al. 2011), NGS technologies have become revolutionary tools. Their application has demonstrated enormous promise for studying gene expression profiles (Wang et al. 2010), genome evolution (Holt et al. 2008), and gene regulation (Pomraning et al. 2009). Over the last 10 years, the number of publications using NGS technologies in fish research has increased 25-fold. By using NGS, fish biologists can better understand how biological processes are regulated at the genome and transcriptome levels, find new biomarkers for ecotoxicological applications, clarify toxicity pathways, and better study evolutionary issues than was previously possible with traditional population genetic markers like DNA microsatellites (Mehinto et al. 2012). Fortunately, sequencing prices are decreasing, allowing for the examination of several biological replicates (i.e., individual genomes) within a single analysis. The use of NGS technology will allow for better links between individual genotype, phenotype, and transcriptome responses under a variety of environmental variables and experimental settings. The integration of NGS into fisheries genomics has resulted in pioneering advances in several critical areas.

14.8 Species Identification Through DNA Barcoding

The accurate identification of species is fundamental to fishery research, with implications across various disciplines including ecology, microbiology, and conservation. Precise species recognition is crucial for comprehending biodiversity, geographical distribution, and ecological relationships, as well as for detecting pathogens and vulnerable species. Nevertheless, species identification can be complex, particularly for microscopic organisms or those exhibiting significant morphological variability. DNA barcoding has emerged as a technique for species identification utilizing a standardized DNA sequence (Chac and Thinh 2023). This method has gained traction due to its high precision and effectiveness in species recognition, especially in instances where morphological identification proves challenging or unfeasible. However, to accurately identify organisms from diverse samples, suitable DNA barcode sequences must be employed at various taxonomic levels. These DNA barcodes can be located in either the nucleus or cytoplasm of an organism's cells and may be utilized independently or in conjunction to enhance identification accuracy. NGS techniques for DNA barcoding and metabarcoding have transformed biodiversity assessments, enabling swift, high-precision species identification, even from compromised DNA samples (Letsiou et al. 2024). To better understand the genetic variety of fish species in India's crucial coastal ecosystem, Chilika Lagoon, researchers used DNA barcoding, which highlighted the remarkable biodiversity of the lagoon by identifying 10 putative new species from 155 DNA barcode sequences that represented 63 morphospecies. Additionally, the study established the genetic evidence for the members of three genera from Chilika Lake: *Hypoatherina*, *Odontamblyopus*, and *Scombrops* (Kundu et al. 2024). A study in Korea identified freshwater fish species by creating a DNA chip that uses mitochondrial cytochrome c oxidase subunit I (COI) gene sequences. With 53 species covered, this molecular identification technology provides a precise and high-throughput approach to species discovery. The DNA chip promotes efforts to conserve biodiversity and makes it easier to monitor freshwater ecosystems effectively (Kim et al. 2011). NGS thus proves particularly valuable in uncovering cryptic species, monitoring illicit fishing activities, and combating seafood mislabelling.

14.9 Adaptive Genomics and Evolutionary Studies

The new field of fisheries genomics promotes the application of genomic methods to fisheries management problems. By making it possible to integrate adaptive variety to understand the basic features of fisheries resources, this method presents a new paradigm. Moreover, the opportunity to contribute to different facets of fisheries management is unmatched by genomic tools. They facilitate surveillance and enforcement (Nielsen et al. 2012), help assess the evolutionary potential of exploited fishing resources to protect their evolutionary history, and shed light on the evolutionary dynamics causing geographical and temporal stock variability (Therkildsen et al. 2013; Valenzuela-Quiñonez 2016). Allozymes, mitochondrial DNA, and microsatellites are among the few potentially neutral markers used in this process (Hauser and Seeb 2008; Ward 2000). Single-nucleotide polymorphisms (SNPs) have been widely discovered employing NGS technology, which has made it possible to analyse both allele-specific (adaptive) and genome-wide (neutral) diversity patterns in a variety of organisms, including non-model species. Through NGS, scientists looked into the genetic underpinnings of the Tibetan Schizothoracine fish's (*Gymnocypris przewalskii ganzihonensis*) adaptation to high altitude. The study used NGS to find genes linked to osmoregulation and hypoxia tolerance, which provided insight into how the species evolved to adapt to the harsh environment of the Tibetan Plateau. Another study from Australia resulted in the discovery of the complete genome of the critically endangered spotted handfish (*Brachionichthys hirsutus*), distinguished by its unusual pectoral and pelvic fin movement, which has been sequenced by researchers. In addition to supporting monitoring and captive breeding initiatives, these genomic data provide comprehensive insights into the genetic diversity and adaptive characteristics of the species. The study described the population genetic diversity and structure for the first time using more than 4000 single-nucleotide polymorphism sites (Appleyard et al. 2021). In addition to all these, NGS also facilitates the investigation of genetic adaptations to environmental stressors through transcriptomics (RNA-Seq) (Liu et al. 2015) and epigenomics.

14.10 Population Genetics and Stock Assessment

The preservation of wild fish genetic resources depends on the assessment of genetic diversity. With little genetic interchange, the majority of species are made up of separate populations or genetic stocks (Hilborn et al. 2003). Because of intraspecies heterogeneity, these populations retain distinctive genetic traits. Evolutionary processes like migration, mutation, selection, and genetic drift all have an impact on this differentiation. Since migration is unlikely to restore the population, overfishing of these units could result in fishery collapse. NGS has significantly improved our capacity to examine fish population structures, genetic variability, and gene flow on an unprecedented scale. Whole-genome sequencing (WGS) and genotyping-by-sequencing (GBS) offer detailed insights into population dynamics of fishes, facilitating the recognition of distinct stocks and guiding sustainable fisheries management approaches (Andersson et al. 2024). Insights about dispersal during the obscure larval stage are provided by evaluating the genetic composition (DNA) of individuals across geographic areas. Numerous molecular methods have been established; preliminary research has compared differences in microsatellite markers (Wenne 2023), mitochondrial or nuclear markers, and allozyme markers. One such study in the Mediterranean Sea, where huge red prawns (*Aristaeomorpha foliacea*) are an important fishing resource, NGS was used to identify microsatellite markers for sustainable management. These markers aid in the identification of genetically distinct units or stocks within a species. Understanding the genetic structure of giant red shrimp populations thus helps in establishing suitable management plans, ensuring the genetic variety and long-term sustainability of the species (Heras et al.

2024). Another study of wild medaka (*Oryzias latipes*) populations in Southern Japan used high-throughput genome sequencing to determine genetic diversity and population structure. The findings aided in the creation of a near-isogenic panel of inbred lines, which will be a great resource for vertebrate genetic research and help us better understand evolutionary processes in real fish populations (Spivakov et al. 2013). These methods improve the capacity to find demographically independent equities. Research using similar techniques on fish species that have been harvested has shown clear isolation patterns well within the species' dispersion ranges (Payet et al. 2024); minor genetic differences between populations separated by 10–100 km; and year-to-year fluctuations in the genetic structure (Payet et al. 2022).

14.11 Using Environmental DNA (eDNA) to Track Ecosystems

In fisheries and aquaculture, eDNA technology has become a game-changer, offering novel approaches to managing and monitoring aquatic ecosystems. Better methods for extracting, amplifying, and sequencing DNA from environmental materials have been developed as a result of the growing interest and development of eDNA technology in recent years (Chouhan et al. 2023). NGS-based eDNA analysis has emerged as a potent, non-invasive method for monitoring aquatic biodiversity. One technical advancement that opens up new avenues for analysing and quantifying biosphere recovery, resilience, and change is the use of eDNA for species identification (Thomsen and Willerslev 2015). In order to discover cryptic, difficult-to-reach, and rare species, advanced monitoring techniques improve accuracy and standardization. Researchers in Japan used eDNA analysis to determine the spread of Japanese Jack Mackerel (*Trachurus japonicus*) in Maizuru Bay. The study successfully discovered its presence by collecting and analysing water samples using NGS, providing a 'snapshot' of its spread. This strategy has proven useful in wide maritime habitats where traditional survey techniques are difficult, providing a non-invasive way to monitor fish populations and inform management tactics (Yamamoto et al. 2016). It is possible to identify specific species or evaluate entire ecosystems using eDNA derived from both ancient and modern origins. Aquatic ecosystem monitoring, conservation biology, and biodiversity evaluation are just a few of the ecological domains where eDNA technology may find use (Thomsen et al. 2012). It helps scientists monitor threatened species, find disease carriers, and evaluate how human activity affects the ecosystem.

14.12 Discovery of Antibiotic Resistance Gene (ARGs) and Disease Resistance

As aquaculture expands rapidly, genomic tools are increasingly employed to enhance selective breeding programmes. NGS enables researchers to pinpoint genetic markers associated with disease resistance, growth rates, and ARGs, facilitating the development of more resilient fish stocks. By making it possible to resolve complex sections, next-generation sequencing technologies present a promising path towards improving genome assemblies (Holt et al. 2008). Moreover, the scientific community has also focused on the prevalence of viruses in aquatic environments. Aquatic viruses can infect a variety of aquatic species, including phytoplankton, prawns, and fish. High-throughput sequencing platforms can thus be used in aquatic virology to discover new viruses, understand pathogenesis mechanisms, and study molecular epidemiology. Fish farms are more susceptible to infection and disease transmission than natural settings because of the fish's close quarters, ongoing stress, and exposure to different contaminants. As a result, NGS helps in keeping an eye on aquatic health and treating illnesses (Potts et al. 2021). With the ease with which scientists can now extract DNA, a thorough assessment of the pathogens found in any aquatic environment or aquaculture facility is now possible (McArthur and Tsang 2017). Bacterial resistance has emerged or developed as a result of aquaculture's overuse of antibiotics. For instance, over nine antibiotic-resistant bacteria (ARB) species were found in water samples from tilapia farms: *Actinobacteria* (11.0%), *Bacteroidetes* (9.8%), *Planctomycetes* (8.3%), *Verrucomicrobia* (6.1%), *Chloroflexi* (2.1%), *Chlorobi* (1.7%), *Firmicutes* (1.0%), *Cyanobacteria* (26.5%), and *Proteobacteria* (26.4%) (Zhou et al. 2021), which had been identified using NGS. Up until now, sulfa medications (including trimethoprim), tetracycline, quinolones, and macrolides have been the primary subjects of studies on ARGs in aquatic environments. Sulfonamide ARGs (sul1, sul2, and sul3), tetracycline ARGs (tetA, tetB, tetC, tetD, tetE, tetG, tetH, tetM, tetO, tetQ, tetS, tetW, tetX, and tetZ), quinolone ARGs (qnrA, qnrD, and qnrS), and macrolide ARGs (ermC, ermA, emf (C), and mph) were the most commonly found ARGs. Another NGS study also showed that *Bacteroidetes*, *Fusobacteria*, and *Firmicutes* dominate the gut ecology of spotted catfish (Yuan et al. 2023). The most prevalent bacteria in Nile tilapia are found to be *Pseudomonas aeruginosa* and *Aeromonas hydrophila* (Sherif et al. 2021). In contrast, the most common organisms found in snow crabs were *Acinetobacter*, *Bacillus*, and *Psychrobacter* (Kim et al. 2013), all of which has been discovered through NGS. Therefore, it can be concluded that different host organisms would have different dominant ARBs, and NGS makes it easier to identify. Thus, compared with conventional sampling methods, this monitoring approach provides a more comprehensive assessment of disease prevalence while lowering stress during surveillance, which contributes to sustainable aquaculture practices and diminishes reliance on antibiotics.

14.13 Integrating Multi-omics in Fisheries Research

Innovative biotechnological tools, such as recently developed omics technologies, such as genomics, transcriptomics, proteomics, and metabolomics, have been used to improve our understanding of aquaculture safety, production, quality, and health to overcome biological limitations and meet growing demands. The immunological response of fish and the underlying molecular mechanisms of fish immunity in response to pathogenic infections can be clarified using NGS technologies. Beyond

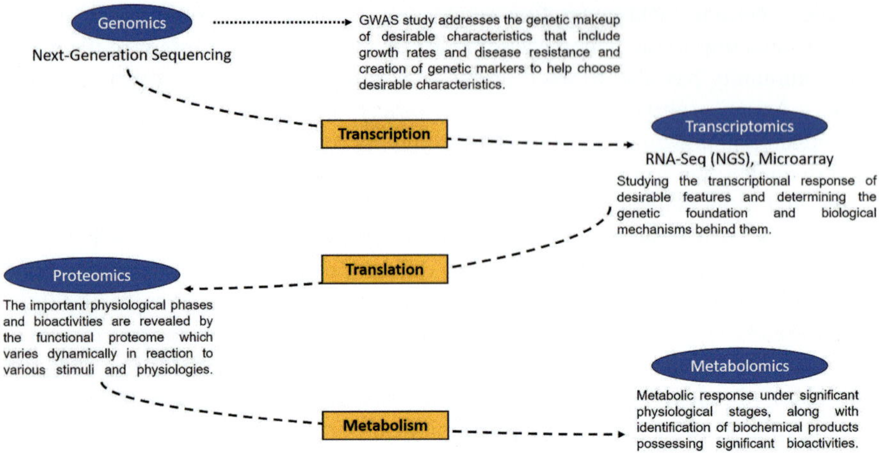

Fig. 14.4 Integration of proteomics, metabolomics, transcriptions, and genomes through multi-omics. (Adapted from Natnan et al. (2021))

genomics, NGS enables the integration of transcriptomics, proteomics, and metabolomics (Fig. 14.4), thereby offering a comprehensive understanding of fish biology (Pathak et al. 2019). These multi-omics approaches help uncover gene regulatory networks, metabolic pathways, and immune responses, providing deeper insights into fish health and adaptation. The most recent improvements in experimental techniques and computational methods used in the processing of omics data can provide new insights and help us understand the complexity of biological processes (Natnan et al. 2021). Scientists have used an integrative multi-omics method to examine how fish respond to acute hypoxia stress in a study on genetically improved farmed tilapia (*Oreochromis niloticus*). This study identified notable changes in glycolipid metabolism pathways, which are essential for energy production in low-oxygen environments, by examining data from transcriptomics, proteomics, and metabolomics. These results advance our knowledge of the mechanisms underlying tilapia's response to hypoxia and offer important insights for breeding initiatives targeted at enhancing the tolerance of aquaculture to stress (Ma et al. 2021). Thus, a deeper understanding of fish genetics and protein function under various settings will result from the use of genomic or transcriptomic technologies to provide genetic information about tissues.

14.14 Challenges in Next-Generation Sequencing (NGS)

Fisheries research has been profoundly changed by Next-Generation Sequencing (NGS), which provides in-depth genetic and environmental insights into conservation, stock management, and aquaculture enhancement. To optimize its advantages, several obstacles must be overcome. These include the need for additional technological developments, ethical and policy problems, computational and bioinformatics limitations, and mechanical limitations (Lu and Luo 2020). Large volumes of sequencing data are produced by NGS, which presents difficulties in data management, storage, processing, and interpretation. High-performance computing (HPC) clusters are necessary for the analysis of raw sequencing data with sophisticated computational capacity, particularly for WGS and metagenomic research. Furthermore, the interpretation of NGS data depends on bioinformatics pipelines that include a number of intricate processes, including variant analysis, genome assembly, and quality control. Genome assembly and annotation are made more difficult by the scarcity of high-quality reference genomes for many fish species. Concerns about biodiversity protection, genetic resource sharing, and the use of genetic changes in aquaculture are among the ethical and legal issues (Martinez-Martin and Magnus 2019) brought up by the incorporation of NGS into fisheries science. Additionally, when studying genetic resources from particular regions, decision-making procedures should involve local communities that depend on fisheries and Indigenous people. NGS is still transforming fisheries science in spite of obstacles pertaining to data management, computational resources, and ethical issues. AI-driven bioinformatics, long-read sequencing, and single-cell sequencing are examples of emerging technologies that have the potential to overcome present constraints (Cheng et al. 2023).

14.15 Conclusion and Future Research

Because of genome duplication events and the existence of numerous isoforms for many genes, fishes pose different hurdles than mammals. However, teleost fishes are valuable model organisms for investigating some human diseases and evaluating the effects of anthropogenic pollutants in the environment (Albertson et al. 2009). The combination of multiple NGS platforms should enable whole-genome sequencing initiatives and increase our understanding of ecologically significant species as the costs of DNA and RNA sequencing decline. Gaining insight into the connections between gene expression and exposure to environmental chemicals can help with environmental risk assessments (ERA). Future studies should concentrate on international cooperation, policy creation, and technology advancements to completely incorporate NGS into fisheries science and management. The efficacy of genome-wide association studies (GWAS) and selective breeding programs is limited by the unavailability of high-quality reference genomes for many economically significant fish species. Furthermore, in order to help aquaculture genetics and conservation, studies should be made on producing high-quality genome assemblies for both wild and farmed fish species. To enhance real time fisheries assessment and biodiversity protection, eDNA should be used more widely for species detection and population monitoring. In conclusion, NGS has emerged as a potent instrument in aquaculture growth, encompassing not only fish health but also fish larval production, nutrition, medicine, and food safety (Fig. 14.5). New perspectives on basic issues for fisheries management are provided by the shift from fisheries genetics to fisheries genomics. An entirely novel approach in fisheries management is

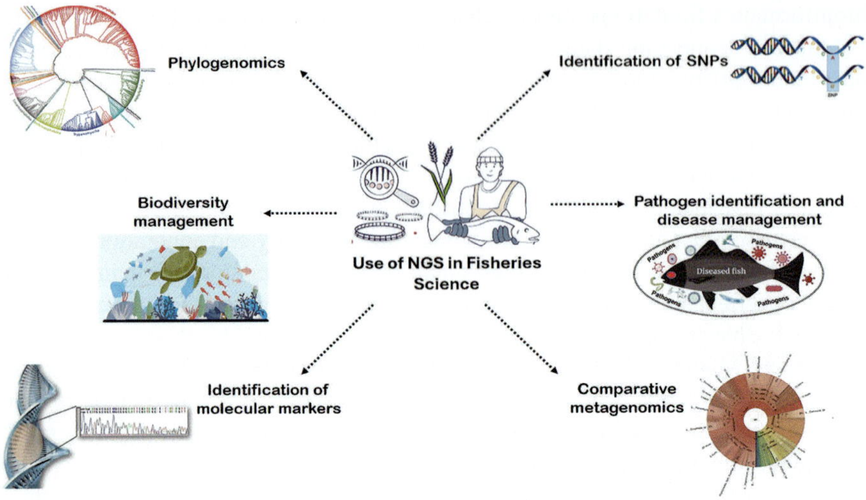

Fig. 14.5 Applications of next-generation sequencing (NGS) in fisheries research

emerging, one that incorporates adaptive diversity to comprehend the basic characteristics of fisheries resources. In the future, new hypotheses in fish study will require a workable framework that incorporates the vast amounts of genomic data collected from transcriptomics, gene control, and evolutionary biology.

References

Adams MD, Celniker SE, Holt RA, Evans CA, Gocayne JD, Amanatides PG, Scherer SE, Li PW, Hoskins RA, Galle RF, George RA, Lewis SE, Richards S, Ashburner M, Henderson SN, Sutton GG, Wortman JR, Yandell MD, Zhang Q et al (2000) The genome sequence of *Drosophila melanogaster*. Science 287(5461):2185–2195

Albertson RC, Cresko W, Detrich HW, Postlethwait JH (2009) Evolutionary mutant models for human disease. Trends Genet 25(2):74–81

Andersson L, Bekkevold D, Berg F, Farrell ED, Felkel S, Ferreira MS, Fuentes-Pardo AP, Goodall J, Pettersson M (2024) How fish population genomics can promote sustainable fisheries: a road map. Annu Rev Anim Biosci 12(1):1–20

Aparicio S, Chapman J, Stupka E, Putnam N, Chia J, Dehal P, Christoffels A, Rash S, Hoon S, Smit A, Gelpke MDS, Roach J, Oh T, Ho IY, Wong M, Detter C, Verhoef F, Predki P, Tay A et al (2002) Whole-genome shotgun assembly and analysis of the genome of *Fugu rubripes*. Science 297(5585):1301–1310

Appleyard SA, Lynch TP, Green MA, Encinas-Viso F (2021) Genetic diversity and restricted genetic connectivity in an endangered marine fish. Mar Freshw Res 72(12):1735–1745

Bergh Ø, BØrsheim KY, Bratbak G, Heldal M (1989) High abundance of viruses found in aquatic environments. Nature 340(6233):467–468

Bräutigam A, Gowik U (2010) What can next generation sequencing do for you? Next generation sequencing as a valuable tool in plant research: next generation sequencing. Plant Biol 12(6):831–841

Brenner S, Elgar G, Sanford R, Macrae A, Venkatesh B, Aparicio S (1993) Characterization of the pufferfish (Fugu) genome as a compact model vertebrate genome. Nature 366(6452):265–268

Chac LD, Thinh BB (2023) Species identification through DNA barcoding and its applications: a review. Biol Bull 50(6):1143–1156
Cheng C, Fei Z, Xiao P (2023) Methods to improve the accuracy of next-generation sequencing. Front Bioeng Biotechnol 11:982111
Chouhan N, Dekari D, Choudhary B, Singh A, Gon Choudhury T (2023) Environmental DNA (eDNA) technology: fisheries and aquaculture perspectives. Indian J Anim Health 62:75–85
Cruz-Cano R, Kolb M, Saldaña-Vázquez RA, Bretón-Deval L, Cruz-Cano N, Aldama-Cervantes A (2024) Existing evidence on the use of environmental DNA as an operational method for studying rivers: a systematic map and thematic synthesis. Environ Evid 13(1):2
FAO Report: Global Fisheries and Aquaculture Production Reaches a New Record High (n.d.) Fao.Org. https://www.fao.org/newsroom/detail/fao-report-global-fisheries-and-aquaculture-production-reaches-a-new-record-high/en
Franco CM, Ambrosio RL, Cepeda A, Anastasio A (2021) Fish intended for human consumption: from DNA barcoding to a next-generation sequencing (NGS)-based approach. Curr Opin Food Sci 42:86–92
Goffeau A, Barrell BG, Bussey H, Davis RW, Dujon B, Feldmann H, Galibert F, Hoheisel JD, Jacq C, Johnston M, Louis EJ, Mewes HW, Murakami Y, Philippsen P, Tettelin H, Oliver SG (1996) Life with 6000 genes. Science 274(5287):546–567
Hauser L, Seeb JE (2008) Advances in molecular technology and their impact on fisheries genetics. Fish Fish 9(4):473–486
Heras S, Abras A, Palahí A, García-Marín J-L, Roldán MI (2024) Genomic analysis of the giant red shrimp (Aristaeomorpha foliacea) using next-generation sequencing: set of tools for population studies. Genes 15(11):1360
Hilborn R, Quinn TP, Schindler DE, Rogers DE (2003) Biocomplexity and fisheries sustainability. Proc Natl Acad Sci 100(11):6564–6568
Holt KE, Parkhill J, Mazzoni CJ, Roumagnac P, Weill F-X, Goodhead I, Rance R, Baker S, Maskell DJ, Wain J, Dolecek C, Achtman M, Dougan G (2008) High-throughput sequencing provides insights into genome variation and evolution in Salmonella Typhi. Nat Genet 40(8):987–993
Kim S, Koo H, Kim J-H, Jung J-W, Hwang SY, Kim W (2011) DNA chip for species identification of Korean freshwater fish: a case study. Biochip J 5(1):72–77
Kim M, Kwon T-H, Jung S-M, Cho S-H, Jin SY, Park N-H, Kim C-G, Kim J-S (2013) Antibiotic resistance of bacteria isolated from the internal organs of edible snow crabs. PLoS One 8(8):e70887
Kundu S, Rath S, Mishra SS, Tyagi K, Banerjee D, Mohapatra A, Kumar V (2024) Reassessment of the fish diversity of Chilika Lagoon by DNA barcoding: genetic evidence and new distribution records for 10 species. Rec Zool Surv India 124(1):79–83
Letsiou S, Madesis P, Vasdekis E, Montemurro C, Grigoriou ME, Skavdis G, Moussis V, Koutelidakis AE, Tzakos AG (2024) DNA barcoding as a plant identification method. Appl Sci 14(4):1415
Liang P, Saqib HSA, Ni X, Shen Y (2020) Long-read sequencing and de novo genome assembly of marine medaka (Oryzias melastigma). BMC Genomics 21(1):640
Liu L, Li Y, Li S, Hu N, He Y, Pong R, Lin D, Lu L, Law M (2012) Comparison of next-generation sequencing systems. J Biomed Biotechnol 2012:1–11
Liu H, Zheng H, Zhang H, Deng L, Liu W, Wang S, Meng F, Wang Y, Guo Z, Li S, Zhang G (2015) A de novo transcriptome of the noble scallop, Chlamys nobilis, focusing on mining transcripts for carotenoid-based coloration. BMC Genomics 16(1):44
Lu G, Luo M (2020) Genomes of major fishes in world fisheries and aquaculture: status, application and perspective. Aquac Fish 5(4):163–173
Ma J-L, Qiang J, Tao Y-F, Bao J-W, Zhu H-J, Li L-G, Xu P (2021) Multi-omics analysis reveals the glycolipid metabolism response mechanism in the liver of genetically improved farmed Tilapia (GIFT, Oreochromis niloticus) under hypoxia stress. BMC Genomics 22(1):105
Martinez-Martin N, Magnus D (2019) Privacy and ethical challenges in next-generation sequencing. Expert Rev Precis Med Drug Dev 4(2):95–104

Maxam AM, Gilbert W (1977) A new method for sequencing DNA. Proc Natl Acad Sci 74(2):560–564

McArthur AG, Tsang KK (2017) Antimicrobial resistance surveillance in the genomic age. Ann N Y Acad Sci 1388(1):78–91

Mehinto AC, Martyniuk CJ, Spade DJ, Denslow ND (2012) Applications for next-generation sequencing in fish ecotoxicogenomics. Front Genet 3:62

Metzker ML (2010) Sequencing technologies—the next generation. Nat Rev Genet 11(1):31–46

Mostovoy Y, Levy-Sakin M, Lam J, Lam ET, Hastie AR, Marks P, Lee J, Chu C, Lin C, Džakula Ž, Cao H, Schlebusch SA, Giorda K, Schnall-Levin M, Wall JD, Kwok P-Y (2016) A hybrid approach for de novo human genome sequence assembly and phasing. Nat Methods 13(7):587–590

Nakamura K, Oshima T, Morimoto T, Ikeda S, Yoshikawa H, Shiwa Y, Ishikawa S, Linak MC, Hirai A, Takahashi H, Altaf-Ul-Amin M, Ogasawara N, Kanaya S (2011) Sequence-specific error profile of Illumina sequencers. Nucleic Acids Res 39(13):e90–e90

Natnan ME, Low C-F, Chong C-M, Bunawan H, Baharum SN (2021) Integration of omics tools for understanding the fish immune response due to microbial challenge. Front Mar Sci 8:668771

Nielsen EE, Cariani A, Aoidh EM, Maes GE, Milano I, Ogden R, Taylor M, Hemmer-Hansen J, Babbucci M, Bargelloni L, Bekkevold D, Diopere E, Grenfell L, Helyar S, Limborg MT, Martinsohn JT, McEwing R, Panitz F, Patarnello T et al (2012) Gene-associated markers provide tools for tackling illegal fishing and false eco-certification. Nat Commun 3(1):851

Nkili-Meyong AA, Bigarré L, Labouba I, Vallaeys T, Avarre J-C, Berthet N (2016) Contribution of next-generation sequencing to aquatic and fish virology. Intervirology 59(5–6):285–300

Nowrousian M (2010) Next-generation sequencing techniques for eukaryotic microorganisms: sequencing-based solutions to biological problems. Eukaryot Cell 9(9):1300–1310

Pareek CS, Smoczynski R, Tretyn A (2011) Sequencing technologies and genome sequencing. J Appl Genet 52(4):413–435

Pathak AK, Rashid I, Nagpure NS, Kumar R, Pati R, Singh M, Murali S, Kushwaha B, Kumar D, Rai A (2019) FisOmics: a portal of fish genomic resources. Genomics 111(6):1923–1928. https://doi.org/10.1016/j.ygeno.2019.01.003

Payet SD, Pratchett MS, Saenz-Agudelo P, Berumen ML, DiBattista JD, Harrison HB (2022) Demographic histories shape the population genomics of the common coral grouper (Plectropomus leopardus). Evol Appl 15(8):1221–1235

Payet SD, Underwood J, Berry O, Saunders T, Travers MJ, Wakefield CB, Miller K, Newman SJ (2024) Population genomics informs the management of harvested snappers across north-western Australia. Sci Rep 14(1):26598

Pomraning KR, Smith KM, Freitag M (2009) Genome-wide high throughput analysis of DNA methylation in eukaryotes. Methods 47(3):142–150

Potts RWA, Gutierrez AP, Penaloza CS, Regan T, Bean TP, Houston RD (2021) Potential of genomic technologies to improve disease resistance in molluscan aquaculture. Philos Trans R Soc Lond Ser B Biol Sci 376(1825):20200168

Powers DA (1989) Fish as model systems. Science 246(4928):352–358

Quail M, Smith ME, Coupland P, Otto TD, Harris SR, Connor TR, Bertoni A, Swerdlow HP, Gu Y (2012) A tale of three next-generation sequencing platforms: comparison of Ion Torrent, Pacific Biosciences, and Illumina MiSeq sequencers. BMC Genomics 13(1):341

Rather MA, Agarwal D, Bhat TA, Khan IA, Zafar I, Kumar S, Amin A, Sundaray JK, Qadri T (2023) Bioinformatics approaches and big data analytics opportunities in improving fisheries and aquaculture. Int J Biol Macromol 233:123549

Rhoads A, Au KF (2015) PacBio sequencing and its applications. Genomics Proteomics Bioinformatics 13(5):278–289

Rothberg JM, Leamon JH (2008) The development and impact of 454 sequencing. Nat Biotechnol 26(10):1117–1124

Rothberg JM, Hinz W, Rearick TM, Schultz J, Mileski W, Davey M, Leamon JH, Johnson K, Milgrew MJ, Edwards M, Hoon J, Simons JF, Marran D, Myers JW, Davidson JF, Branting

A, Nobile JR, Puc BP, Light D et al (2011) An integrated semiconductor device enabling non-optical genome sequencing. Nature 475(7356):348–352

Sherif AH, Gouda M, Darwish S, Abdelmohsin A (2021) Prevalence of antibiotic-resistant bacteria in freshwater fish farms. Aquac Res 52(5):2036–2047

Spivakov M, Auer TO, Peravali R, Dunham I, Dolle D, Fujiyama A, Toyoda A, Aizu T, Minakuchi Y, Loosli F, Naruse K, Birney E, Wittbrodt J (2013) Genomic and phenotypic characterisation of a wild Medaka population: establishing an isogenic population genetic resource in fish (Version 2). arXiv

Therkildsen NO, Hemmer-Hansen J, Hedeholm RB, Wisz MS, Pampoulie C, Meldrup D, Bonanomi S, Retzel A, Olsen SM, Nielsen EE (2013) Spatiotemporal SNP analysis reveals pronounced biocomplexity at the northern range margin of ATLANTIC COD GADUS *morhua*. Evol Appl 6(4):690–705

Thomsen PF, Willerslev E (2015) Environmental DNA – an emerging tool in conservation for monitoring past and present biodiversity. Biol Conserv 183:4–18

Thomsen PF, Kielgast J, Iversen LL, Wiuf C, Rasmussen M, Gilbert MTP, Orlando L, Willerslev E (2012) Monitoring endangered freshwater biodiversity using environmental DNA. Mol Ecol 21(11):2565–2573

Valenzuela-Quiñonez F (2016) How can fisheries management benefit from genomics? Brief Funct Genomics 15(5):352–357

Walker PJ, Winton JR (2010) Emerging viral diseases of fish and shrimp. Vet Res 41(6):51

Wang R-L, Bencic D, Villeneuve DL, Ankley GT, Lazorchak J, Edwards S (2010) A transcriptomics-based biological framework for studying mechanisms of endocrine disruption in small fish species. Aquat Toxicol 98(3):230–244

Ward RD (2000) Genetics in fisheries management. Hydrobiologia 420(1):191–201

Wenne R (2023) Microsatellites as molecular markers with applications in exploitation and conservation of aquatic animal populations. Genes 14(4):808

Yamamoto S, Minami K, Fukaya K, Takahashi K, Sawada H, Murakami H, Tsuji S, Hashizume H, Kubonaga S, Horiuchi T, Hongo M, Nishida J, Okugawa Y, Fujiwara A, Fukuda M, Hidaka S, Suzuki KW, Miya M, Araki H et al (2016) Environmental DNA as a 'snapshot' of fish distribution: a case study of Japanese Jack Mackerel in Maizuru Bay, Sea of Japan. PLoS One 11(3):e0149786

Young AD & Gillung JP (2019) Phylogenomics — principles, opportunities and pitfalls of big-data phylogenetics. Syst Entomol 45(2):225–247. https://doi.org/10.1111/syen.12406

Yuan X, Lv Z, Zhang Z, Han Y, Liu Z, Zhang H (2023) A review of antibiotics, antibiotic-resistant bacteria, and resistance genes in aquaculture: occurrence, contamination, and transmission. Toxics 11(5):420

Zhou A, Xie S, Junaid M, Sun D, Tang H, Chuan J, Li X, Xu G, Zou J (2021) First insight into the environmental microbial communities associated with potentially pathogenic strains in pond cultured tilapia (*Oreochromis niloticus*) at various growth stages based on 16S, 18S, and ITS2 rRNA gene amplicons sequencing. Aquaculture 532:736007

Bioinformatics and Big Data in Marine Conservation

15

Ayushman Gadnayak, Hirak Jyoti Chakraborty, and Smruti Priyambada Pradhan

15.1 Introduction

Marine ecosystems are intricate networks of diverse organisms interacting with their physical environment. These ecosystems range from the sunlit surface waters teeming with phytoplankton and zooplankton to the mysterious depths of the ocean floor, where unique species thrive in extreme conditions. The ocean is a vast reservoir of biodiversity, supporting a vast array of life forms from microscopic bacteria to the largest whales, all of which contribute to the ecological balance. Marine biodiversity is crucial, providing numerous ecosystem services such as carbon sequestration, nutrient cycling, and food supply. Furthermore, marine ecosystems play a pivotal role in regulating the Earth's climate and weather patterns. However, marine ecosystems face significant threats, including climate change, which leads to ocean acidification and rising sea temperatures. These changes in ocean chemistry and temperature can have profound effects on marine life, particularly coral reefs, which are highly sensitive to changes in their environment. Additionally, pollution from plastics and chemicals is a major threat to marine ecosystems. Plastic debris can be ingested by marine animals, causing physical harm or even death, while chemical pollutants can accumulate in the tissues of marine organisms, leading to toxic effects. Overfishing is another critical issue, as it disrupts food chains and depletes fish populations, which are essential for the livelihoods of millions of people around the world. Each of these factors has cascading effects, altering habitats and endangering marine species (Pörtner et al. 2014; Halpern et al. 2008).

Bioinformatics and big data have revolutionized our understanding of marine science. Bioinformatics involves the use of software tools to understand biological data, and when combined with big data techniques, it allows for the analysis of vast

A. Gadnayak (✉) · H. J. Chakraborty · S. P. Pradhan
ICAR-Central Inland Fisheries Research Institute, Barrackpore, Kolkata, India

© The Author(s), under exclusive license to Springer Nature Singapore Pte Ltd. 2026

B. K. Das et al. (eds.), *Bioinformatics Tools for Predictive Ecology and Fisheries*, Sustainable Aquatic Food Systems, https://doi.org/10.1007/978-981-95-2573-7_15

amounts of information generated by next-generation sequencing and other technologies (Kanehisa et al. 2004). Next-generation sequencing technologies have enabled scientists to decode the genomes of various marine organisms rapidly, providing insights into their physiology, adaptation mechanisms, and evolutionary history. For instance, the genome of the Atlantic cod has been sequenced, revealing genes that allow it to tolerate cold waters, which is crucial information for managing fish stocks in the face of climate change. The integration of big data allows for the collection and analysis of data on an unprecedented scale, from tracking oceanographic data to monitoring the health of coral reefs through satellite imagery. This wealth of data provides a more comprehensive understanding of marine ecosystems and their dynamics. For example, big data analytics can help identify patterns and trends in marine biodiversity, such as changes in species distribution due to climate change or human activities. Despite these advancements, challenges remain, such as the need for high computational power, data storage solutions, and the development of robust algorithms to handle and interpret complex datasets (Yau et al. 2014). The application of bioinformatics and big data in marine science presents both challenges and opportunities. One significant challenge is the management and analysis of enormous datasets that require sophisticated computational infrastructure and expertise in data science. Additionally, there are ethical considerations related to data sharing and the use of genetic information. However, these technologies offer immense opportunities for advancing marine conservation efforts, understanding the impacts of climate change, and discovering new species and compounds with potential biotechnological applications. For instance, bioinformatics can be used to identify genes responsible for the production of bioactive compounds in marine organisms, which can lead to the development of new drugs or other products.

By harnessing the power of bioinformatics and big data, scientists can develop more effective strategies for preserving marine biodiversity and ensuring the sustainability of marine resources for future generations (Costello et al. 2010; Glover et al. 2010). For example, predictive models can be developed to forecast the impacts of climate change on marine ecosystems, enabling proactive measures to be taken to mitigate these effects. Moreover, big data analytics can support the enforcement of fishing regulations by providing real-time monitoring of fishing activities, helping to prevent overfishing and illegal fishing practices. Overall, the integration of bioinformatics and big data into marine science holds great promise for enhancing our understanding and stewardship of the marine environment.

15.2 Bioinformatics Tools and Techniques

In the realm of marine bioinformatics, genomics and transcriptomics are critical areas of study. DNA sequencing technologies, such as next-generation sequencing (NGS), have revolutionized our understanding of marine organisms. These technologies enable the sequencing of entire genomes rapidly and at a lower cost, providing insights into genetic diversity, evolutionary processes, and adaptive mechanisms in marine species. Genome assembly and annotation are essential steps that follow sequencing. Assembly involves piecing together short DNA reads into longer sequences, while annotation identifies functional elements like genes and regulatory regions within the genome (Pop and Salzberg 2008). RNA sequencing (RNA-seq) is another powerful tool used to study gene expression in marine organisms. By sequencing RNA molecules, researchers can analyze which genes are active under different environmental conditions, shedding light on how marine species respond to stressors like climate change and pollution (Wang et al. 2009).

Proteomics and metabolomics are complementary approaches that provide a deeper understanding of the functional biology of marine organisms. Proteomics involves the large-scale study of proteins, the workhorses of the cell, using mass spectrometry-based techniques. This approach allows scientists to identify and quantify proteins, as well as to study post-translational modifications and protein–protein interactions (Aebersold and Mann 2003). Metabolomics, on the other hand, focuses on the comprehensive analysis of metabolites, the small molecules involved in metabolism. Techniques such as liquid chromatography-mass spectrometry (LC-MS) and nuclear magnetic resonance (NMR) spectroscopy are commonly used for metabolomic profiling (Fiehn 2002). Together, proteomics and metabolomics provide a holistic view of the molecular mechanisms underlying physiological processes and adaptive responses in marine organisms, offering potential applications in areas like drug discovery and environmental monitoring.

The complexity and scale of data generated in marine bioinformatics necessitate the use of bioinformatics pipelines and workflows. These pipelines automate the analysis of large datasets, ensuring reproducibility and efficiency. Popular bioinformatics tools and software include BLAST for sequence alignment, Clustal Omega for multiple sequence alignment, and R for statistical analysis and visualization (Altschul et al. 1990; Sievers et al. 2011; R Core Team 2020). Additionally, the advent of cloud computing and high-performance computing (HPC) has greatly enhanced the capacity to handle and process massive datasets. Cloud platforms like Amazon Web Services (AWS) and Google Cloud offer scalable computing resources that can be tailored to the needs of bioinformatics projects. These advancements enable researchers to conduct complex analyses, such as metagenomic studies and predictive modeling, more efficiently and cost effectively (Schatz et al. 2010). An overview of tools and techniques is shown in Table 15.1.

15.3 Part II: Big Data Applications in Marine Conservation

15.3.1 Marine Microbial Ecology

Marine microbial ecology is significantly enhanced through big data applications, particularly in the use of metagenomics and metatranscriptomics. Metagenomics involves sequencing the collective genome of microorganisms in a given environment, which allows researchers to identify and characterize the diversity of microbial communities. Metatranscriptomics, on the other hand, focuses on the complete set of RNA transcripts, providing insights into the functional activity of microbial

Table 15.1 Overview of bioinformatics tools and techniques in marine science

Aspect	Description
Genomics and Transcriptomics	DNA sequencing (NGS) provides insights into genetic diversity, evolution, and adaptation. Genome assembly reconstructs DNA sequences, while annotation identifies functional elements. RNA-seq analyzes gene expression under various conditions
Proteomics	Studies proteins using mass spectrometry to identify, quantify, and analyze post-translational modifications and interactions
Metabolomics	Analyzes metabolites using LC-MS and NMR spectroscopy to understand metabolism and adaptive responses
Bioinformatics tools	BLAST (sequence alignment), Clustal omega (multiple sequence alignment), R (statistical analysis and visualization)
Computational resources	Cloud computing (AWS, Google cloud) and high-performance computing (HPC) enhance data processing and analysis
Applications	Conservation, drug discovery, environmental monitoring, predictive modeling, and metagenomics

communities under various environmental conditions (Huson et al. 2016). These techniques enable comprehensive microbial community analysis and functional profiling, revealing the roles of microbes in nutrient cycling, primary production, and other essential ecosystem processes. The integration of big data technologies allows for the analysis of vast datasets, improving our understanding of microbial ecology in marine ecosystems and their responses to environmental changes (Sunagawa et al. 2015).

15.3.2 Population Genetics and Phylogeography

Big data also plays a crucial role in population genetics and phylogeography, which are essential for conservation efforts. Population genetics involves studying the genetic diversity and structure of populations, while phylogeography focuses on the historical processes that shape the geographic distributions of species (Excoffier and Heckel 2006). Next-generation sequencing technologies have facilitated large-scale genetic analyses, enabling researchers to assess population structure and genetic diversity with high resolution. Phylogenetic analysis, which involves reconstructing the evolutionary relationships among species, is enhanced through big data approaches that can handle extensive datasets (Kumar et al. 2016). These insights are vital for conservation genetics and population management, as they help identify genetically distinct populations, assess their vulnerability, and develop strategies for their protection and restoration (Allendorf et al. 2010).

15.3.3 Marine Ecosystem Modeling

Marine ecosystem modeling benefits immensely from big data applications, particularly in ecological niche modeling and species distribution modeling. Ecological niche modeling predicts the suitable habitat conditions for species based on environmental variables, while species distribution modeling forecasts the potential distribution of species under current and future climate scenarios (Elith and Leathwick 2009). These models rely on extensive datasets, including species occurrence records and environmental data, which are increasingly accessible through big data platforms. Ecosystem modeling and forecasting are also crucial for understanding the dynamics of marine ecosystems, such as predator–prey interactions, nutrient fluxes, and the impacts of human activities (Plagányi et al. 2014). By incorporating big data into these models, scientists can make more accurate predictions and inform conservation and management decisions.

15.3.4 Oceanographic Data Analysis

Oceanographic data analysis has been transformed by big data technologies, enabling more sophisticated studies of marine environments. Remote sensing and satellite imagery provide continuous, large-scale observations of oceanographic variables such as sea surface temperature, chlorophyll concentration, and ocean currents (Hansen and Loveland 2010). Oceanographic data assimilation integrates these observations with numerical models to improve the accuracy of ocean state estimates and forecasts (Cummings and Smedstad 2013). Big data also plays a critical role in studying the impacts of climate change on marine ecosystems, as it allows for the analysis of long-term datasets to detect trends and anomalies (Hoegh-Guldberg and Bruno 2010). This comprehensive understanding of oceanographic processes is essential for predicting the effects of climate change and implementing effective conservation measures.

15.4 Part III: Ethical Considerations and Future Directions

15.4.1 Data Sharing and Standardization

In the realm of marine science, ethical considerations around data sharing and standardization are paramount. The FAIR data principles, which stand for Findable, Accessible, Interoperable, and Reusable, are designed to enhance the transparency and utility of scientific data (Wilkinson et al. 2016). Implementing these principles ensures that data can be easily located, accessed, and utilized by others, fostering collaboration and advancing scientific discoveries. Data repositories and databases, such as the Ocean Biogeographic Information System (OBIS) and the Global Biodiversity Information Facility (GBIF), play crucial roles in aggregating and disseminating marine data (Costello et al. 2015). However, there are challenges regarding data privacy and security, particularly when dealing with sensitive ecological or genetic information. Ensuring robust data governance frameworks is essential to protect privacy while promoting open science (Borgman 2015).

15.4.2 Citizen Science and Crowdsourcing

Citizen science and crowdsourcing represent innovative approaches to engaging the public in marine research. Projects like Reef Check and the Great Barrier Reef Census involve volunteers in data collection and monitoring, increasing the scale and scope of research efforts (Thiel et al. 2014). Engaging citizens not only enhances scientific literacy but also fosters a sense of stewardship for marine environments. However, ensuring the quality and reliability of data collected through citizen science is crucial. Implementing rigorous protocols for data validation and incorporating training for volunteers can help mitigate issues related to data accuracy (Kosmala et al. 2016). By leveraging the power of crowdsourcing, marine scientists can gather extensive datasets that might otherwise be unattainable due to resource limitations.

15.4.3 Artificial Intelligence and Machine Learning

Artificial intelligence (AI) and machine learning (ML) are transforming marine conservation efforts. AI-driven approaches, such as automated species identification and habitat mapping, enable researchers to analyze vast amounts of data quickly and accurately (Christin et al. 2019). For example, machine learning algorithms can classify underwater images or acoustic recordings to identify species and track their populations. Despite the promise of AI in marine science, there are challenges and limitations to consider. The development and training of accurate AI models require large, high-quality datasets, which may not always be available. Additionally, the interpretability of AI models remains a challenge, as complex algorithms can be difficult to understand and validate (Samek et al. 2017). Addressing these issues is essential to fully realize the potential of AI in marine conservation. A detailed flow of work is depicted in Fig. 15.1.

15.5 Future Trends and Emerging Technologies

Looking to the future, several emerging technologies hold great promise for advancing marine science and conservation. Synthetic biology and marine biotechnology are at the forefront, with potential applications ranging from developing biofuels to bioremediation (Cameron et al. 2014). CRISPR-Cas9 gene editing technology offers unprecedented opportunities to study gene function and develop targeted interventions for conservation and management (Hsu et al. 2014). Additionally, the establishment of ocean observatories and real-time monitoring systems provides continuous data on ocean conditions, enabling timely responses to environmental changes (Lefevre et al. 2017). These advancements will enhance our ability to protect and preserve marine ecosystems, ensuring their resilience in the face of global challenges.

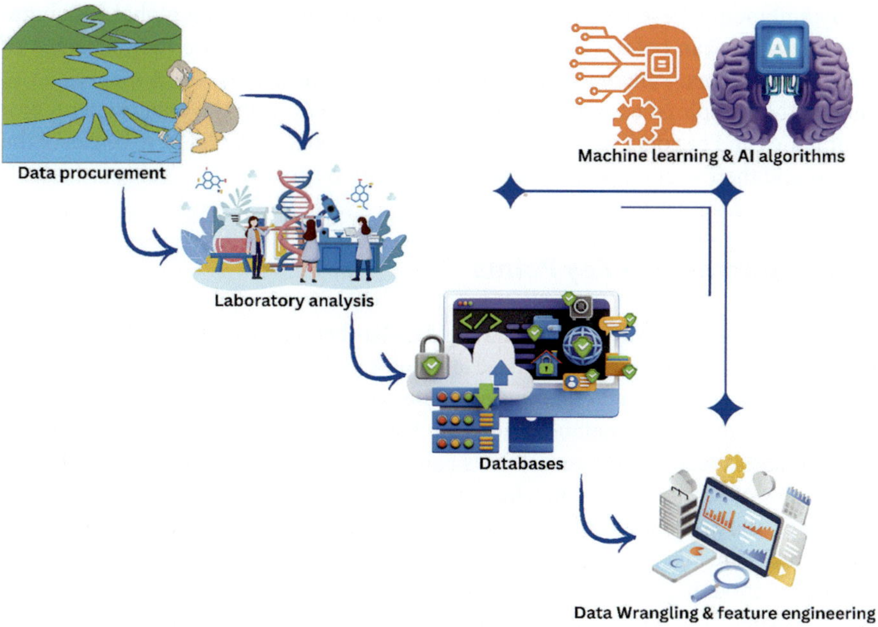

Fig. 15.1 AI and ML applications in marine conservation

15.6 Conclusion

Marine ecosystems are invaluable reservoirs of biodiversity, essential for maintaining ecological balance, supporting human livelihoods, and mitigating climate change impacts. However, they face mounting threats from climate change, pollution, and overfishing, which necessitate innovative and effective conservation strategies. The integration of bioinformatics and big data into marine science has revolutionized our ability to study and understand these complex ecosystems. Through advancements in genomics, proteomics, and metabolomics, scientists can uncover the genetic and functional diversity of marine organisms, shedding light on their adaptive mechanisms and ecological roles. Meanwhile, big data analytics, coupled with tools like AI and machine learning, enable comprehensive monitoring, modeling, and prediction of marine ecosystem dynamics, offering critical insights for conservation efforts. Challenges such as managing vast datasets, ensuring data privacy, and addressing the interpretability of AI models remain significant but are being addressed through innovations in computational infrastructure, adherence to FAIR data principles, and collaborative approaches like citizen science. The future holds immense promise with emerging technologies, including synthetic biology, CRISPR gene editing, and real-time ocean monitoring systems, which will expand our capacity to mitigate threats to marine ecosystems and promote their sustainability.

By embracing these technological advancements and fostering global collaboration, humanity can better protect marine biodiversity and ensure the resilience of ocean ecosystems, safeguarding them for future generations. As we continue to unravel the complexities of marine life and its interplay with the environment, bioinformatics and big data will remain indispensable tools in our quest to preserve the oceans and their irreplaceable contributions to life on Earth.

15.7 Summary of Key Points

The integration of bioinformatics and big data has become indispensable in advancing marine conservation efforts. These technologies have revolutionized the way we study marine ecosystems, enabling more detailed and comprehensive analyses of genetic, proteomic, and metabolomic data. By leveraging next-generation sequencing and mass spectrometry, researchers can uncover the genetic diversity and functional roles of marine organisms, leading to more informed conservation strategies. Moreover, the application of bioinformatics pipelines and high-performance computing facilitates the management and interpretation of vast datasets, allowing for predictive modeling and real-time monitoring of marine environments. These advancements hold significant potential for future research, as they provide the tools necessary to address complex ecological questions and respond to emerging environmental challenges. Looking ahead, several emerging trends and challenges will shape the future of marine science and conservation. The continued development of synthetic biology and gene editing technologies, such as CRISPR-Cas9, offers exciting opportunities for targeted interventions and the discovery of new biotechnological applications. However, these advancements also raise ethical and regulatory considerations that must be carefully addressed. Additionally, the integration of artificial intelligence and machine learning in marine research will likely drive further innovation, though challenges related to data quality, model interpretability, and computational resources remain. To fully realize the potential of these technologies, interdisciplinary collaboration and capacity building are essential. By fostering partnerships between marine biologists, bioinformaticians, data scientists, and policymakers, we can enhance our collective ability to protect and preserve marine ecosystems for future generations.

The future of marine conservation is bright, but it requires a concerted effort to harness the power of cutting-edge technologies and address the complex ethical, technical, and environmental challenges ahead. Through continued innovation and collaboration, we can ensure the health and sustainability of our oceans.

References

Aebersold R, Mann M (2003) Mass spectrometry-based proteomics. Nature 422(6928):198–207
Allendorf FW, Luikart G, Aitken SN (2010) Conservation and the genetics of populations. Wiley
Altschul SF, Gish W, Miller W, Myers EW, Lipman DJ (1990) Basic local alignment search tool. J Mol Biol 215(3):403–410

Borgman CL (2015) Big data, little data, no data: scholarship in the networked world. MIT Press

Cameron DE, Bashor CJ, Collins JJ (2014) A brief history of synthetic biology. Nat Rev Microbiol 12(5):381–390

Christin S, Hervet É, Lecomte N (2019) Applications for deep learning in ecology. Methods Ecol Evol 10(10):1632–1644

Costello MJ, Cheung A, De Hauwere N (2010) Topography statistics for the surface and seabed area, volume, depth, slope, and topographic variation for the world's seas, oceans, and countries. Environ Sci Technol 44(23):8821–8828

Costello MJ, Michener WK, Gahegan M, Zhang ZQ, Bourne PE (2015) Biodiversity data should be published, cited, and peer reviewed. Trends Ecol Evol 28(8):454–461

Cummings JA, Smedstad OM (2013) Variational data assimilation for the global ocean. In: Data assimilation for atmospheric, oceanic and hydrologic applications, vol II. Springer, pp 303–343

Elith J, Leathwick JR (2009) Species distribution models: ecological explanation and prediction across space and time. Annu Rev Ecol Evol Syst 40:677–697

Excoffier L, Heckel G (2006) Computer programs for population genetics data analysis: a survival guide. Nat Rev Genet 7(10):745–758

Fiehn O (2002) Metabolomics–the link between genotypes and phenotypes. Plant Mol Biol 48(1–2):155–171

Glover AG, Higgs N, Horton T (2010) Deep-sea species diversity and the impact of biogeography, isolation and time. Biol Rev 85(3):588–610

Halpern BS, Walbridge S, Selkoe KA, Kappel CV, Micheli F, D'Agrosa C, McMahon G (2008) A global map of human impact on marine ecosystems. Science 319(5865):948–952

Hansen MC, Loveland TR (2010) A review of large area monitoring of land cover change using Landsat data. Remote Sens Environ 122:66–74

Hoegh-Guldberg O, Bruno JF (2010) The impact of climate change on the world's marine ecosystems. Science 328(5985):1523–1528

Hsu PD, Lander ES, Zhang F (2014) Development and applications of CRISPR-Cas9 for genome engineering. Cell 157(6):1262–1278

Huson DH, Auch AF, Qi J, Schuster SC (2016) MEGAN analysis of metagenomic data. Genome Res 17(3):377–386

Kanehisa M, Goto S, Sato Y, Furumichi M, Tanabe M (2004) KEGG for integration and interpretation of large-scale molecular data sets. Nucleic Acids Res 32(suppl_1):D277–D280

Kosmala M, Wiggins A, Swanson A, Simmons B (2016) Assessing data quality in citizen science. Front Ecol Environ 14(10):551–560

Kumar S, Stecher G, Tamura K (2016) MEGA7: molecular evolutionary genetics analysis version 7.0 for bigger datasets. Mol Biol Evol 33(7):1870–1874

Lefevre N, Watson AJ, Olsen A et al (2017) A global database of sea-air CO2 fluxes derived from surface ocean CO2 Atlas (SOCAT) data. Earth Syst Sci Data 9(3):1357–1386

Plagányi ÉE, Skewes TD, Dowling NA, Haddon M, Campbell RA (2014) Risk management tools for sustainable fisheries management under changing climate: a sea cucumber example. Clim Change 124(3):685–698

Pop M, Salzberg SL (2008) Bioinformatics challenges of new sequencing technology. Trends Genet 24(3):142–149

Pörtner HO, Karl D, Boyd PW, Cheung WWL, Lluch-Cota SE, Nojiri Y, Schmidt DN (2014) Ocean systems. In: Field CB, Barros VR, Dokken DJ (eds) Climate change 2014: impacts, adaptation, and vulnerability. Part a: global and sectoral aspects. Cambridge University Press, Cambridge, UK/New York, NY, pp 411–484

R Core Team (2020) R: a language and environment for statistical computing. R Foundation for Statistical Computing, Vienna

Samek W, Wiegand T, Müller K-R (2017) Explainable artificial intelligence: understanding, visualizing and interpreting deep learning models. ITU Jl: ICT Discov 1(1):39–48

Schatz MC, Langmead B, Salzberg SL (2010) Cloud computing and the DNA data race. Nat Biotechnol 28(7):691–693

Sievers F, Wilm A, Dineen D, Gibson TJ, Karplus K, Li W et al (2011) Fast, scalable generation of high-quality protein multiple sequence alignments using Clustal omega. Mol Syst Biol 7(1):539

Sunagawa S, Coelho LP, Chaffron S et al (2015) Structure and function of the global ocean microbiome. Science 348(6237):1261359

Thiel M, Penna-Díaz MA, Luna-Jorquera G, Salas S, Sellanes J, Stotz W, Tala F (2014) Citizen scientists and marine research: volunteer participants, their contributions, and projection for the future. Oceanogr Mar Biol Annu Rev 52:257–314

Wang Z, Gerstein M, Snyder M (2009) RNA-Seq: a revolutionary tool for transcriptomics. Nat Rev Genet 10(1):57–63

Wilkinson MD, Dumontier M, Aalbersberg IJ et al (2016) The FAIR guiding principles for scientific data management and stewardship. Sci Data 3:160018

Yau S, Criddle CS, Knight R (2014) Microbial biogeography of public restroom surfaces. PLoS One 9(1):e89381

Functional Genomics in Fish Immunology

16

Bhagyeswari Behera, Smruti Priyambada Pradhan, Subhasmita Behera, and Animesh Pattnaik

16.1 Introduction

Fish immunology encompasses the defence systems of fish, a taxonomically diverse group of aquatic vertebrates integral to aquatic ecosystems and essential to global human economies (Ghani et al. 2024). As ectothermic vertebrates, fish represent an evolutionary link between invertebrates and higher vertebrates, making their immune systems unique and valuable for comparative immunological studies (Du Pasquier 2001). Their immune systems comprise both innate and adaptive components, offering insights into fundamental immunological processes and their evolutionary origins (Müller et al. 2018).

Fish immunology is essential for multiple reasons, like in aquaculture, where fishes are prone to diseases due to high stocking densities and environmental stresses, which lead to significant economic losses (Mugwanya et al. 2022). Furthermore, as environmental indicators, the immune health of fish reflects the impacts of pollution and climate change on aquatic ecosystems (Dar et al. 2022).

The immune system of fish, like other vertebrates, is a complex network of specialized cells and organs that work together to defend against pathogens (Buchmann 2014). However, fish exhibit distinct immunological features due to their aquatic environment and evolutionary position. Understanding the key immune cells and organs in fish is crucial for deciphering their defense mechanisms and developing strategies to enhance their immunity, particularly in aquaculture.

B. Behera · A. Pattnaik (✉)
Department of Botany, School of Applied Sciences, Centurion University of Technology and Management, Bhubaneswar, India
e-mail: animesh.pattnaik@cutm.ac.in

S. P. Pradhan · S. Behera
ICAR-Central Inland Fisheries Research Institute, Barrackpore, Kolkata, India

© The Author(s), under exclusive license to Springer Nature Singapore Pte Ltd. 2026

B. K. Das et al. (eds.), *Bioinformatics Tools for Predictive Ecology and Fisheries*, Sustainable Aquatic Food Systems, https://doi.org/10.1007/978-981-95-2573-7_16

Genomics is a transformative tool in the study of fish immunology, offering insights into the genetic and molecular mechanisms underpinning immune responses in fish. Fish represent a critical group in aquaculture, biodiversity, and evolutionary studies, making it essential to understand their immune systems for health management and ecological preservation. Fish possess unique immune features, such as the absence of certain mammalian immune components and the presence of alternative pathways (e.g., nucleated erythrocytes with immune functions). Functional genomics aids in identifying and characterizing fish-specific immune genes and pathways. Understanding the molecular basis of immune responses in fish is crucial for developing effective vaccines. Functional genomics identifies antigenic targets and immune-related genes that can improve vaccine design, enhancing fish health and productivity in aquaculture systems. By exploring the structure and function of these immune cells and organs, researchers can better understand fish immunology, contributing to advancements in vaccine development, disease prevention, and sustainable aquaculture practices.

This topic focuses on the rapidly growing field of bioinformatics and its transformative role in predictive ecology and fisheries science. With the increasing availability of large datasets generated by environmental monitoring, genomic studies, and fisheries management, bioinformatics tools have become indispensable for processing, analyzing, and interpreting complex information.

16.2 Fundamentals of Fish Immunology

Fish possess both innate and adaptive immune systems that work together to defend against pathogens. The innate immune system serves as the first line of defense, providing a nonspecific response that is fast-acting but lacks memory. Components of the innate immune system in fish include physical barriers such as skin and mucus, cellular components like macrophages, neutrophils, and natural killer cells, as well as humoral factors like lysozymes, complement proteins, and antimicrobial peptides (Mokhtar et al. 2023). The adaptive immune system, on the other hand, provides a specific response that is slower to develop but includes immunological memory, allowing for an enhanced response upon subsequent exposures to the same pathogen (Natoli and Ostuni 2019). The adaptive immune system in fish includes lymphocytes, particularly T cells and B cells, which produce antibodies. Fish antibodies, often referred to as immunoglobulins (e.g., IgM and IgT/IgZ), are structurally and functionally distinct from those of mammals but play a similar role in pathogen recognition and neutralization (Salinas et al. 2011). Unlike mammals, fish lack bone marrow and lymph nodes, relying on other lymphoid organs such as the spleen, kidney, and mucosa-associated lymphoid tissues to support immune responses. Mucosal immunity is especially critical in aquatic environments, where fish are constantly exposed to pathogens (Tosi 2005; den Haan et al. 2014).

Table 16.1 Major fish microbial pathogens

Pathogen type	Examples	Disease/effects
Bacterial	*Aeromonas* spp.	Hemorrhagic septicemia, ulcerative diseases
Vibrio spp.	Vibriosis (hemorrhages, ulcers, high mortality)	
Edwardsiella spp.	Edwardsiellosis (liver, spleen damage)	
Flavobacterium spp.	Bacterial gill disease, columnaris disease	
Streptococcus spp.	Streptococcosis (septicemia, neurological symptoms)	
Viral	IHNV	Infectious hematopoietic necrosis (high mortality in young fish).
IPNV	Infectious pancreatic necrosis (pancreatic necrosis)	
KHV	Koi herpesvirus (gill damage, high mortality)	
SVCV	Spring viremia of carp (systemic hemorrhagic disease)	
VHSV	Viral hemorrhagic septicemia (hemorrhages, organ damage)	
Fungal	*Saprolegnia* spp.	Saprolegniosis (skin, gills, eggs affected).
Aphanomyces spp.	Epizootic ulcerative syndrome (EUS, deep ulcers)	
Parasitic	*Ichthyophthirius multifiliis* (Ich)	White spot disease (freshwater fish)
Cryptobia spp.	Systemic infections, weight loss	
Dactylogyrus spp.	Monogenean parasites (gill infections)	
Cestodes, nematodes	Intestinal infections, malnutrition	
Other pathogens	Algal toxins	Harmful algal blooms
Rickettsia-like organisms (RLOs)	Intracellular bacteria cause systemic diseases	

The interaction between innate and adaptive systems in fish reflects an evolutionary adaptation to their aquatic environment, showcasing the diversity and complexity of vertebrate immune systems. Understanding these systems is crucial for improving aquaculture practices and developing effective disease management strategies (Boehm 2011) (Table 16.1).

Fish have specialized immune cells and organs that contribute to their defense mechanisms against pathogens. These components are adapted to their aquatic environment and support both innate and adaptive immune responses (Secombes and Wang 2012; Panara et al. 2024; Zapata 2022; Guha et al. 2025; Rathor and Swain 2024).

16.2.1 Innate Immune Cells

In fish, as in other vertebrates, the *innate immune system* serves as the first line of defence against pathogens. It is nonspecific and fast-acting and does not require prior exposure to an invader to initiate a response. Fish possess a range of innate immune cells that play critical roles in recognizing and eliminating pathogens such as bacteria, viruses, and parasites.

Cell type	Function	Features
Macrophages	Phagocytosis of pathogens, antigen presentation, and cytokine production	Found in tissues like the spleen, kidney, and liver
Neutrophils	Rapid responders to infection, phagocytosis, and release of antimicrobial substances	Similar to mammalian neutrophils
Monocytes	Circulate in the blood, differentiate into macrophages in tissues	Important for inflammation and tissue repair
Dendritic-like cells	Antigen-presenting cells that link innate and adaptive immunity	Less well characterized than in mammals
Natural killer (NK)-like cells	Target and destroy virus-infected and abnormal cells	Limited characterization, but functional analogs exist
Eosinophils and basophils	Roles in parasite defense and inflammation	Rare, but present in some species
Mast cells	Release histamine and other mediators, often found in mucosal tissues	More prevalent in mucosal immune defense

16.2.2 Adaptive Immune Cells

The *adaptive immune system* in fish, though less complex than in mammals, is well developed and capable of *antigen-specific responses* and *immunological memory*. This system allows fish to mount a more tailored and long-lasting immune defence upon repeated exposure to pathogens.

Cell type	Function	Fish-specific
B cells	Produce antibodies (mainly IgM, also IgD and IgT in teleosts)	IgT (or IgZ in some species) is unique to mucosal immunity
T cells	Involved in cell-mediated immunity, including helper and cytotoxic functions	$CD4^+$ and $CD8^+$ T-cell subsets exist, though markers vary
Plasma cells	Differentiated B cells that secrete large amounts of antibodies	Play a role in long-term humoral immunity
Memory cells	Persist after an infection to ensure faster responses upon re-exposure	Evidence of memory responses is growing, but less defined than in mammals

16.2.3 Immune Organs in Fish

Immune organs in fish are specialized anatomical structures that are responsible for the development, coordination, and execution of immune responses. These organs produce immune cells, process antigens, and orchestrate defense mechanisms to protect fish from invading pathogens such as bacteria, viruses, fungi, and parasites. Unlike mammals, fish do not possess lymph nodes or bone marrow, but they have evolved a unique set of immune organs that fulfill equivalent roles in maintaining health and immunity in aquatic environments. They are classified into primary and secondary lymphoid organs. These organs play a crucial role in adaptive and innate immune responses, enabling fish to defend against pathogens in their aquatic environments. Primary lymphoid organs (thymus and anterior kidney) are responsible for the development and maturation of lymphocytes (B and T cells). Secondary lymphoid organs (spleen, posterior kidney, MALT) are essential sites for the initiation of immune responses, where lymphocytes encounter antigens and become activated.

(a) *Thymus:* Located bilaterally near the dorsal region of the gill cavity, the thymus is the primary lymphoid organ responsible for the development, selection, and maturation of T lymphocytes. The structure of the thymus is less compartmentalized compared to mammals, but it is functionally analogous and essential for cell-mediated immunity.
(b) *Spleen:* Serves as the main secondary lymphoid organ in teleost fish. It helps in blood filtration, removing damaged erythrocytes and pathogens, provides a site for antigen processing, where immune cells interact with foreign particles and also a center for lymphocyte activation and proliferation, particularly B cells and macrophages.
(c) *Mucosa-Associated Lymphoid Tissues (MALT):* Fish are constantly exposed to waterborne pathogens through their external surfaces. As such, mucosal immunity is a cornerstone of their immune defence, mediated by specialized lymphoid tissues like *GILL-Associated Lymphoid Tissue (GIALT)* which protects the respiratory surfaces of the gills, *SKIN-Associated Lymphoid Tissue (SALT)* which provides frontline defence at the epidermal barrier, and *GUT-Associated Lymphoid Tissue (GALT)* which guards the intestinal tract and is involved in the oral tolerance and mucosal immune responses. These tissues support local immune cell populations, including IgT^+ B cells, phagocytes, and antigen-presenting cells, and are critical for immune responses without systemic inflammation.
(d) ***Liver:*** A secondary lymphoid organ that plays a **supportive role in immunity**. It synthesizes **acute-phase proteins** such as complement components, lysozymes, and C-reactive proteins during infections. It also participates in the **detoxification of immune complexes** and contributes to systemic immune regulation.
(e) Kidney: A bifunctional, divided into two immunologically distinct regions; anterior (Head) kidney (primary hematopoietic tissue in fish, analogous to bone

marrow in mammals) produces various immune cells, including lymphocytes, granulocytes, and macrophages, also involved in endocrine regulation and Posterior (Trunk) kidney that is mainly involved in excretory functions but also plays a role in antigen trapping, inflammatory responses, and immune surveillance during systemic infections.

Organ	Primary functions
Thymus	T-cell development and selection
Spleen	Antigen processing, lymphocyte activation, and blood filtration
Head kidney	Hematopoiesis, immune cell production, and hormonal regulation
Trunk kidney	Excretion, antigen trapping, and inflammatory response
GIALT/SIALT/GALT	Mucosal immunity at gills, skin, and gut surfaces
Liver	Acute-phase protein production, detoxification, and systemic immune support

16.2.4 Bacterial Pathogens

- *Aeromonas* spp.: Causes hemorrhagic septicemia and ulcerative diseases.
- *Vibrio* spp.: Associated with vibriosis, characterized by hemorrhages, ulcers, and high mortality.
- *Edwardsiella* spp.: Leads to diseases like edwardsiellosis, affecting the liver, spleen, and other organs.
- *Flavobacterium* spp.: Causes bacterial gill disease and columnaris disease, affecting the gills and skin.
- *Streptococcus* spp.: Causes streptococcosis, leading to septicemia and neurological symptoms.

16.2.5 Viral Pathogens

- *Infectious Hematopoietic Necrosis Virus (IHNV)*: Affects salmonids, causing severe mortality in young fish.
- *Infectious Pancreatic Necrosis Virus (IPNV)*: Impacts trout and salmon, leading to necrosis in pancreatic tissue.
- *Koi Herpesvirus (KHV)*: Affects koi and common carp, causing high mortality and gill damage.
- *Spring Viremia of Carp Virus (SVCV)*: Causes systemic hemorrhagic disease in carp.
- *Viral Hemorrhagic Septicemia Virus (VHSV)*: Affects various fish species, causing hemorrhages and organ damage.

16.2.6 Fungal Pathogens

- *Saprolegnia* spp.: Causes saprolegniosis, a common water mold infection that affects skin, gills, and eggs.
- *Aphanomyces* spp.: Leads to epizootic ulcerative syndrome (EUS), causing deep tissue ulcers.

16.2.7 Parasitic Pathogens

- *Protozoa*:
  - *Ichthyophthirius multifiliis* (Ich): Causes white spot disease in freshwater fish.
  - *Cryptobia* spp.: Leads to systemic infections and weight loss.
- *Helminths*:
  - *Dactylogyrus* spp.: Monogenean parasites that affect gills, causing respiratory distress.
  - *Cestodes and nematodes*: Infect the intestines, leading to malnutrition and blockage.
- *Crustaceans*:
  - *Argulus* spp. (fish lice and *Lernaea* spp.) (anchor worms cause skin irritation and secondary infections).

16.2.8 Other Pathogens

- *Algal Toxins*: Harmful algal blooms can produce toxins that affect fish health.
- *Rickettsia-like Organisms (RLOs)*: Intracellular bacteria causing systemic diseases.

16.3 Economic and Environmental Impact

These pathogens can lead to mass mortality, reduced growth rates, and secondary infections, with substantial impacts on aquaculture productivity and wild fish populations. Effective management includes vaccination, improved water quality, biosecurity measures, and targeted therapeutics.

16.4 Technological Advances in Functional Genomics

Functional genomics has revolutionized fish biology by facilitating comprehensive investigations into gene functions, interactions, and regulatory pathways. Technological advancements in high-throughput sequencing, transcriptomics, and CRISPR-based genome editing have enabled unprecedented insights into the molecular underpinnings of fish development, physiology, and adaptation. These tools have enhanced the ability to decode complex genetic traits, elucidate immune system dynamics, and understand species-specific responses to environmental challenges. These technologies progress in areas such as disease resistance, growth optimization, and stress tolerance. These are being actively used in aquaculture, conservation genomics, and biotechnological innovation, ensuring more sustainable and resilient fish populations.

16.4.1 High-Throughput Sequencing Technologies

High-throughput sequencing has revolutionized fish genomics by enabling rapid and cost-effective analysis of entire genomes like *Whole Genome Sequencing (WGS)* (provides comprehensive insights into fish genomes, aiding in species-specific genetic improvement and biodiversity studies), *Reduced Representation Sequencing* (like RAD-seq and GBS allow for high-resolution genetic mapping and population genetics studies), and *Epigenomics* (Bisulfite sequencing and ATAC-seq reveal DNA methylation patterns and chromatin accessibility).

16.4.2 Transcriptomics: RNA Sequencing (RNA-Seq)

RNA sequencing (RNA-Seq) has emerged as a cornerstone in functional genomics, enabling both quantitative and qualitative analyses of the transcriptome. It provides high-resolution insights into gene expression patterns under various physiological and environmental conditions. In fish biology, RNA-Seq has facilitated major breakthroughs in understanding immune function, developmental biology, and environmental adaptation. RNA-Seq enables the identification of genes that are differentially expressed across tissues, developmental stages, or in response to external stimuli such as pathogens, temperature fluctuations, or pollutants. Beyond protein-coding genes, it has illuminated the roles of noncoding RNAs, particularly microRNAs (miRNAs) and long noncoding RNAs (lncRNAs) in gene regulation. These ncRNAs are increasingly recognized as key modulators of gene expression, epigenetic control, and cellular homeostasis. In fish, studies have demonstrated that specific miRNAs can influence immune cell signaling, while lncRNAs are involved in gonadal development and environmental stress responses.

16.4.3 Proteomics and Metabolomics Approaches

While genomics and transcriptomics provide valuable information about gene presence and expression, proteomics and metabolomics delve into the actual functional manifestations of molecular changes. In fish biology, these downstream "omics" approaches are critical for linking gene expression with physiological traits and phenotypic outcomes, offering a holistic view of organismal function under both natural and aquaculture conditions.

Proteomics focuses on the comprehensive study of the entire protein complement expressed by a genome under specific conditions. In fish, proteomic analyses have become instrumental in elucidating the molecular mechanisms of immune defense, stress resilience, growth performance, and metabolic regulation.

- Proteomic profiling using tools such as mass spectrometry (MS) and 2D gel electrophoresis has identified immune-related proteins, including cytokines, complement factors, antimicrobial peptides, and pattern recognition receptors. These insights support the development of biomarkers and vaccines for disease management in aquaculture.
- Exposure to stressors such as hypoxia, temperature fluctuations, or contaminants leads to measurable alterations in the proteome. Proteomic approaches help identify stress-responsive proteins such as heat shock proteins (HSPs), antioxidant enzymes, and structural proteins involved in tissue remodeling.
- Techniques like liquid chromatography coupled with tandem mass spectrometry (LC-MS/MS) and protein microarrays now enable high-throughput, high-resolution protein identification and quantification across multiple tissues and time points in fish models such as zebrafish, tilapia, salmon, and carp.

Metabolomics complements proteomics by focusing on small-molecule metabolites, the end products of cellular processes. These compounds offer direct signatures of metabolic activity and physiological states, and their profiles are highly sensitive to environmental, nutritional, and genetic factors.

- Metabolomic studies in fish, especially using Nuclear Magnetic Resonance (NMR) and LC-MS/MS, have revealed how fish metabolite profiles shift in response to salinity, temperature, oxygen deprivation, and exposure to pollutants. Such shifts often reflect changes in energy metabolism, osmoregulation, and oxidative stress response.
- In the context of aquaculture, metabolomics has been used to evaluate the effects of diet composition, probiotics, and functional feed additives. These studies have contributed to improving feed formulations that promote better growth rates, immune competence, and overall fish health.
- Metabolomic fingerprints can serve as early indicators of disease, stress, or nutritional imbalance. This makes them valuable for real-time health assessment and for designing intervention strategies in commercial aquaculture systems.

Together, proteomic and metabolomic analyses provide a system-level understanding of how fish respond to genetic, environmental, and pathological factors. Their integration with genomic and transcriptomic data is driving the development of precision aquaculture, improved breeding programs, and more effective conservation strategies.

16.4.4 Gene Editing Technologies in Fish Functional Genomics

Gene editing has emerged as a revolutionary tool in fish functional genomics, enabling precise and targeted manipulation of the genome. These technologies allow us to validate gene functions, investigate regulatory networks, and engineer desirable traits for aquaculture and conservation.

The CRISPR-Cas9 system has become the most widely adopted gene-editing platform due to its simplicity, versatility, and efficiency. In fish research, it is primarily used to generate targeted gene knockouts or to insert specific genetic modifications, enabling in vivo functional studies at an unprecedented scale. CRISPR-mediated gene knockouts in model species such as zebrafish have been used to dissect gene roles in embryonic development, immune regulation, reproductive biology, and neurobiology. This has significantly accelerated the annotation of gene function in fish genomes. In economically important species like salmon, tilapia, and catfish, CRISPR has been applied to enhance disease resistance by disrupting susceptibility genes or augmenting immune pathways. Edits targeting growth-related genes such as *myostatin* have shown promising results in increasing muscle mass and growth rates, while others have targeted stress-response genes to improve tolerance to high stocking density or low oxygen conditions. CRISPR has also been explored as a method to induce sterility or sex reversal in aquaculture populations, which can prevent uncontrolled breeding, improve meat quality, and reduce the ecological risks of escapees in open-water systems.

Advanced gene editing technologies such as base editing and prime editing allow precise single-nucleotide modifications without introducing double-strand breaks or requiring donor DNA templates. It enables the direct conversion of one nucleotide base to another (e.g., C→T or A→G), allowing fine-tuned genetic edits. This is especially useful for correcting point mutations, studying regulatory elements, or mimicking naturally occurring SNPs in fish populations. Gene drives are engineered genetic elements that bias inheritance to rapidly propagate specific traits through wild or cultured populations. Though still in experimental stages in fish, gene drive systems present novel opportunities. This can be used to suppress invasive fish populations by promoting sterility or sex bias. This can serve as an ecological tool to mitigate biodiversity threats in sensitive habitats.

16.5 Applications of Functional Genomics in Fish Immunology

16.5.1 Identification of Immune-Related Genes

Functional genomics has enabled the discovery and characterization of genes involved in fish immunity. These genes play crucial roles in both innate and adaptive immune responses, helping fish combat pathogens in their aquatic environment. The identification of immune-related genes is achieved using high-throughput sequencing technologies, genome-wide association studies (GWAS, and transcriptomic analyses (Rauta et al. 2012; Ghani et al. 2024)). *Example: (a) Toll-Like Receptors (TLRs).* Toll-like receptors are a family of pattern recognition receptors (PRRs) that play a central role in innate immunity by recognizing pathogen-associated molecular patterns (PAMPs). Functional genomics has identified multiple TLR genes in fish species such as zebrafish (*Danio rerio*), salmon (*Salmo salar*), and common carp (*Cyprinus carpio*) (Rauta et al. 2012). TLRs recognize PAMPs like bacterial lipopolysaccharides (LPS) and viral double-stranded RNA, triggering signaling pathways that lead to the production of pro-inflammatory cytokines and type I interferons. In zebrafish, *TLR3* was identified as a key gene for recognizing viral RNA. Functional studies showed that *TLR3* activation enhances antiviral responses by upregulating interferon-stimulated genes (ISGs), thus protecting the host from viral infections (Sullivan et al. 2021). *(b) Major Histocompatibility Complex (MHC Genes)* essential for antigen presentation in adaptive immunity have been extensively studied in fish. Functional genomics identified variations in *MHC Class I* and *MHC Class II* genes in species like rainbow trout *(Oncorhynchus mykiss)* (Yamaguchi and Dijkstra 2019). These genes are critical for presenting peptides to T cells, enabling specific immune responses against pathogens. In Atlantic salmon (*Salmo salar*), polymorphisms in *MHC Class II* genes were linked to resistance against bacterial infections like furunculosis, highlighting their role in disease resistance.

16.5.2 Host–Pathogen Interactions

Functional genomics provides powerful tools to unravel the complex molecular mechanisms underlying host–pathogen interactions in fish. By studying the genomes, transcriptomes, and proteomics of both fish hosts and their pathogens, researchers can identify how pathogens invade and evade immune responses and how fish mount their defenses (Sudhagar et al. 2018). *Example: (a) Host–Pathogen Interaction in Viral Hemorrhagic Septicemia Virus (VHSV),* VHSV is a deadly virus affecting a wide range of fish species, causing viral hemorrhagic septicemia. Functional genomics techniques, particularly transcriptomics, have been employed to study the immune response of rainbow trout (*Oncorhynchus mykiss*) to VHSV infection. RNA-Seq revealed the upregulation of genes encoding pattern recognition receptors (PRRs) like *TLR3* and *RIG-I*, which recognize viral RNA and initiate antiviral signaling pathways) (Xu et al. 2016). The study identified increased expression of interferon-stimulated genes (ISGs, such as *Mx* and *ISG15*), which inhibit viral replication (Broering et al. 2010). Elevated expression of pro-inflammatory cytokines like *IL-1β* and *TNF-α* was observed, highlighting their role in recruiting immune cells to the site of infection (Kany et al. 2019). VHSV genomic studies identified viral proteins that suppress host immune responses, such as the nonstructural protein NV, which inhibits type I interferon signaling (Pereiro 2017).

16.5.3 Gene Expression Profiling Under Pathogenic Stress in Fish

Gene expression profiling under pathogenic stress reveals the fish's response at the molecular level to infections. Functional genomics, particularly transcriptomics, enables the identification and quantification of genes that are upregulated or down-regulated in response to pathogens. This uncovers immune pathways, stress adaptation mechanisms, and potential biomarkers for disease management (Sudhagar et al. 2018). *Example: Response to Aeromonas hydrophila in Catfish. Aeromonas hydrophila* is a bacterial pathogen causing motile aeromonad septicemia (MAS) in fish, leading to tissue damage and systemic infections. Channel catfish (*Ictalurus punctatus*) is highly susceptible to *A. hydrophila*, making it a model for studying bacterial infections. RNA-Seq analysis of infected catfish showed increased expression of immune-related genes such as *IL-1β*, *IL-6*, and *TNF-α*, which mediate inflammation and recruit immune cells, *Hepcidin* and *β-defensins*, which inhibit bacterial growth and also *C3* and *C5*, which enhance phagocytosis and bacterial lysis. Genes involved in metabolic pathways, such as lipid metabolism, were down-regulated, indicating a shift in energy allocation toward immune responses. Heat shock proteins (*HSP70*, *HSP90*) were significantly upregulated, reflecting cellular stress due to infection. For example: Viral Infection in Atlantic Salmon (Christiansen et al. 2011).

16.5.4 Functional Characterization of Immune Genes in Fish

Functional characterization of immune genes involves studying the roles of specific genes in immune responses to pathogens. By manipulating or modifying these genes using tools such as CRISPR-Cas9, RNA interference (RNAi, or gene overexpression, researchers can assess how they contribute to immune defense). This type of functional genomics helps identify key immune regulators and pathways that fish use to respond to infections, and it provides valuable information for enhancing disease resistance in aquaculture (Zhao et al. 2016). *TLR3* (Toll-like receptor 3) is a critical component of the innate immune system that recognizes viral double-stranded RNA, a molecular signature of many viral infections in Zebrafish (*Danio rerio*) (Yu and Levine 2011). RNA interference (RNAi) was used to knock down the expression of *TLR3* in zebrafish. The results showed that the knockdown fish were more susceptible to viral infections (e.g., vesicular stomatitis virus, VSV) compared to wild-type fish. This demonstrated that *TLR3* is essential for the fish's ability to recognize and mount an immune response to viral pathogens. In contrast, overexpressing *TLR3* in zebrafish led to an enhanced immune response, with upregulation of downstream immune genes such as interferons (*IFN-α* and *IFN-)* and antimicrobial peptides. This indicated that *TLR3* activation boosts antiviral immunity (Dar et al. 2022). Major Histocompatibility Complex (MHC) Class II molecules are involved in adaptive immunity by presenting processed antigens to helper T cells, which are crucial for activating specific immune responses. Atlantic salmon (*Salmo salar*) has been extensively studied for its immune responses to bacterial and viral infections (Zhao et al. 2016).

A study used CRISPR-Cas9 to knock out *MHC Class II* genes in Atlantic salmon. The knockout fish showed impaired antibody production and reduced T-cell responses after infection with *Aeromonas salmonicida*, a bacterium that causes furunculosis. This demonstrated that *MHC Class II* is essential for the adaptive immune response to bacterial pathogens (Zhao et al. 2016). Further analysis revealed that fish lacking *MHC Class II* showed diminished memory responses, indicating the gene's role not only in pathogen detection but also in the development of long-lasting immunity.

C3 is a key protein in the complement system, which enhances the ability of antibodies and phagocytic cells to clear pathogens. Common carp (*Cyprinus carpio*) is often used in complement system studies. Overexpressing *C3* in carp led to enhanced bacterial clearance following infection with *Streptococcus iniae*, a pathogen causing streptococcal infections. The fish with higher levels of *C3* showed increased phagocytosis and more efficient pathogen opsonization, highlighting the importance of *C3* in immune defense. Knockdown of *C3* resulted in reduced complement activity and slower pathogen clearance, demonstrating that *C3* is crucial for an efficient immune response (Christiansen et al. 2011).

16.5.5 Comparative Genomics and Evolution of Immune Responses in Fish

Comparative genomics involves the comparison of genetic material from different species to understand evolutionary relationships and how immune systems have adapted to various environmental challenges. By comparing the genomes of different fish species, researchers can gain insights into how immune responses have evolved and identify conserved and divergent immune mechanisms. This approach helps in understanding how fish species are equipped to deal with pathogens in their specific habitats (Christiansen et al. 2011). Comparative genomics provides valuable evolutionary insights into how immune systems have adapted to different environments and pathogens over time. By comparing the genomes of distantly related species, researchers can identify genes that are conserved across species, providing clues about essential immune functions, and those that have evolved to meet the specific challenges posed by different pathogens (Christiansen et al. 2011). Immune gene families, such as those encoding receptors, cytokines, and complement proteins, are central to host defence mechanisms. By comparing the immune gene families across different fish species, researchers can identify both conserved genes essential for immune responses and divergent genes that have evolved to recognize different pathogens or environmental challenges (Dar et al. 2022). A comparison of *MHC Class I* and *Class II* gene families across species such as zebrafish, rainbow trout, and carp reveals that while the basic structure of these genes is highly conserved, the diversity within the gene families is higher in species exposed to more diverse pathogens. For instance, salmonids (e.g., rainbow trout) have a more diverse

MHC region compared to species with more stable environments, such as goldfish (*Carassius auratus*). This genetic diversity enables salmonids to recognize a broader array of pathogens, highlighting the role of *MHC* in adaptive immune responses (Dar et al. 2022).

16.5.6 Adaptive Evolution of Immune-Related Genes

Adaptive evolution of immune-related genes refers to the process by which immune genes evolve in response to selective pressures from pathogens and environmental conditions. The ability of fish to adapt to pathogens through changes in immune gene expression and function is a key feature of their evolutionary success. Comparative genomics helps identify immune genes that have undergone adaptive evolution, providing insight into how fish immune systems evolve to enhance pathogen defence (Dar et al. 2022). *Immunoglobulin M (IgM)* is a critical antibody in fish immune responses, particularly in adaptive immunity. Studies on the evolution of *IgM* in fish species have shown that certain fish populations, such as those living in freshwater environments, have evolved variants of *IgM* that are better suited to defending against local bacterial pathogens. For example, research on salmon and carp has shown that their *IgM* genes have undergone adaptive evolution in response to the specific pathogens present in their habitats. In salmon, the *IgM* gene has evolved to produce a more diverse antibody repertoire capable of recognizing and neutralizing a broader range of bacterial pathogens in cold water environments (Sudhagar et al. 2018).

16.6 Functional Genomics in Vaccine Development

Functional genomics has emerged as a powerful tool in the field of fish immunology, providing comprehensive insights into the immune responses of fish to various pathogens. By leveraging high-throughput sequencing technologies, researchers can analyze the expression, regulation, and interaction of genes involved in the immune system, leading to the identification of key players in fish immunity (Abdelrahman et al. 2017; Betancor et al. 2015a). This review explores the application of functional genomics in fish immunology, focusing on the identification of immune-related genes, genomic approaches to understanding immune mechanisms, and the evaluation of immune responses using genomic tools. The identification of immune-related genes in fish has been significantly advanced by functional genomics. High-throughput sequencing, such as RNA sequencing (RNA-seq), allows for the comprehensive profiling of gene expression in fish during immune responses. This enables the discovery of novel immune genes and pathways that are activated upon pathogen exposure. Comparative genomics, involving the analysis of genetic material across different fish species, aids in identifying conserved and unique immune genes, providing insights into the evolutionary aspects of fish immunity (Brawand et al. 2014; Abdelrahman et al. 2017). Techniques such as transcriptomics and proteomics are instrumental in identifying differentially expressed genes and proteins during infection, shedding light on the molecular mechanisms underlying fish immune responses. Genomic approaches have revolutionized the understanding of immune mechanisms in fish. By employing tools such as genome-wide association studies (GWAS) and quantitative trait loci (QTL) mapping, researchers can identify genetic loci associated with immune traits, such as disease resistance and susceptibility (Brawand et al. 2014; Calduch-Giner et al. 2012). Functional genomics also facilitates the study of host–pathogen interactions at the molecular level. For instance, single-cell RNA sequencing (scRNA-seq) allows for the dissection of immune cell heterogeneity and the identification of specific cell types involved in immune responses. Additionally, epigenomics, the study of heritable changes in gene function without altering the DNA sequence, provides insights into how environmental factors and pathogen exposure influence gene expression and immune function in fish (Betancor et al. 2015a, b). Evaluating immune responses in fish using genomic tools provides a detailed understanding of the effectiveness of immune responses to infections and vaccinations. High-throughput sequencing technologies enable the monitoring of global gene expression changes during immune responses, offering a comprehensive view of the immune landscape (Calduch-Giner et al. 2012; Carbone and Faggio 2016). Metagenomic sequencing allows for the assessment of microbiome changes in response to infections and treatments, highlighting the interplay between the host immune system and microbial communities. Genomic tools also facilitate the study of immune memory and the identification of biomarkers for disease resistance and vaccine efficacy (Olsvik et al. 2013; Tine et al. 2014). For instance, the identification of specific immune gene expression signatures can serve as indicators of a robust immune response, guiding the development of effective vaccines and therapeutics for fish. Functional genomics has transformed the study of fish immunology, providing deep insights into the molecular mechanisms of immune responses (Tocher and Glencross 2015). By identifying key immune-related genes, elucidating immune mechanisms, and evaluating immune responses using genomic tools, researchers can advance the understanding of fish immunity and develop strategies to enhance disease resistance in aquaculture. The integration of functional genomics with traditional immunological approaches holds great promise for improving fish health and ensuring sustainable aquaculture practices (Tort et al. 1998).

16.7 Role of DNA Methylation and Histone Modifications

DNA methylation and histone modifications play pivotal roles in regulating the immune response in fish. DNA methylation typically involves the addition of a methyl group to the cytosine bases of DNA, particularly at CpG islands, which are regions rich in cytosine and guanine nucleotides (Liu et al. 2016; Lugo-Villarino et al. 2010). This process can either repress or activate gene expression, depending on the context. In fish, DNA methylation patterns in immune cells are crucial for controlling the expression of genes involved in immune responses (Lazado and

Caipang 2014). For instance, the methylation status of promoters in immune genes can determine whether these genes are transcribed and, consequently, how effectively the immune system can respond to pathogens.

Histone modifications, including acetylation, methylation, and phosphorylation, alter the chromatin structure, making it either more or less accessible for transcription (Liu et al. 2016). These modifications play a critical role in the regulation of immune genes in fish. For example, acetylation of histones generally leads to a more open chromatin structure, promoting gene transcription, while methylation can either activate or repress gene expression depending on the specific amino acid residues modified. The interplay between these epigenetic marks and the transcription machinery is essential for the proper functioning of the fish immune system (Star et al. 2011).

16.7.1 Epigenetic Regulation of Immune Genes

The regulation of immune genes through epigenetic mechanisms is a complex and dynamic process in fish. Epigenetic modifications such as DNA methylation, histone modifications, and the action of noncoding RNAs contribute to the fine-tuning of gene expression in response to immune challenges. These mechanisms ensure that immune genes are expressed at the right time and in the appropriate cells, enabling an effective immune response.

Research has shown that specific immune genes in fish are subject to tight epigenetic control. For example, during an infection, certain genes may become demethylated, leading to their activation and the production of immune molecules needed to fight off pathogens. Histone modifications also play a crucial role; for instance, the trimethylation of histone H3 at lysine 4 (H3K4me3) is associated with the activation of immune genes. Comparative studies between fish and other vertebrates reveal both conserved and unique aspects of epigenetic regulation, highlighting the evolutionary adaptations of fish immune systems (Zhao et al. 2016).

16.7.2 Impact of Environmental Factors on Epigenetics

Environmental factors significantly influence the epigenetic landscape in fish, affecting their immune responses and overall health. Pollutants, temperature fluctuations, hypoxia, and other environmental stressors can induce epigenetic changes such as altered DNA methylation patterns and histone modifications. These changes can either enhance or impair the immune function, depending on the nature and duration of the exposure. For example, exposure to pollutants like heavy metals can lead to hypermethylation of immune gene promoters, resulting in decreased gene expression and a weakened immune response. Conversely, beneficial environmental conditions can promote epigenetic modifications that enhance immune function, such as the demethylation of key immune genes or the acetylation of histones associated with these genes. Fish exhibit remarkable epigenetic plasticity, allowing them to adapt to varying environmental conditions. Case studies have shown that specific epigenetic changes induced by environmental factors can lead to increased resistance to diseases, demonstrating the adaptive capacity of fish immune systems (Burggren and Mendez-Sanchez 2023).

16.8 Host–Pathogen Interactions: A Genomic Perspective

Genomic studies have provided profound insights into the mechanisms by which pathogens infect and proliferate within fish hosts. Sequencing and comparative genomics have identified numerous virulence factors, such as toxins, adhesion molecules, and secretion systems, that pathogens use to invade host tissues and evade immune responses. For example, bacterial pathogens like *Vibrio* spp. and *Aeromonas* spp. possess genes encoding for Type III and Type VI secretion systems, which are instrumental in injecting effector proteins into host cells, disrupting cellular processes, and evading immune detection.

Furthermore, viral pathogens such as the infectious salmon anemia virus (ISAV) and the koi herpesvirus (KHV) have been studied extensively at the genomic level, revealing strategies they employ to hijack host cellular machinery for replication. Transcriptomic analyses during infection have shown how these viruses modulate host gene expression to create a favorable environment for their proliferation. Understanding these genomic interactions is crucial for developing targeted interventions, such as vaccines and antimicrobial agents, to mitigate infections.

16.8.1 Host Immune Responses to Different Pathogens

The genomic perspective also extends to understanding how fish hosts respond to various pathogens. Fish possess a complex and adaptive immune system, comprising both innate and adaptive components, which can be studied through high-throughput genomic techniques like RNA sequencing and proteomics. Upon pathogen invasion, fish activate a myriad of immune pathways, including the production of antimicrobial peptides, activation of complement systems, and the expression of cytokines and chemokines that mediate inflammation and recruit immune cells to the site of infection.

For instance, during bacterial infections, genes involved in the recognition of pathogen-associated molecular patterns (PAMPs), such as toll-like receptors (TLRs) and nucleotide binding oligomerization domain-like receptors (NLRs), are upregulated. These receptors trigger signaling cascades that lead to the activation of NF-κB and the production of pro-inflammatory cytokines. Similarly, during viral infections, interferon-stimulated genes (ISGs) play a critical role in antiviral defence by inhibiting viral replication and spread.

Comparative genomics has revealed species-specific immune strategies, highlighting the diversity in immune responses among different fish species. For example, salmonids show robust interferon responses to viral infections, while cyprinids rely more heavily on the rapid induction of heat shock proteins and other stress-related pathways.

16.8.2 Co-evolution of Fish and Pathogens

The co-evolution of fish and their pathogens is a dynamic process driven by continuous genetic adaptation on both sides. Pathogens evolve mechanisms to circumvent host defences, while fish develop more sophisticated immune responses to detect and neutralize these threats. This evolutionary arms race can be studied through genomic analyses, which reveal patterns of positive selection and genetic diversity in both host and pathogen genomes.

In fish, genes involved in immune recognition, such as major histocompatibility complex (MHC) genes, exhibit high levels of polymorphism, reflecting the evolutionary pressure exerted by diverse pathogens. Pathogens, on the other hand, show rapid evolution in genes encoding surface antigens and other virulence factors, enabling them to escape immune surveillance (Betancor et al. 2015a). Studies on co-evolution have also identified instances of gene transfer between pathogens and hosts, which can lead to the emergence of new pathogenic traits or enhanced immune capabilities.

For example, the co-evolution of Atlantic salmon and the salmon louse (*Lepeophtheirus salmonis*) has led to the development of specific immune responses in salmon that target the lice, while the lice have evolved mechanisms to suppress these responses. Similarly, the relationship between zebrafish and *Mycobacterium marinum* provides insights into the evolution of granuloma formation, a hallmark of mycobacterial infections.

16.9 Effects on Fish Immunology and Health

Understanding host–pathogen interactions from a genomic perspective has significant implications for fish health management and aquaculture. Genomic insights can guide the development of more effective vaccines and treatments by identifying key immune pathways and virulence factors (Betancor et al. 2015a). For instance, vaccine development can be informed by identifying conserved antigens across different pathogen strains, ensuring broader protection.

Moreover, genomic tools can be used for breeding programs to enhance disease resistance in fish populations. By identifying genetic markers associated with strong immune responses, selective breeding can produce fish that are more resilient to infections, reducing the reliance on antibiotics and other treatments in aquaculture (*Lepeophtheirus salmonis*).

Additionally, understanding the co-evolution of fish and pathogens helps predict the emergence of new diseases and develop strategies to mitigate their impact. This is particularly important in the context of climate change and environmental stressors, which can alter pathogen dynamics and increase the susceptibility of fish to infections.

16.10 Case Studies in Functional Genomics of Fish Immunology

Functional genomics has revolutionized our understanding of fish immunology by providing comprehensive insights into the molecular mechanisms underlying immune responses. Case studies in functional genomics have utilized high-throughput technologies such as RNA sequencing (RNA-seq), genome-wide association studies (GWAS), and CRISPR/Cas9 gene editing to elucidate the roles of specific genes and pathways in fish immunity (Betancor et al. 2015a). For example, studies on zebrafish (*Danio rerio*) have identified key regulators of the immune response, such as the Toll-like receptors (TLRs) and the Janus kinase/signal transducers and activators of transcription (JAK/STAT) pathway, through loss-of-function and gain-of-function experiments (*Lepeophtheirus salmonis*). These findings have not only advanced our knowledge of the genetic basis of immunity in fish but also provided comparative insights relevant to vertebrate immunology as a whole.

16.11 Transcriptomic Analysis of Fish During Pathogenic Infection

16.11.1 Transcriptomic Analysis During Bacterial Infection

Transcriptomic analysis during bacterial infection has provided valuable insights into the dynamic changes in gene expression associated with the immune response in fish. Utilizing RNA-seq technology, researchers have mapped the transcriptional landscapes of fish species such as Atlantic salmon (*Salmo salar*), rainbow trout (*Oncorhynchus mykiss*), and zebrafish following exposure to bacterial pathogens like *Vibrio anguillarum*, *Aeromonas salmonicida*, *Mycobacterium marinum*, and *Lepeophtheirus salmonis*. These studies have revealed the upregulation of genes involved in pathogen recognition, such as pattern recognition receptors (PRRs), as well as those mediating inflammatory responses, including cytokines, chemokines, and acute-phase proteins. Additionally, the activation of oxidative stress pathways and apoptosis-related genes highlights the multifaceted nature of the immune response to bacterial infections. This comprehensive view of transcriptomic changes aids in identifying potential biomarkers for infection and targets for therapeutic intervention.

16.11.2 Genomic Studies on Viral Immunity in Fish

Genomic studies have significantly enhanced our understanding of viral immunity in fish. By sequencing the genomes of both fish hosts and viral pathogens, researchers have uncovered the genetic determinants of susceptibility and resistance to viral infections (*Lepeophtheirus salmonis*). For instance, the study of the infectious salmon anemia virus (ISAV) in Atlantic salmon has identified genetic variants in interferon-stimulated genes (ISGs) that correlate with enhanced antiviral responses. Similarly, research on the viral *hemorrhagic septicemia* virus (VHSV) in rainbow trout has highlighted the role of major histocompatibility complex (MHC) genes in antiviral immunity. These genomic insights have facilitated the development of selective breeding programs aimed at enhancing disease resistance in aquaculture, thereby reducing the economic impact of viral diseases.

16.11.3 Functional Genomics in Parasitic Interactions

Functional genomics approaches have also been instrumental in dissecting the complex interactions between fish hosts and parasitic organisms. Studies on parasites such as the sea lice (*Lepeophtheirus salmonis*) and the ciliated protozoan (*Ichthyophthirius multifiliis*) have employed transcriptomic and proteomic analyses to identify host genes and proteins that are differentially expressed during parasitic infection (*Lepeophtheirus salmonis*). In Atlantic salmon, for example, the expression of genes involved in epithelial barrier function, immune signalling, and stress responses has been shown to be modulated in response to sea lice infestation. Additionally, functional genomics has revealed the strategies employed by parasites to evade host immune defences, such as the secretion of immunomodulatory proteins and the manipulation of host cellular processes. Understanding these host–parasite interactions at the molecular level provides a foundation for developing novel control measures, including vaccines and antiparasitic agents, to mitigate the impact of parasitic diseases in aquaculture. It has provided a powerful toolkit for unravelling the complexities of fish immunology, offering insights into the molecular underpinnings of immune responses to bacterial, viral, and parasitic infections. These advancements not only enhance our understanding of fish biology but also contribute to the development of more effective strategies for disease management in aquaculture (*Lepeophtheirus salmonis*) (Buchmann 2014).

16.12 Challenges in Functional Genomics of Fish Immunology

One of the primary challenges in the functional genomics of fish immunology is the technical and methodological limitations associated with high-throughput genomic technologies. Despite significant advancements, the complexity and variability of fish genomes pose substantial hurdles. Many fish species have large, complex genomes with high levels of polymorphism and gene duplication, making sequencing and assembly difficult. Additionally, the lack of well-annotated reference genomes for many commercially important or ecologically significant fish species hampers the interpretation of genomic data.

Another challenge is the difficulty in achieving high-resolution temporal and spatial profiling of gene expression during immune responses. Capturing the dynamic changes in gene expression and cellular interactions during the early and late stages of infection requires sophisticated experimental designs and high-sensitivity technologies, which can be cost-prohibitive. Moreover, the inherent variability in immune responses among individual fish further complicates data analysis and interpretation.

Integrating multi-omics data (genomics, transcriptomics, proteomics, and metabolomics) to gain a holistic understanding of fish immune responses is a significant challenge. Each omics layer provides unique insights, but integrating these data sets requires advanced bioinformatics tools and approaches to manage and analyze the large volumes of data. Ensuring data compatibility and consistency across different platforms and experimental conditions is also challenging. Moreover, integrating multi-omics data demands a comprehensive understanding of the underlying biology and the ability to correlate molecular changes across different biological levels. This includes understanding how genetic variations influence transcriptomic responses, how changes in protein expression affect metabolic pathways, and how these alterations ultimately impact immune function and disease resistance. Developing robust computational models to integrate and interpret multi-omics data remains an ongoing challenge.

16.13 Conclusion and Future Prospective

The functional genomics of fish immunology has significantly advanced our understanding of the molecular mechanisms underlying immune responses in fish. Key findings from various studies have highlighted the roles of DNA methylation and histone modifications in regulating immune gene expression, the dynamic transcriptomic changes during bacterial infections, and the genomic basis of viral immunity. Furthermore, functional genomics has elucidated the complex interactions between fish hosts and parasites, revealing both host defence mechanisms and parasitic strategies for immune evasion. These insights have been facilitated by high-throughput technologies such as RNA sequencing, genome-wide association studies, and proteomics, which have provided comprehensive views of the genetic and molecular landscapes of fish immune responses.

The advancements in functional genomics have profound implications for aquaculture and fish health management. Understanding the genetic determinants of disease resistance allows for the development of selective breeding programs aimed at enhancing the resilience of fish populations to infections. This reduces the reliance on antibiotics and other chemical treatments, promoting more sustainable and environmentally friendly aquaculture practices. Additionally, the identification of key immune pathways and biomarkers can inform the development of more effective vaccines and therapeutic interventions, leading to improved disease prevention and management strategies (Guan et al. 2019). The integration of multi-omics data provides a holistic view of fish health, enabling the identification of novel targets for enhancing immune function and overall health. This is particularly important in the face of environmental stressors such as climate change and pollution, which can impact fish immunity and increase susceptibility to diseases. By incorporating environmental factors into functional genomics studies, researchers can develop more comprehensive strategies to mitigate these impacts and ensure the sustainability of aquaculture operations (Foey and Picchietti 2014).

The future of functional genomics in fish immunology holds great promise for further breakthroughs and advancements. Continued efforts to develop more accurate and comprehensive reference genomes for a wider range of fish species will enhance the precision and interpretability of genomic data. The advancement of single-cell sequencing technologies will provide high-resolution insights into the heterogeneity of immune cell populations and their responses to infection, uncovering novel regulatory mechanisms and potential therapeutic targets. Integrating multi-omics data using advanced bioinformatics tools and machine learning approaches will enable the identification of key regulatory networks and predictive biomarkers of disease resistance (Foey and Picchietti 2014). This holistic approach will facilitate the development of targeted breeding programs and therapeutic interventions, enhancing fish health and disease resistance. Furthermore, understanding the epigenetic regulation of immune responses and the impact of environmental factors will be crucial for adapting to changing environmental conditions. Functional genomics will play a key role in developing strategies to mitigate the effects of climate change and pollution on fish health, ensuring the resilience and sustainability of aquaculture operations.

Looking ahead, several research priorities and potential breakthroughs can address these challenges and advance the field of functional genomics in fish immunology. One key priority is the development of more accurate and comprehensive reference genomes for a broader range of fish species. Improved reference genomes will facilitate more precise mapping and interpretation of genomic data, enabling better insights into the genetic basis of immune responses.

Another priority is the advancement of single-cell sequencing technologies, which can provide high-resolution insights into the heterogeneity of immune cell populations and their responses to infection. Single-cell transcriptomics and proteomics can reveal the distinct roles of different cell types in immune responses, uncovering novel regulatory mechanisms and potential targets for intervention. Furthermore, integrating environmental factors into functional genomics studies is crucial for understanding how changing environmental conditions, such as climate change and pollution, impact fish immunity. This includes investigating epigenetic modifications and their role in mediating the effects of environmental stressors on immune function. Potential breakthroughs also lie in the development of advanced bioinformatics tools and machine learning approaches to analyze and integrate multi-omics data. These tools can help identify key regulatory networks and predictive biomarkers of disease resistance, guiding selective breeding programs and the development of targeted therapeutics.

References

Abdelrahman, et al., (2017) Aquaculture genomics, genetics and breeding in the United States: current status,challenges, and priorities for future research. BMC Genom. 18: 191

Betancor MB, Sprague M, Usher S, Sayanova O, Campbell PJ, Napier JA, Tocher DR (2015a) A nutritionally-enhanced oil from transgenic *Camelina sativa* effectively replaces fish oil as a source of eicosapentaenoic acid for fish. Sci Rep 5:8104

Betancor MB, Sprague M, Sayanova O, Usher S, Campbell PJ, Napier JA, Caballero MJ, Tocher DR (2015b) Evaluation of a high-EPA oil from transgenic Camelina sativa in feeds for Atlantic salmon (*Salmo salar* L.): effects on tissue fatty acid composition, histology and gene expression. Aquaculture 444:1e12

Boehm T (2011) Design principles of adaptive immune systems. Nat Rev Immunol 11(5):307–317

Brawand D, Wagner CE, Li YI, Malinsky M et al (2014) The genomic substrate for adaptive radiation in African cichlid fish. Nature 513(7518):375–e381

Broering R, Zhang X, Kottilil S, Trippler M, Jiang M, Lu M et al (2010) The interferon stimulated gene 15 functions as a pro viral factor for the hepatitis C virus and as a regulator of the IFN response. Gut 59(8):1111–1119

Buchmann K (2014) Evolution of innate immunity: clues from invertebrates via fish to mammals. Front Immunol 5:459

Burggren WW, Mendez-Sanchez JF (2023) "Bet hedging" against climate change in developing and adult animals: rolesfor stochastic gene expression, phenotypic plasticity, epigenetic inheritance and adaptation. Frontiers inPhysiology. Oct 6;14:1245875

Calduch-Giner JA, Sitja-Bobadilla A, Davey GC, Cairns MT, Kaushik S, Perez-Sanchez J (2012) Dietary vegetable oils do not alter the intestine transcriptome of gilthead sea bream (*Sparus aurata*), but modulate the transcriptomic response to infection with *Enteromyxum leei*. BMC Genom 13:470

Carbone D, Faggio C (2016) Importance of prebiotics in aquaculture as immunostimulants. Effects on immune system of *Sparus aurata* and *Dicentrarchus labrax*. *Fish Shellfish Immunol* 54:172–e178

Christiansen DH, Østergaard PS, Snow M, Dale OB, Falk K (2011) A low-pathogenic variant of infectious salmon anemia virus (ISAV-HPR0) is highly prevalent and causes a non-clinical transient infection in farmed Atlantic salmon (Salmo salar L.) in The Faroe Islands. J Gen Virol 92(4):909–918

Dar GH, Bhat RA, Qadri H, Al-Ghamdi KM, Hakeem KR (2022) Bacterial fish diseases. Academic

den Haan JM, Arens R, van Zelm MC (2014) The activation of the adaptive immune system: cross-talk between antigen-presenting cells, T cells and B cells. Immunol Lett 162(2):103–112

Du Pasquier L (2001) The immune system of invertebrates and vertebrates. Comp Biochem Physiol B: Biochem Mol Biol 129(1):1–15

Foey A, Picchietti S (2014) Immune defences of teleost fish. In: Aquaculture nutrition: gut health, probiotics and prebiotics, pp 14–52

Ghani MU, Chen J, Khosravi Z, Wu Q, Liu Y, Zhou J et al (2024) Unveiling the multifaceted role of toll-like receptors in immunity of aquatic animals: pioneering strategies for disease management. Front Immunol 15:1378111

Guha R, Beegum S, Kumar N, Nazar A, John N, Lakshmi S et al (2025) Fish immune response: an overview. In: Bacillus probiotics for sustainable aquaculture. CRC Press, pp 27–54

Kany S, Vollrath JT, Relja B (2019) Cytokines in inflammatory disease. Int J Mol Sci 20(23):6008

Lazado CC, Caipang CM (2014) Mucosal immunity and probiotics in fish. Fish. Shellfish Immunol 39(1):78e89

Liu Z, Liu S, Yao J, Bao L, Zhang J, Li Y, Jiang C, Sun L, Wang R, Zhang Y, Zhou T, Zeng Q, Fu Q, Gao S, Li N, Koren S, Jiang Y, Zimin A, Xu P, Phillippy AM, Geng X, Song L, Sun F, Li C, Wang X, Chen A, Jin Y, Yuan Z, Yang Y, Tan S, Peatman E, Lu J, Qin Z, Dunham R, Li Z, Sonstegard T, Feng J, Danzmann RG, Schroeder S, Scheffler B, Duke MV, Ballard L, Kucuktas H, Kaltenboeck L, Liu H, Armbruster J, Xie Y, Kirby ML, Tian Y, Flanagan ME, Mu W, Waldbieser GC (2016) The channel catfish genome sequence provides insights into the evolution of scale formation in teleosts. Nat Commun 7:11757

Lugo-Villarino G, Balla KM, Stachura DL, Bañuelos K, Werneck MB, Traver D (2010) Identification of dendritic antigen-presenting cells in the zebrafish. PNAS 107:15850–e15855

Mokhtar DM, Zaccone G, Alesci A, Kuciel M, Hussein MT, Sayed RK (2023) Main components of fish immunity: an overview of the fish immune system. Fishes 8(2):93

Mugwanya M, Dawood MA, Kimera F, Sewilam H (2022) A review on recirculating aquaculture system: influence of stocking density on fish and crustacean behavior, growth performance, and immunity. Ann Anim Sci 22(3):873–884

Müller V, De Boer RJ, Bonhoeffer S, Szathmáry E (2018) An evolutionary perspective on the systems of adaptive immunity. Biol Rev 93(1):505–528

Natoli G, Ostuni R (2019) Adaptation and memory in immune responses. Nat Immunol 20(7):783–792

Olsvik PA, Hemre G-I, Waagbø R (2013) Exploring early micronutrient deficiencies in rainbow trout (Oncorhynchus mykiss) by next-generation sequencing technology e from black box to functional genomics. PLoS One 8(7):e69461

Panara V, Varaliová Z, Wilting J, Koltowska K, Jeltsch M (2024) The relationship between the secondary vascular system and the lymphatic vascular system in fish. Biol Rev 99(6):2108–2133

Pereiro P (2017) Advances in the knowledge of the antiviral immune response and resistance to viral Haemorrhagic Septicaemia virus (VHSV) in turbot (*Scophthalmus maximus)*. Universidade de Santiago de Compostela

Rathor GS, Swain B (2024) Advancements in fish vaccination: current innovations and future horizons in aquaculture health management. Appl Sci 14(13):5672

Rauta PR, Nayak B, Das S (2012) Immune system and immune responses in fish and their role in comparative immunity study: a model for higher organisms. Immunol Lett 148(1):23–33

Salinas I, Zhang YA, Sunyer JO (2011) Mucosal immunoglobulins and B cells of teleost fish. Dev Comp Immunol 35(12):1346–1365

Secombes CJ, Wang T (2012) The innate and adaptive immune system of fish. In: Infectious disease in aquaculture. Woodhead Publishing, pp 3–68

Star B, Nederbragt AJ, Jentoft S, Grimholt U, Malmstrøm M (2011) The genome sequence of Atlantic cod reveals a unique immune system. Nature 477(7363):207e210

Sudhagar A, Kumar G, El-Matbouli M (2018) Transcriptome analysis based on RNA-Seq in understanding pathogenic mechanisms of diseases and the immune system of fish: a comprehensive review. Int J Mol Sci 19(1):245

Sullivan C, Soos BL, Millard PJ, Kim CH, King BL (2021) Modeling virus-induced inflammation in zebrafish: a balance between infection control and excessive inflammation. Front Immunol 12:636623

Tine M, Kuhl H, Gagnaire PA, Louro B, Desmarais E (2014) European sea bass genome and its variation provide insights into adaptation to euryhalinity and speciation. Nat Commun 5:5770

Tocher DR, Glencross BD (2015) Lipids and fatty acids. In: Lee CS, Lim C, Gatlin DM (eds) Dietary nutrients, additives, and fish health, pp 47–e94

Tort L, Padros F, Rotllant J, Crespo S (1998) Winter syndrome in the gilthead sea bream *Sparus aurata*. Immunological and histopathological features. Fish Shellfish Immunol 8:37–e47

Tosi MF (2005) Innate immune responses to infection. J Allergy Clin Immunol 116(2):241–249

Xu C, Evensen Ø, Mweemba HMA (2016) De novo transcriptome analysis shows that SAV-3 infection upregulates pattern recognition receptors of the endosomal toll-like and RIG-I-like receptor signaling pathways in macrophage/dendritic like TO-cells. Viruses 8(4):114

Yamaguchi T, Dijkstra JM (2019) Major histocompatibility complex (MHC) genes and disease resistance in fish. Cells 8(4):378

Yu M, Levine SJ (2011) Toll-like receptor 3, RIG-I-like receptors and the NLRP3 inflammasome: key modulators of innate immune responses to double-stranded RNA viruses. Cytokine Growth Factor Rev 22(2):63–72

Zapata AG (2022) Lympho-hematopoietic microenvironments and fish immune system. Biology 11(5):747

Zhao C, Fu M, Wang C, Jiao Z, Qiu L. (2016) RNA-Seq analysis of immune-relevant genes in Lateolabrax japonicusduring Vibrio anguillarum infection. Fish & Shellfish Immunology. May 1;52:57–64

Genetic Marker Development for Aquaculture Species

17

Barsha Baisakhi, Jyotirmayee Pradhan, Debasmita Mohanty, Vikash Kumar, and Basanta Kumar Das

17.1 Introduction

Aquaculture, the cultivation of aquatic organisms, has emerged as an essential element of the global food supply, offering a sustainable approach to satisfy the rising need for protein. With the expansion of the aquaculture firm, it is essential to optimize production efficiency, disease resistance, growth rates, and the overall genetic quality of farmed species. The creation of genetic markers is crucial for the exact characterization, selection, and breeding of desirable traits in aquaculture species (Askari et al., 2013). Genetic markers are discernible DNA sequences linked to certain genes or genomic areas of significance. These markers function as instruments for analyzing the genetic structure of phenotypes, assessing genetic diversity, and facilitating marker-assisted selection (MAS) and genomic selection (GS) in breeding initiatives. In aquaculture, the cultivation of varied species with distinct life histories and genetic complexity is enhanced by genetic markers, which provide better control and sustainability of breeding procedures. The advancement of genetic markers has progressed dramatically throughout the years. Initial methodologies depended on protein-based markers such as allozymes, which offered restricted resolution. The emergence of molecular methods transformed the

B. Baisakhi
P.G Department of Zoology, Kuntala Kumari Sabat Women's College, Affiliated to Fakir Mohan University, Balasore, Odisha, India

ICAR-Central Inland Fisheries Research Institute, Barrackpore, Kolkata, India

J. Pradhan
P.G Department of Zoology, Kuntala Kumari Sabat Women's College, Affiliated to Fakir Mohan University, Balasore, Odisha, India

D. Mohanty · V. Kumar · B. K. Das (✉)
ICAR-Central Inland Fisheries Research Institute, Barrackpore, Kolkata, India

© The Author(s), under exclusive license to Springer Nature Singapore Pte Ltd. 2026
B. K. Das et al. (eds.), *Bioinformatics Tools for Predictive Ecology and Fisheries*, Sustainable Aquatic Food Systems, https://doi.org/10.1007/978-981-95-2573-7_17

discipline by adding DNA-based markers, including restriction fragment length polymorphisms (RFLPs), amplified fragment length polymorphisms (AFLPs), and microsatellites. These markers provided enhanced resolution and repeatability, enabling more accurate genetic analysis. The most significant advancement occurred with the emergence of single-nucleotide polymorphisms (SNPs) and high-throughput sequencing technology, enabling genome-wide analysis with unparalleled precision and scale.

Aquaculture species provide distinct problems in genetic marker development owing to their varied biological traits, such as polyploidy in some species (e.g., sturgeon and salmonids), significant genetic diversity, and intricate life cycles. Notwithstanding these obstacles, the use of genetic markers has produced substantial advancements. Genetic markers have effectively identified quantitative trait loci (QTLs) associated with growth, disease resistance, and stress tolerance, which are essential for enhancing aquaculture production and resilience. In addition to breeding, genetic markers are crucial for conservation genetics and population control in aquaculture. They provide the evaluation of genetic diversity within and between populations, monitor inbreeding rates, and safeguard the genetic integrity of both captive and wild populations. Furthermore, genetic markers are essential for traceability and certification, offering a molecular method to authenticate species, confirm the provenance of cultivated items, and prevent illicit trading.

Recent breakthroughs in genomic technology, including whole-genome sequencing (WGS), genotyping-by-sequencing (GBS), and CRISPR-based gene editing, have significantly improved the production and use of genetic markers. These instruments enable researchers to uncover new markers, analyze intricate attributes, and incorporate genetic data into breeding methodologies. Aquaculture is at the forefront of a genomics-driven revolution, with the discovery of genetic markers as a fundamental aspect of this advancement. This chapter examines the methodology, applications, and future possibilities of genetic marker development for aquaculture species. By comprehending the ideas and methodologies of genetic marker application, stakeholders in the aquaculture sector may use these tools to tackle difficulties, enhance productivity, and foster sustainability in this essential business.

The first stage of developing genetic markers is the careful selection of the target species and the establishment of specific objectives for marker production. Aquaculture includes several species; each has distinct ecological and economic significance. The selection of species is often determined by their significance in aquaculture, production capacity, and particular obstacles encountered in respective farming methods. Tilapia is a commonly cultivated freshwater fish recognized for its swift growth and flexibility, making it an ideal subject for the development of genetic markers to enhance growth rate and feed efficiency. Likewise, animals like Atlantic salmon are prioritized for markers that target disease resistance because of the economic losses incurred from illnesses like sea lice infestation and infectious salmon anemia.

Marker development corresponds with commercial aims and the biological traits of the species that also include increasing growth rates, enhancing illness resistance, enabling sex determination, or preserving genetic variety (Gopikrishna, 2023). In shrimp farming, genetic markers facilitate the development of disease-resistant strains of Pacific white shrimp (*Litopenaeus vannamei*), a species significantly impacted by viral illnesses such as white spot syndrome virus (WSSV). In molluscs such as oysters, indicators of stress tolerance have proved essential in alleviating the effects of ocean acidification and temperature variations on the output.

Each species has distinct obstacles that affect marker development tactics. Crustaceans, including crabs and shrimp, possess extensive and complex genomes, sometimes characterized by repeated sequences, complicating the selection of appropriate markers. In contrast, the less complex genomic design of fish such as tilapia facilitates the finding of markers. Identifying these distinctions is essential for customizing marker development strategies to address species-specific requirements.

Moreover, sex determination is a significant application of genetic markers in some animals. In tilapia aquaculture, male specimens are favored for their superior development rates relative to females. The advancement of sex-linked markers has facilitated the creation of mono-sex populations, enhanced productivity, and minimized undesired reproduction. In sturgeon aquaculture, where caviar production is the primary objective, genetic markers for sex determination facilitate the early identification of females, hence decreasing maintenance expenses for non-productive males.

17.2 Sample Collection

Samples should ideally be taken from diverse populations stretching various geographic regions, ecological environments, or aquaculture facilities to include a wide range of genetic variety. In shrimp aquaculture, researchers may get samples from both wild populations in coastal regions and farmed populations kept under controlled conditions. This variety aids in the identification of genetic markers beneficial for local breeding programs and applicable over the species' broad geographical area.

Non-invasive or minimally intrusive sampling methods are often favored to guarantee organisms' ethical treatment and mitigate stress or fatalities. For example, fin clips, scales, or mucus swabs are often used in fish species, while tissue samples may be obtained from shell clippings of molluscs like oysters and mussels. Hemolymph extraction in shrimp and other crustaceans yields genetic materials with little damage. Non-invasive sampling techniques, such as the collection of shed skin, feathers, or environmental DNA (eDNA) from water samples, are increasingly being used for endangered species or those of significant conservation significance. These methodologies guarantee that sample collection adheres to scientific objectives and ethical principles. Efficient sample collection requires meticulous preparation and strict compliance with guidelines to prevent contamination and degradation of DNA. Accurate labeling, storage, and transportation are essential procedures. Samples are often kept in ethanol or held at ultra-low temperatures to ensure DNA integrity until laboratory analysis. GPS mapping tools may be used during sampling to document precise locations, facilitating spatial genetic analyses and the identification of population-specific markers.

17.3 Extraction of DNA/RNA

The basis of genetic marker development is the extraction of high-quality DNA or RNA from the biological specimens of the target species. The quality and amount of extracted nucleic acids directly affect the precision and efficacy of subsequent applications, including sequencing, marker identification, and genotyping. The selection of tissue for extraction often relies on the species and the study's purpose. Frequently used sample categories include fin clips, blood, gill tissue, or muscle, each necessitating customized extraction techniques to enhance yield and purity. For instance, in species such as tilapia or carp, fin clips are often used because of their ease of acquisition with minimum injury to the organism and their provision of sufficient DNA. Conversely, blood samples may be favored in animals such as catfish, producing high-quality nucleic acids. Gill tissue is a prevalent option, especially in marine animals like salmon, owing to its availability and abundance of nucleic acids.

Extraction techniques differ based on the kind of tissue and the intended use. Manual techniques, such as the phenol-chloroform extraction approach, continue to be favored for their cost-effectiveness and capacity to provide high-purity DNA. Commercially accessible kits are often used for high-throughput processing, providing standardized and consistent results. For RNA extraction, essential for transcriptomics or expression investigations, approaches that reduce RNase contamination, such as TRIzol-based procedures, are used. The quality of extracted DNA or RNA is evaluated using spectrophotometers (e.g., Nanodrop) to determine purity and concentration, and agarose gel electrophoresis to verify integrity. High-quality DNA or RNA often has an A260/A280 ratio of 1.8–2.0 and demonstrates low degradation when examined on an agarose gel. Ensuring appropriate extraction is essential since damaged or contaminated materials may induce mistakes in sequencing and genotyping, hence undermining the dependability of genetic marker creation. Specialized methodologies are used in certain applications to address issues. For instance, when dealing with species that generate mucopolysaccharides (e.g., shrimp or shellfish), additional purification procedures may be required to eliminate inhibitory compounds. In polyploid animals like sturgeon, methods must guarantee that the intricate genome is sufficiently represented in the retrieved nucleic acids.

Ultimately, proficient and dependable DNA/RNA extraction establishes the foundation for the following procedures in genetic marker creation, including sequencing, PCR amplification, and genotyping. This is an essential first phase in the examination of genetic variety, the identification of markers associated with desirable features, and the advancement of genetic enhancement and sustainability in aquaculture species.

17.4 Methods for Marker Discovery (Fig. 17.1)

The identification of genetic markers entails recognizing DNA sequences that reliably indicate genetic diversity. Various marker types have been created, each tailored to distinct research and breeding goals:

17.4.1 Microsatellites

Also known as simple sequence repeats (SSRs), are short, repeating DNA sequences characterized by significant polymorphism. They are essential for examining genetic diversity, population structure, and parentage analysis in aquaculture species. SSR markers have been extensively used in shrimp species like *Litopenaeus vannamei* to assess genetic diversity and enhance breeding operations. Their significant diversity makes them ideal for monitoring genetic alterations within and across populations throughout time.

17.4.2 Single-Nucleotide Polymorphisms (SNPs)

Are prevalent variations in the genome, characterized by alterations in individual DNA bases. Their stability and high-throughput detection capabilities have established them as a fundamental component in contemporary genomics. SNP markers have been used in salmon (*Salmo salar*) to pinpoint quantitative trait loci (QTLs)

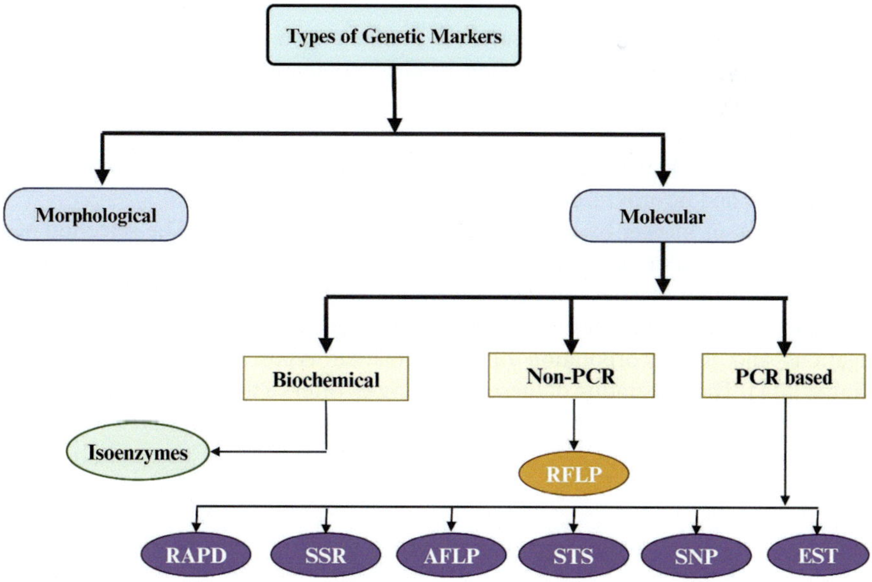

Fig. 17.1 Molecular markers used in fisheries

linked to illness resistance and growth characteristics. SNP arrays and genotyping-by-sequencing (GBS) methodologies enable extensive SNP identification and use in aquaculture breeding.

17.4.3 Amplified Fragment Length Polymorphisms (AFLPs)

Are DNA fragments produced by the use of restriction enzymes and selective PCR amplification. These markers are very valuable for evaluating genetic diversity and developing linkage maps. AFLP markers have been used in tilapia (*Oreochromis* spp.) to investigate hybridization and genetic differences across populations. AFLPs, although less precise than SNPs, are adaptable and do not need previous genome sequence data.

17.4.4 Expressed Sequence Tags (ESTs)

Originating from expressed genes, EST markers serve as functional markers associated with gene expression and certain features. They are crucial in finding markers linked to economically significant features, such as stress tolerance or growth. In catfish (*Ictalurus* spp.), EST markers have been used to identify genes associated with immune response, facilitating the creation of disease-resistant variants.

17.4.5 Mitochondrial Markers

Including mitochondrial DNA (mtDNA) markers such as cytochrome oxidase I (COI) and D-loop sections, are often used in population genetics, phylogenetic research, and species identification. These markers are especially efficacious for elucidating genetic links in organisms with intricate evolutionary histories. Mitochondrial markers have been used in clams (*Meretrix* spp.) to elucidate population patterns and inform resource management.

Each form of marker has distinct benefits and disadvantages. The selection of a marker is contingent upon the study objective, the species of interest, and the resources at hand. Progress in next-generation sequencing (NGS) technology has markedly expedited marker discoveries, facilitating high-resolution genomic investigations and the incorporation of many marker types in aquaculture species (Table 17.1). The integration of these technologies offers a thorough comprehension of genetic diversity and improves the accuracy of breeding strategies.

Table 17.1 Incorporation of different genetic markers in aquaculture species (Liu and Cordes, 2004; Maqsood and Ahmad, 2017; Tanya and Kumar 2010; Vos et al., 1995; Wright and Bentzen, 1995)

Genetic markers	Description	Application	Species
Allozyme markers	Protein-based markers, identified via electrophoresis.	Genetic diversity, population structure, and stock identification.	Salmonids, Tilapia
Microsatellites	Short tandem repeats (STRs), highly polymorphic and codominant markers.	Parentage analysis, linkage mapping, QTL identification, and stock management	Shrimp, Carp, Salmon
RAPD	Random Amplified Polymorphic DNA, which detects polymorphisms without prior sequence information.	Genetic fingerprinting, genetic diversity studies, and strain identification.	Catfish, Sea bass
AFLP	Amplified Fragment Length Polymorphism, combining restriction enzymes and PCR to detect polymorphisms.	Genetic diversity analysis, genome mapping, and hybrid identification.	Prawns, Tilapia
SNPs	Single-nucleotide polymorphisms, highly abundant, codominant markers.	High-resolution genome-wide association studies (GWAS), genomic selection, and breeding programs.	Salmon, Shrimp, Oyster
mtDNA Markers	Mitochondrial DNA markers are maternally inherited.	Phylogenetics, population structure, and evolutionary studies.	Tuna, Clams, Lobsters
ESTs	Expressed Sequence Tags derived from mRNA are useful for identifying coding sequences.	Gene discovery, functional genomics, and marker-assisted selection (MAS).	Trout, Shrimp
RAD-Seq	Restriction site-associated DNA sequencing, a high-throughput method to identify SNPs and other polymorphisms.	Genomic diversity studies, population genomics, and adaptive trait studies.	Salmon, Scallops, Sea cucumbers
SSR-EST	Microsatellites derived from ESTs, which are functional and species-specific.	Functional genomics, QTL mapping, and molecular breeding.	Catfish, Oysters
CRISPR-based Markers	Genome editing technologies to target specific genomic loci for mutation or tagging.	Functional gene analysis, genetic improvement, and disease resistance studies.	Salmon, Zebrafish

17.5 Marker Development and Validation

The first phase of genetic marker development involves discovering DNA sequences that function as dependable markers of gene diversity. High-throughput sequencing technologies, including Next-Generation Sequencing (NGS), have transformed this procedure by allowing researchers to sequence whole genomes or transcriptomes swiftly. Next-generation sequencing (NGS) facilitates the identification of single-nucleotide polymorphisms (SNPs), microsatellites, and other genetic markers throughout the whole genome. For example, in species such as Atlantic salmon (*Salmo salar*), next-generation sequencing (NGS) has been used to find hundreds of single-nucleotide polymorphisms (SNPs) linked to growth rate and disease resistance, establishing a solid basis for selective breeding initiatives.

Following identifying possible markers, the subsequent stage is validation to confirm their reliability and correlation with the traits of interest. Validation entails verifying the marker's consistency across several samples and populations, ensuring it is polymorphic and genetically informative. Methods such as polymerase chain reaction (PCR) are often used for this objective. PCR amplifies particular DNA sequences, enabling researchers to assess markers precisely and economically. Microsatellite markers in tilapia (*Oreochromis niloticus*) have been verified by PCR to assess genetic diversity and monitor breeding lines.

Advanced techniques such as genotyping-by-sequencing (GBS) and microarrays are essential for marker validation. GBS utilizes NGS technology to concurrently identify genotype markers across many individuals, making it especially advantageous for species with limited genetic resources. Genome-wide association studies (GBS) in Pacific oysters (*Crassostrea gigas*) have identified single-nucleotide polymorphisms (SNPs) linked to shell thickness and disease resistance, features essential for aquaculture production. Microarrays, conversely, include the hybridization of DNA to a chip embedded with probes for known markers, facilitating high-throughput genotyping. SNP arrays have been created for shrimp species such as *Penaeus vannamei*, enabling extensive genomic research and breeding initiatives.

The marker validation method includes evaluating the markers' association with quantitative trait loci (QTLs) or candidate genes related to characteristics of interest. Validated SNP markers in rainbow trout (*Oncorhynchus mykiss*) have been associated with QTLs for growth rate and fillet production, offering practical guidance for selective breeding initiatives. These verified indicators enhance the efficacy of breeding techniques while preserving genetic diversity and mitigating the danger of inbreeding.

The combination of high-throughput sequencing and rigorous validation methods has established genetic marker creation as a fundamental aspect of aquaculture genetics. These developments allow researchers and breeders to accurately discover, confirm, and use markers, hence enhancing commercially significant features in aquaculture species.

17.6 Categories of Data Analysis Conducted

Different types of data analysis are essential for advancing genetic markers, particularly for aquaculture species. The following are several forms of data analysis often used in the development of genetic markers.

17.6.1 Detection of Polymorphism

Sequence data is processed to identify single-nucleotide polymorphisms (SNPs), insertions/deletions (indels), and microsatellites. Applications such as GATK (Genome Analysis Toolkit), SAMtools, or variant callers like FreeBayes are applicable for variant calling and polymorphism identification.

17.6.2 Analysis of Heterozygosity and Allele Frequencies

The allele frequency and heterozygosity (the ratio of heterozygous individuals at a certain locus) are computed for each polymorphic locus. Software like as PLINK, Arlequin, or TASSEL may be used to calculate these statistics among genotyped individuals.

17.6.3 Analysis of Linkage Disequilibrium (LD)

Linkage disequilibrium analysis elucidates the inheritance patterns of genetic variants across distinct loci. This is often executed using software like Haploview or PLINK, which computes metrics like D' and r^2 to assess the intensity of linkage disequilibrium between marker pairs.

17.6.4 Quantitative Trait Locus (QTL) Mapping

QTL mapping entails correlating genetic markers with phenotypic characteristics (such as growth rate, disease resistance, or temperature tolerance). Statistical methodologies like interval mapping, composite interval mapping (CIM), and marker regression are used. Software such as QTL Cartographer, MapQTL, and R/qtl are often used for QTL analysis.

17.6.5 Genome-Wide Association Studies (GWAS)

Genome-wide association studies (GWAS) include evaluating statistical correlations between genetic markers, often single-nucleotide polymorphisms (SNPs), and phenotypic characteristics throughout extensive populations. It employs regression models or mixed linear models to address population structure and kinship. In GWAS analysis, tools such as PLINK, TASSEL, and GEMMA are often used.

17.6.6 Analysis of Population Structure and Genetic Diversity

Population structure may be assessed with programs such as STRUCTURE, ADMIXTURE, and PCA-based techniques like EIGENSOFT. These technologies categorize individuals into separate groups according to genetic similarities. Furthermore, measurements like FST (fixation index) may assess genetic divergence across populations.

17.6.7 Genomic Selection and Prediction

Genomic selection employs genome-wide markers to forecast the breeding value of individuals. Statistical methods, including genomic best linear unbiased prediction (GBLUP), and machine learning techniques such as random forests, support vector machines, and deep learning are used. Tools such as rrBLUP, BGLR, and GEneS are often used in genomic selection.

17.6.8 Phylogenetic and Evolutionary Examination

Phylogenetic analysis uses genetic markers to elucidate the evolutionary history of species. Software like MEGA, RAxML, and BEAST may create phylogenetic trees using genetic distance matrices. This may assist in tracing the origin of aquaculture species and evaluating their genetic integrity.

17.6.9 Transcriptomic Expression Analysis

RNA sequencing (RNA-seq) data may be used to discern genes that exhibit differential expression under certain situations, such as illness stress or temperature variations. Tools for differential expression analysis, such as DESeq2 or edgeR, are often used for this objective.

17.6.10 Genetic Mapping and Physical Mapping

For the development of comprehensive genetic maps and correlate markers with specific physical positions on chromosomes, genetic mapping may be constructed, illustrating the relative placements of markers throughout chromosomes by using genetic and physical markers. Physical mapping methods may amalgamate molecular cytogenetics with high-resolution sequencing.

17.7 Application of Genetic Markers in Aquaculture

The uses of genetic markers are many, including selective breeding, disease resistance, population management, and sex determination. Below, we elaborate on the various applications of genetic markers and the methodologies used to implement them in aquaculture. Below, we elaborate on the various applications of genetic markers and the methodologies used to implement them in aquaculture (Okumus and Ciftci, 2003; Lo Presti, Lisa and Di et al., 2009; Yue et al., 2024).

17.7.1 Selective Breeding

Selective breeding with genetic markers enhances the efficiency of aquaculture species improvement by decreasing generation time and facilitating more precise genetic enhancements. Genetic markers are used in marker-assisted selection (MAS), using markers associated with advantageous features to identify individuals for breeding purposes. This strategy is often associated with genomic selection (GS), a more sophisticated method that utilizes genome-wide data to forecast breeding parameters. In both MAS and GS, markers linked to attributes including growth rate, disease resistance, and stress tolerance are found via techniques like QTL mapping or GWAS. SNP markers linked to accelerated development or enhanced resistance to illnesses such as vibriosis or viral hemorrhagic septicemia may be used to identify better breeding stock (Fig. 17.2).

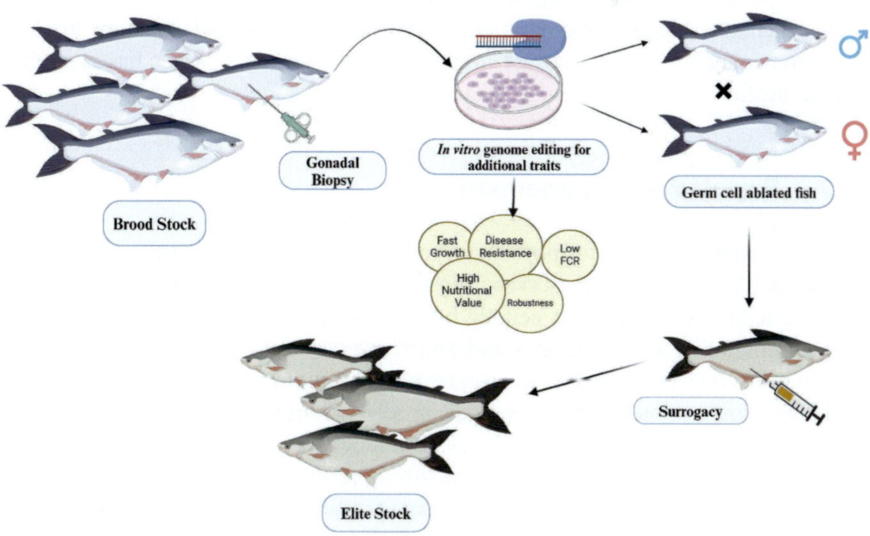

Fig. 17.2 Genome editing in aquaculture

17.7.2 Parentage and Pedigree Analysis

A significant difficulty in aquaculture is preserving genetic variety while preventing inbreeding. Genetic markers are essential for lineage tracing, monitoring breeding lines, and preventing the loss of genetic diversity or the accumulation of undesirable alleles in breeding programs. Parentage analysis utilizes genetic markers (e.g., microsatellites, SNPs) to accurately allocate offspring to their progenitors. These indicators facilitate the establishment of pedigrees that trace the lineage of individuals. Parentage analysis is often conducted using software such as Cervus or Colony, which use probabilistic algorithms to ascertain the most probable parents based on marker data. Alongside parentage analysis, pedigree reconstruction is often used in aquaculture breeding programs to prevent the mating of animals with a shared lineage, hence mitigating inbreeding (Fig. 17.3).

17.7.3 Disease Resistance

Disease outbreaks pose significant issues in aquaculture, and selecting for disease-resistant livestock may decrease death rates and reduce the need for antibiotics or other treatments. Genetic indicators linked to disease resistance enable breeders to identify individuals with enhanced natural immunity or resistance to certain diseases. Genetic markers associated with disease resistance are often found by QTL mapping or GWAS, wherein the genetic variations correlated with resistance features are localized to particular genomic areas. Markers associated with resistance to bacterial infections (e.g., *Aeromonas hydrophila*) or viral illnesses (e.g., infectious pancreatic necrosis in salmonids) are chosen. These markers are then used in MAS or GS to select individuals for breeding that possess the resistance-associated alleles. genetic selection methodologies such as GBLUP and machine learning algorithms may be used to forecast an individual's disease resistance based on their genetic profile.

17.7.4 Population Management

Utilizing genetic markers for population management guarantees the enduring sustainability of aquaculture, preserves robust wild populations, and enables informed breeding techniques that avert genetic bottlenecks and the loss of variety. Genetic markers, including microsatellites and SNPs, are used to evaluate genetic diversity within and across populations. Researchers may monitor genetic variation in both wild and farmed populations by comparing allele frequencies and computing metrics such as heterozygosity and FST (fixation index). In aquaculture, it is crucial to maintain genetic variety in farmed stocks throughout time to preserve their capacity to adjust to environmental changes and disease outbreaks. The use of genomic data from next-generation sequencing (NGS) technology facilitates more thorough

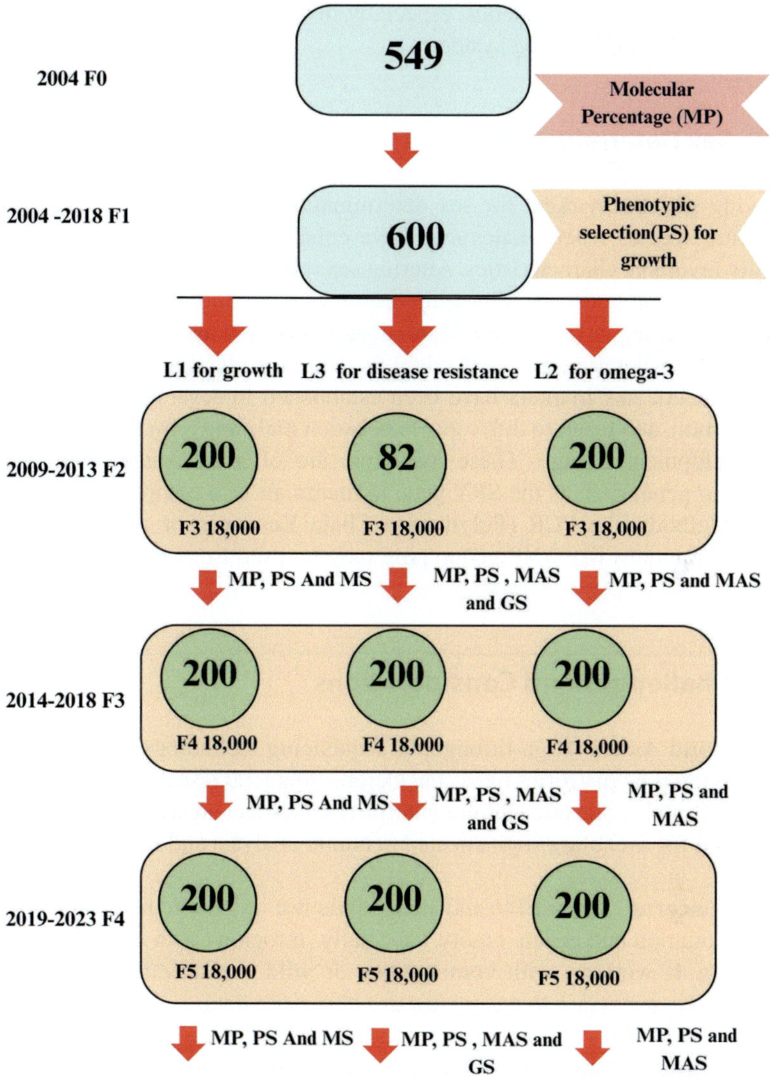

Fig. 17.3 Overview of selecting approaches for the 20-year Asian seabass breeding program in Singapore. The numbers in each cycle are the number of selected elite broodfish. F0, F1, F2, F3, and F4: Generations from F0 to F4, and the numbers behind F0, F1, F2, F3, and F4 are the numbers of fish raised for selection in each generation. L1, L2, and L3: Lines 1, 2, and 3. *MS* Marker-assisted selection, *GS* Genomic selection

evaluations of population genetics, especially in species characterized by extensive genomes or intricate breeding systems.

17.7.5 Sex Determination

Employing genetic markers for sex determination is especially advantageous in aquaculture species where males may have enhanced growth rates or other commercially favorable characteristics. Altering sex ratios may enhance production efficiency and decrease costs by allocating resources to the more productive gender. In livestock exhibiting sexual dimorphism, genetic markers associated with sex chromosomes or sex-determining areas might facilitate the identification of an individual's sex. Genetic sex markers have been established in several species, including tilapia, salmon, and trout, to differentiate between male and female individuals at an early developmental stage. These indicators are often associated with the sex-determining gene, such as the SRY gene in mammals or a comparable gene in fish species. Methods like PCR (Polymerase Chain Reaction) or more sophisticated genomic techniques like SNP genotyping may be used to ascertain sex with great precision.

17.8 Challenges and Considerations

Expenses and Assets High-throughput sequencing technology and the advancement of genotyping platforms need considerable financial resources. These expenses can hamper small-scale aquaculture enterprises. Competent staff and infrastructure are essential for executing sophisticated genomic analysis and data management.

Ethical Concerns Sustainable and responsible use of genetic resources is vital to avoid exploitation and ensure equity, especially in regions with indigenous or wild populations. If working with communities or wild stocks, ethical approval and agreements are necessary to ensure fair use of genetic data.

Environmental Impact Over-reliance on genetically improved strains can reduce the genetic diversity of farmed populations, potentially leading to inbreeding depression or vulnerability to diseases. Escapees from aquaculture systems can interbreed with wild populations, potentially altering their genetic makeup. Proper containment and management strategies are essential to mitigate this risk.

17.9 Conclusion and Future Prospects

To conclude, genetic markers are instrumental in advancing aquaculture by enhancing growth rates, disease resistance, and overall sustainability. They enable more precise and efficient selective breeding, population management, and species conservation. The transition from protein-based markers to DNA-based technologies such as SNPs and high-throughput sequencing has significantly expanded the precision and scope of genetic analysis. Despite challenges like high costs, ethical concerns, and the complexity of certain genomes, genetic markers remain vital for addressing key issues in aquaculture, including genetic diversity preservation and traceability. As genomic technology progresses, the integration of novel methods like whole-genome sequencing and CRISPR will likely further optimize genetic marker development, ensuring continued innovation and sustainability in aquaculture practices. This ongoing advancement supports the sector's critical role in meeting global food demands responsibly.

References

Askari GH, Shabani A, Miandare HK (2013) Application of molecular markers in fisheries and aquaculture. Anim Sci 2(4):82–88

Gopikrishna G (2023) Application of molecular markers in aquaculture. In: Frontiers in aquaculture biotechnology. Academic, pp 47–53

Liu ZJ, Cordes JF (2004) DNA marker technologies and their applications in aquaculture genetics. Aquaculture *238*(1-4):1–37

Lo Presti R, Lisa C, Di Stasio L (2009) Molecular genetics in aquaculture. Ital J Ani Sci *8*(3):299–313

Maqsood HM, Ahmad SM (2017) Advances in molecular markers and their applications in aquaculture and fisheries. Genet Aquat Org *1*(1):27–41

Okumuş İ, Çiftci Y (2003) Fish population genetics and molecular markers: II-molecular markers and their applications in fisheries and aquaculture. *Turk J Fish Aquat Sci 3*(1):51–79

Tanya C, Kumar R (2010) Molecular markers and their applications in fisheries and aquaculture. Adv Biosci Biotechnol *1*:281–291

Vos P, Hogers R, Bleeker M, Reijans M, Lee TVD, Hornes M et al (1995) AFLP: a new technique for DNA fingerprinting. Nucleic Acids Res *23*(21):4407–4414

Wright JM, Bentzen P (1995) Microsatellites: genetic markers for the future. In: Molecular genetics in fisheries. Springer, Dordrecht, pp 117–121

Yue GH, Wang L, Sun F, Yang ZT, Wong J, Wen YF et al (2024) Improving growth, omega-3 contents, and disease resistance of Asian seabass: status of a 20-year family-based breeding program. *Rev Fish Biol Fish 34*(1):91–110

Predictive Modelling in Aquatic Ecosystems Using Bioinformatics

18

Maheswata Sahoo, Ananya Nayak, Ankita Pati, and Dibyajyoti Uttameswar Behera

18.1 Introduction

In aquatic environments, predictive modelling is a crucial tool for comprehending and predicting the behaviour of intricate environmental systems. From freshwater environments like rivers, lakes, and wetlands to marine habitats like oceans and coral reefs, aquatic ecosystems are complex webs of interdependent biological, chemical, and physical elements. Natural processes like water flow, nutrient cycling, and climate change, as well as human activities like pollution, overfishing, and habitat destruction, have a continuous impact on these systems (Shavalieva et al. 2022).

A way to model these interactions and predict how aquatic ecosystems might react to management changes, environmental stressors, and climate change is through predictive modelling. Predictive models can provide information about species distributions, ecosystem health, biodiversity loss, and the effects of human-induced changes by combining large datasets and applying sophisticated computational techniques. This information can then be used to inform conservation and management decisions (Niinemets et al. 2017; Häder and Barnes 2019).

Bioinformatics has become a potent instrument in recent years to improve aquatic ecosystem predictive modelling (Pais 2022). The analysis of biological data, including genetic sequences, protein structures, and metabolic pathways, has historically been the main focus of bioinformatics. But in recent decades, its purview has broadened to encompass the analysis of environmental data, including that gathered from ecological data repositories, microbial community profiling, and environmental DNA (eDNA) (Kim et al. 2003; Pittman et al. 2004). These days, bioinformatics tools are essential for processing vast amounts of biological data,

M. Sahoo (✉) · A. Nayak · A. Pati · D. U. Behera
Centre for Biotechnology, Siksha 'O' Anusandhan (Deemed to be University), Bhubaneswar, India

© The Author(s), under exclusive license to Springer Nature Singapore Pte Ltd. 2026

B. K. Das et al. (eds.), *Bioinformatics Tools for Predictive Ecology and Fisheries*, Sustainable Aquatic Food Systems, https://doi.org/10.1007/978-981-95-2573-7_18

identifying significant patterns, and offering insights into how ecosystems function. Bioinformatics can enhance ecological forecasts' precision and scope when paired with predictive modelling, producing more reliable predictions that take ecosystem-level, microbial, and genetic dynamics into account (Jones et al. 2006).

In aquatic ecosystems, the combination of bioinformatics and predictive modelling is particularly effective in gaining insight into species interactions, habitat suitability, and biodiversity (Ummah et al. 2025). For instance, direct sampling, which can be time-consuming and environmentally disruptive, is no longer necessary for the detection and monitoring of species presence or absence in water bodies, thanks to DNA barcoding and eDNA techniques (Nørgaard et al. 2021). These genetic methods offer real-time data for predictive models and a non-invasive way to measure biodiversity and monitor changes over time. Bioinformatics tools can enhance species distribution models by integrating genomic data with environmental factors (such as temperature, pH, and salinity) and identifying regions at risk from changing environmental conditions or areas with high biodiversity. This is especially helpful in comprehending how aquatic organisms are affected by invasive species, habitat degradation, and climate change. A more thorough understanding of how these changes will impact individual species as well as entire ecosystems is made possible by predictive models that integrate bioinformatics-based data. This offers the possibility of predicting and reducing ecological harm before it becomes irreparable (Pais 2022).

Furthermore, by making it possible to investigate microbial communities in aquatic environments, bioinformatics aids in predictive modelling. Although microorganisms are essential to the cycling of nutrients, the control of water quality, and the health of ecosystems, traditional modelling techniques frequently ignore their dynamic interactions. The composition and functional potential of microbial communities can be analysed by bioinformatics tools using metagenomic data, which aids in identifying important species that propel ecological processes like organic matter decomposition or nitrogen fixation. Incorporating microbial data into predictive models enhances the precision of forecasts concerning water quality, nutrient cycling, and ecosystem resilience to environmental stressors while offering a more thorough understanding of ecosystem processes (Tucker and Duplisea 2012; Pais 2022).

Predictive models in aquatic ecosystems are further improved by the combination of bioinformatics, machine learning (ML), and artificial intelligence (Tucker and Duplisea 2012). When it comes to analysing massive, complicated datasets that incorporate biological, chemical, and environmental elements, machine learning methods are especially well-suited. ML models can produce forecasts on future ecosystem dynamics, such as changes in species distributions or modifications in water quality brought on by pollution or climate change, by finding patterns and linkages in these datasets. Additionally, by identifying important variables that affect ecosystem health, these algorithms offer important insights into how ecosystems may react to particular management techniques or interventions. AI and ML approaches will continue to propel the creation of more precise, scalable, and effective models as predictive modelling advances in complexity (Jamialahmadi et al. 2024).

Even though predictive modelling shows great promise in aquatic environments, there are still a number of obstacles to overcome. One of the main challenges is the intricacy of aquatic habitats, which are impacted by a wide range of interrelated elements that can be challenging to measure or simulate. For instance, the effects of contaminants like plastics and medications on aquatic life are not well understood, and it takes complex models that take into consideration a wide range of factors to forecast the long-term effects of these contaminants. Predictive models' efficacy may also be constrained by the quality and accessibility of the data. For models to be calibrated and validated, high-resolution, accurate data is required; otherwise, predictions may not be accurate. This problem can be solved by combining bioinformatics and predictive modelling, which makes it possible to gather and examine massive amounts of ecological, environmental, and genomic data (Quazi 2022; Jamialahmadi et al. 2024).

This chapter illustrates how the integration of bioinformatics and predictive modelling in aquatic ecosystems has great promise for enhancing our comprehension of biodiversity, ecosystem dynamics, and the effects of natural and human stresses. Scientists and researchers can produce more precise predictions of ecosystem behaviour, direct conservation efforts, and influence policy decisions by combining cutting-edge computer approaches with ecological and biological data. Predictive modelling and bioinformatics both have the potential to become vital tools for managing aquatic ecosystems in the face of a rapidly changing environment, ultimately assisting in the preservation of the ecosystem services and biodiversity that these systems offer.

By using statistical, computational, and machine learning methods, predictive modelling makes predictions about future events or patterns based on available data. These models are used to examine intricate relationships between biological, chemical, and physical elements in water systems in aquatic ecology studies. The objective is to model, forecast, and control aquatic environments under a range of man-made and natural forces.

- Statistical models correlating environmental variables with observed patterns, such as species distribution, are among the key techniques (Arhonditsis et al. 2008, 2014).
- Mechanistic models predict ecosystem behaviour by using biological processes and physical principles (e.g., nutrient cycling models).
- Machine learning models manage big datasets and nonlinear interactions to forecast events like fish population dynamics or toxic algal blooms.
- Hybrid models improve prediction accuracy, combine data-driven and mechanistic methods.

18.2 The Value of Predictive Modelling in Aquatic Environments

- Being Aware of Ecosystem Dynamics: The distribution of species in aquatic habitats is influenced by biotic and abiotic parameters, such as temperature, salinity, and dissolved oxygen, which are explained by predictive models.
- Evaluation of Environmental Effects: Models help us comprehend the long-term effects of pollution, habitat degradation, and climate change on aquatic ecosystems.
- Preservation and Diversity in Nature: By predicting changes in habitat or species distribution brought on by climate change, these tools help guide conservation priorities and initiatives.
- Management of Resources: Through the estimation of population trends, breeding cycles, and the effects of overfishing, predictive models maximise fisheries management.
- Reducing the Impact of Climate Change: To create adaptation strategies for the resilience of aquatic ecosystems, models evaluate scenarios such as increasing temperatures and shifting precipitation patterns.
- Handling Negative Occurrences: Predictive techniques that provide early warnings about invasive species, hypoxia, or dangerous algal blooms help reduce hazards to human health and biodiversity.
- Sustainable Growth: Models help policymakers strike a balance between ecological preservation and development by offering insights into ecosystem functions and possible disruptions.

18.3 Application in Aquatic Environment

- Predicting pollution levels or eutrophication concerns is known as water quality management.
- Finding susceptible species or hotspots for intervention is known as biodiversity monitoring.
- Habitat restoration: evaluating the results of restoration projects in water bodies that have deteriorated.
- Fish stock trends are modelled in fisheries under various harvesting and environmental situations.
- Ecosystem services: comprehending and forecasting the provision of ecosystem services, including the sequestration of carbon in wetlands.

18.4 Bioinformatics in the Aquatic Environment: An Overview

In aquatic environmental research, bioinformatics has emerged as a vital tool. By combining biological and computational methods, it aids in analysing aquatic biodiversity, unravelling intricate ecological relationships, and tackling environmental issues. Proteomic, metagenomic, transcriptomic, and genomic data from aquatic species and their habitats are analysed in this discipline to help understand aquatic ecosystems.

18.5 Definition and Scope

Bioinformatics is the use of computational tools to analyse biological data about aquatic animals, ecosystems, and processes in aquatic environments. It encourages research into genetic diversity, population dynamics, and how aquatic life is affected by environmental changes.

18.6 The Scope of Aquatic Bioinformatics Includes

- Decoding the genomes of aquatic species—such as fish, corals, and algae—to comprehend their evolution and adaptation is known as aquatic genomics.
- Investigating the ecological functions and possible biotechnological uses of the microbial communities found in lakes, rivers, and seas is known as marine metagenomics.
- Monitoring fish stocks, migration trends, and conservation status through the analysis of genetic variation is known as population genetics.
- Ecosystem monitoring is the process of detecting changes in ecosystem health and biodiversity brought on by pollution or climate change.
- Aquatic pathogen detection is the process of identifying and monitoring illnesses that impact aquatic species, such as harmful algal blooms (HABs), using bioinformatics technologies.
- Aquaculture: Understanding disease resistance in farmed species and optimising breeding programs through genomic selection (Table 18.1).

18.7 Machine Learning and Artificial Intelligence

Artificial intelligence (AI) tools are being used more and more to monitor water quality, predict species distributions, and identify dangerous environmental changes like HABs. In order to fully understand the intricacies of aquatic ecosystems, bioinformatics is essential. It facilitates comprehension of ecological interactions, biodiversity, and environmental issues such as habitat loss, pollution, and climate change. Researchers can support aquatic resource conservation and sustainable management by utilising state-of-the-art computational tools and technologies (Arhonditsis and Brett 2004).

Table 18.1 Key tools and technologies in aquatic bioinformatics

Category	Tools/Database	Purpose
Microbial diversity analysis	QIIME	Analyses microbial diversity in aquatic ecosystems
	MEGAHIT/ MetaPhlAn	For metagenomic data assembly and profiling microbial communities
	Anvi'o	Visualises and analyses metagenomic and microbial genomics data
Genomics and Transcriptomics	Trinity	For de novo transcriptome assembly in aquatic species
	Salmon and Kallisto	Quantify transcript abundances in aquatic organisms
	Genome browsers (UCSC, Ensembl)	Explore aquatic species genomes
Phylogenetics and evolutionary	BEAST	For evolutionary and phylogenetic analysis
	PhyloBayes	Handles complex evolutionary models in aquatic species studies
Population genetics and biodiversity	STRUCTURE	Analyses population structure using genetic markers
	Arlequin	Conducts population genetics and diversity analysis
	iBOL	DNA barcoding to identify and monitor aquatic species
Environmental DNA (eDNA)	OBITools and DADA2	Analyse eDNA to monitor aquatic diversity and detect invasive species
	EcoPCR	For DNA sequence-based taxonomic assignment in ecological studies
Proteomics and functional analysis	MaxQuant	For proteomic analysis of aquatic species
	InterProScan	Identifies protein domains and functions in aquatic species
Databases specific to aquatic research	FISH-BOL	A global initiative for identifying fish species
	Ocean Gene Atlas	Analyses ocean-related gene expression and biodiversity data
	MAR databases (Marine metagenomics)	Explore functional and taxonomic data of marine microbiomes
Modelling and visualisation	Cytoscape	For network visualisation in aquatic ecosystems
	Ecopath with Ecosim	Models' aquatic trophic interactions and energy flows

18.8 Aquatic Ecosystems and Their Complexity

Freshwater (lakes, rivers, wetlands) and marine (oceans, coral reefs, estuaries) habitats are both part of the dynamic and varied aquatic ecosystems. Because of their complex physical, chemical, and biological interactions, these ecosystems rank among the most complex natural systems, despite being essential to human well-being, ecosystem services, and biodiversity worldwide (Arhonditsis and Brett 2004).

18.8.1 Types of Aquatic Ecosystems

Freshwater Ecosystems

Water bodies that flow at different rates, such as rivers and streams, have an impact on biodiversity and sediment transport. Standing water ecosystems with distinct zones (littoral, limnetic, profundal, and benthic) are lakes and ponds. Wetlands are areas of transition between terrestrial and aquatic ecosystems that are abundant in ecosystem services and biodiversity.

Marine Ecosystems

About 71% of the Earth's surface is made up of oceans, which are known for their rich biodiversity and vital roles in regulating the planet's climate. Coral reefs are biodiverse marine environments that are frequently referred to as the "rainforests of the sea." Estuaries are areas of transition between fresh and salt water that serve as nurseries for a variety of species.

18.8.2 Key Features of Aquatic Ecosystems

- High Biodiversity: Aquatic ecosystems harbour diverse organisms, including phytoplankton, zooplankton, fish, amphibians, and aquatic plants, many of which are interdependent.
- Zonation: Aquatic environments are stratified based on depth, light penetration, and temperature, such as the photic and aphotic zones in oceans.
- Dynamic Physical and Chemical Factors: Water Flow: Influences nutrient distribution and habitat structure.
- Temperature and Light: Affect metabolic rates and primary productivity. Dissolved Oxygen and pH: Critical for organism survival and ecosystem health.
- Nutrient Cycling: In global biogeochemical cycles like the nitrogen, phosphorus, and carbon cycles, aquatic ecosystems are essential.
- Connectivity: Rivers that feed into oceans or wetlands that serve as buffers frequently create intricate networks among aquatic ecosystems.

18.8.3 Complexity in Aquatic Ecosystems

- Multiscale Interactions: Aquatic ecosystems operate at multiple scales, ranging from microscopic interactions between plankton to large-scale ocean currents that impact the climate globally.
- Food Web Dynamics: The intricate food webs present in aquatic environments involve primary producers (phytoplankton), primary consumers (zooplankton), and top predators (fish, marine mammals). Any change in the trophic level has the potential to have a significant impact on the system as a whole.
- Temporal and Spatial Variability: Aquatic ecosystems are impacted by tidal cycles, seasonal changes, and long-term climate fluctuations. Spatial heterogeneity in water depth, temperature, and salinity adds to their complexity.
- Human Impacts: Pollution, overfishing, habitat destruction, and climate change all contribute to the instability of ecosystems and add layers of complexity.
- Anthropogenic factors alter nutrient cycles and species interactions, leading to issues like eutrophication or biodiversity loss.
- Emergent Properties: The interactions between organisms and their environment give rise to emergent properties such as resilience to shocks and ecosystem self-regulation.
- Evolution and Adaptations: The ecosystems of aquatic species are highly specialised but also extremely sensitive to change due to their unique adaptations, such as temperature tolerance, osmoregulation, and bioluminescence.

18.8.4 Challenges in Studying Aquatic Ecosystems

1. Data collection: Monitoring aquatic ecosystems can be logistically challenging due to their size and accessibility (e.g., deep oceans).
2. Multidisciplinary Approach: Understanding aquatic environments requires knowledge from a number of disciplines, such as hydrology, biology, chemistry, and physics. Dynamic Interactions: Because the environment changes so quickly, it is difficult to predict long-term trends in ecosystem health.

18.9 Data Collection and Management

Obtaining physical, chemical, and biological data from aquatic systems is necessary to comprehend biodiversity, ecosystem health, and environmental changes. Important approaches include remote sensing techniques for extensive evaluations of water bodies and in situ monitoring with sensors and probes to measure parameters like temperature, pH, dissolved oxygen, and salinity. Techniques like net sampling for fish and plankton, environmental DNA (eDNA) analysis for species identification, and sediment sampling for benthic community research are all part of the biological data collection process. For accuracy, accessibility, and interoperability to be guaranteed, proper data management is essential. This entails data digitisation, safe database storage, and analysis using bioinformatics tools. Effective visualisation and sharing are made possible by integrated platforms such as cloud-based storage systems and GIS (Geographic Information Systems), which support cooperative research projects and long-term monitoring (Beng and Corlett 2020).

18.10 Modelling Approaches in Aquatic Ecosystems

In aquatic ecosystems, modelling techniques are crucial for comprehending their intricate dynamics, forecasting changes, and guiding management choices. These models simulate ecosystem interactions, analyse ecological responses to environmental changes, and evaluate the effects of humans by integrating biological, physical, and chemical processes. Process-based models, which mimic mechanistic interactions like energy transfer, primary production, and nutrient cycling in food webs, are important strategies. Biogeochemical models, which monitor nutrient dynamics and greenhouse gas emissions, and hydrodynamic models, which examine water flow and stratification, are two examples. In order to evaluate trophic cascades, migration patterns, and biodiversity, population and community models forecast the dynamics of species populations and their interactions.

Ecosystem models such as Ecopath with Ecosim (EwE), which assess species interactions and energy flows within food webs to aid in fisheries management, are another important strategy. Using massive datasets, machine learning and data-driven models are being used more and more to forecast changes in habitat suitability, biodiversity, and water quality. In order to track and forecast land–water interactions, coastal erosion, and habitat changes, remote sensing data and Geographic Information Systems (GIS) are frequently incorporated into spatially explicit models.

These modelling techniques help with conservation planning and policy-making by providing predictive capabilities to model future scenarios, such as the effects of pollution, habitat destruction, and climate change. But to ensure realistic and useful results, their success hinges on precise data inputs, interdisciplinary integration, and validation against observed data (Jorgensen 2002).

18.10.1 Types of Predictive Models

- Statistical Models: Regression analysis and time-series forecasting are two methods that use historical data to find trends and connections.
- Machine Learning Models: Use algorithms such as neural networks, decision trees, and random forests to forecast results from intricate datasets.
- Using hydrodynamic models, one can forecast changes in aquatic systems by simulating water flow and movement.
- Examine primary production, nutrient cycling, and ecosystem reactions to environmental changes using biogeochemical models.
- Species Distribution Models (SDMs) use occurrence data and environmental factors to forecast a species' geographic range.

18.11 Machine Learning and AI in Predictive Modelling

Predictive modelling is being revolutionised by machine learning (ML) and artificial intelligence (AI), which makes it possible to analyse large, complicated datasets in order to find patterns, generate predictions, and inform choices. AI and machine learning are used in aquatic ecosystems to forecast things like species distributions, water quality, and the effects of climate change. Machine learning (ML) algorithms are very effective for ecosystem-level predictions because, in contrast to traditional modelling approaches, they can handle high-dimensional data and non-linear relationships. Common approaches include unsupervised techniques like clustering and anomaly detection, as well as supervised learning techniques like decision trees, support vector machines, and neural networks (Battineni et al. 2020).

Predictive models driven by AI are especially useful for combining various datasets, such as real-time sensor measurements, environmental DNA (eDNA) data, and remote sensing imagery. These tools are used to forecast pollution events, monitor harmful algal blooms, and evaluate habitat vulnerability. Furthermore, a subset of artificial intelligence called deep learning models is used for underwater monitoring image and video analysis, automating processes like species identification. Predictive modelling with AI improves ecological research's accuracy and scalability by enabling real-time forecasts and adaptive management. To optimise their potential in environmental applications, however, issues like data accessibility, model interpretability, and computational requirements need to be resolved (Yang 2022).

18.12 Applications of Machine Learning in Bioinformatics

Large, complex biological datasets can now be analysed and interpreted, thanks to machine learning (ML), which has revolutionised bioinformatics. ML is utilised in genomics to identify functional elements like splice sites and regulatory sequences, as well as to predict genes and annotate the genome. In proteomics, machine learning algorithms aid in the prediction of protein interactions, folding patterns, and structures—all of which are essential for comprehending molecular processes and developing new drugs. By evaluating RNA-seq data for gene expression profiling and finding disease biomarkers, machine learning (ML) also plays a crucial part in transcriptomics. Reconstructing and analysing biological networks, including gene regulatory and metabolic networks, using machine learning models, offers insights into cellular functions in the field of systems biology (Sharma et al. 2024). Precision medicine is another important area where machine learning is being used to predict patient reactions to medications, find genetic variations linked to diseases, and create individualised treatment regimens. To speed up drug discovery, ML techniques are used in structural bioinformatics to predict protein–ligand interactions.

Furthermore, metagenomics has been transformed by machine learning, which makes it possible to classify and functionally annotate microbial communities from environmental samples. Relevant information is extracted from biomedical literature through the use of natural language processing (NLP), a subset of machine learning. These uses demonstrate the adaptability and revolutionary power of machine learning in bioinformatics, propelling advances in both basic and applied biomedical sciences (Sharma et al. 2024).

18.13 Ecological Factors Influencing Predictions

Environmental Variables

In aquatic systems, environmental factors play a crucial role in influencing ecological model predictions. Temperature, light availability, nutrient levels, pH, salinity, and dissolved oxygen are important factors that have a big impact on aquatic organisms' physiological functions and the dynamics of the ecosystem as a whole. For example, light availability affects phytoplankton primary production, while temperature affects metabolic rates, reproduction, and species distributions. Eutrophication is caused by nutrient levels, especially nitrogen and phosphorus, which change species compositions by causing algal blooms and oxygen depletion. From freshwater species to marine life, habitat suitability is influenced by salinity gradients and pH variations. External elements that impact nutrient transport, sediment deposition, and water temperature include hydrodynamic forces (such as currents, tides, and stratification) and climatic fluctuations. Since these variables frequently show both spatial and temporal variability, accurate predictions necessitate ongoing monitoring of these variables and their interactions (Peng et al. 2017).

18.14 Interactions Within Aquatic Food Webs

Aquatic food web interactions determine population dynamics, nutrient cycling, and energy flow, all of which have a substantial impact on ecological predictions. Aquatic ecosystems have extremely complex food webs that include apex predators, secondary consumers (small fish), primary producers (phytoplankton), and primary consumers (zooplankton). Understanding trophic interactions, competition, and predation is necessary to forecast changes within these webs. For instance, trophic cascades brought on by the overfishing of predators can upset the equilibrium of an ecosystem. By outcompeting native organisms or introducing new predation pressures, invasive species can also change the dynamics of the food web. Seasonal variations that cause shifts in consumer populations, like phytoplankton blooms, make forecasting even more difficult. Because species are interdependent, even small changes at one trophic level can have a ripple effect throughout the system, increasing unpredictability. To capture the dynamic complexity of aquatic food webs and increase prediction accuracy, sophisticated modelling techniques that incorporate species interactions and environmental factors are crucial (Sun et al. 2019).

18.15 Validation and Calibration of Models

In order to create trustworthy ecological models for aquatic systems, validation and calibration are crucial processes. Validation checks that a model accurately depicts real-world conditions by comparing its predictions to observed data. In order to improve the model's predictive power, calibration entails adjusting model parameters to match simulations with empirical data. When combined, these procedures increase the validity and usefulness of models, allowing for precise predictions of species distributions, environmental changes, or ecosystem reactions. Models may produce inaccurate predictions if they are not properly validated and calibrated, which would undermine conservation and management initiatives (Sun et al. 2019).

18.16 Techniques for Model Validation

Several methods are used in model validation to assess the precision and dependability of predictions. In order to assess the model's performance on unseen data, a popular technique is cross-validation, which divides the dataset into training and testing subsets. Goodness-of-fit tests use statistical metrics such as R-squared, mean squared error, or correlation coefficients to evaluate how well model outputs match observed data. Sensitivity analysis identifies important factors that impact predictions by examining how changes in input variables affect model outputs. Furthermore, independent data validation provides an objective evaluation of the model's predictive power by comparing it to a different dataset that was not used in its creation. Scenario testing is another aspect of validation, in which models are tested in fictitious or extreme scenarios to gauge their dependability and robustness. These techniques ensure that models are not only statistically sound but also ecologically meaningful (Landry et al. 1983).

18.17 Importance of Calibration in Predictions

In order to improve the precision and dependability of predictive models, calibration is essential. In order to reduce disparities and enhance overall performance, it entails modifying model parameters to match predictions with observed data. Parameters like growth rates, nutrient uptake, or mortality rates, for example, are adjusted during calibration in aquatic ecosystem modelling to account for regional environmental conditions. Models can more accurately depict intricate ecological processes and improve their capacity to forecast results in a range of situations when they are calibrated accurately. Calibration's significance goes beyond its capacity to enhance judgement. Realistic projections from well-calibrated models can direct resource management, conservation tactics, and policy decisions. For instance, determining nutrient concentration thresholds in advance of algal blooms guarantees prompt actions to lessen their effects. Calibration also improves the model's transferability between systems, which makes it a useful instrument for ecological assessments on a global scale. Calibration increases confidence in the model's application to ecological challenges by guaranteeing that its outputs are based on observed realities (Staartjes and Kernbach 2020).

18.18 Case Studies

Successful Applications of Predictive Modelling in Aquatic Ecosystems

Numerous issues in aquatic ecosystems have been successfully resolved through the use of predictive modelling, which has informed conservation and management initiatives and offered insights into ecosystem dynamics. Models that forecast harmful algal blooms (HABs), for instance, have been essential in reducing their negative effects on public health and fisheries. To predict bloom events and enable prompt interventions, these models incorporate environmental factors like nutrient levels, water temperature, and light availability. Another noteworthy use is in fisheries management, where trophic interactions, fish stock dynamics, and the impact of fishing pressure are evaluated using models like Ecopath with Ecosim (EwE).

These instruments support the preservation of marine biodiversity and the establishment of sustainable harvest levels. Similarly, to help with habitat restoration and protection efforts, critical habitats for endangered aquatic species have been identified using habitat suitability models, which are frequently paired with remote sensing data. Studying how climate change affects aquatic ecosystems—including rising sea levels, ocean acidification, and changed hydrological cycles—has also benefited greatly from predictive models (Corrales et al. 2020).

18.19 Lessons Learned and Challenges Faced

Although predictive modelling has shown impressive results, it has also brought to light a number of difficulties and lessons. One important lesson is that accurate predictions require the integration of long-term, high-quality datasets. The quality of the data used to develop and validate a model determines its dependability. Significant errors can result from incomplete or biased datasets, which emphasises the necessity of regular monitoring and data exchange amongst management and research institutions. Since ecological modelling frequently necessitates knowledge of biology, hydrology, physics, and computer science, interdisciplinary collaboration is another important lesson. Nonetheless, there are still issues, such as the complexity and unpredictability of aquatic ecosystems. Accurately capturing every ecological process is challenging due to nonlinear interactions like trophic cascades and climate feedback mechanisms. Scaling problems also occur because models created for particular areas or circumstances might not work well in larger or different contexts. Logistical challenges also arise from the computational demands of high-resolution modelling and the requirement for real-time predictions. It is still very difficult to strike a balance between model complexity for accuracy and simplicity for usability. In order to overcome these challenges and guarantee the ongoing effectiveness and usefulness of predictive modelling in aquatic ecosystem management, improvements in data collection, processing power, and teamwork are needed (Corrales et al. 2020).

18.20 Future Directions in Predictive Modelling

Improving the accuracy, scalability, and applicability of predictive modelling to new environmental challenges is key to its future in aquatic ecosystems. The temporal and spatial resolution of models will be greatly enhanced by incorporating real-time data streams from satellite remote sensing and sophisticated sensor networks. By incorporating artificial intelligence (AI) and machine learning into conventional ecological models, it will be possible to analyse intricate datasets, uncover nonlinear interactions, and generate dynamic predictions. The creation of coupled models, which concurrently address hydrodynamic, biogeochemical, and ecological processes and provide a more comprehensive understanding of ecosystem dynamics, is another exciting avenue. In order to inform adaptive management tactics, future models will increasingly concentrate on climate change scenarios, such as the effects of ocean warming, acidification, and extreme weather events.

18.21 Emerging Trends in Bioinformatics

Advances in technology and the availability of more data are driving the rapid evolution of bioinformatics. The use of AI and deep learning methods to analyse intricate biological datasets, like multi-omics data integration, is one new trend. The emergence of single-cell sequencing technologies is another trend that gives researchers insights into the diversity and function of microorganisms by enabling them to examine individual cellular responses within ecosystems. Environmental DNA (eDNA) analysis is also being used more and more in bioinformatics to track biodiversity in aquatic systems. Big data analytics and cloud-based platforms are revolutionising the processing, sharing, and storing of bioinformatics data, promoting international cooperation. Another growing field is the creation of specialised bioinformatics tools for systems biology and metagenomics, which allow for a better understanding of microbial ecosystems and how they interact with their surroundings.

18.22 Potential for Future Research

Future studies in bioinformatics and predictive modelling have a lot of potential, especially where these two domains converge. Research in aquatic ecosystems might concentrate on combining predictive models and bioinformatics tools, like eDNA-based biodiversity monitoring, to improve predictions of species distribution and conservation tactics. Another crucial area is examining the effects of microplastics and newly discovered pollutants on aquatic life, where bioinformatics and predictive modelling can work together to clarify the effects at the molecular and ecosystem levels. Furthermore, there are chances for innovation in the study of microbial communities' roles in biogeochemical cycling and how they react to climate change. By facilitating quicker simulations of intricate ecological processes, developments in quantum computing have the potential to completely transform predictive modelling. In the end, interdisciplinary approaches will probably be the focus of future research, utilising state-of-the-art technologies and teamwork to tackle the urgent issues confronting aquatic ecosystems in a changing global environment.

18.23 Summary

In aquatic ecosystems, predictive modelling is essential to comprehending and controlling the complexity of these settings. These models assist in forecasting important occurrences such as detrimental algal blooms, species distributions, and the effects of climate change by combining ecological processes and environmental data. Predictive modelling's future rests in improving accuracy through the use of AI, machine learning, and real-time data to handle nonlinear interactions and increase scalability. Coupled models that incorporate cloud computing and a variety of ecological processes will increase the models' applicability and accessibility across geographical boundaries.

Along with advancements like single-cell sequencing and eDNA analysis to track biodiversity, new trends in bioinformatics are concentrating on using AI and deep learning to analyse complex biological datasets. The creation of specialised tools for combining multi-omics data and investigating microbial ecosystems is part of the future of bioinformatics. With opportunities to investigate microplastics, microbial communities, and the wider effects of climate change, future research is probably going to concentrate on fusing bioinformatics and predictive modelling to gain a better understanding of aquatic ecosystems. These multidisciplinary developments have a great deal of promise to enhance conservation initiatives, manage ecosystems better, and solve the problems aquatic ecosystems face as a result of environmental change on a worldwide scale.

References

Arhonditsis GB, Brett MT (2004) Evaluation of the current state of mechanistic aquatic biogeochemical modeling. Mar Ecol Prog Ser 271:13–26

Arhonditsis GB, Perhar G, Zhang W, Massos E, Shi M, Das A (2008) Addressing equifinality and uncertainty in eutrophication models. Water Resour Res 44(1)

Arhonditsis GB, Stow CA, Rao YR, Perhar G (2014) What has been accomplished twenty years after the Oreskes et al. (1994) critique? Current state and future perspectives of environmental modeling in the Great Lakes. J Great Lakes Res 40:1–7

Battineni G, Sagaro GG, Chinatalapudi N, Amenta F (2020) Applications of machine learning predictive models in the chronic disease diagnosis. J Pers Med 10(2):21

Beng KC, Corlett RT (2020) Applications of environmental DNA (eDNA) in ecology and conservation: opportunities, challenges and prospects. Biodivers Conserv 29(7):2089–2121

Corrales X, Katsanevakis S, Coll M, Heymans JJ, Piroddi C, Ofir E, Gal G (2020) Advances and challenges in modelling the impacts of invasive alien species on aquatic ecosystems. Biol Invasions 22(3):907–934

Häder DP, Barnes PW (2019) Comparing the impacts of climate change on the responses and linkages between terrestrial and aquatic ecosystems. Sci Total Environ 682:239–246

Jamialahmadi H, Khalili-Tanha G, Nazari E, Rezaei-Tavirani M (2024) Artificial intelligence and bioinformatics: a journey from traditional techniques to smart approaches. Gastroenterol Hepatol Bed Bench 17(3):241

Jones MB, Schildhauer MP, Reichman OJ, Bowers S (2006) The new bioinformatics: integrating ecological data from the gene to the biosphere. Annu Rev Ecol Evol Syst 37(1):519–544

Jorgensen SE (2002) Recent trends in the development of ecological models applied on aquatic ecosystems. Sci World J 2(1):387–406

Kim SY, Imoto S, Miyano S (2003) Inferring gene networks from time series microarray data using dynamic Bayesian networks. Brief Bioinform 4(3):228–235

Landry M, Malouin JL, Oral M (1983) Model validation in operations research. Eur J Oper Res 14(3):207–220

Niinemets Ü, Kahru A, Mander Ü, Nõges P, Nõges T, Tuvikene A, Vasemägi A (2017) Interacting environmental and chemical stresses under global change in temperate aquatic ecosystems: stress responses, adaptation, and scaling. Reg Environ Chang 17:2061–2077

Nørgaard L, Olesen CR, Trøjelsgaard K, Pertoldi C, Nielsen JL, Taberlet P et al (2021) eDNA metabarcoding for biodiversity assessment, generalist predators as sampling assistants. Sci Rep 11(1):6820

Pais RJ (2022) Predictive modelling in clinical bioinformatics: key concepts for startups. Biotech 11(3):35

Peng Y, Wang Q, Fan M (2017) Identification of the key ecological factors influencing vegetation degradation in semi-arid agro-pastoral ecotone considering spatial scales. Acta Oecol 85:62–68

Pittman J, Huang E, Dressman H, Horng CF, Cheng SH, Tsou MH et al (2004) Integrated modeling of clinical and gene expression information for personalized prediction of disease outcomes. Proc Natl Acad Sci 101(22):8431–8436

Quazi S (2022) Artificial intelligence and machine learning in precision and genomic medicine. Med Oncol 39(8):120

Sharma A, Lysenko A, Jia S, Boroevich KA, Tsunoda T (2024) Advances in AI and machine learning for predictive medicine. J Hum Genet 69(10):487–497

Shavalieva G, Papadokonstantakis S, Peters G (2022) Prior knowledge for predictive modeling: the case of acute aquatic toxicity. J Chem Inf Model 62:4018–4031

Staartjes VE, Kernbach JM (2020) Importance of calibration assessment in machine learning–based predictive analytics. J Neurosurg Spine 32(6):985–987

Sun Z, Sokolova E, Brittain JE, Saltveit SJ, Rauch S, Meland S (2019) Impact of environmental factors on aquatic biodiversity in roadside stormwater ponds. Sci Rep 9(1):5994

Tucker A, Duplisea D (2012) Bioinformatics tools in predictive ecology: applications to fisheries. Philos Trans R Soc B Biol Sci 367(1586):279–290

Umma SB, Agbeja YE, Iyiola AO, Ogwu MC, Izah SC, Vazquez-Arenas JG, Feleke ST, Wei X (2025) Sustainable Bioeconomy Development in the Global South Volume II Bioeconomy Techniques Sustainable Extraction and Production of Raw Materials for Bio-based Products in the Global South Springer Nature Singapore Singapore 1–38

Yang CC (2022) Explainable artificial intelligence for predictive modeling in healthcare. J Healthc Inform Res 6(2):228–239

Ethical and Legal Considerations in Aquatic Genomics Research

19

Pradyut Malakar, Soumya Prasad Panda, Dhananjay Soren, Vikash Kumar, and Basanta Kumar Das

19.1 Introduction

Demand for fish and marine products for human and animal use is rising gradually. Due to this demand, aquaculture has grown quickly, sometimes with adverse ecological effects and financial losses from diseases and fish escape. Application of high-quality feed sources are now limiting issues for aquaculture's further growth (Naylor et al. 2000). Some of these issues may have technological solutions using genetic engineering (GE) techniques, such as transgenic fish, genetically modified (GM) plants used as edible vaccines or GM feed, and DNA vaccines. Since it may lower production costs per fish and improve feed conversion efficiency, the majority of transgenic fish development has been linked to growth promotion in commercially viable fish (FAO 2003; Melamed et al. 2002). Genetically modified plants are being developed with enhanced nutritional properties, including higher vitamin content, altered oil composition, and expression of chemicals that promote growth (Fisheries and Oceans Canada 2003; Ye et al. 2000). Furthermore, there are opportunities to immunize fish against a variety of diseases using edible vaccines, DNA vaccines, GE vaccines, and vaccines based on recombinant proteins (Hew and Fletcher 2002; Lorenzen et al. 1999a, b). Therefore, it is likely that GE will be used in commercial aquaculture in the coming years.

The knowledge on the possible ecological impacts of GM feed and DNA vaccine is less spread worldwide. Genetically modified feed may impact animal welfare, by introduction of DNA in feed or through DNA vaccines which may transfer genes to the host system (NRC 2002, 2004; Expert Panel of the Royal Society of Canada

P. Malakar · S. P. Panda · V. Kumar · B. K. Das (✉)
ICAR-Central Inland Fisheries Research Institute, Barrackpore, Kolkata, West Bengal, India

D. Soren
Department of Zoology, Ravenshaw University, Cuttack, Odisha, India

© The Author(s), under exclusive license to Springer Nature Singapore Pte Ltd. 2026

B. K. Das et al. (eds.), *Bioinformatics Tools for Predictive Ecology and Fisheries*, Sustainable Aquatic Food Systems, https://doi.org/10.1007/978-981-95-2573-7_19

2001). The primary advantages and risks of introducing genetically engineered organisms (GE) into aquaculture are outlined in this paper, along with areas that require further investigation. We concentrate on the regulatory and scientific facets of risk assessment. By offering a normative standard by which harm can be evaluated in ecological, socioeconomic, and ethical terms, we propose that the use of sustainability as a criterion may have an impact on the scope and framework of risk assessment. "Sustainable development is development that satisfies the demands of the present without compromising the ability of future generations to satisfy their own needs," according to the Brundtland Commission's definition (WCED 1987). The 1992 Rio Declaration made it clear that governments have an obligation to settle environmental disputes in a way that safeguards both human and nature's interests. Therefore, while using and releasing genetically modified organisms(GMOs), the Norwegian Gene Technology Act of 1993 mandates that their contribution to sustainability be evaluated. The Holmenkollen guidelines for sustainable aquaculture (1998) and the EU's 2002 message, "A strategy for the sustainable development of European aquaculture," offer suggestions for applying sustainability in aquaculture. To ensure an environmentally sound sector, the EU commission plan naturally incorporates a long-term view, a presumption of democratic decision-making processes, and a prospect for safety measures for both environment and human health. This is suggested to identify the high stake decisions which can influence a proactive policy, stakeholders such as biotechnology companies and research institutions, fish breeders and farmers, government offices, and non-governmental organisations (NGOs) like consumer and thus environmental activist groups must be involved from the conceptualization to the commercial stage.

Aquatic genomics research holds immense potential for biodiversity conservation, sustainable fisheries, and biotechnological advancements. However, it also raises critical ethical and legal challenges, including genetic privacy, bio-prospecting, and environmental impact. This manuscript explores the ethical dilemmas associated with genomic data collection, the legal frameworks governing genetic resources, and the implications of emerging technologies in marine and freshwater ecosystems which can encourage the researchers to ensure responsible research practices that balance scientific progress with ecological and societal well-being by addressing these concerns.

19.2 Advantages and Dangers of DNA Vaccinations

According to estimates, infectious illnesses claim the lives of 10% of farmed fish (Leong and Fryer 1993). Neither live, attenuated viruses nor vaccines containing recombinant viral antigens have proven effective in preventing various viral illnesses. By protecting fish from viral infections, DNA vaccines can enhance animal welfare by lowering the need for antibiotics, and stop the spread of disease. A bacterial plasmid containing the target gene, a polyadenylation/transcriptional termination sequence, and a potent viral promoter make up DNA vaccines. To start protein expression in vivo and eventually trigger the generation of antibodies, the plasmid is first isolated from bacteria, purified, dissolved in a saline solution, and then directly injected into the muscle as naked DNA. Muscle cells often express various cellular proteins following DNA injection. The plasmid expressing a glycoprotein of the infectious hematopoeitic necrosis virus (IHNV), which is transcriptionally controlled by a cytomegalovirus promoter (pCMV), is an example of a DNA vaccine which was injected into Atlantic salmon to produce resistance to IHNV (Traxler et al. 1999). Low cost, simplicity in production and quality control, heat stability, a single production procedure for all vaccines, and the potential to create multivalent vaccinations using plasmid cocktails are some of the benefits of DNA vaccines (Hew and Fletcher 2002; Liu 1998; Heppell and Davis 2000). The IM delivery method is the recommended approach for DNA vaccination of fish, according to Heppell and Davis (2000) and Evensen and Leong (2013). This can be done manually with an injection needle or, alternatively, with the use of tools such as compressed air (Dhar et al. 2014) (Fig. 19.1).

19.2.1 DNA Vaccine's Effectiveness

Numerous studies have been conducted on fish with encouraging outcomes, including protection against viral infections, after early trials on DNA vaccination in mammalian species (Romøren 2003). Furthermore, security is offered for a few days following vaccination by protection against heterologous viruses brought on by stimulating innate defense mechanisms (such as Mx protein) (LaPatra et al. 2000; Lorenzen et al. 2001). The effectiveness of the vaccines has been demonstrated by Corbeil et al. (2000) that a protective immune response may be elicited with a single dosage of nanogram amounts of IHNV DNA. The possible adverse effects of using DNA in vaccinations are currently less understood. It has been demonstrated that in certain vertebrates, the expression and synthesis of several proinflammatory chemicals are altered after DNA vaccination; elevated expression of IL-1 beta and IFN-gamma may rise (Lawson et al. 1997; Stacey et al. 1996). As with other polyanions, it has been proposed that the scavenger receptors may bind the majority of the plasmid DNA after injection before internalization. It was discovered that additional receptors bind DNA through the use of competitive ligand binding properties. CpG oligodeoxynucleotides have been demonstrated to bind to Toll-like receptor 9 (TLR9), which signs that plasmid DNA containing these patterns can also binds to TLR9 (Bauer et al. 2001; Zhao et al. 2004). Intracellular signaling is triggered by activating TLR9; this may trigger inflammatory processes apart from cytotoxic reactions (Kandimalla et al. 2003). Although fish have the TLR9 gene (Meijer et al. 2003), it is unclear if plasmid DNA could bind to Fish TLR9.

Fig. 19.1 DNA vaccination in aquaculture practices

19.2.2 Triggering Specific Immunological Reactions

Following DNA vaccination, the host cells may produce the target antigen. Antigen-presenting cells (APC) may endocytose these antigens and present them to T-cells through MHC class-II molecules, which may also causes activation of plasma B cells to make antibodies. As a result, administering DNA vaccines is a successful method of producing humoral immune responses that are specific to the encoded antigen or antigens (Russell et al. 1998; Lorenzen et al. 1999a; Fernandez-Alonso et al. 2001; Nusbaum et al. 2002; Verri et al. 2003). Thus to elicit a broader range of antibody responses and production, combination DNA vaccine consisting of several distinct plasmids encoding several disease antigens may be used. This might work well for immunization against viruses that change their antigens, including the infectious salmon anemia virus (ISAV) and the infectious pancreatic necrosis virus

(IPNV) (Lee et al. 1996; Wang and Nicholson 1996; Kibenge et al. 2001). When compared to standard vaccines, the potency (number of specific antibodies created) may be low despite the possibility of great vaccine efficacy (improved survival). Applying larger doses or boosting or co-administering plasmids expressing cytokines or co-stimulatory molecules can increase the effectiveness of DNA vaccines. On the other hand, opposite outcomes could also occur with adverse effect. Xiang and Ertl (1995) noted that the antibody response was inhibited when plasmids encoding IFN-α were co-injected.

19.2.3 Tolerance

According to Joosten et al. (1997), fish develop immune tolerance due to recurrent antigen stimulation. According to Dijkstra et al. (2001), long-term expression of recombinant DNA vaccine and its proteins could affects the host systems where chances of introducing new genes in to chromosome has yet to be investigated.

19.2.4 Additional Immunological Reactions After DNA Immunization

It has been proposed that the induction of antiviral systems, including Mx and IFN gamma/IL 1beta expression, is responsible for the early protection that follows DNA vaccination (LaPatra et al. 2000; Lorenzen et al. 2001). Mx protein was expressed in the kidney, liver, and muscle after DNA immunization using a plasmid encoding a viral glycoprotein (Kim et al. 2000; Boudinot et al. 1998). It is unknown if other immune-related molecules are also synthesized after DNA vaccination, besides the activation of Mx and IFN production and the cytokines crucial for the polarization of T cell response. In a recent experiment on Japanese flounders an injection of a DNA vaccine that encodes the viral G protein were given to prevent VHS (viral hemorrhagic septicemia). NK (natural killer cell), Kupffer cell receptor, macrophage inflammatory factor 1-alpha, CD20 receptor, CD8 alpha chain, and CD40 were among the numerous up-regulated genes after the DNA vaccine injection. These genes were demonstrated to be induced at levels greater than those for control plasmid injection 1–3 days after the injection (Byon et al. 2005). Unfortunately, the analysis of the up-regulation of genes by the empty plasmid DNA compared to fish injected with PBS was not the main focus. Microarray technology, which may be useful for evaluating global gene expression, was employed in this study to track gene expression. These microarrays, which include a huge number of genes, are also being created for salmon, carp, rainbow trout, pufferfish, and zebra fish. Microarray technology can potentially track the effects of DNA vaccination on fish over both short- and long-term periods.

19.2.5 Plasmid DNA Degradation and Tissue Distribution

Numerous investigations have demonstrated that in mice plasmid DNA was degraded quickly upon intramuscular injection (Dupuis et al. 2000). It was demonstrated that radiolabeled plasmid DNA gradually left the muscle (injection site) and began to disintegrate as early as 5 min after injection. At the injection site, the majority of the radioactivity was found in the interfibrillar gaps of a significant section of the muscle. A little portion of the injected DNA "escaped" to another compartment where it was shielded from endogenous DNAs, but the majority of it was quickly broken down by the host DNAase. Plasmid DNA was found in degradative compartments of cells that were thought to be antigen-presenting cells (APCs) that lived in between muscle cells (Dupuis et al. 2000). Three hours following the injection, DNA was also discovered in the rear limb's draining superficial and deep lymph nodes. The isolated cells from these nodes resembled APC. Only the muscle cells expressed the transgene, despite these cells containing plasmid DNA. Similarly, following an intravenous infusion, plasmid DNA was also discovered in the liver endothelial cells of mice (Kobayashi et al. 2001). Nevertheless, no discernible gene expression was seen and reported. This suggests that transgenes may not be expressed even while cells possess its own DNA construct. The location of receptors and cells involved in fluid pinocytosis and endocytosis significantly impacts the tissue distribution of both endogenous and exogenous DNA (Dalmo et al. 1997). The DNA delivery method also influences the plasmid DNA's tissue distribution in fish. Fluorescence-labeled plasmid DNA constructs were primarily detected in cells situated between myofibrils at the injection site following intramuscular injection. Other tissues like the kidney, gills, and heart also contained trace of levels (R.A. Dalmo, University of Tromsø, unpublished). After intraperitoneal injection, plasmid DNA was significantly stained in the peritoneal membrane, hematopoietic organs, and connective tissues covering internal organs (R.A. Dalmo, University of Tromsø, unpublished). Similarly, using a luciferase expression system, it was demonstrated that following intramuscular and intraperitoneal injections, the reporter gene was expressed by the rainbow trout liver and spleen (Romøren et al. 2003). The authors hypothesized that mono-nuclear cells were expressing the reporter, but regrettably, the study did not characterize the cells producing luciferase. Some of these particles may contain membrane-disrupting chemicals that facilitate the cytoplasmic delivery of plasmid DNA from endosomes and lysosomes. APC, such as monocyte-derived immature dendritic cells, may ingest these DNA-coated vehicles and express the transgenes by introducing DNA-binding vehicles (liposomes, cationic polymers) (Strobel et al. 2000) which could then cause these cells to produce more antigens.

19.2.6 DNA Vaccines Released into the Environment

Due to the relative absence of physical and physiological barriers in aquatic habitats, DNA can spread over large areas, distances, and phyla. Furthermore, it was not until recently discovered that DNA is far more resilient to degradation in ecosystems. Aquatic habitats are no exception (Tappeser et al. 2002; Heinemann and Roughan 2000). Experiments must be started to investigate on the uptake of DNA constructs in aquatic animals, such as marine fishes and mammals, through horizontal gene transfer (HGT), and DNA stability in aquatic settings.

19.2.7 Transparency Issues and Regulatory Challenges

An experimental veterinary DNA vaccine created by the CDC and produced by Aldevron (Fargo, ND) was expedited in 2003 by the U.S. Centers for Disease Control (CDC) and Prevention (Bouchie 2003). The California condor was the objective of the immunization, which was intended to prevent West Nile virus infection in this endangered animal. Novartis's IHNV DNA vaccine is being tested for commercial usage in aquaculture in Canada (M. Horne, Novartis, personal communication). Nevertheless, neither the risk assessments nor the data underlying these apps are openly accessible. In the EU, where GM pharmaceuticals lack the public transparency essential to other GMO applications, is an evident of lack of transparency. To safeguard the proprietary interests of the company that created the medical device, they flouted the need of transparency. However, the public's trust and confidence in future judgments pertaining to DNA vaccinations may be diminished by the proper evaluation information's of lack of transparency and accessibility. On the contrary, by allowing stakeholders to examine the data, transparency may contribute to a wise choice. Revealing the authorities' assessment will guarantee the process's integrity and offer a way to enhance the knowledge used to inform policy decisions (Pew Initiative on Food and Biotechnology 2003). Directives governing GMOs are in place in the United States and Europe, and the debate over how to regulate DNA vaccines is currently just being started (British Agriculture and Environment Biotechnology Commission 2002; Foss and Rogne 2003). The regulatory definitions of "genetic alteration" and "medicinal products," as well as which regulatory system should be involved, are at the heart of this debate. According to the U.S. Food and Drug Administration, genetic constructions given to animals are legally considered pharmacological substances in the United States. This is in line with European rules; the European Agency of Medicinal Products uses a centralized process to authorize medications based on GE. Nonetheless, national GMO agencies are involved in assessing the environmental dangers of the medications and the animals that get them as part of the European process. According to the Norwegian Directorate for Nature Management, for example, an animal that has received a DNA vaccination is deemed genetically changed as long as the additional DNA remains in the animal (Foss and Rogne 2003). This could have an impact on the requirement for traceability and labeling. According to the EU's 2002/0173(COD) regulation on genetically modified food and feed, items made from animals treated with GM medications are exempt from the requirement to be labeled as such. It is unclear, though, if this labeling exemption applies only to medications that have been approved as GMOs or if DNA-vaccinated goods are also covered. Therefore, policy implications arise from the current limited scientific understanding of the fate of DNA vaccines. It also highlights the need for research that focuses on the stability of the DNA construct, any immunological effects, and whether the DNA construct may integrate into the recipient organism's chromosomes.

19.3 Genetically Engineered (GE) Fish

Genetically engineered (transgenic) fish contain recombinant DNA sequences developed in laboratories. The International Council for Exploration of the Seas (ICES) classifies GMOs to include both transgenic fish and those with modified chromosome sets, excluding those improved via traditional breeding. Early transgenic fish were produced in rainbow trout (Maclean and Talwar 1984) and goldfish (Zhu et al. 1985). GE fish have been developed primarily for faster growth (Stokstad 2002) and, to a lesser extent, disease resistance. Targeted species include aquaculture fish (e.g., carp, tilapia, salmon, channel catfish) and model species (e.g., zebrafish, medaka, goldfish) (Alestrom 1996).

19.3.1 Techniques for Producing Transgenic Fish

Microinjection remains the most common method for producing transgenic fish, despite the development of alternatives like electroporation, retroviral vectors, lipofection, and particle bombardment (Muir and Hostetler 2001). Microinjection involves introducing genetic material into embryos using fine glass needles (Winkler and Schartl 2022) but is labor-intensive. Electroporation, which uses electrical pulses to create temporary pores in cell membranes, offers a more efficient alternative (FAO/WHO 2003; Muir and Hostetler 2001). Other methods, such as particle bombardment and retroviral vectors, have also been explored for gene transfer (Dunham 2003; FAO/WHO 2003). Lipofection, which uses lipophilic particles for DNA uptake and delivery, has shown promise in transfecting both in prokaryotic and eukaryotic cells (Volckaert and Ollevier 1998).

19.3.2 Current Status

Global seafood demand is expected to double by 2040 (Stokstad 2002), making the development of genetically superior broodstocks crucial (Hew and Fletcher 1997). No transgenic animals have been approved for human consumption, though Aqua Bounty's transgenic salmon is under regulatory review (Van Eenennaam and Olin 2006). The only commercially available transgenic fish is the GloFish, a genetically modified zebra danio, sold in the U.S. under pet trade since 2003 (Van Eenennaam 2005; Muir 2004). Regulation of aquatic GMOs is overseen by several U.S. agencies, including the Fish and Wildlife Service (FWS) and the National Marine

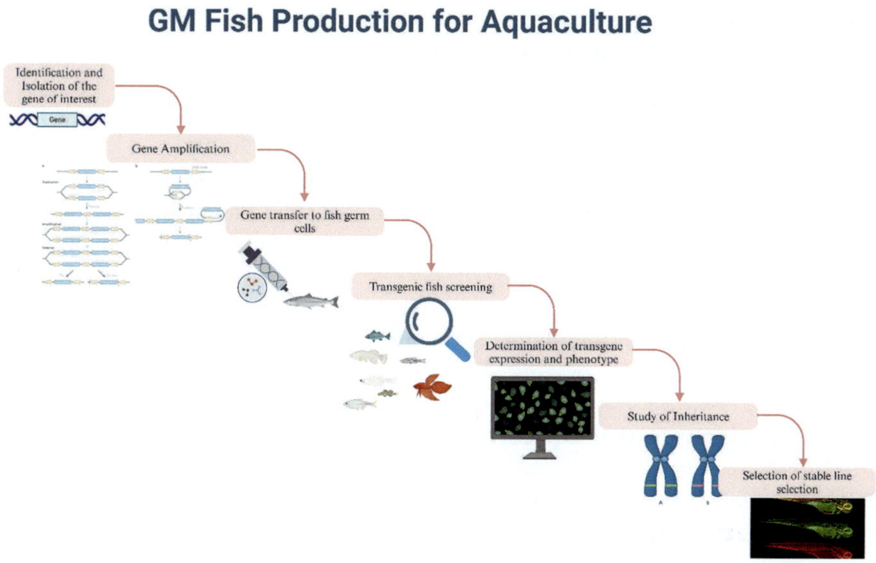

Fig. 19.2 Genetically modified fish production for aquaculture practices

Fisheries Service (NMFS), with international bodies like the OECD and involvement of FAO (Kapuscinski and Hallerman 1994) (Fig. 19.2).

19.3.3 Advantages and Disadvantages of GMOs in Aquaculture

The impact of genetic engineering is typically assessed in terms of long-term effects on future generations (Frewer and Shepherd 1995).

19.3.3.1 Benefits

Transgenic fish offer potential applications, including:

- Pollution detection using genetically modified zebrafish in Singapore (Pew Initiative on Food and Biotechnology 2003)
- Faster-growing fish to enhance productivity and reduce feed waste (Borgatti and Buck 2004)
- Disease-resistant fish to reduce the need for antibiotics and pesticides (Muir 2004)
- Improved cold resistance in fish for expanded cultivation (Borgatti and Buck 2004)
- The use of fluorescent proteins as biosensors for water contamination (Afanassiev et al. 2000)

19.3.3.2 Risks

The introduction of GE fish may lead to ecological risks, such as:

- Interbreeding with native populations, potentially increasing risk of genetic load or causing extinction (Knibb 1997; Maclean and Laight 2000)
- GM fish outcompeting wild species, disrupting ecosystems (Muir 2004)
- GM fish being perceived as alien species (Maclean and Laight 2000)
- Dependency on specialized broodstocks for fish farming (Beardmore and Porter 2003)
- Potential for new, harmful proteins or toxins, posing human health risks, including allergic reactions (Beardmore and Porter 2003)

19.3.4 Environmental Concerns on GE Fish Biotechnology

Environmental concerns surrounding transgenic fish are speculative due to limited commercialization. Some argue that more efficient animal feed production could reduce resource use and environmental impact (Cowan and Becker 2006). Primary concerns include competition with wild populations, transgene movement into the wild gene pool, and ecological disruptions resulting from altered prey and niche requirements (FDA 2006). A central issue is whether transgenic fish will interbreed with wild populations and whether GMOs will be selected for superior mate choice (Volckaert and Ollevier 1998).

Sustainability and biodiversity are critical issues in GM product introduction (Cockburn 2002; Arvanitoyannis and Exadactylos 2006). European opposition to GMOs is driven by concerns about their perceived risks and unnatural nature (Eurobarometer 2000; Hallman et al. 2002). Myhr and Dalmo (2005) highlighted the challenge of balancing environmental protection with economic benefits in aquaculture. The introduction of transgenic fish into wild populations could alter genetic makeup and reduce natural genetic variation (Galli 2002; Regal 1994). Muir and Howard (2001) noted that transgenes could spread even with high juvenile viability costs if they offer other fitness advantages.

Escaped GM fish could disrupt ecosystems by causing local extinction of conspecific populations, leading to cascading effects on other species (Muir and Howard 2002). Studies indicate that transgenic fish may have competitive advantages in certain environments (Devlin et al. 2004; Jönsson et al. 1998). Additionally, transgenic fish could invade new habitats and potentially alter ecosystem dynamics, especially if they spread transgenes through interbreeding (Muir 2004; Muir and Howard 2001).

Research by Hedrick (2001) suggests that transgenes with male-mating advantages could invade natural populations, reducing their viability and increasing the risk of extinction. Devlin et al. (1999) observed that growth hormone (GH)-transgenic salmon exhibited increased feeding motivation, which could give them a competitive edge in the wild. Additionally, the release of transgenic fish could lead to irreversible environmental changes, particularly if the GM fish become invasive species (Dunham 2003).

While the environmental risks of transgenic fish may be low in some contexts (Van Eenennaam 2005), the safest approach is land-based farming to avoid unintended ecological impacts, though the added costs may limit its feasibility (Stokstad 2002). Detecting GM fish in the environment would require DNA-based markers and adequate sampling protocols (FAO/WHO 2003).

19.3.5 Ethical Concerns in Genetic Engineering

Ethics, defined as a set of standards to regulate behavior (Godown 1987), is central to debates over animal genetic engineering (GE). Ethical concerns can be categorized as extrinsic and intrinsic (Kaiser 2005). Extrinsic concerns focus on the perceived risks and consequences of GE applications, which remain speculative (Frewer and Shepherd 1995). Intrinsic concerns relate to moral issues about the process itself, such as views that GE is unnatural or conflicts with religious beliefs (http://www.agbioworld.org/biotech-info/articles/agbio-articles/critical.html). However, as GE allows precise DNA changes, it is considered safer than natural mutations. If modifications are confined to within-species gene transfers, these concerns may be reduced (Macer 2003).

Key categories of concern include:

(i) Potential risks beyond observable traits of the organism
(ii) Unknown long-term human health and environmental impacts
(iii) Inadequate regulation and testing
(iv) Social and economic inequities
(v) Uncertain future benefits of transgenic technologies (Vollmer et al. 2007)

Public opinion on GE in agriculture is less favorable than for medical applications. Even if GE fish are FDA-approved, market success will likely depend on consumer and retailer reactions (Van Eenennaam 2005; FDA 2006). Ethical considerations heavily influence consumer behavior, with those perceiving risks to health and the environment is less likely to accept GM products (Han and Harrison 2007). In contrast, consumers who see benefits may be more willing to purchase GM foods (Han and Harrison 2007). This has made biotechnology a significant social issue, influencing public debates (Liakopoulos 2002).

Factors influencing consumer behavior toward GM foods include:

(i) Awareness and attitudes towards food technology
(ii) Perceived benefits
(iii) Sociodemographic profile
(iv) Social and moral consciousness
(v) Perceived food quality and trust (Arvanitoyannis and Exadactylos 2006).

A study by Han and Harrison (2007) found that food safety concerns were pivotal in consumer decisions regarding GM foods. Religious and ethical concerns were prominent factors influencing attitudes toward cloning and GM animals (Pew Initiative on Food and Biotechnology 2004, 2005, 2006).

In Europe, consumer acceptance of GM foods varies significantly by country. In the United Kingdom, 36.5% of consumers had no opinion on transgenic fish, while 34.8% rejected it, and 28.7% supported it (Penman et al. 1995). Consumer education is seen as key to overcoming initial skepticism (Moses 1999). However, some countries, like Portugal and Spain, remain largely unaware of transgenic issues.

Surveys indicate that 82% of Irish consumers express concern over GM food safety (Morris and Adley 2001). In contrast, only 28% of Americans support the genetic modification of animals, with over half expressing concerns about its risks (Hallman et al. 2002). The right to know how food is produced has led to support for labeling GM products, with 74% of EU respondents in favor (CEC 1997). Clear labeling policies would respect consumer autonomy and enhance informed choice (FAO 2001; Power 2003).

19.4 Advantages and Disadvantages of Distributing Feed Made from Plants

The availability of wild fish species constrains future aquaculture growth utilized as feed in fish farming (Naylor et al. 2000). The possibility of using plant-based feed materials in place of marine fish meal could lower aquaculture expenses and provide a sustainable substitute for fish feed. The current method, which involves feeding fish with marine fish meals rather than turning them into human food, has been considered to be inconsistent with sustainability (Hannesson 2003). Fish feed in Norwegian aquaculture currently consists of two distinct fish species: Chilean and Peruvian anchovies (*Engraulis ringens*) and Indian blue whiting (*Micromesistus poutassou*) (Tuominen and Esmark 2003). Despite not being suitable for direct human consumption, these fish species are essential to food webs that produce other commercial fish species. In other words, eliminating forage fish from an ecosystem may affect higher trophic levels, such as reducing the number of carnivorous fish that are crucial for human consumption (Naylor et al. 2000).

Additionally, because anchovies are exported to India and other countries as fish feed, they are not economically competitive in Chile as a local human food resource. This impacts the supply of fish for human consumption in Chile (Tuominen and Esmark 2003). Therefore, if GE plant-derived feed is introduced into Norwegian aquaculture, it could improve the sustainable use of marine resources, and will raise the value of commercial fisheries (by increasing the number of fish of higher trophic levels) and enhance the sustainability of fishing-dependent livelihoods in Chile.

19.4.1 Fish Consumption of Diet Generated from Plants

Aquaculture, including carnivorous fish species, is paying more and more attention to alternative feed ingredients derived from vegetable sources. However, there are issues with the usage of vegetables in fish feed. Most plants include anti-nutritional elements that hinder the use of feed and negatively impact fish well-being (Makkar

1993). For salmonids, feeding them soy or canola results in impaired digestion and stomach problems (Fisheries and Oceans Canada 2003; Jystad 2001). The difficulty of obtaining soy, maize, canola, or rape seed from the worldwide market that is assured to be free of genetically modified components is another issue with the usage of vegetable sources in fish feed. This brings up new questions and challenges like possibilities of transgene(s) be absorbed by intestinal cells or be susceptible to horizontal gene transfer (HGT) (i.e., nonsexual absorption) to bacteria in the fish intestine. Could the genetic alteration process in the GM plant cause unintended changes?

19.4.2 Unintentional Modifications to the GM Plant

Through genetic editing, the recipient plant acquires new characteristics. The GM plant may exhibit unexpected alterations due to the expression of a previously unidentified component. Therefore, characterization of the transgenic product or products must include immunological activity, sequence homology with proteins known to cause allergy or pathogenic consequences, and examination of degradation under circumstances of ingestion, processing, and storage (Aumaitre et al. 2002; Kuiper and Kleter 2003). Unintentional pleiotropic effects could result from the introduction of transgenes expressing novel proteins or other components, as variations in the amounts and activities of natural enzymes, nutrients, and metabolites (Haslberger 2003; Kuiper et al. 2002; the Expert Panel of the Royal Society of Canada 2001). The transgenic rice with soybean glycinin gene showed 50% increase in vitamin B6 and 20% increase in protein content because of increased glycine (Momma et al. 1999). The genetic insertion might also experience expression instability, such as the inserted transgene becoming silent (Qin et al. 2003). The integrity of the transgene or transgenes, the location of genomic integration, and any pleiotropic effects on the host must therefore be examined to characterize the GM plants (Kuiper et al. 2002; the Expert Panel of the Royal Society of Canada 2001). Most countries' regulatory bodies want this kind of characterization before commercializing a genetically modified plant. As a result, it has been recommended that five pertinent levels be examined to assess any possible ecological effects of administering GM feed. Genome, transcript, protein, metabolite, health impact, and environmental impact are under these levels (Aumaitre et al. 2002; Kuiper et al. 2002; Kuiper and Kleter 2003) (Table 19.1).

Furthermore, a GM food/feed safety evaluation must include feeding tests as a component. Only ten scientific (peer-reviewed) papers on animal nutrition have been published, and four of these were written by the group surrounding Pusztai and Ewen (Domingo 2000; Pryme and Lembcke 2003). There is only one reported feeding research involving fish (Sanden et al. 2004). This study was aimed to find out about the GM soy DNA fragments after feeding Atlantic salmon as a diet that included 17.2% GM soybean for 6 weeks. By using PCR technology on the content of the stomach, mid-intestine, and distal intestine, the authors were only able to identify a little DNA fragments (120 and 195 bp). No GM soy DNA fragments were

Table 19.1 Safety issues regarding GM feed

Issues	Study methods
Characteristics of the genetic modification	DNA analysis (Southern blotting, sequencing)
Change in expression of/by genetic modification, effect on toxicological properties, levels of nutrients and allergens	mRNA and protein analysis (Western blotting, proteomics), amino acid sequence comparisons to toxins and allergens, animal feeding studies
Potential alterations in expression levels	Compositional analysis, metabolomics
Formation of new metabolites	Residue analysis, metabolomics, bioassays
Integration and gene transfer between GM feed and bacteria in intestines	Screening for insertion of recombinant DNA, and study of gene transfer in microcosms

Modified after Kuiper et al. (2002)

found in the brain, muscle, or liver. Therefore, further research is required for this discovery, including extended feeding durations and examination of additional tissues, including blood, spleen, and gills, utilizing both PCR technology and other techniques like Southern blotting.

19.4.3 Intestinal Cells' Uptake of the Transgene or Genes

Foreign DNA may be absorbed by intestinal cells, moved via the circulatory system, and then dispersed to various tissues and organs. The effects of GM feed consumption on the gut microbial system are unknown (NRC 2002). The liver, muscle, blood, kidney, and spleen of pigs, as well as the liver, muscle, kidney, and spleen of chickens and cattle, have been found to contain gene fragments from non-GM plants (N200 bp) (Reuter and Aulrich 2003; Einspanier et al. 2001). Small pieces of GM soy DNA (180 bp) have been discovered in the small intestine of human ileostomies (with colostomy bag), and low-frequency gene transfer to the gut microbiota of gene fragments has been seen (Netherwood et al. 2004). However, in the big intestine of animals the data on GM DNA soya did not lasts. Low amounts of DNA may be taken up by specialized intestinal mucosal cells called M cells in the gastrointestinal tract of higher vertebrates. These cells can endocytose and transfer antigens to expert antigen-presenting cells (Nicoletti 2000). However, the existence of a comparable endocytosis mechanism in fish intestines is still unknown.

19.4.4 Gene Transfer to Bacteria Horizontally

To the best of our knowledge, no research has been done on the possibility of HGT in bacterial communities in fish that are given genetically modified feed or to which DNA constructs are administered as DNA vaccines or by gene therapy. A wide variety of microorganisms colonize fish intestines. These bacteria might have come into contact with the recombinant DNA constructions in fish digestive tracts.

The frequency of horizontal gene transfer from GM plants to soil microorganisms has frequently been regarded as having little effect or being negligible when field trial monitoring has included this monitoring. This is because, particularly when examining low-frequency or low-probability events, it is simple to prove "no effect" under extremely complicated and diverse circumstances. Therefore, it is argued in two recent papers (Heinemann and Traavik 2004; Nielsen and Townsend 2004) that more thorough monitoring may be the only practical method of detecting HGT, and that current methods for sampling and monitoring of HGT from GM plants to soil microorganisms are too insensitive. They also pointed out that, in comparison to the selection pressures operating on the host that receives the new DNA, the frequency of HGT is most likely very slightly significant. Antimicrobial drug resistance and the development of virulence features by bacteria are two examples of how this has been demonstrated (Ochman et al. 2000). Therefore, research on the stability of the DNA construct in fish must be started, and methods for sampling and tracking possible HGT from GM feed to bacteria in fish intestines must be developed.

19.5 Governance of Risks

Identification, evaluation, management, and communication of risks are all included in risk governance. Hazard identification, risk characterization, and risk estimation are other subcategories of risk assessment (Covello and Merkhofer 1993). While risk management and communication involve value judgement with respect to acceptability, trade-off criteria, and the adoption of coping mechanisms for uncertainty, risk assessment has been regarded as a purely "scientific" approach. However, knowledge from the fields of science, ethics, economics, society, and politics influences risk evaluations (Mayo and Hollander 1991; Wynne 1992). Value judgments about the outcomes that should be avoided and the methodology to be used for risk identification and characterization, for example, are part of risk assessment. Numerous articles have been written about the environmental problems linked with transgenic fish as well as recommendations for mitigating such risks (ABRAC 1995; Hallerman et al. 1999; Hallerman and Kapuscinski 1995; Hedrick 2001; Kapuscinski et al. 1999; Muir 2004; Muir and Howard 1999, 2002; NRC 2002; Pew Initiative on Food and Biotechnology 2003). One must be mindful that uncertainties will always be associated with extrapolating these predictions and studies to actual settings, whether based on mathematical or biological models. The problem becomes more complicated when considering the potential ecological effects of genetically modified feed and DNA vaccines because it is difficult or impossible to assign probability values to the positive and negative effects due to the systems' complexity and a lack of biological and ecological knowledge. In this regard, it is crucial to understand that when uncertainty grows, more basic forms of uncertainty, such as ignorance and indeterminacy, start to manifest (Wynne 1992; Stirling 1999).

Situations when the kind of hazard to measure is unknown, when entirely unforeseen and unprecedented hazards may arise, are represented by ignorance. This has happened with pesticides, dioxins, and bovine spongiform encephalopathy (BSE). These instances highlight the necessity of investigating the possible direct and indirect impacts that the dissemination of genetically modified feed and DNA vaccinations may have on aquatic creatures.

The inherent gap between constrained experimental circumstances and reality, where the outcomes of action can never be fully predicted, is referred to as indeterminacy or "huge uncertainty." The intricate functional linkages are reliant on innate biological variety and may be impacted by environmental and climatic factors. Thus, it is imperative that techniques for detecting and tracking the effects of GE introduction in aquaculture be started in order to map out the real ecological effects, uncover unanticipated negative effects, and follow up on the completed risk assessment. Using a monitoring system that incorporates adaptive management, repeated cycles of goal-setting, programmed design, execution, and evaluation would enhance risk assessment and facilitate learning (Kapuscinski et al. 1999).

The topic of whose expertise should be acknowledged and applied in policy processes is one that arises when comprehension gaps are acknowledged. Government agencies, policy makers, scientific reviewers and regulators will need to address this policy decision and decide which scientific methods will be considered sufficiently enough to handle it. Nonetheless, the safety framework that the regulatory bodies deem suitable has an impact on the amount and caliber of scientific knowledge needed to manage risk and uncertainty, as well as the techniques and models to be used. Applying sustainability criteria entails safety standards for the long-term health and environment, founded on the Precautionary Principle (Myhr and Traavik 2003a, b). In order to handle ambiguous and difficult topics, it is required to involve a wide range of scientific disciplines, including pathologists, ecologists, immunologists, molecular biologists, marine biologists, philosophers, and others (Funtowicz and Ravetz 1990). It is possible to view the numerous approaches and models that represent the many disciplines as complementary sources of data and models for research in this issue or system. More data will be produced and more replies will be provided to comprehend shifting conditions if the technique is more diverse. Such an approach will increase the expense of research, introduce redundancies into the process, and challenge the various disciplines involved to comprehend and show respect for alternative interpretations and perspectives. On the other hand, an integrative approach will enhance the body of knowledge and lead to the creation and consensus-building of frameworks for assessing risk and uncertainty related to genetically engineered organisms (GE) in aquaculture.

19.5.1 GE in Aquaculture Is Accepted by the Public

The use of GE in aquaculture has clear commercial benefits. Still, the dissemination of GM feed and DNA vaccinations also carries the risk of negative ecological repercussions and moral quandaries for society. The introduction of GE has generally sparked intense public and scientific discussion. Transgenic animals face the most resistance, but GE used for medical purposes is more widely accepted than GM plants (Anonymous 2003). In addition to technological and financial hazards, the public's perception of GE's dangers also takes into account on social, cultural, and religious values (NRC 2002; Power 2003; Sagar et al. 2000; Wynne 2001). As a result, there can be a number of differing concerns about how customers will react to GE in aquaculture. The consumer needs safe, dependable, high-quality food that has been produced sustainably, as demonstrated by the replies to the publication of Hiter et al. (2004). A global survey of salmon purchased in shops in March 2002 was published by Hiter et al. (2004). According to the authors, customers would be foolish to eat fish produced in Britain and other northern European nations more than six times a year since they are tainted with toxins that cause cancer. Sales of salmon were immediately lowered as a result of this disclosure, and the salmon farming sector is still in danger. Public acceptance of genetically engineered organisms (GE) in aquaculture may also be impacted by the increasing exploitation of the aquatic environment, the looming conflicts between different users of marine resources, and the significance of coastal areas in meeting human needs. The public also cares about the naturalness of GE use and animal welfare. For example, the idea that animals have inherent worth and, thus, rights at the ethical and legal levels in addition to their utilitarian or instrumental value is becoming more and more popular. Animal integrity violations in a genetically engineered environment can occur at three different levels: the individual animal level (phenotypic integrity), the species-specific level (genotypic integrity), and the animality level (self-maintenance, self-organization, independence, animal-type integrity) (Verhoog 2001). Luddism and anti-scientific sentiments have been blamed for the emotive and subjective nature of references to inherent values, such as species integrity and natural values (Davies 2001; Nuffield Council on Bioethics 1999; Tave 1993). Public sentiment, however, can serve as a barometer of moral principles and pose significant queries about "what we are entitled to do?". The question of whether GE use is a continuation of conventional breeding and medicine and is therefore natural or a sign of discontinuity is crucial in this regard.

The "natural" argument, which holds that one should be cautious about altering nature because "nature knows best," can also be interpreted as a means of expressing concerns about negative outcomes. Although ethical concerns are not specifically addressed by public policy in the majority of nations (Cayford 2004; Power 2003; NRC 2002), there are issues that need to be addressed because they may influence public acceptance of the commercial introduction of genetically engineered organisms in aquaculture.

19.5.2 Involving Stakeholders

We propose that expanding the scientific peer community may meaningfully spark discussion among the impacted parties, incorporating many points of view and permitting a broader assessment of risk. The general goals and objectives (consultation, actual decision-making, input to an ongoing process, satisfaction of free informed consent), the participants (professionals, specialists, laypeople, ethical committees, stakeholders), and their representative status (self, organisations, unions, future generations, or animals) comprise the "model" for any participatory decision-making process, as outlined by Rowe and Frewer (2000). For example, Kapuscinski et al. (2003) started a "Safety First Initiative," a collaboration between industry and society, with the goal of creating proactive, socially sound safety standards that can foresee and address safety concerns before GE is used commercially. This leads to a framework that can be used to analyze GE in aquaculture; it defines "decision stakes," involves stakeholders, and aims to influence policies that apply safety criteria, such as sustainability, by weighing advantages against risk and uncertainty.

19.6 Conclusion and Future Perspective

The biggest obstacle to applying GE in a sustainable setting is figuring out possible scientific strategies to safeguard the environment while also reaping financial rewards. Actively examining the suggested advantages and possible drawbacks is a part of this challenge, which calls for a research and policy agenda that promotes long-term, comprehensive thinking. Socioeconomic and ethical concerns are also influenced by sustainability; these factors emphasize the necessity of involving stakeholders in the risk governance of genetically engineered organisms in aquaculture. Such a thorough peer review could assist in determining the kinds of GE applications that might be acceptable at the concept stage, with evaluation of kinds of risk-associated in research should be narrated during the development stage, and with specifying the socioeconomic, ethical, and policy implications significant to the stakeholders during the commercialization stage. The future may see the approval of specific GM animals for large-scale production, with or without their inclusion in the human food supply. In such cases, post-market monitoring may be necessary to track potential environmental spread and food safety risks. Public acceptance of GM and GE foods in Europe, compared to the United States, will depend on shifting perceptions of health and environmental risks, which rely on clear communication from scientists, policymakers, industry, and the media. Greater public involvement in agricultural research planning may be essential.

Ethical concerns about aquatic GMOs often question humanity's right to alter the nature. While genetic modification accelerates and amplifies these changes, humans have been modifying plants, animals, and habitats for millennia. The first GM products on the market are likely to shape consumer attitudes significantly. Labeling can offer individuals some control over their choices, but it is ineffective without public understanding of its logical values and meaning. Effective communication strategies are crucial towards enhancing public awareness of this technology. The success of these advancements ultimately depends on active public engagement in discussions about and on their development.

References

ABRAC (Agricultural Biotechnology Research Advisory Committee–U.S. Department of Agriculture) (1995) Performance standards for safely conducting research with genetically modified Fish and shellfish. http://www.isb.vt.edu/perfstands/psmain.cfm

Afanassiev V, Sefton M, Anantachaiyong T, Barker G, Walmsley R, Wolfl S (2000) Application of yeast cells transformed with GFP expression constructs containing the RAD54 or RNR2 promoter as a test for the genotoxic potential of chemical substances. Mutat Res Genet Toxicol Environ Mutagen 464:297–308

Alestrom P (1996) Biotechnology seminar paper. Genetically modified fish in future aquaculture: technical, environmental and management considerations. International Service for National Agricultural Research

Anonymous H (2003) Europeans and biotechnology in 2002. European Commission. Eurobarometer, p 58

Arvanitoyannis IS, Exadactylos A (2006) Aquaculture biotechnology aiming at enhanced fish production for human consumption. In: Ray RC (ed) Microbial biotechnology in agriculture and aquaculture, vol II. Science Publishers, NeHampshire, pp 453–510

Aumaitre A, Aulrich K, Chesson A, Flachowsky G, Piva G (2002) New feeds from genetically modified plants: substantial equivalence, nutritional equivalence, digestibility, and safety for animals and the food chain. Livest Prod Sci 74:223–238

Bauer S, Kirschning CJ, Häcker H, Redecke V, Hausmann S, Akira S, Wagner H, Lipford GB (2001) HumanTLR9 confers responsiveness to bacterial DNA via species-specific CpG motif recognition. Proc Natl Acad Sci U S A 98:9237–9242

Beardmore JA and Porter JS (2003) Genetically Modified Organisms and Aquaculture. FAO Fisheries Circular No. 989 FIRI/C989(E). FAO, ISSN 0429–9329. ftp://ftp.fao.org/docrep/fao/006/y4955e/Y4955E00.pdf

Borgatti R, Buck EH (2004) Genetically engineered fish and seafood. CRS Report for Congress, vol 7, p 2004

Bouchie A (2003) DNAvaccine deployed for endangered condors. Nat Biotechnol 21:11

Boudinot P, Blanco M, deKinkelin P, Benmansour A (1998) Combined DNA immunisation with the glycoprotein gene of viral hemorrhagic septicaemia virus and infectious hematopoietic necrosis virus induces double-specific protective immunity and non-specific response in rainbow trout. Virology 12:297–306

British Agriculture and Environment Biotechnology Commission (2002) Animals and biotechnology. BAEBC, London

Byon JY, Ohira T, Hirono I, Aoki T (2005) Use of cDNA microarray to study immunity against viral hemorrhagic septicaemia (VHS) in Japanese flounder (*Paralichthys olivaceus*) following DNAvaccination. Fish Shellfish Immunol 18:135–147

Cayford J (2004) Breeding sanity into the GM food debate. Issues Sci Technol 20:49–56

CEC (1997) The Europeans and modern biotechnology. Eurobarometer 46.1. European Commission, Luxemburg

Cockburn A (2002) Assuring the safety of GM food. J Biotechnol 98:79–106

Corbeil S, LaPatra SE, Anderson ED, Kurath G (2000) Nanogram quantities of a DNAvaccine protect rainbow trout fry against heterologous strains of infectious hematopoietic necrosis virus. Vaccine 8:2817–2824

Covello VT, Merkhofer MW (1993) Risk assessment methods; approaches for assessing health and environmental risks. Plenum Press, New York

Cowan T, Becker GS (2006) Biotechnology in animal agriculture: status and current issues. CRS Report for Congress 27, 2006

Dalmo RA, Ingebrigtsen K, Bøgwald J (1997) Non-specific defence mechanisms in fish, with particular reference to the reticulo-endothelial system. J Fish Dis 20:241–273

Davies K (2001) What makes genetically modified organisms so distasteful? Trends Biotechnol 19:424–427

Devlin RH, Johnsson JI, Smailus DE, Biagi CA, Jonsson E, Bjornsson BT (1999) Increased ability to compete for food by growth hormone-transgenic coho salmon *Oncorhynchus kisutch* (Walbaum). Aquac Res 30:479–482

Devlin RH, D'Andrade M, Uh M, Biagi CA (2004) Population effects of growth hormone transgenic coho salmon depend on food availability and genotype by environment interactions. Proc Natl Acad Sci USA 101(25):9303–9308

Dhar AK, Manna SK, Allnutt FCT (2014) Viral vaccines for farmed fnfsh. Virus Dis 25(1):1–17

Dijkstra JM, Okamoto H, Ototake M, Nakanishi T (2001) Luciferase expression 2 years after DNA injection in glass catfish (*Kryptopterus bicirrhus*). Fish Shellfish Immunol 11:199–202

Domingo JL (2000) Health risks of GM foods: many opinions but few data. Science 288:1748–1749

Dunham RA (2003) Status of genetically modified (transgenic) fish: research and application. Working paper topic 2. Food and Agriculture Organization/World Health Organization expert hearings on biotechnology and food safety

Dupuis M, Denis-Mize K, Woo C, Goldbeck C, Selby MJ, Chen M, Otten GR, Ulmer JB, Donnelly JJ, Ott G, McDonald D (2000) Distribution of DNA vaccines determines their immunogenicity after intramuscular injection in mice. J Immunol 165:2850–2858

Einspanier R, Klotz A, Kraft J, Aulrich K, Poser R, Schwägle F, Jahreis B, Flachowsky G (2001) The fate of forage plant, DNA in farm animals, a collaborative case-study investigating cattle and chicken fed recombinant plant material. Eur Food Res Technol 212:129–134

Eurobarometer (2000) Europeans and biotechnology. Eurobarometer 52.1. Luxemburg, Office for Official Publications of the European Communities. Brussels, European Commission, Research DG, 2000

Evensen Ø, Leong JAC (2013) DNA vaccines against viral diseases of farmed fsh. Fish Shellfsh Immunol 35(6):1751–1758

Expert Panel of the Royal Society of Canada (2001) Elements of precaution; recommendations for the regulation of food biotechnology in Canada http://www.rsc.ca

FAO (2001) Genetically modified organisms, consumers, food safety and the environment. Food and Agriculture Organization of the United Nations, Rome

FAO (Food and Agriculture Organization of the United Nations) (2003) Genetically modified organisms and aquaculture. FAO Fish Circ, vol 989. FAO, Rome

FAO/WHO (2003) Safety assessment of foods derived from genetically modified animals, including fish. Rome, 17–21 November 2003

Fernandez-Alonso M, Rocha A, Coll JM (2001) DNA vaccination by immersion and ultrasound to trout viral haemorrhagic septicaemia virus. Vaccine 19:3067–3075

Fisheries and Oceans Canada (2003) Biotechnology to help develop better salmon feeds http://www.dfo-mpo.gc.ca/science.aquaculture/biotech/fact3_e.htm

Food and Drug Administration (FDA) (2006) Questions and answers about transgenic fish. Center for Veterinary Medicine

Foss GS, Rogne S (2003) Gene medification or genetic modification? The devil is in the details. Nat Biotechnol 21:1280–1281

Frewer LJ, Shepherd R (1995) Ethical concerns and risk perceptions associated with different applications of genetic engineering: interrelationships with the perceived need for regulation of the technology. Agric Hum Values 12:50–57

Funtowicz SO, Ravetz JR (1990) Uncertainty and quality in science for policy. Kluwer, Dordrecht

Galli L (2002) Genetic modification in aquaculture. A review of potential benefits and risks. Bureau of Rural Sciences, Canberra

Gene Technology Act (1993) The act relating to the production and use of genetically modified organism. Act no. 38 of 2 April 1993, Oslo

Godown RD (1987) The science of biotechnology. In: Batra LR, Klassen W (eds) Public perceptions of biotechnology. Agricultural Research Institute, Maryland

Hallerman E, King D, Kapuscinski A (1999) A decision support software for safely conducting research with genetically modified fish and shellfish. Aquaculture 73:309–318

Hallman WK, Adelaja AO, Schilling BJ, Lang JT (2002) Public perceptions of genetically modified foods. Americans know not what they eat. Food Policy Institute

Hallerman E, Kapuscinski A (1995) Incorporating risk assessment and risk management into public policies on genetically modified finfish and shellfish. Aquac 173 309–318
Han J-H, Harrison RW (2007) Factors influencing urban consumers' acceptance of genetically modified foods. Rev Agric Econ 29(4):700–719
Hannesson R (2003) Aquaculture and fisheries. Mar Policy 27:169–178
Haslberger AG (2003) Codex guidelines for GM foods include the analysis of unintended effects. Nat Biotechnol 21:739–741
Hedrick PW (2001) Invasion of transgenes from salmon or other genetically modified organisms into natural populations. Can J Fish Aquat Sci 58(5):841–844
Heinemann JA, Roughan PD (2000) New hypotheses on the material nature of horizontally transferred genes. Ann NY Acad Sci 906:169–186
Heinemann JA, Traavik T (2004) Problems in monitoring horizontal gene transfer in field trials of transgenic plants. Nat Biotechnol 22:1105–1109
Heppell J, Davis HL (2000) Application of DNA vaccine technology to aquaculture. Adv Drug Rev 43:29–43
Hew CL, Fletcher GL (1997) Transgenic fish for aquaculture. Chem Ind 21:311–314
Hew CL, Fletcher GL (2002) The role of aquatic biotechnology in aquaculture. Aquaculture 197:191–204
Hiter RA, Foran JA, Carpenter DO, Hamilton MC, Knuth BA, Schawn SJ (2004) Global assessment of organic contamination in farmed salmon. Science 303:226–229
Holmenkollen guidelines for sustainable aquaculture (1998) Proceedings of the second international symposium on sustainable aquaculture. A.A. Balkema, Rotterdam/Brook field, Oslo
Jönsson E, Johnsson JI, Bjornsson BT (1998) Growth hormone increases aggressive behavior in juvenile rainbow trout. Horm Behav 33:9–15
Joosten PHM, Engelsma MY, van der Zee MD, Rombout JHWM (1997) Induction of oral tolerance in carp (*Cyprinus carpio L.*) after feeding protein antigens. Vet Immunol Immunopathol 60:187–196
Jystad PT (2001) An Intrafish.com industry report on fishmeal and oil or vegetable alternatives: will high volume prediction spoil premium fish products? Intrafish, Bodø
Kaiser M (2005) Assessing ethics and animal welfare in animal biotechnology for farm production. Rev Sci Tech Off Int Epiz 24(1):75–87
Kandimalla ER, Zhu F-G, Bhagat L, Yu D, Agrawal S (2003) Toll-like receptor 9: modulation of recognition and cytokine induction by novel synthetic CpG DNAs. Biochem Soc Trans 31:654–658
Kapuscinski AR, Hallerman EM (1994) Benefits, environmental risks, social concerns, and policy implications of biotechnology in aquaculture. Contract report prepared for the Office of Technology Assessment, U.S. Congress. National Technical Information Service, Washington
Kapuscinski AR, Nega T, Hallermann EM (1999) Adaptive biosafety assessment and management regimes for aquatic genetically modified organisms in the environment. In: Pullin RSV, Bartley DM, Kooimans J (eds) Towards policies for conservation and sustainable use of aquatic genetic resources. Conference proceedings, vol 59. ICLARM (International Center for Living Aquatic Resources Management), Manila, pp 225–251
Kapuscinski AR, Goodman RM, Hann SD, Jacobs LR, Pullins EE, Johnson CS, Kinsey JD, Krall RL, La Vin ˜a AGM, Mellon MG, Ruttan VW (2003) Making safety first a reality for biotechnology products. Nat Biotechnol 21:599–601
Kibenge FSB, Kibenge MJT, McKenna PK, Stothard P, Marshall R, Cusack RR, McGeachy S (2001) Antigenic variation among isolates of infectious salmon anaemia virus correlates with genetic variation of the viral haemagglutinin gene. J Gen Virol 82:2869–2879
Kim CH, Johnson MC, Drennan JD, Simon BE, Thomann E, Leong JC (2000) DNA vaccination encoding viral glycoproteins induce non-specific immunity and Mx protein synthesis in fish. J Virol 74:7048–7054
Knibb W (1997) Risk from genetically engineered and modified marine fish. Transgenic Res 6:59–67

Kobayashi N, Kuramoto T, Yamaoka K, Hashida M, Takakura Y (2001) Hepatic uptake and gene expression mechanisms following intravenous administration of plasmid DNA by conventional and hydrodynamic-based procedures. J Pharma Col Exp Ther 297:853–860

Kuiper HA, Kleter GA (2003) The scientific basis for risk assessment of regulation of genetically modified foods. Trends Food Sci Technol 14:277–293

Kuiper HA, Noteborn HPJM, Kok EJ, Kleter GA (2002) Safety aspects of novel food. Food Res Int 35:267–270

LaPatra S, Corbeil S, Jones GR, Shewmaker WD, Kurath G (2000) The dose-dependent effect on protection and humoral responses to a DNA vaccine against infectious hematopoietic necrosis virus (IHN) virus in subyearling rainbow trout. J Aquat Anim Health 12:181–188

Lawson CM, Yeow WS, Lee CM, Beilharz MW (1997) In-vivo expression of an interferon-alpha gene by intramuscular injection of naked DNA. J Interferon Cytokine Res 17:255–261

Lee MK, Blake SL, Singer JT, Nicholson BL (1996) Genomic variation of aquatic birnaviruses analyzed with restriction fragment length polymorphism. Appl Environ Microbiol 62:2513–2520

Leong JC, Fryer JL (1993) Viral vaccines for aquaculture. Annu Rev Fish Dis 3:225–240

Liakopoulos M (2002) Pandora's box or panacea? Using metaphors to create the public representations of biotechnology. Public Underst Sci 11:5–32

Liu MA (1998) Vaccine developments. Nat Med 4:515

Lorenzen N, Olesen NJ, Koch C (1999a) Immunity to VHS virus in rainbow trout. Aquaculture 172:41–61

Lorenzen N, Lorenzen E, Einer-Jensen K, Heppell J, Davis HL (1999b) Genetic vaccination of rainbow trout against viral haemorrhagic septicaemia virus: small amounts of plasmid DNA protect against a heterologous serotype. Virus Res 63:19–25

Lorenzen N, Lorenzen E, Einer-Jensen K, LaPatra S (2001) Immunity induced shortly after DNA vaccination of rainbow trout against rhabdoviruses protects against heterologous virus but not against bacterial pathogens. Dev Comp Immunol 26:173–179

Macer D (2003) Ethical, legal and social issues of genetically modified disease vectors in public health. TDR/STR/SEB/ST/03.1. World Health Organisation, Geneva

Maclean N, Talwar S (1984) Injection of cloned genes with rainbow trout eggs. J Embryol Exp Morphol 82:187

Maclean N and Laight RJ (2000) Transgenic fish: an evaluation of benefits and risks. Fish 1:146–172

Makkar HPS (1993) Anti-nutritional factors in foods for livestock. In: Gill M, Owen E, Pollot GE, Lawrence TLJ (eds) Animal production in developing countries, vol 16. BSAP Occas Publ, pp 69–85

Mayo DG, Hollander RD (1991) Acceptable evidence: science and values in risk management. Oxford University Press, Oxford

Meijer AH, Gabby Krens SF, Medina Rodriquez IA, He S, Bitter W, Snaar-Jagalska BE, Spaink HP (2003) Expression analysis of the toll-like receptor and TIR domain adaptor families of zebrafish. Mol Immunol 40:773–783

Melamed P, Gong Z, Fletcher G, Hew CL (2002) The potential impact of modern biotechnology on fish aquaculture. Aquaculture 204:255–269

Momma K, Hashimoto W, Ozawa S, Kawai S, Katsube T, Takaiwa F, Kito M, Utsumi S, Murata K (1999) Quality and safety evaluation of genetically engineered rice with soybean glycinin: analyses of the grain composition and digestibility of glycinin in transgenic rice. Biosci Biotechnol Biochem 63:314–318

Morris SH, Adley CC (2001) Irish public perceptions and attitudes to modern biotechnology: an overview with a focus on GM foods. Trends Biotechnol 19(2):43–48

Moses V (1999) Biotechnology products and European consumers. Biotechnol Adv 17:647–678

Muir W (2004) The threats and benefits of GM fish. EMBO Rep 5(7):654–659

Muir WM, Hostetler HA (2001) Transgenic fish: production, testing, and risk assessment. Biotechnol Anim Husb 5:261–281

Muir WM, Howard RD (1999) Possible ecological risks of transgenic organisms release when transgenes affect mating success, sexual selection and the Trojan gene hypothesis. Proc Natl Acad Sci USA 96:13853–13856

Muir WM, Howard RD (2001) Fitness components and ecological risk of transgenic release: a model using Japanese medaka (*Oryzias latipes*). Am Nat 158:1–16

Muir WM, Howard RD (2002) Assessment of possible ecological risks and hazards of transgenic fish with implications for other sexually reproductive organisms. Transgenic Res 11(2):101–114

Myhr AI, Dalmo RA (2005) Introduction of genetic engineering in aquaculture: ecological and ethical implications for science and governance. Aquaculture 250:542–554

Myhr AI, Traavik T (2003a) Genetically modified (GM) crops: precautionary science and conflicts of interests. J Agr Environ Ethic 16:227–247

Myhr AI, Traavik T (2003b) Sustainable development and Norwegian genetic engineering regulations: applications, impacts, and challenges. J Agric Environ Ethics 16:317–335

Naylor RL, Goldburg RJ, Primavera JH, Kautsky N, Beveridge MC, Clay J et al (2000) Effect of aquaculture on world fish supplies. Nature 405(6790):1017–1024

Netherwood T, Martin-Orue SM, O'Donnell AG, Gockling S, Graham J, Mathers JC, Gilbert HJ (2004) Assessing the survival of transgenic plant DNA in the human gastrointestinal tract. Nat Biotechnol 22:204–209

Nielsen KM, Townsend JP (2004) Monitoring and modelling horizontal gene transfer. Nat Biotechnol 22:110–1114

Nicoletti C (2000) Unsolved mysteries of intestinal M cells. Gut 47(5):735–739

NRC (National Research Council) (2002) Animal Biotechnology: Science-Based Concerns. National Academy Press, Washington, DC

NRC (National Research Council) (2004) Biological Confinement of Genetically Engineered Organisms. National Academy Press, Washington, DC

Nuffield Council on Bioethics (1999) Genetically modified crops: the ethical and social issues. Nuffield Council on Bioethics, London. http://www.nuffield.org/bioethics

Nusbaum KE, Smith BF, DeInnocentes P, Bird RC (2002) Protective immunity induced by DNA vaccination of channel catfish with early and late transcripts of the channel catfish herpesvirus (IHV-1). Vet Immunol Immunopathol 84(3–4):151–168

Ochman H, Lawrence JG, Groisman EA (2000) Lateral gene transfer and the nature of bacterial innovation. Nature 405(6784):299–304

Penman DJ, Woodwark M, McAndrew BJ (1995) Genetically modified fish populations. In: Environmental Impacts of Aquatic Biotechnolqy. OECD documents 22–27

Pew Initiative on Food and Biotechnology (2003) Future fish: issues in science and regulation of transgenic fish. Pew Initiative on Food and Biotechnology, Washington. http://pewagbiotech.org/research/fish/fish.pdf

Pew Initiative on Food and Biotechnology (2004) Americans' Opinions About Genetically Modified Foods Remain Divided, But Majority Want a Strong Regulatory System. Pew Initiative on Food and Biotechnology, Washington http://pewagbiotech.org/newsroom/releases/112404.php3

Pew Initiative on Food and Biotechnology (2005) Public Sentiments About Genetically Modified Foods. Pew Initiative on Food and Biotechnology, Washington http://pewagbiotech.org/research/2005update/

Pew Initiative on Food and Biotechnology (2006) Pew initiative finds public opinion about genetically modified foods remains 'up for grabs' ten years after introduction of ag biotech. Pew Initiative on Food and Biotechnology, Washington

Power M (2003) Lots of fish in the sea: salmon aquaculture, genomics and ethics, Electronic workshop papers series, vol 84. W. Maurice Young Centre for Applied Ethics, UBC. http://www.ethics.ubc.ca/workingpapers/deg/deg004.pdf

Pryme IF, Lembcke R (2003) In vivo studies on possible health consequences of genetically modified food and feed—with particular regard to ingredients consisting of genetically modified plant materials. Nutr Health 17(1):1–8

Qin H, Dong Y, von Arnim AG (2003) Epigenetic interactions between Arabidopsis transgenes: characterization in light of transgene integration sites. Plant Mol Biol 52:217–231

Regal PJ (1994) Scientific principles for ecologically based risk assessment of transgenic organisms. Mol Ecol 3(1):5–13

Reuter T, Aulrich K (2003) Investigation on genetically modified maize (BT-maize) in pig nutrition: fate of feed-ingested foreign DNA in pig bodies. Eur Food Res Technol 216:185–192.2Romøren K, (2003) Liposome and chitosan-based formulations used for gene delivery to fish: -in vivo and in vitro studies of possibilities and limitations. Ph.D. dissertation, Faculty of Mathematics and Natural Sciences, Dep Pharm School of Pharmacy, University of Oslo, Norway, ISSN 1501–7710

Romøren K, Thu BJ, Evensen J (2003) Distribution of luciferase expression after intraperitoneal administration of lipoplex to rainbow trout (Oncorhynchus mykiss) fry. J Liposom Res 13:44–55

Rowe G, Frewer LJ (2000) Public participation methods: a framework for evaluation. Sci Technol Human Values 25:3–21

Russell PH, Kanellos T, Sylvester ID, Chang KC, Howard CR (1998) Nucleic acid immunisation with a reporter gene results in antibody production in goldfish (Carassius auratus). Fish Shellfish Immunol 8:121–128

Sanden M, Bruce IJ, Rahman MA, Hemre G (2004) The fate of transgenic sequences present in genetically modified plant products in fish feed; investigating the survival of GM soybean DNA fragments during field trials in Atlantic salmon, Salmo salar L. Aquac 237:391–405

Sagar A, Daemmrich A, Ashiya M (2000) The tragedy of the commoners: biotechnology and its public. Nat Biotechnol 18:2–4

Stacey KJ, Sweet MJ, Hume DA (1996) Macrophages ingest and are activated by bacterial DNA. J Immunol 157:2116–2122

Stirling A (1999) On science and precaution on risk management of technological risk. An ESTO project report prepared for the European Commission- JRC Institute for Prospective Technological Studies, Seville http://www.sussex.ac.uk/Units/gec/gecko/r9e-prc-.htm

Strobel I, Berchtold S, Gftze A, Schulze U, Schuler G, Steinkasserer A (2000) Human dendritic cells transfected with either RNA or DNA encoding influenza matrix protein M1 differ in their ability to stimulate cytotoxic T lymphocytes. Gene Ther 7:2028–2035

Tappeser B, Jager M, Eckelkamp C (2002) Survival, persistence, transfer: the fate of genetically modified microorganisms and recombinant DNA in different environments. In: Letourneau, D.K., Burrows, B.E. (Eds.), Genetically Engineered Organisms; Assessing Environmental and Human Health Effects. CRC Press, Boca Raton, FL, pp. 223–251

Tuominen TR, Esmark M (2003) Food for thought; the use of marine resources in fish feed. Report 02/03. WWF, Norway http://www.panda.org/downloads/marine/foodforthoug.pdf

Stokstad E (2002) Engineered fish: friend or foe of the environment? Science 297:1797

Tave D (1993) Genetics for Fish Hatchery Managers, 2nd ed., Van Nostrand Reinhold, New York

Traxler GS, Anderson E, LaPatra SE, Richard J, Shewmaker B, Kurath G (1999) Naked DNAvaccination of Atlantic salmon, Salmon salar against IHNV. Dis Aquat Org 38:183–190

Van Eenennaam AL (2005) Genetic engineering fact sheet 8: genetic engineering and fish. In: University of California, Division of Agriculture and Natural resources, Agricultural Biotechnology in California Series, Publication

Van Eenennaam AL, Olin PG (2006) Careful risk assessment needed to evaluate transgenic fish. Calif Agric 60(3):126

Verri T, Ingrosso L, Chiloiro R, Danieli A, Zonno V, Alifano P, Romano N, Scapigliatti G, Vilella S, Storelli C (2003) Assessment of DNA vaccine potential for gilthead sea bream (Sparus aurata) by intramuscular injection of a reporter gene. Fish Shellfish Immunol 15:283–295

Verhoog H (2001) The intrinsic value of animals: its implementation in governmental regulations in the Netherlands and its implications for plants. In: Heaf, D., Wirz, J. (Eds.), Intrinsic Value and Integrity of Plants in the Context of GE. Proceedings of Ifgene Workshop, Dornach, Switzerland 15–18

Volckaert F, Ollevier F (1998) Transgenic fish. The future of fish with novel genes. Gen Aquac Africa 326:33–58

Vollmer E, Creamer N, Mueller P (2007) Sustainable Agriculture and Transgenic Crops. In: The Second National Conference on Facilitating Sustainable Agriculture Education, Cornell University, July 11–14

Wang WS, Nicholson BL (1996) Antigenic and in vitro host range variety in populations of aquatic birnaviruses demonstrated by selection for neutralization resistance to monoclonal antibodies. Zool Stud 35:93–104

WCED (World Commission on Environment and Development) (1987) Our Common Future. Oxford University Press, Oxford, UK

Winkler C, Schartl M (2022) Gene transfer in laboratory fish: model organisms for the analysis of gene function. In: Transgenic animals. CRC Press, pp 387–395

Wynne B (1992) Uncertainty and environmental learning: reconceiving science and policy in the preventive paradigm. Glob Environ Change 2:111–127

Wynne B (2001) Creating public alienation: expert cultures of risk and ethics of GMOs. Sci Cult 10:445–481

Xiang ZQ, Ertl H (1995) Manipulation of the immune response to a plasmid-encoded viral antigen by coinoculation with plasmids expressing cytokines. Immunity 2:129–135

Ye X, Al-Babili S, Kløti A, Zhang J, Lucca P, Beyer P, Potrykus I (2000) Engineering the provitamin A (beta-carotene) biosynthetic pathway into (carotenoid-free) rice endosperm. Sci 287:303–305

Zhao H, Hemmi H, Akira S, Cheng SH, Scheule RK, Yew NS (2004) Contribution of toll-like receptor 9 signalling to the acute inflammatory response to nonviral vectors. Molec Ther 9:241–248

Zhu Z, He L, Chen S (1985) Novel gene transfer into the fertilized eggs of gold fish (Carassius auratus L. 1758). J Appl Ichthyol 1(1):31–34

Future Trends in Bioinformatics for Aquatic Sciences

20

Hirak Jyoti Chakraborty, Sejyoti Chakraborty, and Basanta Kumar Das

20.1 Introduction

Bioinformatics has revolutionized aquatic sciences by enabling the comprehensive analysis of biological data to better understand ecosystems, biodiversity, and environmental changes. With advancements in sequencing technologies, computational tools, and big data analytics, bioinformatics provides critical insights into the genetic, proteomic, and metagenomic composition of marine and freshwater environments. It plays a key role in monitoring aquatic biodiversity, tracking species' evolution, and assessing the impact of environmental stressors such as pollution, climate change, and habitat destruction (Das et al. 2024).

The integration of bioinformatics in aquatic research allows for efficient management and interpretation of vast datasets, supporting conservation efforts and sustainable resource management. Future advancements in high-throughput sequencing, artificial intelligence, and data-sharing frameworks are expected to further enhance our ability to analyze and predict ecosystem dynamics. By leveraging these innovations, researchers can develop data-driven policies aimed at preserving aquatic life and maintaining ecological balance (Behera et al. 2020) (Fig. 20.1).

Bioinformatics has transformed the study of marine and freshwater ecosystems by integrating computational biology with vast genomic, transcriptomic, proteomic, and metagenomic datasets. This interdisciplinary approach enables a deeper understanding of aquatic biodiversity, species interactions, and ecosystem functions while

H. J. Chakraborty · B. K. Das (✉)
ICAR-Central Inland Fisheries Research Institute, Barrackpore, Kolkata, India

S. Chakraborty
BioCompin, 43/1, Nawab Abdul Latif Street, Belgharia, Kolkata, India

© The Author(s), under exclusive license to Springer Nature Singapore Pte Ltd. 2026

B. K. Das et al. (eds.), *Bioinformatics Tools for Predictive Ecology and Fisheries*, Sustainable Aquatic Food Systems, https://doi.org/10.1007/978-981-95-2573-7_20

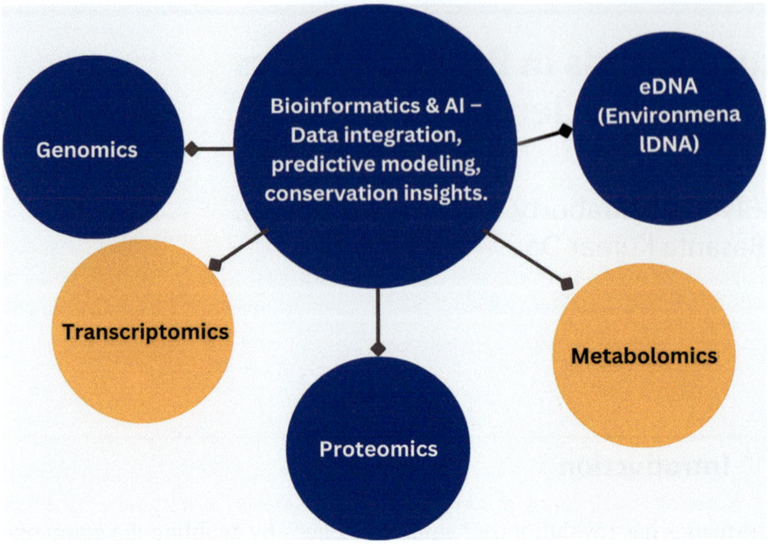

Fig. 20.1 Circular diagram to understand different segments representing various omics layers

addressing critical challenges such as habitat degradation, climate change, pollution, and invasive species proliferation.

Applications of bioinformatics in aquatic research range from studying the genetic diversity of aquatic organisms to deciphering the molecular mechanisms underlying their responses to environmental stressors (Behera et al. 2020). Metagenomic studies, for example, have revolutionized microbial ecology by uncovering the hidden diversity of microorganisms in oceans, rivers, and lakes. Meanwhile, transcriptomic and proteomic analyses shed light on how aquatic organisms adapt to changing environmental conditions.

Furthermore, bioinformatics plays a crucial role in fisheries management, aquaculture, and conservation biology by aiding in species authentication, population genetics studies, and disease diagnostics. The ability to analyze and interpret large-scale datasets is essential for making informed decisions that ensure sustainable fisheries and biodiversity conservation.

Looking ahead, the integration of machine learning algorithms with ecological and environmental datasets will enhance predictive modeling of ecosystem changes, helping researchers anticipate the impact of climate change and human activities on aquatic life. As bioinformatics continues to evolve, it will play an increasingly vital role in shaping policies and conservation strategies, ensuring the health and stability of marine and freshwater ecosystems.

20.2 Omics Integration

Advancements in genomics, transcriptomics, and proteomics have significantly enhanced our understanding of aquatic species, enabling deeper insights into their genetic diversity, physiological adaptations, and responses to environmental stressors. The integration of multi-omics approaches combining genomic, transcriptomic, proteomic, and metabolomic data has become a powerful strategy for studying complex biological systems in fish, invertebrates, and microorganisms. However, challenges such as the vast genetic diversity of aquatic species, limited reference genomes, and the influence of environmental factors on molecular data make multi-omics integration particularly complex. The emergence of long-read sequencing technologies has provided a breakthrough in resolving highly repetitive and complex genomes of marine organisms, allowing for more accurate assembly and functional annotation. Additionally, single-cell sequencing has opened new avenues for studying cellular heterogeneity, enabling researchers to investigate cell-specific functions and adaptations in aquatic organisms (Ghosh et al. 2024). Environmental DNA (eDNA) has further revolutionized aquatic research by facilitating non-invasive monitoring of biodiversity, detecting invasive species, and assessing the impact of pollution on aquatic ecosystems. Coupled with advanced bioinformatics tools, real-time and remote monitoring systems using eDNA are now being developed, providing high-resolution data on species distribution and ecosystem health with minimal disruption to natural habitats. These innovations in omics and environmental monitoring are paving the way for more precise and predictive models of aquatic ecosystem dynamics (Table 20.1).

Table 20.1 An overview of various tools and types of NGS analysis

Omics type	Tool name	Function
Metagenomics	MEGAHIT	De novo metagenomic assembly
	MetaSPAdes	Metagenomic assembler for short reads
	Kraken2	Taxonomic classification of metagenomic reads
	MetaPhlAn3	Microbial community profiling
Genomics	BWA	Read alignment
	GATK	Variant calling
	SAMtools	Data manipulation for high-throughput sequencing
Transcriptomics	HISAT2	Fast RNA-seq alignment
	StringTie	Transcript assembly
	DESeq2	Differential expression analysis
Proteomics	MaxQuant	Quantitative proteomics
	PEAKS	Peptide sequencing
	OpenMS	Open-source proteomics pipeline
Metabolomics	XCMS	Feature detection & metabolite identification
	MetaboAnalyst	Statistical and functional analysis of metabolomics data
	GNPS	Metabolite networking and annotation

20.3 Computational Advances and Big Data Analytics

The rapid expansion of high-throughput sequencing and multi-omics studies in aquatic sciences has led to an explosion of complex datasets, necessitating advanced computational approaches for efficient analysis and interpretation. Machine learning and artificial intelligence (AI) are playing a transformative role in predicting species distributions, modelling ecosystem responses to climate change, and identifying disease outbreaks in marine and freshwater environments. Emerging AI-driven algorithms tailored for aquatic data analysis enable automated pattern recognition, anomaly detection, and predictive modeling, improving our ability to manage and conserve aquatic biodiversity. High-performance computing (HPC) has become essential for handling large-scale bioinformatics workflows, facilitating genome-wide association studies (GWAS), and complex simulations of aquatic ecosystem dynamics (Das et al. 2020). Additionally, cloud computing solutions are revolutionizing data storage and accessibility, allowing researchers to process vast marine metagenomics and oceanographic datasets in real time. Open-access data repositories and collaborative databases are further enhancing knowledge-sharing efforts, promoting interdisciplinary research, and global conservation strategies (Sharma et al. 2024). The integration of these computational advancements is driving a new era of big data analytics in aquatic sciences, enabling more precise and scalable approaches for monitoring and preserving marine and freshwater ecosystems.

20.4 Metagenomics and Microbiome Analysis in Aquatic Ecosystems

Microbial communities play a fundamental role in shaping aquatic ecosystems, influencing nutrient cycles, primary production, and toxin degradation (Behera et al. 2021). Advances in metagenomics have revolutionized our ability to study these microbial interactions and functions, providing insights into the diversity and metabolic potential of aquatic microbiomes. Functional metagenomics is particularly valuable in linking microbial activities to broader ecosystem processes, with applications in bioremediation, aquaculture, and environmental monitoring (Roy et al. 2018). However, studying aquatic microbiomes presents several technical and bioinformatics challenges, including difficulties in sampling dynamic water environments, sequencing complex microbial consortia, and interpreting vast metagenomic datasets (Kumar et al. 2024).

To address these challenges, novel bioinformatics pipelines and analytical frameworks are being developed to improve microbial identification, functional annotation, and ecological modeling (Das er al. 2025a, b). As sequencing technologies and computational tools advance, future research will enable a deeper understanding of microbial-driven processes in marine and freshwater ecosystems, enhancing conservation efforts and sustainable resource management (Fig. 20.2).

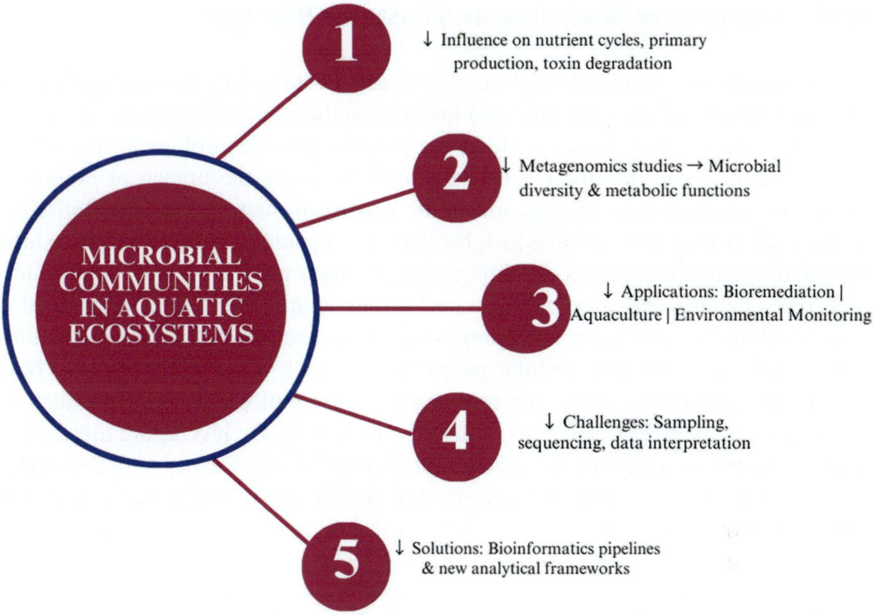

Fig. 20.2 Metagenomics and microbiome analysis in aquatic ecosystems

20.5 Conservation Genomics and Biodiversity Monitoring

Bioinformatics-driven genomic advancements have significantly improved species identification and biodiversity conservation efforts in aquatic ecosystems (Kang et al. 2022). The use of genomic markers allows for precise species identification, population genetics studies, and assessments of genetic diversity, which are crucial for monitoring endangered species and maintaining ecological balance. Conservation genomics plays a vital role in understanding how aquatic species adapt to environmental changes, aiding in the development of strategies for habitat protection and sustainable fisheries management. Additionally, genomic tools are being employed to enhance the resilience of fish stocks and aquaculture species by identifying genetic traits linked to stress tolerance, disease resistance, and climate adaptation. These insights are increasingly shaping conservation policies and resilience planning, helping to mitigate the impacts of climate change on marine and freshwater biodiversity (Arendt et al. 2016). With continuous advancements in sequencing technologies and bioinformatics, conservation genomics is expected to play an even greater role in preserving aquatic ecosystems and ensuring the long-term sustainability of global biodiversity.

20.6 Integrative Modeling and Systems Biology

Advancements in computational modelling and evolutionary bioinformatics are transforming our ability to predict and understand the responses of aquatic ecosystems to environmental changes. By integrating large-scale ecological and omics data, researchers can develop predictive models to assess the impact of pollution, habitat loss, and climate change on marine and freshwater species. Evolutionary bioinformatics provides valuable insights into how aquatic organisms adapt to shifting environmental conditions, tracing genetic changes associated with stress tolerance, reproduction, and survival. Systems biology approaches further enhance this understanding by combining multi-omics data to construct comprehensive models of metabolic networks and cellular pathways (Nayfach and Pollard 2016). These models are particularly useful for studying complex biological processes such as fish behaviors, reproductive cycles, and stress responses. By leveraging these computational tools, researchers can gain a more holistic view of ecosystem dynamics, facilitating informed conservation efforts and sustainable resource management in aquatic environments.

20.7 Synthetic Biology and Aquatic Biotech Applications

The integration of synthetic biology and bioinformatics in aquatic sciences is unlocking new opportunities for genetic engineering, marine bioprospecting, and biotechnological applications. CRISPR and genome-editing technologies are being employed in fish and shellfish to enhance disease resistance, improve growth rates, and increase resilience to environmental stressors, making aquaculture more sustainable (Roy et al. 2018). However, these advancements also raise regulatory and ethical considerations regarding biodiversity conservation, ecological risks, and food safety (Fig. 20.3).

Beyond aquaculture, marine bioprospecting leveraging bioinformatics to explore marine biodiversity has led to the discovery of novel bioactive compounds, enzymes, and biomolecules with applications in pharmaceuticals, biofuels, and industrial biotechnology (Arendt et al. 2016). Computational tools play a crucial role in screening vast marine metagenomic datasets, identifying candidate genes, and predicting the function of novel biomolecules. As bioinformatics and synthetic biology continue to evolve, these technologies hold the potential to revolutionize marine biotechnology, offering sustainable solutions for health, energy, and industry.

20.8 Challenges and Future Prospects

As bioinformatics continues to advance aquatic sciences, several challenges and opportunities shape its future trajectory. One of the key concerns is data privacy and accessibility, particularly in managing aquatic genetic resources. Ensuring ethical data-sharing practices while balancing intellectual property rights remains a

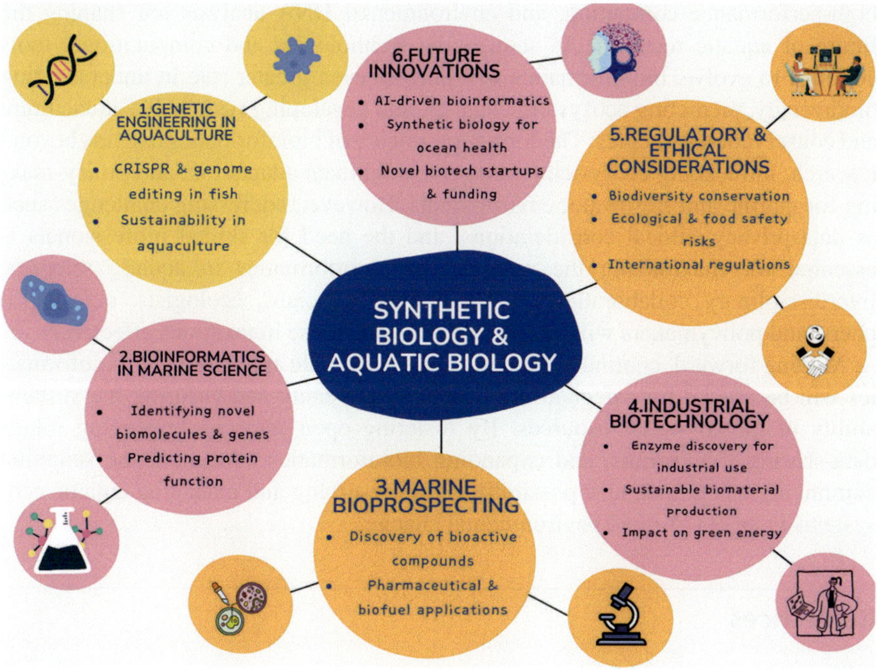

Fig. 20.3 Synthetic biology and aquatic biotech applications

significant challenge. However, global data-sharing networks and open science initiatives are crucial for fostering collaboration and accelerating discoveries in marine and freshwater research. Another pressing issue is the growing need for bioinformatics training among researchers, conservationists, and policymakers. Expanding education through online courses, workshops, and international collaborations will be essential to equip professionals with the skills needed to analyze and interpret large-scale biological data (Nayfach and Pollard 2016). Additionally, the integration of bioinformatics with other disciplines, such as ecology, oceanography, and environmental policy, will be vital for developing holistic solutions to aquatic conservation and resource management. By addressing these challenges and fostering interdisciplinary collaboration, the future of bioinformatics in aquatic sciences promises to drive innovation, sustainability, and a deeper understanding of aquatic ecosystems.

20.9 Conclusion

Bioinformatics has become an indispensable tool in aquatic sciences, driving advancements in species identification, ecosystem modelling, and conservation efforts. Key trends such as multi-omics integration, machine learning, high-performance computing, and environmental DNA analysis are shaping the future of aquatic research. As sequencing technologies and computational tools continue to evolve, bioinformatics will play an even greater role in understanding biodiversity, predicting ecosystem changes, and developing sustainable aquaculture and conservation strategies. The long-term impact of bioinformatics extends beyond research, influencing ecosystem management, climate adaptation, and policy-making for marine and freshwater environments. However, addressing challenges such as data privacy, ethical considerations, and the need for skilled professionals is essential for maximizing the potential of bioinformatics in aquatic sciences. Interdisciplinary collaboration among bioinformaticians, ecologists, oceanographers, and policymakers will be key to harnessing these innovations effectively.

Moving forward, continued research and responsible application of bioinformatics will be crucial for safeguarding aquatic biodiversity and ensuring the sustainability of global water resources. By fostering open science, developing robust data-sharing frameworks, and expanding bioinformatics education, the scientific community can unlock new possibilities for preserving and managing aquatic ecosystems in an era of rapid environmental change.

References

Arendt D, Musser JM, Baker CV, Bergman A, Cepko C, Erwin DH et al (2016) The origin and evolution of cell types. Nat Rev Genet 17(12):744–757

Behera BK, Chakraborty HJ, Patra B, Rout AK, Dehury B, Das BK et al (2020) Metagenomic analysis reveals bacterial and fungal diversity and their bioremediation potential from sediments of river ganga and Yamuna in India. Front Microbiol 11:556136

Behera BK, Dehury B, Rout AK, Patra B, Mantri N, Chakraborty HJ et al (2021) Metagenomics study in aquatic resource management: recent trends, applied methodologies, and future needs. Gene Rep 25:101372

Das BK, Behera BK, Chakraborty HJ, Paria P, Gangopadhyay A, Rout AK et al (2020) Metagenomic study focusing on antibiotic resistance genes from the sediments of river Yamuna. Gene 758:144951

Das BK, Chakraborty HJ, Kumar V, Rout AK, Patra B, Das SK, Behera BK (2024) Comparative metagenomic analysis from Sundarbans ecosystems advances our understanding of microbial communities and their functional roles. Scientific Reports 14(1):16218

Das BK, Kumar V, Roy S, Mohanty D, Bisai K, Chakraborty HJ (2025) De novo assembly of a potential fish pathogen Lactococcus lactis IFLLLBEK1 isolated from Labeo bata. Microbiology Resource Announcements pp.e00455–25

Ezziane Z (2006) Applications of artificial intelligence in bioinformatics: a review. Expert Syst Appl 30(1):2–10

Ghosh D, Sinha N, Sen M, Khatun I, Mohapatra PKD, Saoud J (2024) Advances in microbial ecology illustration using metagenomics and bioinformatics approaches. In: Applications of metagenomics. Academic, pp 335–358

Kang J, Zhang Z, Chen Y, Zhou Z, Zhang J, Xu N et al (2022) Machine learning predicts the impact of antibiotic properties on the composition and functioning of bacterial community in aquatic habitats. Sci Total Environ 828:154412

Kumar R, Raghavendra VC, Gupta S (2024) Mining metagenomic data for novel functional genes using AI. In: Genomic intelligence. CRC Press, pp 239–256

Nayfach S, Pollard KS (2016) Toward accurate and quantitative comparative metagenomics. Cell 166(5):1103–1116

Roy S, Chakraborty HJ, Kumar V, Behera BK, Rana RS, Babu G (2018) *In silico* structural studies and molecular docking analysis of Delta6-desaturase in HUFA biosynthetic pathway. Anim Biotechnol 29(3):161–173

Sharma N, Das BK, Bhattacharjya BK, Chaudhari A, Behera BK, Kumar AP, Chakraborty HJ (2024) Metagenomic insights into microbial community, functional annotation, and antibiotic resistance genes in Himalayan Brahmaputra River sediment, India. Front Microbiol 15:1426463

Empowering Fisheries with AI, ML, and Deep Learning: A Big Data Approach

21

Ananya Nayak, Dibyajyoti Uttameswar Behera, Ankita Pati, and Maheswata Sahoo

21.1 Introduction

Aquaculture and fisheries are paramount to global food security and support livelihoods and the world economy. The sector directly employs more than 58 million people worldwide, while billions more depend upon fish as a primary source of protein, especially in places where access to alternative protein is limited (FAO 2010). Nonetheless, the very same industry faces an increasing range of challenges, including, but not limited to, overfishing, habitat degradation, climate change, and poor resource management. These threats not only endanger the very fabric of marine ecosystems but also jeopardize the livelihood and food security of millions of people worldwide (Lincoln et al. 2022). Building such novel, data-driven solutions is essential to address such complex and entwined challenges, which require insights generated from processes and analyses of a growing amount of highly diverse datasets that arise from fisheries operations.

The emergence of AI, ML, and DL in fisheries science and management has opened an entirely new chapter in this domain. These technologies work with exceptionally high accuracy in the big data era, making it easy for researchers and practitioners to manage and analyze these huge datasets (Kühn et al. 2025). From stock assessment to resource allocation and ecosystem monitoring, it is also AI-based tools that may help reshape the management of fisheries, encouraging

A. Nayak · A. Pati · M. Sahoo
Centre for Biotechnology, Siksha 'O' Anusandhan Deemed to be University, Bhubaneswar, India

D. U. Behera (✉)
Centre for Biotechnology, Siksha 'O' Anusandhan Deemed to be University, Bhubaneswar, India

Regional Centre, ICAR–Central Tuber Crops Research Institute, Bhubaneswar, India

© The Author(s), under exclusive license to Springer Nature Singapore Pte Ltd. 2026

B. K. Das et al. (eds.), *Bioinformatics Tools for Predictive Ecology and Fisheries*, Sustainable Aquatic Food Systems, https://doi.org/10.1007/978-981-95-2573-7_21

sustainable behaviors by striking a balance between social and environmental, and economic goals (Vivekanandan 2017). These technologies provide real-time insights into fish populations and their dynamics to encourage dynamic and better management options.

This review lays out the revolutionary possibilities that AI, ML, and DL technologies could have towards changing fisheries science and management. In addition, these tools have worked miracles where collective challenges like species identification, predicting fish stock dynamics, and tackling illegal fishing are concerned (Yang et al. 2021). AI-based systems allow processing and integration of heterogeneous datasets, including remote sensing data, acoustic monitoring, and socioeconomic indicators, which results in a holistic framework for informed decision-making (Alotaibi and Nassif 2024). Synthesis from multivariate data streams, such as those found in marine resource governance, is central to their very effective management, particularly in climate change contexts and the ecological uncertainties they pose (Wang et al. 2024).

The use of emerging technologies involving AI and others falls in line with global demands of sustainable development, and in particular, the UN Sustainable Development Goal (SDG) 14. SDG 14 provides for the conservation and sustainable utilization of oceans, seas, and marine resources, and can make significant contributions through advanced analytical tools (ICS 2017; Nahar 2024). Solutions powered by AI will help to address the twin concerns of ecological conservation and economic growth in balancing the dual prospects of allocating and deploying resources efficiently and sustainably in the future management of marine resources.

These reviews will introduce an opportunity to explore the present trends, challenges, and ethical considerations of AI, ML, and DL in fisheries and will elucidate the integration of "big data" and its predictive capabilities, thus bringing them on par for better management decision-making with multi-dimensional datasets, and forecast different pathways and ways of development of technology for fisheries management corresponding to emerging paradigms of big data use. Harnessing these forward-looking approaches, the fisheries sector will arrive at agreements constituting the response to new realities that will secure the sustainable use of marine ecosystems while confronting the blossoming global challenges ahead.

21.2 Overview of the Importance of Fisheries in the Global Context

Fisheries have played very important roles in attaining global food security and economic development, and are crucial in ecological sustainability (Belton and Thilsted 2014). The sector plays a critical role in the food supply in the world and supports millions of livelihoods while providing a global source of essential proteins for over one billion people (Tigchelaar et al. 2022). More to this, fisheries are very vital to coastal countries' economies, particularly in the developing world, supporting local economies and trade (Watson et al. 2021). Fish have enormous economic impacts, but they are also an important sector in the marine ecosystem that supports biodiversity and ecological balance (Tidwell and Allan 2001).

The sector faces several major issues that include overfishing, climate change effects, habitat destruction, and illegal fishing (Singh et al. 2024). This jeopardizes the sustainability of fisheries; hence, novel methods to tackle them effectively must be developed. The emergence of various technologies, in particular big data, is integrating to become the solution that would facilitate management, ensuring the sustainability of fisheries on a global basis (Zhang et al. 2023).

21.3 Emergence of AI, ML, and Deep Learning in Fisheries Management

Artificial intelligence, machine learning, and deep learning are transforming the face of various industries and surely fisheries management. It is possible to analyze large datasets that are coming from different sources, like satellite imagery, vessel monitoring systems, underwater cameras, and sensors, while extracting actionable insights that empower stakeholders to make informed decisions, improve operations, and address pressing issues by virtue of AI-driven solutions (Salhaoui et al. 2020).

The application of AI and ML is very useful to predict the volume of fish, analyze migration activities, and also detect illegal catches. Deep Learning, in its analysis of unstructured data such as pictures and videos, helps in species identification, monitors fish behavior, and streamlines aquaculture activities. Improved accuracy and efficacy of data-based methods also work towards sustainability and resilience in fisheries (Liu et al. 2023; Yang et al. 2021).

It represents a big opportunity to develop enhanced fisheries management practices by the integration of artificial intelligence, machine learning, and deep learning into large datasets. Stakeholders may employ these technologies in proactive and adaptive ways to secure long-term sustainability for marine resources with ecological, economic, and social priorities in mind (Rowan 2023).

21.4 Big Data in Fisheries

Big datasets have become the foundation in modern fisheries management, permitting the collection, analysis, and interpretation of large and complicated information (Swain et al. 2021). Insightful analysis of massive data supports informed decision-making, refined resource allocation, and enhanced understanding of marine ecosystems (Smit et al. 2021). In fisheries, wide data includes a variety of types, comprising biological, environmental, economic, and operational data. Such datasets originate from a multitude of sources, which include satellite observations, acoustic sensors, e-logbook records, and social media platforms (Lan et al. 2017).

The application of various types of big data in fisheries includes, among others, stock evaluation improvements, supply chain optimization, and sustainability.

However, for big data to deliver on its promise, the very challenges that undermine its perceived potential must be addressed, with all the other opportunities properly seized.

21.4.1 Challenges and Opportunities of Big Data in Fisheries

The implementation of big data in fisheries presents a distinct array of challenges:

- Data Flow and Quality: A constant struggle to maintain regularity, authenticity, and completeness of data collection. Many domains lack the infrastructure or resources required to maintain high-quality datasets.
- Data Integration: To combine diverse datasets from separate sources, highly advanced approaches and an emphasis on standardization are required.
- Financial Implications and Availability: The adoption of big data technologies can incur significant costs, creating challenges for smaller participants.
- Privacy and Security: In other words, such as ensuring the privacy and safety of sensitive information (e.g., fishing vessel location) to maintain trust and compliance.
- Skill Gaps: The effective utilization of big data depends on a strong data science capability, which is not typically present in traditional fisheries management agencies.

Considering these obstacles, big data has substantial opportunities:

- Decision-Making: Big data analytics provides real-time monitoring and predictive modeling, so people can make well-informed decisions and take action before something goes wrong.
- Sustainability: Understanding broad temporal patterns can help understand overfishing hotspots, guide conservation or sustainable efforts.
- Efficiency Improvements: Big data may streamline processes across the fisheries value chain, reducing waste and increasing profitability.
- Promote interdisciplinary research: Sharing and cross-consolidation of datasets within corporations and across geographies can foster creativity and interdisciplinary research.

21.4.2 Integration of Big Data Technologies in Fisheries Research

Big data technologies for fisheries research are revolutionizing the field through comprehensive and rapid analysis across space and time. Cloudera and ESRI are just two examples of the types of advanced instruments, like cloud computing, machine learning algorithms, and GIS technology, that have become critical to fisheries research (Cheng et al. 2023; Kamilaris et al. 2017).

- Cloud Computing: Cloud systems provide scalable resources for the storage and processing of large-scale data and enable the access of data for global users. This is particularly useful in cases where many institutions are working together (Berisha et al. 2022).
- Machine Learning and Predictive Analytics: Machine learning algorithms can look for trends and associations in datasets; help stock assessment, habitat modeling, or migration predictive (Gladju et al. 2022).
- Geospatial Analysis: Geographical Information Systems (GIS), remote sensing technologies provide place-based analytics; from mapping of fishing efforts to marine protected areas and habitat changes monitoring (Kapetsky and Aguilar-Manjarrez 2007).
- IoT and Sensor Networks: Via Internet of Things (IoT), sensor technologies to monitor parameters in real time, the environment, vessel performance, and aquaculture systems (Bachtiar et al. 2022).

Engineers and fisheries researchers will have access to a deep understanding of ecological processes, human impacts on the ecosystem, and opportunities for creative management practices through the use of this technology. The Inauguration of big data technologies and approaches is capable of big progress in fisheries science and sustainability.

21.5 Artificial Intelligence (AI) in Fisheries

Artificial intelligence swiftly revolutionizes the fishing business by facilitating intelligent and adaptive management strategies. AI improves efficiency, accuracy, and sustainability in fisheries operations with improved algorithms and data processing skills (Mana et al. 2024).

21.5.1 Applications of AI in Fisheries Monitoring and Management

Artificial intelligence possesses extensive applications in the monitoring and management of fisheries:

- Fish Stock Assessment: AI models can analyze historical and real-time data to estimate fish numbers and predict stock changes. This enables the setting of sustainable catch limits (Rowan 2023).
- IUU Fishing Detection: AI-based algorithms analyze vessel tracking data to identify unusual activity, including illegal fishing in protected areas (Welch et al. 2024).
- Species Identification: AI-driven image recognition algorithms can accurately identify fish species from images or videos, aiding biodiversity research and monitoring initiatives (Mifsud Scicluna et al. 2024).

- Aquaculture Automation: Artificial intelligence facilitates the automation of feeding schedules, water quality assessment, and disease identification in aquaculture systems, enhancing production and sustainability (Ashraf Rather et al. 2024).
- Environmental Impact Assessment: AI technologies evaluate the ecological consequences of fishing activities by simulating habitat alterations and monitoring bycatch occurrences (Cheng et al. 2023).

21.5.2 Case Studies of AI Implementation in Fisheries

The pragmatic application of AI in fisheries management has revealed its transformational capabilities:

- *Global Fishing Watch*: This program uses artificial intelligence to scrutinize vessel movement data from the Automatic Identification System (AIS) to monitor worldwide fishing activity. Recognizing patterns indicative of illicit fishing fosters transparency and adherence to regulations (Merten et al. 2016; Taconet et al. 2017).
- *FishFace Project*: Developed by The Nature Conservancy, this program utilizes AI-powered facial recognition software to identify fish species. It helps in tracking catches and imposing regulations effectively (Li et al. 2022).
- *SmartFish Project*: The SmartFish H2020 Project is a European Union-funded project using artificial intelligence in the development of smart monitoring equipment to reduce bycatch and enhance fishing gear for sustainable fisheries (Alsharabi et al. 2024).
- *AI in Aquaculture*: Industries are applying AI-based solutions to improve feeding systems, monitor fish health, and minimize waste in the aquaculture industry (Saad et al. 2024).

All these show the flexibility and relevance of AI to the major concerns in fisheries management. With the encouragement of innovation and cooperation, AI is currently shaping the future of sustainable fishing.

21.6 Machine Learning (ML) in Fisheries

Machine Learning (ML) has become an effective instrument for tackling intricate issues in fisheries management and research. Utilizing extensive datasets, machine learning algorithms facilitate the creation of predictive models and automated systems that improve decision-making and operational efficiency.

21.6.1 Supervised and Unsupervised ML Techniques in Fisheries Research

Machine learning approaches are primarily classified into supervised and unsupervised learning, each serving unique applications in fisheries.

- Supervised Learning: This methodology entails training machine learning models on labeled datasets, rendering it appropriate for applications such as species identification, stock assessment, and bycatch detection. Convolutional neural networks (CNNs) have been employed to classify fish species using picture data, attaining excellent accuracy.
- Unsupervised Learning: Analyzing unlabeled data, unsupervised learning methods like clustering and dimensionality reduction allow patterns and relationships to be identified. The approaches can help define migration patterns, habitat preference, and fishing effort distribution.

With both supervised and unsupervised methodologies, the scientist can investigate multiple aspects of fisheries, including the understanding of ecological dynamics up to the enhancement of management practices.

21.6.2 ML Models for Stock Estimation and Prediction

To meet the most significant requirements in fisheries management, machine learning models are gaining importance rapidly in various activities related to stock estimation and prediction.

- Time-Series Analysis: These recurrent neural networks and long short-term memory networks are useful models in modeling temporal data, such as fish population trends and seasonal migrations. This type of analysis is known as time-series analysis (Ishfaque et al. 2022).
- Species Distribution Modeling (SDM): Machine learning algorithms, such as random forests and gradient boosting, are employed to forecast species distribution based on environmental data, facilitating habitat conservation initiatives (Agrawal and Konno 2016).
- Catch Prediction: Machine learning models amalgamate data from vessel monitoring systems, oceanographic conditions, and historical catch records to project capture volumes, facilitating improved planning and sustainability (Christiani et al. 2019; Watson et al. 2018).
- Ecological modeling: Sophisticated machine learning methodologies, including deep reinforcement learning, mimic ecological dynamics and anthropogenic effects, thus allowing adaptive management (Zuccotto et al. 2024).

21.7 Deep Learning in Fisheries

Deep Learning (DL), a branch of machine learning, has demonstrated a transformative impact on fisheries research and management. Utilizing artificial neural networks (ANNs) and other sophisticated designs, deep learning systems analyze extensive and intricate datasets, providing actionable insights and automating labor-intensive processes. The implementation of deep learning in fisheries has notably advanced in picture and video analysis, facilitating effective species identification, behavioral research, and stock evaluations (Mana et al. 2024; Taye 2023).

21.7.1 Neural Networks and Deep Learning Architectures in Fisheries Applications

Deep learning architectures, especially convolutional neural networks (CNNs), are extensively employed in fisheries for image identification and spatial modeling applications. Recurrent neural networks (RNNs) and Long Short-Term Memory (LSTM) networks are utilized for temporal data analysis, facilitating predictive modeling of fish stock dynamics and migration patterns (Hosseinnia Shavaki and Ebrahimi Ghahnavieh 2023; Yang et al. 2021).

Convolutional Neural Networks (CNNs) have exhibited exceptional precision in categorizing fish species from aquatic images, aiding biodiversity evaluations and regulatory adherence (Alzubaidi et al. 2021). Concurrently, LSTMs enable the examination of seasonal patterns in catch volumes and environmental alterations, assisting managers in proactively addressing changing maritime circumstances (Choi et al. 2020). Recent breakthroughs encompass the utilization of generative adversarial networks (GANs) to model fish populations and environmental scenarios, providing innovative instruments for fisheries research (Wang and Tabeta 2023).

The scalability and adaptability of deep learning architectures enable academics to address several difficulties in the field, ranging from enhancing aquaculture efficiency to monitoring illicit fishing activities. As computer resources become increasingly accessible, the range of deep learning applications in fisheries continues to broaden.

21.7.2 Image Recognition and Classification for Fish Species Identification

Within the realm of fish species identification, the capability of deep learning to analyze visual data has been a game-changer. In tough underwater situations with limited visibility, researchers are able to obtain high accuracy in recognizing species by training CNNs on vast datasets of tagged fish photos. This allows them to identify species with more precision (Alzubaidi et al. 2021; Hamzaoui et al. 2023).

The following case studies show how DL is indeed useful in this domain:

- Algorithms for image recognition were utilized by the Fish4Knowledge Project in monitoring populations of coral reef fish, thus acquiring substantial knowledge about marine biodiversity (Fisher et al. 2016).
- The use of convolutional neural networks (CNNs) for real-time species identification in aquaculture settings was accomplished by Salman et al. (2020) in their research (Salman et al. 2020). This allowed for more accurate and expedient stock monitoring.
- To classify deep-sea species captured in film from remotely operated vehicles, scientists at the Monterey Bay Aquarium Research Institute used deep learning. This allowed them to greatly cut down the amount of time needed for the traditional manual analysis (Lopez-Vazquez et al. 2023).

Apart from species identification, deep learning enables the detection of anomalies, such as invasive species or signs of disease in aquaculture systems. These skills enhance the precision and efficiency of fisheries management while promoting conservation goals.

21.8 Data Collection and Pre-processing in Fisheries

Gathering and preparing vast amounts of data is the foundation of effective fisheries management and research. Researchers have the potential to derive relevant insights, maximize resource use, and address difficulties related to sustainability when they have access to datasets that are accurate, comprehensive, and well-processed. The process of collecting, cleaning, and preparing data for analysis in the field of fisheries is undergoing a transformation as a result of developments in sensor technology and data preparation techniques.

21.8.1 Sensor Technologies for Data Collection in Fisheries

Modern sensor technologies have changed the way fishers collect data by giving high-resolution data in real-time across many dimensions. Significant progress encompasses:

- Acoustic Sensors: Commonly employed in stock evaluation, acoustic sensors evaluate fish abundance and biomass by monitoring sound waves reflected from schools of fish (Yassir et al. 2023).
- Remote Sensing and Satellite Imagery: Satellites collect oceanographic data such as sea surface temperature, chlorophyll content, and ocean currents. This enables habitat modeling and fish zone forecasts (Belkin 2021).
- Underwater Cameras and Drones: These tools collect visual data for species identification, behavioral research, and environmental assessment. Automated systems, equipped with AI, speed up the process of processing photographs and videos in real time (Salhaoui et al. 2020).

- Internet of Things (IoT): IoT-enabled sensors utilized in aquaculture systems assess water quality indicators such as temperature, pH, and oxygen levels, thereby providing optimal conditions for fish farming (Bachtiar et al. 2022; Rastegari et al. 2023).
- Electronic Monitoring Systems (EMS): Employed aboard fishing vessels, EMS document actions such as gear deployment, catch composition, and bycatch occurrences, thereby enhancing compliance and transparency (Gilman et al. 2020).

21.8.2 Data Cleaning and Preprocessing Techniques

Data collected needs to be cleaned and pre-processed to ensure its validity and usability for analysis. Preprocessing involves a number of critical steps:

- Data Cleaning: Removing errors, outliers, and inconsistencies from raw datasets is important. Techniques include noise filtration in sensor data and rectification of misclassifications in species identification records (Hosseinzadeh et al. 2023).
- Normalization and Standardization: Datasets are transformed to a standardized scale or form to allow comparisons. This is particularly critical if data is collected from several sources (Institute of Medicine 2013).
- Dealing with Missing Data: Missing data in fisheries datasets occurs mostly due to broken equipment or faults in the communication of data. Interpolation, imputation, or data augmentation methods are applied for this purpose.
- Feature Engineering: Essential characteristics pertinent to the study aims are retrieved or generated. Environmental variables such as temperature and salinity can be extracted from raw oceanographic data.
- Data Reduction: Extensive datasets are streamlined by dimensionality reduction methods such as Principal Component Analysis (PCA), rendering research computationally manageable while preserving essential information (Rastogi et al. 2023).

Efficient preprocessing improves data quality and guarantees that models and analytics produce precise and actionable insights. Preprocessing underwater imaging data enhances species detection by reducing noise and enhancing contrast.

21.9 Ethical and Environmental Concerns

The most vital aspects to be considered when applying recent innovations, such as artificial intelligence and machine learning, that continue to shape the administration of fisheries, have become ethical and environmental considerations. Considering both the societal effect on humankind that remains reliant on fishing, as well as the environmental impact of an AI-based innovation introduced into the marine environment (Rowan 2023).

21.9.1 Ethical Dimensions of AI and ML in Fisheries Management

When it comes to fisheries management, the integration of artificial intelligence and machine learning throws up a myriad of ethical questions, particularly relating to equity, transparency, and data privacy:

- Influence on the Communities in the Area: The digital gap is sometimes caused by the fact that small-scale fishermen do not have access to more advanced technologies. Using artificial intelligence and machine learning could have a disproportionately positive impact on industrial fisheries, which would result in the marginalization of traditional fishing communities and an increase in economic disparities (Cheng et al. 2023).
- Decision-Making and Algorithmic Bias: Artificial intelligence systems rely on previous data, which may contain the possibility of biases or mistakes. These biases can result in an unfair distribution of resources, such as an overestimation of stocks in certain regions while an underestimation of stocks in other regions. To reduce the impact of biases like these, it is essential to build algorithms that are both transparent and inclusive (Ferrara 2023).
- Concerns Over Surveillance and Monitoring: The use of technologies like electronic monitoring systems and vessel tracking raises questions about surveillance and the implications it has for the privacy and autonomy of fishermen (Rowan 2023).

21.9.2 Sustainable Practices in Implementing AI Technologies in Fisheries

The integration of AI technologies in fisheries aims to achieve sustainability, necessitating alignment with environmental conservation principles and long-term ecological objectives (Alsharabi et al. 2024). Essential practices consist of:

- Energy-Efficient AI Model: Energy-efficient AI models necessitate substantial computational resources for training and deployment, which in turn contribute to carbon emissions. The implementation of energy-efficient algorithms and the utilization of green data centers can reduce the environmental impact of these technologies (Olawade et al. 2024).
- Minimizing Environmental Disruption: Underwater cameras and acoustic sensors need to be developed in ways that have less interference with marine ecosystems. Non-invasive acoustic devices are preferred because they pose fewer threats to aquatic fauna (Salhaoui et al. 2020).
- Sustainable Resource Allocation: AI-driven decision support systems can align and rationalize fishery activities to avoid overexploitation by placing controls on the sustainable utilization of resources. It helps balance the ecological needs and economic necessities as it ensures accurate predictions of stocks and advice on optimal harvest limits (Rowan 2023).

- Incorporation of Indigenous Knowledge: Integrating indigenous knowledge in the use of AI technologies integrates traditional ecological practices with data-driven insights, creating sustainability in their fusion (Rowan 2023).
- Lifecycle Assessment of AI Tools: An assessment of hardware, such as sensors and drones, is to be made regarding environmental impact throughout its lifecycle-from production to disposal, in order to prevent marine pollution or resource depletion during its utilization.

Implementing sustainable practices in AI technologies aligns with environmental objectives and enhances the credibility and acceptance of these tools among various stakeholders. A collaborative approach that includes policymakers, researchers, and local communities is crucial for achieving these outcomes.

21.10 Future Prospects and Developing Trends

Technological development changes the fisheries sector at lightning speed. Recent discoveries in artificial intelligence, machine learning, and quantum computing offer major potentialities to solve highly complex problems within sustainable fisheries management. These enable more accurate, scalable, and effective solutions to the benefit of greater sustainability and well-being of marine ecosystems.

21.10.1 Advancements in Artificial Intelligence and Machine Learning to Sustain Fisheries

The future of artificial intelligence and machine learning for fisheries management can be characterized as the advancement of specialized algorithms that integrate multiple data sources into actionable insights. Some major developments include:

- Integration of Multimodal Data Sources: Future AI systems will combine data sources from different areas, including satellite images, acoustic sensors, and environmental databases, to provide a rich understanding of marine ecosystems. This combination enhances the accuracy of stock assessments and habitat models (Alotaibi and Nassif 2024).
- Real-Time Monitoring and Decision Support Systems: Real-time monitoring and decision support systems will benefit from innovations in edge computing and IoT, facilitating the real-time processing of data gathered from fishing vessels and aquaculture systems. This capability enables swift decision-making, minimizes bycatch, and ensures adherence to sustainable practices (Al-Mutairi and Al-Aubidy 2023; Bachtiar et al. 2022).
- AI-Driven Predictive Analytics: Advanced predictive models will integrate climate change projections to forecast changes in fish distribution and productivity. These models will facilitate adaptive management strategies that proactively address environmental changes (Chen et al. 2023; Fatima et al. 2023).

- Automated Species Identification and Behavioral Studies: Future machine learning algorithms will help accelerate automated species identification and behavioral studies, enhancing image recognition capabilities and identifying rare or cryptic species (Bartlett et al. 2022). The tools will allow for biodiversity conservation, effectively decreasing manual labor in data analysis.
- Ethical AI Systems: Ethical AI Systems will increasingly focus on addressing equity and transparency issues, ensuring that small-scale fishers benefit inclusively while minimizing algorithmic biases (Rowan 2023).

Fisheries management can strike a balance between financial objectives and environmental sustainability by utilizing these developments.

21.10.2 The Potential Impact of Quantum Computing on Fisheries Research

Quantum computing is a cutting-edge technology that could greatly enhance fisheries research by tackling complex computational challenges with remarkable speed and accuracy (Du et al. 2015; Zhu and Jiang 2010). Here are some potential impacts:

- Accelerated Data Processing: Quantum computing enables the swift analysis of large datasets from fisheries, allowing researchers to examine intricate patterns in fish populations, migration paths, and ocean conditions in real time. Quantum algorithms can optimize fishing routes and resource distribution more effectively than traditional methods (Zhu and Jiang 2010).
- Improved Stock Assessment Models: The ability of quantum computing to handle high-dimensional datasets will refine stock assessment models by incorporating various factors, such as environmental influences, fish behavior, and socio-economic data. This will lead to more accurate and adaptable forecasts.
- Climate Change Impact Analysis: Quantum simulations can effectively model the interactions between climate change factors and marine ecosystems. These insights will help predict long-term effects on fisheries and guide policy-making.
- Enhanced Machine Learning Models: Quantum machine learning (QML) significantly boosts the training of AI models, enhancing accuracy in image recognition, predictive analytics, and anomaly detection within fisheries.
- Sustainability Optimization: Quantum computing can aid in formulating strategies for sustainable fishing, aquaculture, and marine resource management by solving complex optimization challenges.

While the practical use of quantum computing in fisheries research is currently limited by technological constraints, ongoing advancements suggest it could play a crucial role in the coming decades.

21.11 Conclusion

The fascinating world of bioinformatics has emerged as an important tool in fisheries science that will lead to critical insights and solutions for achieving sustainability and resilience. Employing modern computational techniques and genomic data is resulting in the discovery of new routes toward understanding and preserving genetic diversity, optimizing aquaculture practices, and combating pressing challenges such as disease outbreaks and the impacts of climate change.

This inclusion of bioinformatics into fisheries enables scientists and practitioners to draft more effective strategies for sustainable management. These range from monitoring fish stocks and developing breeding programs to the protection of aquatic ecosystems to maintain biodiversity. Such efforts will ensure not only the long-term viability of fisheries but also a strengthened level of contribution toward global food security and the health of ecosystems around the planet. Bioinformatics-based fisheries-pesca provides the ability to counteract trends arising from warping trends of change that beset the sector, especially now with domestic and foreign dimensions of environmental and socio-economic challenges opening wide. The Navy offers a leap toward greater affordance to all living beings within the systems while ensuring the maintenance of productivity for the betterment of humanity.

References

Agrawal AA, Konno K (2016) Species distribution models: ecological explanation and prediction across space and time. Annu Rev Ecol Evol Syst 40(2009):311–331

Al-Mutairi AW, Al-Aubidy KM (2023) IoT-based smart monitoring and management system for fish farming. Bull Electr Eng Inform 12(3):1435–1446

Alotaibi E, Nassif N (2024) Artificial intelligence in environmental monitoring: in-depth analysis. Discov Artif Intell 4(1):84

Alsharabi N, Ktari J, Frikha T, Alayba A, Alzahrani AJ, Jadi A, Hamam H (2024) Using blockchain and AI technologies for sustainable, biodiverse, and transparent fisheries of the future. J Cloud Comput 13(1):135

Alzubaidi L, Zhang J, Humaidi AJ, Al-Dujaili A, Duan Y, Al-Shamma O, Santamaría J, Fadhel MA, Al-Amidie M, Farhan L (2021) Review of deep learning: concepts, CNN architectures, challenges, applications, future directions. J Big Data 8(1):53. https://doi.org/10.1186/s40537-021-00444-8

Ashraf Rather M, Ahmad I, Shah A, Ahmad Hajam Y, Amin A, Khursheed S, Ahmad I, Rasool S (2024) Exploring opportunities of artificial intelligence in aquaculture to meet increasing food demand. Food Chem X 22:101309

Bachtiar MI, Hidayat R, Anantama R (2022) Internet of things (IoT) based aquaculture monitoring system. MATEC Web Conf 372:04009

Bartlett P, Eberhardt U, Schütz N, Beker HJ (2022) Species determination using AI machine-learning algorithms: hebeloma as a case study. IMA Fungus 13(1):13

Belkin IM (2021) Review remote sensing of ocean fronts in marine ecology and fisheries. Remote Sens 13(5):1–22

Belton B, Thilsted SH (2014) Fisheries in transition: food and nutrition security implications for the global South. Glob Food Sec 3(1):59–66

Berisha B, Mëziu E, Shabani I (2022) Big data analytics in cloud computing: an overview. J Cloud Comput 11(1):24

Chen Y, Shan X, Gorfine H, Dai F, Wu Q, Yang T, Shi Y, Jin X (2023) Ensemble projections of fish distribution in response to climate changes in the Yellow and Bohai Seas, China. Ecol Indic 146:109759

Cheng X, Zhang F, Chen X, Wang J (2023) Application of artificial intelligence in the study of fishing vessel behavior. Fishes 8(10):516

Choi E, Cho S, Kim DK (2020) Power demand forecasting using long short-term memory (LSTM) deep-learning model for monitoring energy sustainability. Sustainability 12(3):1109

Christiani P, Claes J, Sandnes E, Stevens A (2019) Precision fisheries: navigating a sea of troubles with advanced analytics. McKinsey Insights, pp 1–19. https://www.mckinsey.com/industries/agriculture/our-insights/precision-fisheries-navigating-a-sea-of-troubles-with-advanced-analytics

Du T, Hu Y, Ke X (2015) Improved quantum artificial fish algorithm application to distributed network considering distributed generation. Comput Intell Neurosci 2015:1–13

FAO (2010) The state of world fisheries and aquaculture. Food and Agriculture Organization of the United Nations. Aquaculture 218. http://www.fao.org/docrep/013/i1820e/i1820e.pdf

Fatima N, Shuaib SE, Kong JD (2023) Predicting adaptations of fish and fishing communities to rapid climate velocities in Canadian waters: a systematic review. Environ Adv 14:100452

Ferrara E (2023) Fairness and bias in artificial intelligence: a brief survey of sources, impacts, and mitigation strategies. Sci 6(1):3

Fisher RB, Shao KT, Chen-Burger YH (2016) Overview of the fish4knowledge project. Intell Syst Ref Libr 104:1–17

Gilman E, De Ramón Castejón V, Loganimoce E, Chaloupka M (2020) Capability of a pilot fisheries electronic monitoring system to meet scientific and compliance monitoring objectives. Mar Policy 113:103792

Gladju J, Kamalam BS, Kanagaraj A (2022) Applications of data mining and machine learning framework in aquaculture and fisheries: a review. Smart Agric Technol 2:100061

Hamzaoui M, Ould-Elhassen Aoueileyine M, Romdhani L, Bouallegue R (2023) An improved deep learning model for underwater species recognition in aquaculture. Fishes 8(10):514

Hosseinnia Shavaki F, Ebrahimi Ghahnavieh A (2023) Applications of deep learning into supply chain management: a systematic literature review and a framework for future research. Artif Intell Rev 56(5):4447–4489

Hosseinzadeh M, Azhir E, Ahmed OH, Ghafour MY, Ahmed SH, Rahmani AM, Vo B (2023) Data cleansing mechanisms and approaches for big data analytics: a systematic study. J Ambient Intell Humaniz Comput 14(1):99–111

ICS (2017) A guide to SDG interactions: from science to implementation, vol 33. International Council for Science, Paris

Institute of Medicine (2013) Standardization to enhance data sharing - sharing clinical research data - NCBI bookshelf. In: Sharing clinical research data: workshop summary, pp 43–55

Ishfaque M, Dai Q, Haq N u, Jadoon K, Shahzad SM, Janjuhah HT (2022) Use of recurrent neural network with long short-term memory for seepage prediction at Tarbela Dam, KP, Pakistan. Energies 15(9):3123

Kamilaris A, Kartakoullis A, Prenafeta-Boldú FX (2017) A review on the practice of big data analysis in agriculture. Comput Electron Agric 143:23–37

Kapetsky JM, Aguilar-Manjarrez J (2007) Geographic information systems, remote sensing and mapping for the development and management of marine aquaculture. FAO Technical Paper. Food & Agriculture Org, p 125

Kühn B, Cayetano A, Fincham JI, Moustahfid H, Sokolova M, Trifonova N, Watson JT, Fernandes-Salvador JA, Uusitalo L (2025) Machine learning applications for fisheries—at scales from genomics to ecosystems. Rev Fish Sci Aquacult 33:334–357

Lan KW, Shimada T, Lee MA, Su NJ, Chang Y (2017) Using remote-sensing environmental and fishery data to map potential yellowfin tuna habitats in the Tropical Pacific Ocean. Remote Sens 9(5):444

Li D, Su H, Jiang K, Liu D, Duan X (2022) Fish face identification based on rotated object detection: dataset and exploration. Fishes 7(5):219

Lincoln S, Andrews B, Birchenough SNR, Chowdhury P, Engelhard GH, Harrod O, Pinnegar JK, Townhill BL (2022) Marine litter and climate change: inextricably connected threats to the world's oceans. Sci Total Environ 837:155709
Liu H, Ma X, Yu Y, Wang L, Hao L (2023) Application of deep learning-based object detection techniques in fish aquaculture: a review. J Mar Sci Eng 11(4):867
Lopez-Vazquez V, Lopez-Guede JM, Chatzievangelou D, Aguzzi J (2023) Deep learning based deep-sea automatic image enhancement and animal species classification. J Big Data 10(1):37
Mana AA, Allouhi A, Hamrani A, Rahman S, el Jamaoui I, Jayachandran K (2024) Sustainable AI-based production agriculture: exploring AI applications and implications in agricultural practices. Smart Agric Technol 7:100416
Merten W, Reyer A, Savitz J, Amos J, Woods P, Sullivan B (2016) Global Fishing Watch: Bringing transparency to global commercial fisheries. arXiv preprint arXiv:1609.08756
Mifsud Scicluna B, Gauci A, Deidun A (2024) AquaVision: AI-powered marine species identification. Information (Switzerland) 15(8):437
Nahar S (2024) Modeling the effects of artificial intelligence (AI)-based innovation on sustainable development goals (SDGs): applying a system dynamics perspective in a cross-country setting. Technol Forecast Soc Chang 201:123203
Olawade DB, Wada OZ, David-Olawade AC, Fapohunda O, Ige AO, Ling J (2024) Artificial intelligence potential for net zero sustainability: current evidence and prospects. Next Sustain 4:100041
Rastegari H, Nadi F, Lam SS, Ikhwanuddin M, Kasan NA, Rahmat RF, Mahari WAW (2023) Internet of things in aquaculture: a review of the challenges and potential solutions based on current and future trends. Smart Agric Technol 4:100187
Rastogi AK, Taterh S, Kumar BS (2023) Dimensionality reduction algorithms in machine learning: a theoretical and experimental comparison. Eng Proc 59(1):82
Rowan NJ (2023) The role of digital technologies in supporting and improving fishery and aquaculture across the supply chain – Quo Vadis? Aquac Fish 8(4):365–374
Saad A, Baikas A, Remen M, Bjørnson FO (2024) Optimizing feeding strategies in aquaculture using machine learning: ensuring sustainable and economically viable fish farming practices. Procedia Comput Sci 246(C):4712–4721
Salhaoui M, Molina-Molina JC, Guerrero-González A, Arioua M, Ortiz FJ (2020) Autonomous underwater monitoring system for detecting life on the seabed by means of computer vision cloud services. Remote Sens 12(12):1981
Salman AS, Salman OS, Katz GE (2020) Extending CNN classification capabilities using a novel feature to image transformation (FIT) algorithm. In: Advances in intelligent systems and computing, vol 1229. AISC, pp 198–213
Singh H, Ranjan D, Verma P, Upadhyay AK, Kumar P, Singh A (2024) Current issues with fish and fisheries sector: challenges and solutions. Biotica Res Today 6(1):39–45
Smit KP, Bernard ATF, Lombard AT, Sink KJ (2021) Assessing marine ecosystem condition: a review to support indicator choice and framework development. Ecol Indic 121:107148
Swain PR, Parida PK, Behera BK, Swain HS, Sahu SK, Das BK (2021) Big data application in fisheries with special reference to inland fisheries sector in India. Indian J Fish 68(4):164–170
Taconet M, Fernandes JA, Miller NA, Kroodsma D (2017) Conclusions and overview of global atlas of AIS-based fishing activity
Taye MM (2023) Understanding of machine learning with deep learning: architectures, workflow, applications and future directions. Computers 12(5):91
Tidwell JH, Allan GL (2001) Fish as food: aquaculture's contribution. Ecological and economic impacts and contributions of fish farming and capture fisheries. EMBO Rep 2(11):958–963
Tigchelaar M, Leape J, Micheli F, Allison EH, Basurto X, Bennett A, Bush SR, Cao L, Cheung WWL, Crona B, DeClerck F, Fanzo J, Gelcich S, Gephart JA, Golden CD, Halpern BS, Hicks CC, Jonell M, Kishore A et al (2022) The vital roles of blue foods in the global food system. Glob Food Secur 33:100637

Vivekanandan E (2017) Importance of fish stock assessment to fisheries management. Summer School on Advanced Methods for Fish Stock Assessment and Fisheries Management, pp 123–131
Wang J, Tabeta S (2023) Four-channel generative adversarial networks can predict the distribution of reef-associated fish in the South and East China Seas. Eco Inform 78:102321
Wang Z, Wang T, Zhang X, Wang J, Yang Y, Sun Y, Guo X, Wu Q, Nepovimova E, Watson AE, Kuca K (2024) Biodiversity conservation in the context of climate change: facing challenges and management strategies. Sci Total Environ 937:173377
Watson JT, Haynie AC, Sullivan PJ, Perruso L, O'Farrell S, Sanchirico JN, Mueter FJ (2018) Vessel monitoring systems (VMS) reveal an increase in fishing efficiency following regulatory changes in a demersal longline fishery. Fish Res 207:85–94
Watson B, Reimer MN, Guettabi M, Haynie A (2021) Commercial fisheries & local economies. J Environ Econ Manag 106:102419
Welch H, Ames RT, Kolla N, Kroodsma DA, Marsaglia L, Russo T, Watson JT, Hazen EL (2024) Harnessing AI to map global fishing vessel activity. One Earth 7(10):1685–1691
Yang X, Zhang S, Liu J, Gao Q, Dong S, Zhou C (2021) Deep learning for smart fish farming: applications, opportunities and challenges. Rev Aquac 13(1):66–90
Yassir A, Jai Andaloussi S, Ouchetto O, Mamza K, Serghini M (2023) Acoustic fish species identification using deep learning and machine learning algorithms: a systematic review. Fish Res 266:106790
Zhang R, Chen T, Wang Y, Short M (2023) Systems approaches for sustainable fisheries: a comprehensive review and future perspectives. Sustain Prod Consum 41:242–252
Zhu K, Jiang M (2010) Quantum artificial fish swarm algorithm. In: 2010 8th world congress on intelligent control and automation, pp 1–5
Zuccotto M, Castellini A, La Torre D, Mola L, Farinelli A (2024) Reinforcement learning applications in environmental sustainability: a review. Artif Intell Rev 57(4):88

Tools and Techniques in Bioinformatics Research

22

Vikash Kumar and Basanta Kumar Das

22.1 Introduction

Aquatic resources face threats from various human activities. The level of pollution in rivers, reservoirs, and lakes can be somewhat predicted by considering several factors, including industrialization, population density, pollution control methods, and specific features of the receiving water body, such as exchange rates, depth, and others. The effect of pollutants depends on factors like their amount, toxicity, and the sensitivity or productivity of the recipient ecosystem. When sufficient knowledge of these factors is available, several conclusions can be drawn about the pollution status of aquatic ecosystems. Due to pollution, the biodiversity of fisheries is also declining daily in the aquatic ecosystems (Kumar et al., 2024).

Conventional approaches for evaluating biological quality often need experts to visually identify a small number of distinct taxonomic categories, which is a time-consuming task and may provide a restricted perspective on ecological conditions. The field of bioinformatics has fundamentally transformed the manner in which we analyze data associated with predictive ecology. Utilizing molecular techniques coupled with bioinformatics to evaluate biodiversity and ecosystem function has many benefits compared to conventional approaches. These techniques are now being used across a range of organisms, from microorganisms to mammals, following the "barcodes to biomes" strategy. The advancements made in these areas and ongoing enhancements in high-throughput sequencing using the bioinformatics approach provide opportunities for using sequence data in a formal integrated ecosystem assessment. Over the last 10 years, there has been a significant surge in the sequencing of fish genomes, which has had a transformative impact on the fields of comparative and evolutionary genomics. This has resulted in a wealth of new

V. Kumar · B. K. Das (✉)
ICAR-Central Inland Fisheries Research Institute, Barrackpore, Kolkata, India

© The Author(s), under exclusive license to Springer Nature Singapore Pte Ltd. 2026

B. K. Das et al. (eds.), *Bioinformatics Tools for Predictive Ecology and Fisheries*, Sustainable Aquatic Food Systems, https://doi.org/10.1007/978-981-95-2573-7_22

knowledge on the biology of vertebrate genomes (Yoshida et al. 2014). The field of fish genomics has undergone a fast transformation due to the emergence of high-quality genome assemblies at the chromosomal level and extensive sequencing datasets (Roy et al., 2024, 2025). These resources serve as valuable roadmaps for making significant discoveries (Behera et al. 2021).

Omics technology, representative of collective technologies, is used to explore the function, mechanisms, interactions, and relationships between genes, transcripts, proteins, lipids, and other biomolecules in a specific biological sample in an unbiased and non-targeted manner (Horgan et al. 2011). The completion of the human genome project marked the start of high-throughput technologies. From this point onward, various comprehensive "omic" approaches—such as genomic, transcriptomic, proteomic, metabolomic, and others—were progressively developed to address specific questions in biological research. These high-throughput screening technologies facilitate the exploration of the complete spectrum of biological molecules, including genes (genomics), mRNA (transcriptomics), proteins (proteomics), lipids (lipidomics), and metabolites (metabolomics), which collectively form and sustain cellular life (Horgan et al. 2011; Westerhoff et al. 2011). The main reasons behind the increasing popularity of "omics" technology in biological research are its capability to give the following satisfactory information to the researcher: understandable view of a complex biological system as a whole, exploration of the global status of biological molecule/s inside the organelle/cell/tissue/organ under specific condition, identification of key molecule/s in the biological system/process/sample being studied, revelation of the reason behind the different behavior of biomolecules under the condition being examined. The potential application diversity of "omic" approaches is another cause of their increasing importance in biological research. It is used for exploring the normal biological process as well as for understanding the etiology, screening, diagnosis, and prognosis of a disease (Horgan et al. 2011). In the future, biological research will be dependent on integrative "omic" technologies or system biology for the development of predictive, preventive, and personalized new approaches for fighting against the diseases (Kumar et al. 2025a, 2025b; Kell et al. 2006; Evans et al. 2011).

22.2 Genomics Application in Fisheries and Ecosystems

Genomics involves the comprehensive study of gene expression worldwide. While the genome contains the organism's genetic information, it does not reveal functional details. The genome remains static, whereas the proteome is highly dynamic (see Fig. 22.1). The genetic data in the genome manifests in biological functions, primarily through proteins. Genomics can be subdivided into structural and functional genomics. Structural genomics deals with structure, organization, and evolution, whereas functional genomics deals with expression analysis.

Living organism comprises the largest number of non-tetrapod vertebrates and have significantly advanced our knowledge of evolution, ecology, and genetics. The concept of "genomics" pertains to the entirety of heritable information contained

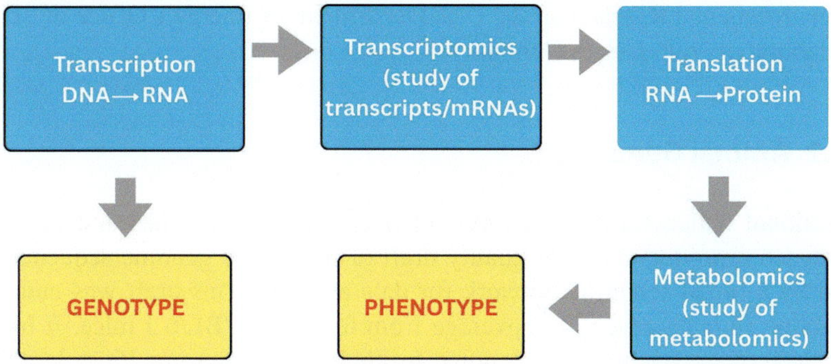

Fig. 22.1 Flow of omics technology and central dogma of molecular biology

within the DNA of an organism. Contemporary genomic technologies offer the necessary tools to analyze this data and reveal variations in genome composition that yield valuable insights regarding specific individuals, populations, and species (Ahmad et al. 2021). Genomics plays a significant role in enhancing our understanding of genome function, evolution, speciation, selection, adaptation, and species diversity. It is crucial to sequence the genomes of an organism to investigate their detailed genomic evolution and to employ various comparative genomics methodologies to examine the genomes of other organisms. The genomic sequences of an organism are compared with those of other organisms in order to track the evolutionary history and determine the origin of species by using Molecular Evolutionary Genetics Analysis (MEGA) software (Tamura et al. 2021). For instance, the taxonomic makeup of prokaryotes is determined using a collection of 16S rRNA genes, while fungi are analyzed using ITS sequences. Animals, on the other hand, are studied using either 18S rRNA or a portion of the mitochondrial COI gene. The technique is referred to as metabarcoding, drawing an analogy to species barcoding. The use of comparative genomics, molecular cytogenetics, in situ hybridization of sequences, and chromosomics presents new opportunities for the progression of genome study. Given the significance of bioinformatics coupled with genomic methods, it is widely thought that studying Zebrafish (Danio rerio) would pave the way to deepen our knowledge of many disease-causing genes and their activities in human diseases.

22.2.1 Human Genome Project: Revolution of Genomics

The completion of the human genome project marked the beginning of the genomics revolution. It involved identifying and mapping the human genome, which comprises 3.2 billion bases and roughly 30,000 to 40,000 protein-coding genes, from both physical and functional perspectives (Bench 1998). In 2005, a complete draft of the genome was announced by Armand Patrinos, Head of the Office of Biological and Environmental Research at the U.S. Department of Energy's Office of Science, in collaboration with Celera Genomics.

22.2.2 Animal Genomes

International collaboration between centers in the United States and the United Kingdom has produced a high-quality draft of the mouse genome sequence, supported by a broad scientific network for data analysis. This draft was created by assembling roughly sevenfold coverage from female C57BL/6 J mice. It includes about 96% of the euchromatic genome (excluding chromosome Y) in large, contiguous sequence contigs, often over 50 megabases (Mb). The mouse (Mus musculus) genome is approximately 14% smaller than the human genome (2.5 Gb vs. 2.9 Gb) (Waterston et al. 2002), likely due to a higher deletion rate in mice. The chicken (Gallus gallus) genome (~1.2 Gb), sequenced by the Washington University Genome Sequencing Center (WUGSC), had its first draft released in March 2004. Notably, avian genomes show large variability in chromosome size: besides sex chromosomes, chickens have 38 pairs of autosomes—5 macro-, 5 intermediate, and 28 microchromosomes. An international consortium found that chickens and humans share over half of their genes, yet their DNA sequences diverge in ways that might explain some key differences between birds and mammals (Bellott et al. 2010). The initial bovine (Bos taurus) genome assembly and analysis and the study of cattle genetic history were published on April 24, 2009 (Zimin et al. 2009). The DNA for the whole-genome shotgun sequences was provided by Dr. Timothy Smith's laboratory at the US Meat Animal Research Centre, Clay Centre. The western clawed frog (*Xenopus tropicalis*) serves as an important model for studying vertebrate development, offering the experimental benefits of the African clawed frog (*Xenopus laevis*) along with more manageable genetics. Its genome contains over 20,000 protein-coding genes, including orthologs of at least 1700 human disease genes (Hellsten et al. 2010). More than one million expressed sequence tags have confirmed the annotation. The genome shares significant synteny with human and chicken genomes across large parts of chromosomes, although this is disrupted by lineage-specific chromosome fusions and fissions, primarily within the mammalian lineage.

22.2.3 Fish Genomes

Sydney Brenner and his colleagues, Greg Elgar, Sam Aparicio, and Byrappa Venkatesh, initiated the fugu (*Takifugu rubripes* or *Fugu rubripes*) genome project in Cambridge in 1989. The draft genome of fugu was published in 2002 by the International Fugu Genome Consortium through whole-genome shotgun Sequencing (Aparicio 2002). Fugu is the first vertebrate genome made publicly available after the human genome. More than 30,000 genes have been identified in the fugu genome. Compared to the human genome, they have observed that approximately the same number of genes are present in both the human and fugu genomes, with few exceptions in genes related to the immune system, metabolic pathways, and other physiological activities. However, the genome size of fugu is seven times smaller than the human genome, which might be due to the absence of large repeat-filled tracts and the lower number of chromosomes (22 chromosomes) in fugu. This compactness in the genome helps in the detection of gene sequences and regulatory elements. More than 1000 unrecognized human genes are predicted in comparison with the genome, chromosomal segments from the common vertebrate ancestor, but with considerable scrambling of gene order. *Tetraodon nigroviridis* is a freshwater puffer fish known for having the smallest vertebrate genome. Its draft genome was published in 2004, featuring long-range linkage and significant anchoring to the 21 chromosomes of *Tetraodon*. The genome was sequenced via whole-genome shotgun (WGS) technology, covering 342 Mb, with the largest scaffold reaching 7.6 Mb, approximately the length of a typical *Tetraodon* chromosome arm (Jaillon et al. 2004) (Table 22.1). The freshwater teleost medaka (Oryzias latipes) is native to East Asia and is an excellent model organism for a wide range of biology, including ecotoxicology, carcinogenesis, sex determination, and developmental genetics. The draft genome sequence of medaka consists of 20,141 genes, including 2900 novel genes. The whole-genome shotgun covers 700 megabases of Medaka's genome (Kasahara et al. 2007).

Danio rerio is a commonly used model organism for studying vertebrate development and gene functions. In 2009, researchers from the Institute of Genomics and Integrative Biology in Delhi announced they had sequenced the genome of a wild zebrafish strain, which contains 1.7 billion base pairs (see Table 22.1). This genome was sequenced with 39-fold coverage (Patowary et al. 2013). The International Collaboration to Sequence the Atlantic Salmon Genome (ICSASG) published a genome sequence that locates and maps all genes in Atlantic salmon, serving as a reference for other salmonids (Davidson et al. 2010; Berthelot et al. 2014). Atlantic cod (*Gadus morhua*), a large, cold-water teleost pivotal to longstanding commercial fisheries and emerging aquaculture, has a genome assembled mainly through 454 sequencing of shotgun and paired-end libraries. Automated annotation identified 22,154 genes (see Table 22.1), and the draft genome was published in 2011 (Star et al. 2011). The genome of the three-spined stickleback, at 675 megabases, is quite compact for vertebrates—less than a quarter the size of a typical mammal genome. Sticklebacks have 21 cytologically visible chromosomes (Jones et al. 2012). Rainbow trout (Oncorhynchus mykiss), a species of major ecological significance, had its entire genome published in 2014, with a size of 2.0 Gb (Berthelot et al. 2014; see Table 22.1). A draft genome of domesticated common carp (*Cyprinus carpio*, strain *Songpu*) reveals 52,610 protein-coding genes and about 92.3% coverage of its paleotetraploid genome ($2n = 100$), with a total genome size of 1700 Mbp (see Table 22.1). The most recent whole-genome duplication event is estimated to have occurred around 8.2 million years ago. Genome resequencing of 33 individuals from global populations indicates a single origin for *C. carpio*, with two subspecies: *C. carpio haematopterus* and *C. carpio carpio* (Xu et al. 2014).

Table 22.1 The tools and packages utilized for the metagenomics study

Sl. no.	Computational pipeline	Resource/web tools
1.	Read quality control (QC)	Meta-QC-chain FitGCP NGS-eval StreamingTrim FastQC MultiQC FastQ screen BBDuk Khmer
2.	Read assembly	MEGAHIT RAY MetaVelvet-SL SOAPdenovo ABySS Genovo CLC GenomicsWorkbench
3.	Binning and analysis	MetaBAT2 MG-RAST Eragatis DIYA Kaiju
4.	Visualizing metagenomic data	AmphoraVizu Elviz Krona MGAviewer
5.	Functional Annotation	COGNIZER RTMg SALT UProC
6.	Gene prediction	MetaGeneAnnotator MetaGeneTack TransGeneScan Glimmer FragGeneScan Prodigal Prokka

22.3 Post-genomic Era

The first and overwhelming trend of genomics is that it is going functional. The genomics research is going ahead to the next steps, such as gene expression analysis, correlating the genome expression and functions, understanding networking among genes, and proteomics and protein–protein interactions. Potential application areas include clinical diagnostics, agrobiotechnology, environmental biotechnology, and pharmaco genomics. The functional aims are to simultaneously study the expression pattern of all genes present in the genome. This can be done at the DNA (metagenomics), RNA (transcriptomics), or protein level (Proteomics).

22.3.1 Metagenomics in Fisheries and Ecosystems

The metagenome refers to the whole collection of genomes from all species present in a given environmental sample, including extracellular DNA. Metagenomics, with the help of advanced next-generation sequencing techniques, offers a comprehensive understanding of the dynamics of the biome in various aquatic resources. This valuable information helps to unravel the physiological, metabolic, and ecological functions of the communities. Metagenomics can effectively analyze various types of microbiomes and their associated functional pathways, making it a valuable tool for monitoring different aquatic resources. By doing an analysis of sequence and function-driven metagenomics, it becomes possible to uncover novel pathways that may lead to the identification of new gene sequences with certain desired functionalities that can have significant ecological, industrial, and pharmacological value. Amplicon and shotgun approaches are the two main categories of metagenomic technologies based on the methods that are currently used (Table 22.1). The first strategy involves using the metagenome of a sample just as a template for amplifying a specific gene using the polymerase chain reaction (PCR). The sequencing process is performed on the PCR products acquired, rather than on the original metagenomic DNA template. On the other hand, shotgun sequencing examines the whole metagenome. There are no discrepancies or errors related to the selection of primers and the faulty PCR in this scenario. Nevertheless, this approach requires more effort and costs more (Vecherskii et al. 2021).

The integration of metagenomics with advanced bioinformatic techniques enabled the detection of microbial communities and provided access to genomic data without the need to isolate and culture microorganisms. Prolonged antibiotic use has resulted in the emergence of antimicrobial resistance (AMR) and alterations in its genetic determinants (Pillay et al. 2022). Metagenomics is a frequently used method for effectively monitoring a diverse range of AMR in environmental samples, including aquatic ecosystems. This metagenomics also allows us to study the beneficial probiotics, prebiotics, and bioremediation bacteria that have the natural tendency to establish a healthy environment (Brar et al. 2023).

Over the past 10 years, metagenomics has emerged as one of the most significant areas of microbial ecology research. Starting from 16S rRNA marker gene analysis to identify the community compositions to entire metagenome shotgun sequencing, which in turn permits functional analysis and metagenomics applications in a wide spectrum of study topics.

(i) Amplicon sequencing or targeted metagenomic sequencing: The most widely used method for characterizing microbial populations is targeted sequencing. Since 16S rRNA gene amplicon sequencing can be completed faster than other taxonomic markers, it is regarded as the most conserved taxonomic marker (bacterial). It is, therefore, the gold standard for in-depth phylogenetic study. The same technique is utilized for metagenomic research to identify the sample's taxonomic profile of microbial populations. ITS internal transcribed spacer sequencing of the nuclear ribosomal DNA is used to identify the eukaryotes, particularly the fungal community, in the metagenomic samples.

(ii) Whole genome shotgun metagenomic sequencing is a technique used to analyze the abundance of microorganisms in a specific environment. It not only identifies microbial species but also provides data on the genes within the metagenomic sample. This method enables functional characterization of microbial genes and is PCR-independent, avoiding biases from primer binding. As a result, it is useful for detecting unknown microbes that might be missed by targeted sequencing. Additionally, it helps in discovering novel viruses in the environment.

According to the microorganism studied, metagenomics can be divided into

1. Bacterial metagenomics
2. Fungal metagenomics
3. Viral metagenomics

Currently, bacterial metagenomics is the most extensively researched, while the development of sequence analysis metagenomics is advancing rapidly. Fungal and viral metagenomics are primarily utilized within sequence analysis metagenomics (Figs. 22.2 and 22.3).

22.3.2 Transcriptome and Metatranscriptome

Transcriptomics is a method used to study mRNA, messages, and transcripts. It involves analyzing gene expression by detecting and quantifying transcripts relative to each other. The goals of transcriptomics extend beyond measuring changes in gene expression levels across different samples; they also include mapping and annotating the transcriptome and understanding the functional structure of each gene within the genome.

Popular tools such as FastQC, Trimmomatic/Fastx Toolkit, GSNAP/Hisat2, HTSeq2/FeatureCounts, and EdgeR/DESeq2 are used for conducting transcriptomic analysis. Transcriptomic analysis involves the study of gene expression patterns in a biological sample. This manual outlines the step-by-step process from raw sequencing data quality control to differential expression analysis (Fig. 22.4).

22.3.2.1 Initial Quality Control with FastQC

Before starting the analysis, install FastQC and configure it for your system. FastQC is a widely used tool for initial quality control in genomics research. It assesses the quality of raw sequencing data by providing detailed visualizations and metrics, allowing researchers to quickly identify potential issues such as adapter contamination, overrepresented sequences, and poor base quality. Its user-friendly interface and comprehensive output make it an essential step in the preprocessing of next-generation sequencing data.

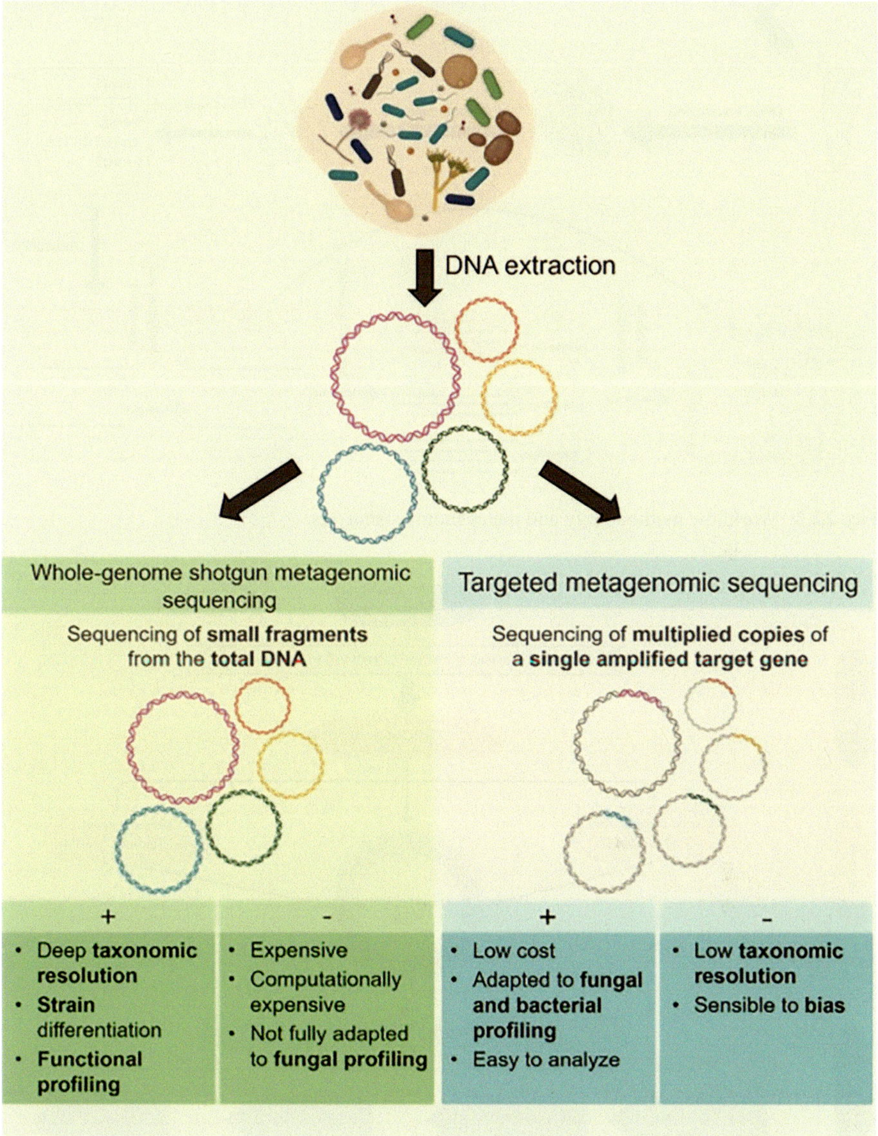

Fig. 22.2 Classification of metagenomics based on sequencing

22.3.2.2 Pre-processing and Quality Filtering

Trimmomatic and configure it for trimming and filtering raw data: Pre-processing and quality filtering are crucial steps in preparing raw sequencing data for downstream analysis. This involves trimming adapters, removing low-quality bases, and filtering out sequences that do not meet defined quality thresholds. These processes

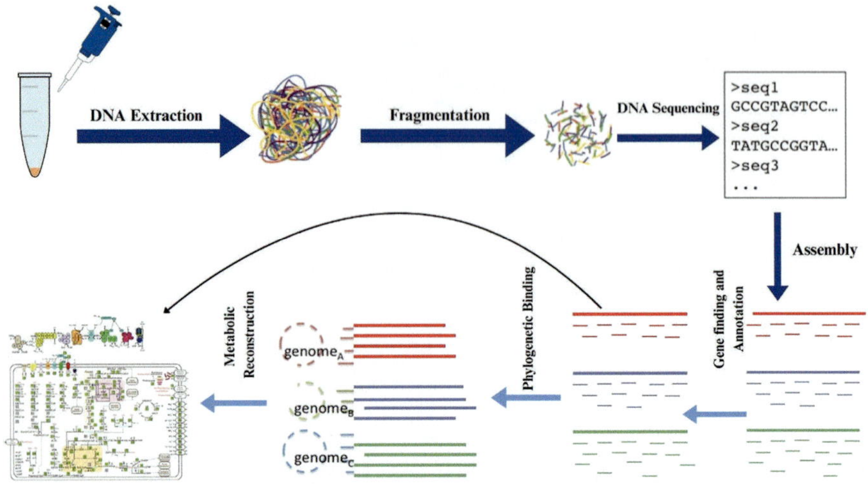

Fig. 22.3 Workflow methodology and metagenomics analysis

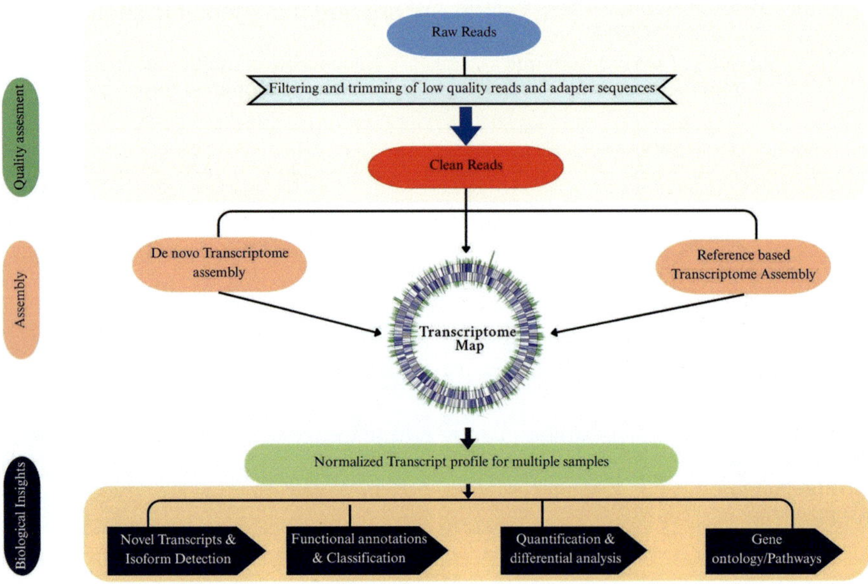

Fig. 22.4 Steps involved in transcriptome analysis

enhance the accuracy and reliability of subsequent analyses, ensuring that only high-quality data is used for further bioinformatics applications.

22.3.2.3 Alignment of Sequencing Reads

GSNAP for read alignment: GSNAP (Genomic Short-read Nucleotide Alignment Program) is a specialized tool for aligning high-throughput sequencing reads to a reference genome. It excels in handling spliced alignments, making it particularly useful for mapping RNA-seq data, where reads may span intron–exon junctions. GSNAP incorporates advanced algorithms to align reads accurately in the presence of splice junctions, contributing to more precise transcriptomic analyses in genomics research.

HISAT2 for read alignment: HISAT2 (Hierarchical Indexing for Spliced Alignment of Transcripts) is a widely used alignment tool in genomics research, particularly for RNA-seq data. It employs an innovative indexing strategy, enabling efficient and sensitive alignment of sequencing reads to a reference genome, especially across splice junctions. HISAT2's speed, accuracy, and compatibility with large eukaryotic genomes make it a popular choice for transcriptome analysis in diverse biological studies.

22.3.2.4 Counting Reads with HTSeq2 or Feature Counts

HTSeq2 and Feature Counts are commonly used tools for counting sequencing reads aligned to genomic features, such as genes or exons. They take aligned reads and assign them to specific features, facilitating the quantification of gene expression levels. These tools play a crucial role in transcriptome analysis, providing essential input for downstream differential expression and functional interpretation studies in genomics research.

22.3.2.5 Differential Expression Analysis

EdgeR and DESeq2: EdgeR and DESeq2 are widely used bioinformatics tools for conducting differential expression analysis in RNA-seq data. EdgeR employs a negative binomial distribution model to assess differential expression, while DESeq2 utilizes a negative binomial generalized linear model. Both tools help identify genes that are significantly differentially expressed between experimental conditions, providing valuable insights into biological processes and pathways affected by experimental perturbations in genomics research.

- Normalization and Transformation
- Perform normalization and transformation of count data.
- Statistical Testing for Differential Expression
- Conduct statistical tests to identify differentially expressed genes

22.3.2.6 Metatranscriptomics Analysis

Metatranscriptomics examines the functions and activities of all transcripts in environmental samples. It uncovers metagenomic transcripts and helps understand transcriptional regulation under specific conditions or times, offering a holistic perspective. An important benefit is that it eliminates the need to isolate and culture individual microbial species, thereby broadening resource availability. The typical workflow involves four main steps: sample preparation, library construction, high-throughput sequencing, and integrated bioinformatics analysis. We utilize a variety of sequencing platforms, such as Sanger, Illumina MiSeq/HiSeq, Roche 454, and

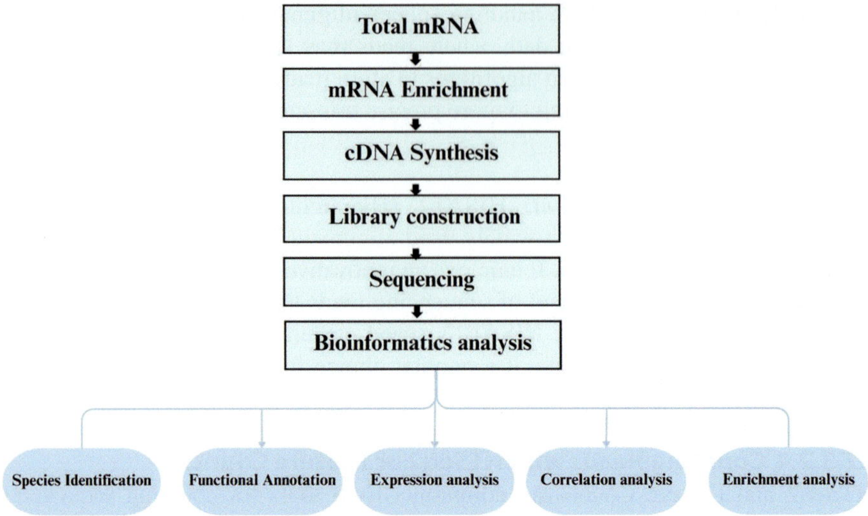

Fig. 22.5 Schematic workflow of metatranscriptomics

PacBio SMRT systems. The choice of sequencing strategy depends on your sample type and research aims. Our experienced team conducts comprehensive and tailored bioinformatics analyses to support your studies (Fig. 22.5).

22.3.3 Proteome and Proteomics

Proteins are biological macromolecules comprised of one or more long-chain polypeptides composed of amino acid residues linked through peptide bonds. As the meaning indicates, it is the "workhorse" of a biological system, essential for almost every structure, function, and regulation of life. The diversity of its function is expanded in every single step, from giving the shape of a cell to its division for generating a new one. Beyond the cellular level, it is also involved in each step necessary for maintaining the life of an organism. Proteome is the set of expressed proteins in a given type of cell or organism, at a given time, under defined conditions. The term proteomics was first coined in 1997 (James 1997), and it is defined as "the large-scale study of proteins" (Pandey and Mann 2000). "Proteomics includes not only the identification and quantification of the proteins, but also the determination of their localization, modifications, interactions, activities, and, ultimately, their function" (Fields 2001), "proteomics is a comprehensive, quantitative description of protein expression and its changes under the influence of biological perturbations such as stress, ageing, disease or drug treatment" (Anderson and Anderson 1998). Although existence of so many different definitions among researchers exists, the more complete one is "the study of all the protein forms expressed within an organism as a function of time, age, state, external factors, etc". The proteomics field can be divided into three major types: expression proteomics, structural proteomics, and functional proteomics (Graves and Haystead 2002; Monteith et al. 2005). The expression proteomics is classified when the quantitative differential proteome expression status of the biological samples is studied to answer biological queries. The comparative expression status can be checked at the total or sub-proteome level of the cell, organelle, tissue, or body fluids of normal and diseased samples. Depending on the study conditions and sample, the specific answer will be coming out like disease-specific proteins, novel signal transduction proteins, disease markers, and therapeutic targets (Graves and Haystead 2002). Structural proteomics deals with the three-dimensional structure of proteins or protein complexes, present or formed in specific cellular compartments under specified conditions. This proteomic field helps to understand the overall architecture of cells and explain how the specific structure of certain proteins or protein complexes gives a cell its unique characteristics (Graves and Haystead 2002; Christendat et al. 2000; Burley et al. 1999). Functional proteomics is an emerging field of biological research which "used to monitor and analyze the spatial and temporal properties of the molecular networks and fluxes involved in living cells" or "used to identify the molecular species that participate in such networks via functional stimulation, perturbation, or isolation of these networks" (Godovac-Zimmermann and Brown 2001).

22.3.3.1 Homology Modeling

Homology modeling, also known as comparative modeling, is a detective workhorse in the world of protein structures. It's a computational technique used to predict the 3D structure of a target protein by comparing it to a known structure of a template protein. Think of it like this: imagine you have a protein with an unknown shape (the target) and you find a picture of a similar protein (the template) with its shape revealed. By meticulously comparing the amino acid sequences of both proteins, like a detective comparing fingerprints, you can infer how the target protein is likely folded based on the similarities between them.

Here's how it works:

- Identify a template: You start by finding a protein with a known 3D structure that shares significant similarity in its amino acid sequence to your target protein. This "template" acts as a blueprint for your model.
- Align the sequences: Next, you carefully align the amino acid sequences of the target and template proteins. This ensures that corresponding residues in both proteins are matched up correctly.
- Build the model: Using the alignment and the template structure, you can then build a model of the target protein's 3D structure. This involves fitting the target sequence onto the template's backbone and then adjusting the side chains and loops to optimize the fit.
- Refine and evaluate: Finally, you refine the model to improve its accuracy and assess its reliability using various biophysical and statistical methods (Fig. 22.6).

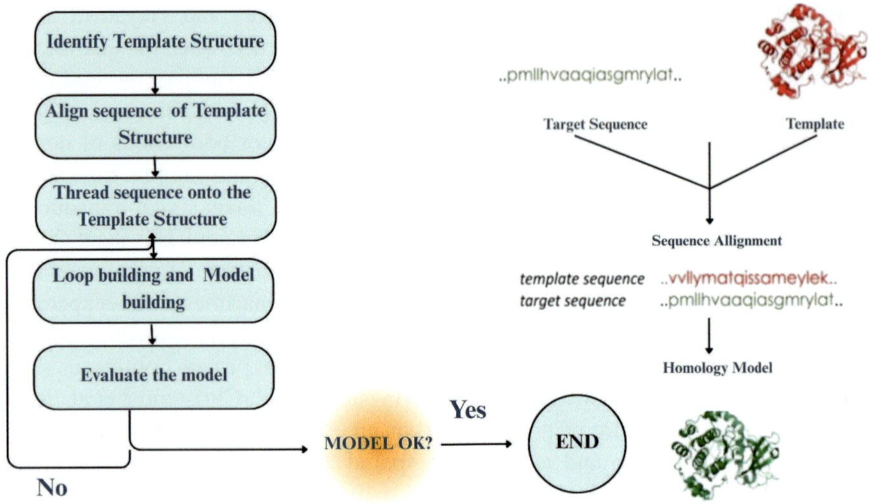

Fig. 22.6 Schematic workflow of protein modeling

22.3.3.2 Steps of Homology Modeling

Step 1: Template recognition and initial alignment:

- To identify the template, the program compares the query sequence to all the sequences of known structures in the PDB (e.g., BLAST).
- Usually, the template structure with the highest sequence identity and coverage will be the first option.
- Other considerations:
  - Conformational state (i.e., active or inactive)
  - Present co-factors
  - Other molecules or multimeric complexes
- It is possible to choose multiple templates and build multiple models.
- It is possible to combine multiple templates into one structure that is used for modeling.

Step 2: Alignment correction

- Having identified one or more possible modeling templates using the initial screen described above, more sophisticated methods are needed to arrive at a better alignment.

Step 3: Backbone generation

- When the alignment is ready, the actual model building can start
- Creating the backbone is trivial for most of the model: one simply transfers the coordinates of those template residues that show up in the alignment with the model

- If two aligned residues differ, the backbone coordinates for N, Cα, C and O and often also the Cβ can be copied
- Conserved residues can be copied completely to provide an initial guess

Step 4: Loop modeling

- For the majority of homology model-building cases, the alignment between the model and template sequence contains gaps
- Gaps in the model sequence are addressed by omitting residues from the template
- Gaps in the template sequences are treated by inserting missing residues into the continuous backbone
- Changes in loop conformation are notoriously hard to predict
- Loop modelling: a search is made through the PDB for known loops containing endpoints that match the residues between which the loop is to be inserted.

Step 5: Side chain modeling

- Libraries of common rotamers extracted from high-resolution X-ray structures are often used to position side chains
- The various rotamers are subsequently explored and scored with a variety of energy functions

Step 6: Model optimization

- To predict the side chain rotamers with high accuracy, we need the correct backbone, which in turn depends on the rotamers and their packing
- The common approach to address this problem is to iteratively model the rotamers and backbone structure
- First, we predict the rotamers, then remodel the backbone to accommodate rotamers, followed by a round of refitting the rotamers to the new backbone
- This process is repeated until the solution converges
- This boils down to a series of rotamer prediction and energy minimization steps

22.3.4 Proteogenomics

The completion of whole genome sequencing of several prokaryotes and eukaryotes, most importantly the human genome, guided the biological research towards a new era of genomic research. At the same time, limitations of genomic research present a challenge to biological researchers. To address the limitations, researchers started applying their knowledge beyond genomics. Accordingly, proteogenomics studies have been applied, combining genomics with proteomics to fill the gap between genomics and cellular mechanisms and unravel the underlying mechanisms of physiological conditions (Dove 1999) (Fig. 22.7). Although complexity increases from genomics to proteomics, it also facilitates researchers to a greater extent in discovering the scientific reasons behind the altered cellular, tissue-level, and organismal phenomena as a result of biological perturbation or external factors.

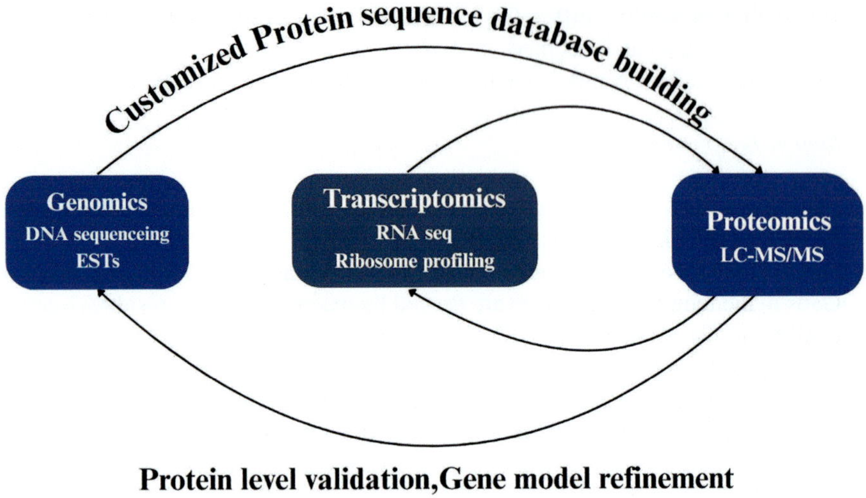

Fig. 22.7 Flow diagram of proteogenomics strategy

Proteomics can accurately explore almost every step of intercellular communication, which mediates complex intracellular signaling pathways through which cell maintains their fates as proliferation, migration, recognition, and differentiation (Tian et al. 2014; Choudhary and Mann 2010). Therefore, various types of proteomic approaches and technologies have been developed to address specific biological queries under a given set of conditions.

In a previous study, the muscle proteome map of *Labeo catla* revealed 70 spots on a 2-D gel, which were compared to muscle proteins identified in other species. Since the genome sequence data for this species is unavailable, a proteogenomic approach was used to generate partial gene sequences for the identified muscle proteins (Mohanty 2015). These methods effectively utilize proteomic data and adapt it to the organism's existing genomic information (Mann et al. 2001).

22.3.5 Phenomics

Phenome refers to the set of physical and biochemical traits belonging to a given organism. Phenomes get changed in response to genetic mutations and in response to gene-environment interaction. Phenomics is the study of measuring phenomes precisely. Specific phenotypes are the result of the combined effects of genetic and environmental factors and their interaction. The phenotypic differences between individuals could be attributed to their genetic makeup or environmental variation. Phenomics is increasingly used to understand disease mechanisms in humans and plants. Several phenomics databases for different organisms like the mouse phenomics database (http://phenome.jax.org); Knockout Mouse Project (http://www.mousephenotype.org), zebrafish (Zebrafish Phenome Project, and Zebrafish

Mutation Project, http://www.sanger.ac.uk/ resources/zebrafish/zmp; Cheng et al. 2011), and dog (Canine Phenome Project, http://www.caninephenome.org) have been developed.

22.3.6 Metabolomics

Unlike genes, transcripts, and proteins, metabolites are not directly encoded in the genome. They exhibit high chemical diversity, including carbohydrates, amino acids, lipids, nucleotides, and others. The metabolome represents the final product of gene transcription, meaning that changes here are often amplified compared to shifts in the transcriptome and proteome (Horgan and Kenny 2011). One key advantage of metabolomics is its ability to encompass a broad array of chemical structures that are highly variable and change over time. This approach provides a comprehensive view by enabling the unbiased and simultaneous measurement of thousands of metabolites, which helps in understanding biological responses to environmental factors (both natural and human-made), sex, gender, lifecycle stages, and biotic interactions such as infection or herbivory. Metabolomics involves analyzing the global metabolite profile within a system—whether it is a cell, tissue, or organism—under specific conditions. However, analyzing the metabolome is particularly challenging due to the diverse chemical nature of metabolites.

22.3.7 Lipidomics

Lipidomics is the study of the structure and function of the complete set of lipids (the lipidome) produced in a given cell or organism, as well as their interactions with other lipids, proteins, and metabolites. Lipidomics is classified into two categories: membrane lipidomics and mediator lipidomics. Membrane lipidomics involves the quantitative study of lipids that are constituents of membranes, and mediator lipidomics involves the study of low-abundant lipids.

22.4 Interpretation of Omics Technology Through Computational Biology

High-throughput technologies generate vast amounts of data, including numerous genes, gene expression profiles, and protein abundances, often obtained through single or combined methods. Computational biology and bioinformatics tools are essential for analyzing these large datasets. This approach enables the integration of unstructured multiomics data into a coherent framework, facilitating the exploration of common biological phenomena (Kuo et al. 2013). Mathematical modelling establishes links between molecules identified by omic technologies and relevant biological mechanisms, helping researchers understand how complex biological processes interact under specific conditions and how their alterations may influence disease development. Furthermore, mathematical models assist in systematically analyzing system perturbations, generating new hypotheses and experimental designs, and evaluating the potential of specific molecules as novel therapeutic targets. Several publicly available databases and data management tools have been developed for standardized data handling, analysis, and reporting. Consequently, systems biology approaches effectively bridge the gap between in vitro systems and in vivo human biology, enabling the analysis of complex biological processes.

22.4.1 Public Database, Genome Browsers, Sequence Retrieval, File Formats, BLAST Analysis, and Primer Designing

Bioinformatics research has been revolutionized by the advent of various tools and resources that facilitate the exploration and analysis of genetic information. In this write-up, we will delve into the essential components of molecular biology research, including public databases, genome browsers, sequence retrieval, file formats, BLAST analysis, and primer design.

National Center for Biotechnology Information (NCBI): NCBI is a complex center that provides a wealth of biological knowledge and is crucial to bioinformatics. NCBI, which houses databases like as GenBank, PubMed, and BLAST, is critical to the preservation and dissemination of genetic data. GenBank, a storehouse of nucleotide sequences, stimulates research efforts, whereas PubMed is an extensive resource for literature searches. The NCBI's BLAST (Basic Local Alignment Search Tool) supports sequence similarity searches, which help in gene identification and annotation.

Uniprot: Uniprot, a collaboration between the European Bioinformatics Institute (EBI), the Swiss Institute of Bioinformatics (SIB), and the Protein Information repository (PIR), serves as a comprehensive repository for protein sequence and functional data. Its enormous collection of protein data, which includes both published (Swiss-Prot) and unreviewed (TrEMBL) entries, facilitates the study of protein structure, function, and evolution.

ExPasy: The Expert Protein Analysis System (ExPasy) supplements Uniprot by offering tools and resources for analyzing protein sequences. ExPasy helps researchers understand protein biology by predicting post-translational modifications and modeling 3D structures.

Protein Data Bank (PDB): The PDB is the primary archive for three-dimensional structures of biological macromolecules. PDB, which has a large library of protein, nucleic acid, and complex structures, allows for the investigation of molecular structures at the atomic level. Researchers use this resource to obtain insight into the spatial organization of biomolecules, which aids drug development and structural biology research.

Database Search: Efficient investigation of primary sequence databases requires advanced database search strategies. Researchers may use tools like BLAST to compare their sequences to enormous databases in search of homologous sequences, revealing evolutionary linkages and functional insights.

Genome Databases: Genome databases contain entire genome sequences, which provide a comprehensive picture of an organism's genetic composition. Ensembl and Genomes Online Database (GOLD) are resources for comparative genomics that help researchers comprehend genome diversity and evolution.

BLAST (Basic Local Alignment Search Tool): BLAST, a key component of the bioinformatics toolbox, enables the fast comparison of nucleotide or protein sequences against large databases. This tool facilitates the identification of sequence similarities, allowing for the inference of functional and evolutionary links.

File Format: Standardized file formats enable efficient data interchange in bioinformatics. Common formats like as FASTA and GenBank enable the smooth exchange and analysis of sequence data, fostering global collaboration among researchers.

Primer Design: Primer design is an important step in molecular biology and PCR research. Researchers may use tools like Primer3, primer BLAST, APE, etc., to generate primers with optimum PCR properties, assuring specificity and efficiency in experimental settings.

22.4.2 ClustalW, Phylogenetic Tree Construction Using MEGA and iTOL

Understanding the evolutionary links between various species or genes is a critical component of molecular biology and genomics. In this write-up, we'll look at three key components in this field: ClustalW, a multiple sequence alignment tool; MEGA (Molecular Evolutionary Genetics Analysis), software for building phylogenetic trees; and iTOL (Interactive Tree of Life), a platform for visualizing and annotating phylogenetic trees.

22.4.2.1 ClustalW: Multiple Sequence Alignment Tools

ClustalW is a commonly used bioinformatics program that aligns multiple nucleotide or protein sequences. Des Higgins and colleagues developed ClustalW, which uses progressive pairwise alignment algorithms to align sequences based on their similarities. The technology is essential in bioinformatics and molecular biology, allowing researchers to compare and analyze the evolutionary links between distinct sequences. The basic stages in using ClustalW are to enter the sequences in a suitable format (for example, FASTA), then pick the relevant settings and run the software. The result generally consists of a multiple sequence alignment, where homologous areas are positioned against one another, offering insight into conserved regions and variances throughout the sequences.

22.4.2.2 Phylogenetic Tree Construction: Understanding Evolutionary Relationships

Phylogenetic trees are graphical representations of evolutionary relationships among different biological entities, such as species or genes. These trees illustrate the divergence and common ancestry of these entities based on their molecular or morphological characteristics. In the context of molecular biology, phylogenetic tree construction involves analyzing molecular data, such as DNA or protein sequences, to infer the evolutionary history of the organisms or genes in question. Phylogenetic tree construction typically involves three main steps: data collection (gathering molecular sequences), alignment (arranging sequences based on similarity), and tree inference (using algorithms to construct the tree). The resulting tree helps researchers understand the evolutionary relatedness, divergence times, and evolutionary patterns among the analyzed entities.

22.4.2.3 Phylogenetic Tree Construction Using MEGA

MEGA (Molecular Evolutionary Genetics Analysis) is a complete software program for performing molecular evolutionary analyses, including creating phylogenetic trees.

How it works:

- Input: MEGA requires an aligned dataset, typically generated using tools like ClustalW.
- Model Selection: Users choose an appropriate evolutionary model that best fits their dataset, considering factors like nucleotide substitution rates and sequence evolution characteristics.
- Tree Construction: MEGA employs various algorithms, such as Neighbor-Joining or Maximum Likelihood, to construct phylogenetic trees based on the evolutionary model and the input data.
- Bootstrapping: MEGA can perform bootstrap analysis to assess the robustness of the tree by generating multiple trees from resampled datasets.
- Visualization: The software provides visualization tools for exploring and annotating the constructed phylogenetic tree.

22.4.2.4 iTOL (Interactive Tree of Life)

The Interactive Tree of Life (iTOL) is a web-based tool for viewing and annotating phylogenetic trees. It enables users to explore and personalize the presentation of their trees interactively.

How it works:

- Input: iTOL accepts phylogenetic tree files, often in Newick format, generated by programs like MEGA.
- Visualization: iTOL provides an interactive and customizable interface for visualizing phylogenetic trees with branch coloring, labeling, and annotation features.

- Annotation: Users can add supplementary information to the tree, such as taxonomic labels, sequence data, or environmental metadata.
- Export: The final tree can be exported in various formats for publication or further analysis.

22.5 Applications of Omics Technology in Fisheries and Aquaculture

The application of Omics technology has the potential to provide commercial and non-commercial fisheries with the tools to support strong, competitive growth based on product quality and security, technological soundness, economic viability, environmental integrity, and social license. Omics technology can be an important part of the toolset required for such studies and for developing aquaculture practices that guarantee welfare and health (Fig. 22.8).

22.5.1 Disease Monitoring

A disease is a process that is characterized by "any impairment that interferes with or modifies the performance of normal functions, including responses to environmental factors such as toxicants and climate, nutrition, infectious agents, inherent or congenital defects, or any combination of these factors" (Das et al. 2025a, b, c; Mohanty et al. 2025). Fish diseases are major obstacles to aquaculture, impacting economic growth and sector revenue worldwide. They are a primary limiting factor for production, with disease outbreaks causing at least US$1400 billion in losses in developing Asian countries in 1990 (ADB/NACA 1991). Fish health is

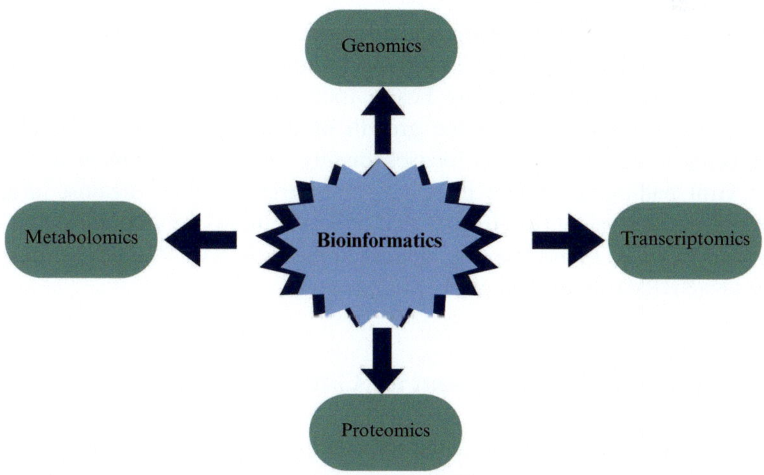

Fig. 22.8 Interpretation of omics technology by bioinformatics

closely linked to environmental conditions, as diseases are inherent in aquatic ecosystems. Often, issues with fish diseases in open inland waters are overlooked for years. Many factors contribute to disease development, but their interactions form a complex system. Causes include environmental factors, genetic host factors, and pathogen virulence (e.g., microbial). Diseases affect performance, predation risk, reproduction success, and survival. DNA-based diagnostics have been employed to detect important viral pathogens in farmed finfish and shrimp (Kumar et al. 2025a, 2025b; Das et al. 2024a, 2024b). For example, in finfish, detection methods target infectious hematopoietic necrosis virus, viral hemorrhagic septicemia virus, viral nervous necrosis virus (VNNV), and *Renibacterium salmoninarum*. In Japan, PCR screening of striped jack (*Pseudocaranx dentex*) broodstock for VNNV helps prevent vertical transmission by selecting PCR-negative spawners. However, concerns about this diagnostic method exist (FAO Fisheries). Phenomics is a new approach transitioning from traditional genotype/phenotype studies to comprehensive genome/phenome analyses, enhancing understanding of disease mechanisms and therapy. Historically, studying disease phenotypes involved forward genetics, linking phenotypes to genotypes, or reverse genetics, linking genotypes to phenotypes. Advances in "Omics" sciences now challenge existing definitions of complex diseases and drug combinations, as they reveal more detailed genomic, proteomic, and metabolomic variations. This has created major challenges for a complete understanding of the relationship between the genomic/proteomic/metabonomic variability and the phenotypic vascular disease in individual patients, and therefore, newer approaches are needed for redefining the disease using not only the traditional signs and symptoms but also the underlying genomic/proteomic causes and other factors.

22.5.2 Management Measure Development

One of the main difficulties that aquaculture farms face is the outbreak of infectious diseases that seriously impact their socio-economic development in several parts of the world. Therefore, approaching health issues in aquaculture has become an urgent requirement to maintain the growth of this activity. Immunization with nucleic acids in aquaculture has focused mainly on the development of vaccines against viral pathogens, where there has been great progress against infectious hematopoietic necrosis virus (IHNV) that affects salmon, viral hemorrhagic septicemia virus (VHSV) in trout, spring viremia of the carp (SVCV), and the snakehead rabdovirus (SHRV), among others.

22.5.3 Biomarker Development for Clinical Diagnosis

Arsenic is a highly carcinogenic environmental contaminant, and arsenic toxicity (arsenicosis) is a major public health problem in many parts of the world. Early diagnosis would help in the better management of patients. Banerjee et al. (2017) studied the impact of arsenic on lower vertebrates with the aim of identifying potential biomarkers of chronic arsenic toxicity using an Indian major carp, *Labeo rohita*, experimental organism. In this study, they have analyzed the plasma proteome of arsenic-exposed (15 ppm) carp by 2D gel electrophoresis and identified ApoA1, A2ML, Wap65, and Transferin as potential biomarkers.

Heat stress or high temperature stress has become a cause of concern because of the rise in global temperature in recent times. It has been predicted that around 20–30% of fish species assessed so far are at risk of extinction if the global temperature rises above 1.5–2 °C. Therefore, it is necessary to look for biomarkers of high-temperature stress that could be useful in monitoring stress in fish and taking necessary ameliorative steps. Attempts have been made by various researchers to use the heat shock proteins (Hsps), especially the Hsp70, to monitor heat stress in fish. However, Hsp70 is a protein the expression of which could increase in several stress conditions. Moreover, there is a great range of heat shock protein families. Thus, attempts have been made to identify the Hsp families that respond well to heat stress. A study in the murrel *Channa striatus* exposed to high-temperature stress for different time periods showed that hsp60, hsp70, and hsp78 responded well and could be used to monitor heat stress in C. striatus (Purohit et al. 2014). Similarly, in another fish, Puntius sophore, it was found that hsp47 and hsp90 play an important role in heat stress tolerance (Mohanty et al. 2016a). To find out what other proteins are in cross-talk with the Hsps, a 2D gel-based proteomics study was carried out in the liver tissues of C. striatus. This study showed that apart from Hsps, there are other chaperones like protein disulphide isomerase (PDI) and the antioxidative enzymes SOD, Ferritin, and glutathione-s-transferase that help the fish withstand heat stress. Further, pathway analysis revealed that both the antioxidative enzymes and heat shock proteins are regulated by Nrf-2 (Mohanty et al. 2016b). This study led to another feeding trial study where curcumin, a potential Nrf-2 inducer, was supplemented to fish to ameliorate heat stress, and positive results were obtained (Mohanty et al. 2018).

22.5.4 Nutrigenomics Studies

Nutrigenomics is a functional genomics discipline that investigates the effect of food and food constituents on gene expression. Nutrigenomics study aims to recognize the nutrient as well as diet-mediated modification on gene expression, alterations in normal metabolism, and effect on health status. Ngoh et al. (2015) have studied the effect of pelleted feed on Asian Sea bass, they have fed three different pelleted feeds to juvenile Sea bass, and comparative nutrigenomics analysis has been carried out relative to fish fed with frozen batfish through fatty acid composition, histology, and liver transcriptome analysis. They have reported that biosynthesis of unsaturated fatty acids and steroid biosynthesis pathways are significantly affected, and a reduced number of goblet cells were observed in fish fed with pelleted feed. Ganguly et al. (2018) reported the nutrigenomic information on hilsa (*Tenualosa ilisha*). They have studied different flesh quality attributes of hilsa in terms of nutritive value, pH, water holding capacity, expression of genes associated with fatty acid metabolism, and flesh quality. Additionally, a comparative analysis has been carried out in hilsa from the rivers Hooghly and Padma. Based on their findings, they have reported that Hooghly hilsa (medium size category, 500–700 g size) is superior to Padma hilsa in terms of oil content, ù-3 PUFAs, EPA and DHA, and essential amino acids. Based on these studies, it could be said that nutrigenomics studies help in the selection of feed and feed constituents for better growth and production in fisheries and the aquaculture sector.

22.5.5 Aspects Relating to Environmental Monitoring

Aquatic pollution with natural and anthropogenic contaminants is one of the major environmental concerns (Das et al. 2025d; Kumar et al. 2023). Water resources act as the ultimate sink for pollutants. Aquatic organisms, especially fish, have attracted considerable attention in pollution impact assessment studies. Biomonitoring is the most acceptable way to assess the pollution status in large open water ecosystems, and some biomonitoring tools are already incorporated into environmental monitoring programs (Viarengo et al. 2007), including CYP1A, HSP, and Metallothionine (MT). Several studies have reported the induction of CYP1A as an indicator of organic xenobiotics like polyaromatic hydrocarbons (PAH), polychlorinated biphenyls (PCB), organochlorine phosphates (OCP), persistent organic pollutants (POP), drugs, and pharmaceutical products (Al-Arabi and Goksøyr 2002; Tripathi et al. 2014). Similarly, MT is known as a biomarker of heavy metal exposure. Previously, Porte et al. (2001) have reported the detection of three isofroms of Hsp70 (78 kDa, 72 kDa and 70 kDa) from gill tissues of marine mussels (*Mytilus galloprovincialis*); in which 72 kDa band was the inducible omits in *M. galloprovincialis* collected from polluted sites of the Galician coast, NW Spain. Similarly, three spots were identified from white muscle tissues of *Rita rita* collected from polluted stretches of the river (Mohanty et al. 2018). A recent study have analyzed the gene expression of battery of heat shock proteins (Hsp27, Hsp47, Hsp60, Hsp70 and Hsp90) in liver and gill tissues of *R. rita* collected from different stretches of river Ganga and revealed that up-regulation of hsp47 and appearance of hsp70 isoform (Hsp70b) correlated well with environmental organic pollutants like HCH and aldrin (Mitra et al. 2018).

22.6 Conclusion and Future Perspective

Research involves not only collecting results and data but also interpreting that data, which is crucial for achieving the study's objectives. Bioinformatics, a branch of computer science, focuses on discoveries in biological science. It is an emerging field advancing in various areas of life sciences, such as biotechnology, molecular biology, agriculture, and medicine. Bioinformatic tools can facilitate the organization, analysis, and interpretation of extensive DNA sequence datasets and are also employed to examine gene expression and predict protein structures. In the future, the development and combined use of metabolomics, phenomics, genomics, and proteomics will enhance both the quality and quantity of aquaculture production through improved feed formulation, disease management, and vaccine production. The future looks promising as proteomic databases grow and more advanced, updated proteome techniques emerge, offering new chances to enhance, expand, and develop related products in aquaculture. Achieving this will need interdisciplinary collaboration among various fields such as physiology, cell biology, computer science, aquaculture, and the food industries.

References

Ahmad SF, Jehangir M, Srikulnath K, Martins C (2021) Fish genomics and its impact on fundamental and applied research of vertebrate biology. Rev Fish Biol Fish 32(2):357–385

Al-Arabi SAM, Goksøyr A (2002) Cytochrome P4501A responses in two tropical fish species, riverine catfish (*Rita rita*) and marine mudfish (*Apocryptes bato*). Comp Biochem Physiol C Toxicol Pharmacol 131:61–71

Anderson NL, Anderson NG (1998) Proteome and proteomics: new technologies, new concepts, and new words. Electrophoresis 19:1853–1851

Aparicio S (2002) Whole-genome shotgun assembly and analysis of the genome of Fugu rubripes. Science 297:1301–1310

Banerjee S, Mahanty A, Mohanty S, Mazumder DG, Cash P, Mohanty BP (2017) Identification of potential biomarkers of hepatotoxicity by plasma proteome analysis of arsenic-exposed carp *Labeo rohita*. J Hazard Mater 336:71–80

Behera BK, Dehury B, Rout AK, Patra B, Mantri N, Chakraborty HJ, Sarkar DJ, Kaushik NK, Bansal V, Singh I, Das BK, Rao AR, Rai A (2021) Metagenomics study in aquatic resource management: recent trends, applied methodologies and future needs. Gene Rep 25:101372

Bellott DW, Skaletsky H, Pyntikova T, Mardis ER, Graves T, Kremitzki C, Brown LG, Rozen S, WarrenWC, Wilson RK, Page DC (2010) Convergent evolution of chicken Z and human X chromosomes byexpansion and gene acquisition. Nature, 466(7306), pp.612–616

Bench AJ, Aldred MA, Humphray SJ, Champion KM, Gilbert JG, Asimakopoulos FA, Deloukas P, Gwilliam R, Bentley DR, Green AR (1998) A detailed physical and transcriptional map of the region ofchromosome 20 that is deleted in myeloproliferative disorders and refinement of the common deleted region. Genomics, 49(3), pp.351–362

Berthelot C, Brunet F, Chalopin D, Juanchich A, Bernard M, Noël B, Bento P, Da Silva C, Labadie K, Alberti A, Aury JM (2014) The rainbow trout genome provides novel insights into evolution after wholegenomeduplication in vertebrates. Nat Commun, 5(1), p.3657

Brar B, Kumar R, Sharma D, Sharma AK, Thakur K, Mahajan D, Kumar R (2023) Metagenomic analysis reveals diverse microbial community and potential functional roles in Baner rivulet, India. J Genet Eng Biotechnol 21(1):1–12

Burley SK, Almo SC, Bonanno JB, Capel M, Chance MR, Gaasterland T, Lin D, Šali A, Studier FW, Swaminathan S (1999) Structural genomics: beyond the human genome project. Nat Genet 23(2):151–157

Cheng PJ, Wang TH, Huang SY, Kao CC, Lu JH, Hsiao CH, Steven Shaw SW (2011) Differential proteomics analysis of amniotic fluid in pregnancies of increased nuchal translucency with normal karyotype. Prenat Diagn 31(3):274-281

Choudhary C, Mann M (2010) Decoding signalling networks by mass spectrometry-based proteomics. Nat Rev Mol Cell Biol 11(6):427–439

Christendat D, Yee A, Dharamsi A, Kluger Y, Savchenko A, Cort JR, Booth V, Mackereth CD, Saridakis V, Ekiel I, Kozlov G (2000) Structural proteomics of an archaeon. Nat Struct Biol 7(10):903–909

Das BK, Kumar V, Roy S, Malick RC, Bisai K, Jana AK, Dhar S (2024a) Pathological effects and immune modulation in host during Tilapia Parvovirus (TiPV) outbreak in cage and wetland Tilapia farms. Sci Rep, 14(1), p.28689

Das BK, Chakraborty HJ, Kumar V, Rout AK, Patra B, Das SK, Behera BK (2024b) Comparative metagenomic analysis from Sundarbans ecosystems advances our understanding of microbial communities and their functional roles. Sci Rep, 14(1), p.16218

Das BK, Kumar V, Roy S, Mohanty D, Jana AK, Gadnayak A (2025a) Whole-genome sequencing of novel pathogenic and multidrug-resistant Kytococcus sedentarius, causing mortality in fish species (Labeo rohita). Microbiology Resource Announcements, 14(1), pp.e01127–24

Das BK, Kumar V, Roy S, Mohanty D, Dhar S, Pradhan S.P (2025b) Genome assembly of multidrug-resistant Enterococcus gallinarum IFEGHNEK1 isolated from bighead carp, Hypophthalmichthys nobilis. Microbiology Resource Announcements, 14(4), pp.e01063–24

Das BK, Kumar V, Roy S, Mohanty D, Bisai K, Chakraborty HJ (2025c) De novo assembly of a potential fish pathogen Lactococcus lactis IFLLLBEK1 isolated from Labeo bata. Microbiology Resource Announcements, pp.e00455–25

Das BK, Mandal B, Kumar V, Roy S, Dalal S, Samanta S (2025d) Molecular traces of invasion: eDNA-based high-resolution mapping of Nile tilapia (Oreochromis niloticus) across eastern India's freshwater habitats of West Bengal, India. Frontiers in Marine Science, 12, p.1614086

Davidson WS, Koop BF, Jones SJ, Iturra P, Vidal R, Maass A, Jonassen I, Lien S, Omholt SW(2010) Sequencing the genome of the Atlantic salmon (Salmo salar). Genome biol, 11(9), p.403

Dove R (1999) Knowledge management, response ability, and the agile enterprise. J Knowl Manage 3(1):18–35

Evans JP, Meslin EM, Marteau TM, Caulfield T (2011) Deflating the genomic bubble. Sci, 331(6019), pp.861–862

Fields S (2001) Proteomics in genomeland. Science 291(5507):1221–1224

Ganguly S, Mahanty A, Mitra T, Mohanty S, Das BK, Mohanty BP (2018) Nutrigenomic studies on hilsa to evaluate flesh quality attributes and genes associated with fatty acid metabolism from the rivers Hooghly and Padma. Food Res Int 103:21–29

Godovac-Zimmermann J, Brown LR (2001) Perspectives for mass spectrometry and functional proteomics. Mass Spectrom Rev 20(1):1–57

Graves PR, Haystead TA (2002) Molecular biologist's guide to proteomics. Microbiol Mol Biol Rev 66:39–63

Hellsten U, Harland RM, Gilchrist MJ, Hendrix D, Jurka J, Kapitonov V, Ovcharenko I, Putnam NH,Shu S, Taher L, Blitz IL (2010) The genome of the Western clawed frog Xenopus tropicalis. Sci, 328(5978), pp.633–636

Horgan RP, Kenny LC (2011) 'Omic'technologies: genomics, transcriptomics, proteomics and metabolomics.Obstet Gynaecol, 13(3)

Jaillon O, Aury JM, Brunet F, Petit JL, Stange-Thomann N, Mauceli E, Bouneau L, Fischer C, Ozouf-Costaz C, Bernot A, Nicaud S (2004) Genome duplication in the teleost fish Tetraodon nigroviridis revealsthe early vertebrate proto-karyotype. Nat, 431(7011), pp.946–957

James P (1997) Protein identification in the post-genome era: the rapid rise of proteomics. Q Rev Biophys 30(4):279–331

Jones FC, Chan YF, Schmutz J, Grimwood J, Brady SD, Southwick AM, Absher DM, Myers RM,Reimchen TE, Deagle BE, Schluter D (2012) A genome-wide SNP genotyping array reveals patterns ofglobal and repeated species-pair divergence in sticklebacks. Curr biol, 22(1), pp.83–90

Kasahara M, Naruse K, Sasaki S, Nakatani Y, Qu W, Ahsan B, Yamada T, Nagayasu Y, Doi K, KasaiY, Jindo T (2007) The medaka draft genome and insights into vertebrate genome evolution. Nat, 447(7145), pp.714–719

Kell DB (2006) Systems biology, metabolic modelling and metabolomics in drug discovery and development. Drugdiscovery today, 11(23–24), pp.1085–1092

Kumar V, Bera T, Roy S, Vuong P, Jana C, Sarkar DJ, Devi MS, Jana AK, Rout AK, Kaur P, Da BK (2023) Investigating bio-remediation capabilities of a constructed wetland through spatial successional study of the sediment microbiome. NPJ Clean Water, 6(1), p.8

Kumar V, Roy S, Parida SN, Bisai K, Dhar S, Jana AK, Das BK (2024) Deciphering the impact of endoparasitic infection on immune response and gut microbial composition of Channa punctata. Frontiers in Cellular and Infection Microbiology, 14, p.1296769

Kumar V, Das BK, Parida PK, Dhar S, Adhikari A, Baisakhi B, Majumder A, Sahoo R, Jana AK, Samanta S (2025a) Co-infection of Aeromonas veronii and Megasporoporia sp. associated with mass mortality in Labeo bata: pathological effects and immune modulation in the host. Microbial Pathogenesis, p.107997

Kumar V, Das BK, Roy S, Dhar S, Bisai K, Adhikari A, Majumder A, Jana AK (2025b) Environmental transition navigates phenotype switching, affecting the virulence and multidrug-resistant profile of pathogenic Morganella morganii. Microbial Pathogenesis, 202, p.107430

Kuo TC, Tian TF, Tseng YJ (2013) 3Omics: a web-based systems biology tool for analysis, integration and visualization of human transcriptomic, proteomic and metabolomic data. BMC Syst Biol 7(1):64

Mann M, Hendrickson RC, Pandey A (2001) Analysis of proteins and proteomes by mass spectrometry. Ann Rev Biochem 70(1):437–473

Mitra T, Mahanty A, Ganguly S, Purohit GK, Mohanty S, Parida PK, Behera PR, Raman RK. Mohanty BP (2018) Expression patterns of heat shock protein genes in Rita rita from natural riverine habitat as biomarker response against environmental pollution. Chemosphere 211:535–546

Mohanty BP (2015) Proteomics in fisheries and aquaculture. Proteomics Society of India (PSI). PSI Newsletter

Mohanty BP, Ganguly S, Mahanty A, Sankar TV, Anandan R, Chakraborty K, Paul BN, Sarma D, Syama Dayal J, Venkateshwarlu G, Mathew S (2016a) DHA and EPA content and fatty acid profile of 39 food fishes from India. BioMed Res Int 2016(1):4027437

Mohanty BP, Sankar TV, Ganguly S, Mahanty A, Anandan R, Chakraborty K, Paul BN, Sarma D, Dayal JS, Mathew S, Asha KK (2016b) Micronutrient composition of 35 food fishes from India and their significance in human nutrition. Biol Trace Elem Res 174(2):448-458

Mohanty BP, Mahanty A, Mitra T, Parija SC, Mohanty S (2018) Heat shock proteins in stress in teleosts. In Regulation of Heat Shock Protein Responses (pp. 71–94). Cham: Springer International Publishing

Mohanty D, Kumari M, Kumar V, Kumari P, Behera B, Das BK (2025) Genome sequence of prodigiosin-producing and tetracycline-resistant Serratia marcescens (IFSMLMEK1), identified from a freshwater bivalve Lamellidens marginalis. Microbiology Resource Announcements, 14(4), pp.e01115–24

Monteith DT, Hildrew AG, Flower RJ, Raven PJ, Beaumont WRB, Collen P, Kreiser AM, Shilland EM, Winterbottom JH (2005) Biological responses to the chemical recovery of acidified fresh waters in the UK. Environ Pollut 137(1):83–101

Ngoh SY, Tan D, Shen X, Kathiresan P, Jiang J, Liew WC, Thevasagayam NM, Kwan HY, Saju JM, Prakki SR, Goh CH (2015) Nutrigenomic and nutritional analyses reveal the effects of pelleted feeds on Asian seabass (Lates calcarifer). PloS One 10(12):e0145456

Pandey A, Mann M (2000) Proteomics to study genes and genomes. Nature 405(6788):837–846

Patowary A, Purkanti R, Singh M, Chauhan R, Singh AR, Swarnkar M, Singh N, Pandey V, Torroja C, Clark MD, Kocher JP (2013) A sequence-based variation map of zebrafish. Zebrafish, 10(1), pp.15–20

Pillay S, Calderón-Franco D, Urhan A, Abeel T (2022) Metagenomic-based surveillance systems for antibiotic resistance in non-clinical settings. Front Microbiol 13:1066995

Porte H, Siat J, Guibert B, Lepimpec-Barthes F, Jancovici R, Bernard A, Foucart A, Wurtz A (2001) Resection of adrenal metastases from non-small cell lung cancer: a multicenter study. Ann Thorac Surg 71(3):981–985

Purohit GK, Mahanty A, Suar M, Sharma AP, Mohanty BP, Mohanty S (2014) Investigating hsp gene expression in liver of Channa striatus under heat stress for understanding the upper thermal acclimation. BioMed Res Int 2014(1):381719

Roy S, Parida PK, Ramya VL, Kumar V, Bhakta D, Behera BK, Das BK (2024) Whole mitochondrial genome sequencing and phylogenetic analysis of Gangetic mystus (Mystus cavasius). Mitochondrial DNA Part B, 9(11), pp.1539–1543

Roy S, Parida PK, Ramya VL, Bhakta D, Kumar V, Behera BK, Das BK (2025) The Complete Mitochondrial Genome of Indigenous Catfish Mystus gulio: Shotgun Assembly, Insight into Structural Characterization and Phylogenetic Relationships. Agricultural Research, pp.1–10

Star B, Nederbragt AJ, Jentoft S, Grimholt U, Malmstrøm M, Gregers TF, Rounge TB, Paulsen J, Solbakken MH, Sharma A, Wetten OF (2011) The genome sequence of Atlantic cod reveals a uniqueimmune system. Nat, 477(7363), pp.207–210

Tamura K, Stecher G, Kumar S (2021) MEGA11: molecular evolutionary genetics analysis version 11. Mol Biol Evol 38(7):3022–3027

Tian Y, Koganti T, Yao Z, Cannon P, Shah P, Pietrovito L, Modesti A, Aiyetan P, DeLeon-Pennell K, Ma Y, Halade GV (2014) Cardiac extracellular proteome profiling and membrane topology analysis using glycoproteomics. PROTEOMICS–CA 8(7-8):595–602

Vecherskii MV, Semenov MV, Lisenkova AA, Stepankov AA (2021) Metagenomics: a new direction in ecology. Biol Bull 48(3):S107–S117

Viarengo A, Lowe D, Bolognesi C, Fabbri E, Koehler A (2007) The use of biomarkers in biomonitoring: a 2-tier approach assessing the level of pollutant-induced stress syndrome in sentinel organisms. Comp Biochem Physiol C Toxicol Pharmacol 146(3):281–300

Waterston RH, Lander ES, Sulston JE (2002) On the sequencing of the human genome. Proceedings ofthe National Academy of Sciences, 99(6), pp.3712–3716

Westerhoff HV (2011) Systems biology left and right. In Methods in Enzymology (Vol. 500, pp. 3–11). AcademicPress.

Xu P, Zhang X, Wang X, Li J, Liu G et al (2014) Genome sequence and genetic diversity of the common carp, Cyprinus carpio. Nat Genet 46:2012–2019

Yoshida S, Date Y, Akama M, Kikuchi J (2014) Comparative metabolomic and ionomic approach for abundant fishes in estuarine environments of Japan. Sci Rep 4:7005

Zimin AV, Delcher AL, Florea L, Kelley DR, Schatz MC (2009) A whole-genome assembly of the domestic cow, *Bos taurus*. Genome Biol 10:R42

23 Scope of Molecular Modeling in Aquatic Animal Health Management

Gulshan Kumar, Pravesh Kumar, Roshan Kumar Ram, Rajive Kumar Brahmchari, and P. P. Srivastava

23.1 Introduction

Aquaculture has emerged as a vital sector in global food production, with fish contributing significantly to the animal protein intake of populations worldwide. In 2022, the global production of aquatic animals reached 185 million tonnes, with an increasing rate of 4% from 2020 (FAO 2024). Farming of aquatic animals contribute 94 million tonnes, representing 51% of the total production. The first sale value of the 2022 global production of aquatic animals was estimated at USD 452 billion, of which the contribution of aquaculture is USD 296 billion. Disease outbreak is a major challenge to global aquaculture due to huge environmental and economic losses. As the demand for aquatic animal protein is increasing day by day, the intensity of aquaculture is also increasing. In the current time, aquaculture most often witnesses regular disease outbreaks. These diseases also include many global pathogenic diseases caused by bacteria, viruses, fungi, and parasites. These diseases occur both in fish and shellfish, causing huge environmental and economic losses globally. These problems underscore the pressing need for effective disease management strategies in aquaculture. Traditional diagnostic methods, while foundational, often fall short in sensitivity and speed, hindering timely disease management and control. Advancements in molecular diagnostics have revolutionized the detection and understanding of aquatic pathogens (Abdelsalam et al. 2023; Jaies et al. 2024). Techniques such as polymerase chain reaction (PCR), real-time PCR, and loop-mediated isothermal amplification (LAMP) offer rapid and accurate

G. Kumar (✉) · P. Kumar · R. K. Ram · R. K. Brahmchari · P. P. Srivastava
College of Fisheries, Dr. Rajendra Prasad Central Agricultural University, Dholi, Muzaffarpur, Bihar, India
e-mail: gulshankumarcof35@rpcau.ac.in

© The Author(s), under exclusive license to Springer Nature Singapore Pte Ltd. 2026

B. K. Das et al. (eds.), *Bioinformatics Tools for Predictive Ecology and Fisheries*, Sustainable Aquatic Food Systems, https://doi.org/10.1007/978-981-95-2573-7_23

identification of pathogens, even in asymptomatic carriers. These methods have enhanced our ability to monitor disease outbreaks and implement effective biosecurity measures.

Molecular modeling encompasses computational techniques that simulate molecular structures and behaviors (Karkoutly et al. 2021; Yalcin-Ozkat 2021). It includes methods like structure prediction of biomolecules, molecular docking, molecular dynamics, and quantum mechanics to predict molecular interactions and properties. These techniques are pivotal in drug discovery and understanding biomolecular functions. Recent advancements in molecular modeling tools have enhanced their applications in pharmaceutical and biomedical research, facilitating the exploration of ligand-receptor interactions and protein conformations. Prediction of three-dimensional (3D) structure of biomolecules, especially proteins, is an essential part of molecular modeling. Traditional methods for determining protein structures, such as X-ray crystallography and nuclear magnetic resonance (NMR) spectroscopy, are time-consuming and resource-intensive. The advent of computational approaches has revolutionized this field with AlphaFold, developed by DeepMind, marking a significant breakthrough. This tool has achieved a precision comparable to experimental techniques, enabling researchers to predict protein structures with high confidence (Jumper et al. 2021; Tunyasuvunakool et al. 2021; Chen et al. 2024). These techniques can be very helpful in not only understanding pathogenesis but also mitigating the disease through novel therapeutics.

23.2 Basics of Molecular Modeling

23.2.1 Protein Structure Prediction

Protein structure prediction plays a crucial role in understanding the molecular basis of disease progression and in designing interventions for aquatic animal health. Since the function of a protein is closely linked to its three-dimensional (3D) structure, accurately predicting protein conformations helps researchers identify functional domains, model host-pathogen interactions, and screen therapeutic molecules. With limitations in experimental methods like X-ray crystallography and cryo-EM, due to cost, time, or difficulty in crystallizing proteins, computational approaches have become indispensable (Zhang 2007; Pevsner 2015; Bertoline et al. 2023). These methods fall into several categories: homology modeling, threading (fold recognition), ab initio (de novo) modeling, and the more recent deep learning-based prediction, each differing in their underlying principles, accuracy, and application contexts (Table 23.1).

23.2.1.1 Homology-Based Modeling (Comparative Modeling)

Homology modeling, also known as comparative modeling, is a widely used method for predicting the three-dimensional structure of a protein based on its sequence similarity to proteins with known structures (Zhang 2007; Pevsner 2015; Bertoline et al. 2023). The underlying assumption is that homologous proteins share similar

Table 23.1 Commonly used tools for molecular modeling of biomolecules

S. No.	Name	Uses	Mode	URL
1.	Modeller	Homology modeling of protein	Local installation	https://salilab.org/modeller/
2.	I-TASSER	Threading or fold recognition of protein	Online server	https://zhanggroup.org/I-TASSER/
3.	AlphaFold 2	Neural network-based template-free protein modeling	Online server	https://g.co/kgs/4tg1NWx
4.	AutoDock	Protein–ligand (small molecule) docking	Local installation	https://autodock.scripps.edu/
5.	AutoDock Vina	Faster, more accurate docking of protein and small molecule ligand with improved scoring and speed	Local installation	https://vina.scripps.edu/
6.	AutoDock tools	GUI for setting up and analyzing AutoDock and Vina docking runs	Local installation	https://autodocksuite.scripps.edu/adt/
7.	GROMACS	Molecular dynamic simulation of individual biomolecules and complexes	Local installation	https://www.gromacs.org/
8.	CHARM	Molecular dynamic simulation of individual biomolecules and complexes	Local installation	https://academiccharmm.org/
9.	CHARM GUI	Input generation for molecular dynamic simulation	Online server	https://www.charmm-gui.org/
10.	NAMD	Molecular dynamic simulation of individual biomolecules and complexes	Local installation	https://www.ks.uiuc.edu/Research/namd/
11.	UCSF-Chimera	Structural visualization and analysis	Local installation	https://www.cgl.ucsf.edu/chimera/
12.	PyMol	Structural visualization and analysis	Local installation	https://pymol.org/

structures. This approach involves identifying a suitable template protein, aligning the target sequence with the template, and constructing a model based on this alignment. Homology modeling is most effective when the target and template share a sequence identity greater than 30%, leading to models with reasonable accuracy. Tools like SWISS-MODEL and MODELLER are commonly used for this purpose (Gupta et al. 2016). However, the accuracy of the model heavily depends on the quality of the sequence alignment and the availability of suitable templates.

23.2.1.2 Threading Based Modeling (Fold Recognition)

Threading, or fold recognition, is employed when homologous structures are not available (Zhang 2007; Pevsner 2015). This method involves aligning the target sequence to a library of known protein folds to identify the best-fitting structure, even in the absence of significant sequence similarity. Threading algorithms evaluate the compatibility of the target sequence with existing folds using scoring functions that consider factors like hydrophobicity, secondary structure propensity, and residue-residue contacts. Tools such as Phyre2 and I-TASSER utilize threading techniques. Threading is particularly useful for detecting remote homologs and can provide structural insights when traditional homology modeling is not feasible.

23.2.1.3 Ab Initio Modeling (De Novo)

Ab initio modeling predicts protein structures solely from the amino acid sequence, without relying on homologous templates. This method uses physical principles and statistical potentials to search for the most energetically favorable conformation. Due to the vast conformational space, ab initio modeling is computationally intensive and traditionally limited to small proteins. However, advancements in computational power and algorithms have improved its applicability. Tools like Rosetta (Simons et al. 1999) and QUARK (Xu and Zhang 2012) are prominent in this domain. Ab initio methods are invaluable for modeling novel proteins and those lacking homologous structures.

23.2.1.4 Deep Learning-Based Modeling

Recent breakthroughs in deep learning have revolutionized protein structure prediction. AlphaFold, developed by DeepMind, employs advanced neural networks to predict protein structures with remarkable accuracy (Jumper et al. 2021; Tunyasuvunakool et al. 2021; Bertoline et al. 2023). By integrating multiple sequence alignments and structural templates, AlphaFold can model complex protein folds, often rivaling experimental methods in accuracy. In the Critical Assessment of Structure Prediction (CASP14), AlphaFold achieved a near-experimental accuracy. The AlphaFold Protein Structure Database has since released over 350,000 predicted structures, significantly expanding the structural coverage of the protein universe.

23.2.2 Molecular Docking

Molecular docking is a foundational technique in molecular modeling that aims to predict the optimal binding pose of two interacting molecules—typically a receptor protein and a protein or non-protein ligand (Xiong 2006; Bonvin 2006; Morris et al. 2009). The method has gained significant attention due to its ability to provide insights into the structural basis of molecular recognition, which is critical for drug discovery, functional annotation of proteins, and vaccine design. Molecular docking is not limited to small-molecule drug discovery but also includes modeling interactions between proteins (protein–protein docking), which is essential for understanding signaling pathways, host-pathogen interactions, and enzyme-substrate dynamics. As a predictive approach, docking helps to estimate the binding affinities of protein-ligand complexes, thereby narrowing down the number of compounds to be screened during drug discovery. So, it significantly reduces the cost and time of modern drug discoveries. With the increasing availability of high-resolution protein structures and genome data, docking has become indispensable in both basic and translational research.

The process of molecular docking involves two primary components: the search algorithm and the scoring function (Bonvin 2006; Morris et al. 2009; Moreira et al. 2010; Eberhardt et al. 2021). The search algorithm explores the potential orientations and conformations of the ligand within the binding site of the target molecule, while the scoring function ranks these poses based on their predicted binding affinity. In protein–ligand docking, the focus is typically on identifying small molecules that fit precisely within a defined binding pocket. In protein–protein docking, the challenge is greater due to larger and more flexible interaction surfaces. Here methods such as FFT-based global search (e.g., in ZDOCK) and interface refinement (as in HADDOCK) are used. Scoring functions in both cases combine empirical energy terms, such as van der Waals forces, electrostatic interactions, hydrogen bonding, and desolvation energy, to approximate the binding free energy. Commonly used tools for protein-small molecule docking include AutoDock, AutoDock Vina, Glide (Schrödinger), and SwissDock (Table 23.1). For protein–protein docking, the commonly used tools include ZDOCK, HADDOCK, ClusPro, and PatchDock. Recent methods also integrate machine learning and deep learning for improved accuracy and speed.

Molecular docking can be classified into rigid and flexible docking. In rigid docking, both the ligand and receptor are treated as static structures, which makes computations faster but less realistic. In contrast, flexible docking allows movement in either or both molecules, accommodating conformational changes that often occur upon binding. This flexibility is particularly important in modeling induced fit interactions, which are common in biological systems. Another important type is blind docking, where the binding site is not specified in advance, allowing the software to search the entire surface of the receptor. The accuracy of docking is influenced by the quality of the receptor structure, the resolution of input data, and the reliability of scoring functions.

The applications of molecular docking span across disciplines, from drug discovery and protein engineering to vaccine design and host–pathogen interaction analysis. In drug design, docking is used for virtual screening to prioritize compounds with high predicted affinity for a biological target. In vaccine design, particularly in reverse vaccinology and immunoinformatics, docking helps predict binding of T-cell epitopes to MHC molecules or antibodies to antigens (Kardani et al. 2020; Parvizpour et al. 2020). Protein–protein docking, on the other hand, is extensively used to model signaling complexes, host–pathogen interfaces, and immune receptors. This in silico strategy reduces reliance on animal trials and accelerates the development of disease control measures.

23.2.3 Molecular Dynamics Simulation

Molecular dynamics (MD) simulation is a computational technique that models the physical movements of atoms and molecules over time (Hollingsworth and Dror 2018; Wu et al. 2022; Hénin et al. 2022). By numerically solving Newton's equations of motion, MD simulations provide detailed insights into the structural and dynamic behavior of biomolecules at atomic resolution. This method allows researchers to observe how molecular systems evolve, offering a dynamic perspective that complements static experimental structures. Molecular dynamics simulations have become indispensable in fields like structural biology, biophysics, and drug discovery, enabling the exploration of molecular interactions and conformational changes that are often challenging to capture experimentally.

In molecular modeling, MD simulations play a crucial role by providing time-resolved information about biomolecular systems (Hollingsworth and Dror 2018). They allow for the exploration of conformational flexibility, stability, and interactions within proteins, nucleic acids, and other macromolecules. This dynamic information is essential for understanding biological processes such as enzyme catalysis, signal transduction, and ligand binding. Moreover, MD simulations aid in refining structural models, validating experimental data, and predicting the effects of mutations or environmental changes on molecular function. Their ability to simulate realistic physiological conditions makes them a powerful tool for hypothesis testing and mechanism elucidation in molecular biology.

Molecular dynamics simulations employ various methods and tools to accurately model molecular systems. Key components include force fields, which define the potential energy functions governing atomic interactions, and integration algorithms that compute the trajectories of particles over time. Popular software packages for MD simulations include GROMACS (Páll et al. 2020), AMBER (Salomon-Ferrer et al. 2013), NAMD (Nelson et al. 1996), and CHARMM (Brooks et al. 2009), each offering specialized features for different types of systems and research objectives (Table 23.1). Advanced techniques such as enhanced sampling methods (e.g., meta dynamics, replica exchange) have been developed to overcome energy barriers and sample rare events more efficiently. These tools and methodologies have significantly expanded the scope and accuracy of MD simulations in studying complex biomolecular phenomena.

The typical workflow of an MD simulation involves several key steps that ensure accurate and meaningful results (Páll et al. 2020; Fig. 23.1). First, the molecular system is prepared by obtaining a starting structure (often from experimental data or homology modeling) and adding missing atoms or residues. This is followed by the selection of a suitable force field, solvation of the system (usually in a water box), and addition of ions to neutralize the charge. The system then undergoes energy minimization to relieve any steric clashes or bad contacts, followed by equilibration steps under constant volume (NVT) and constant pressure (NPT) ensembles. After equilibration, the production MD run is performed, typically for nanoseconds to microseconds, depending on the study. Finally, trajectory analysis is conducted to assess structural stability (RMSD, RMSF), interactions (hydrogen bonds, salt bridges), and conformational changes. Visualization and interpretation of these results are essential for drawing biologically relevant conclusions.

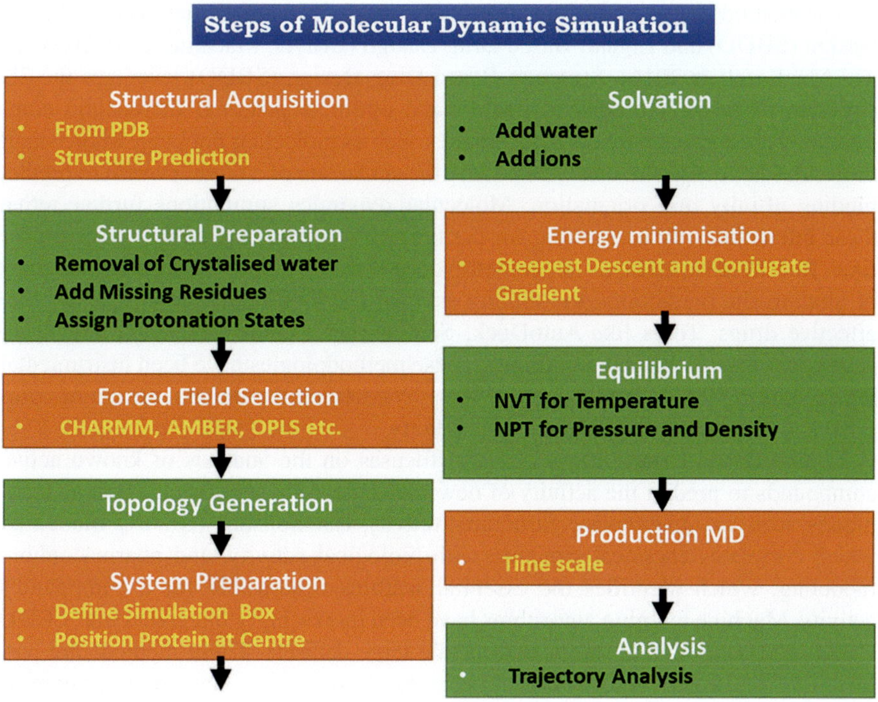

Fig. 23.1 Molecular dynamics simulation procedure for protein

23.2.4 Computer-Aided Drug Design

Computer-Aided Drug Design (CADD) encompasses computational techniques that facilitate the identification and optimization of potential therapeutic compounds (Macalino et al. 2015; Bassani and Moro 2023; Hessler and Baringhaus 2018; Schneider et al. 2020). By simulating and analyzing molecular interactions, CADD streamlines the drug discovery process, reducing both time and cost compared to traditional experimental methods. CADD integrates various disciplines, including bioinformatics, chemistry, and pharmacology, to predict how small molecules interact with biological targets, thereby aiding in the rational design of new drugs. Recent advancements in computational power and algorithms have significantly enhanced the accuracy and efficiency of CADD approaches. The significance of CADD in modern drug discovery lies in its ability to efficiently screen vast chemical libraries and predict the biological activity of compounds. By employing in silico models, researchers can prioritize candidates with optimal pharmacokinetic and pharmacodynamic properties, thereby focusing experimental efforts on the most promising leads. This approach not only accelerates the development pipeline but also reduces the attrition rate of drug candidates in clinical trials. Moreover, CADD facilitates the exploration of chemical space beyond existing compounds, enabling the discovery of novel therapeutics for various diseases.

The methodologies of CADD are broadly categorized into Structure-Based Drug Design (SBDD) and Ligand-Based Drug Design (LBDD; Macalino et al. 2015; Yu and MacKerell Jr 2016). Structure-Based Drug Design (SBDD) leverages the 3D structure of target proteins to identify and optimize potential therapeutic compounds. This approach involves techniques such as molecular docking, where small molecules are computationally fitted into the active site of a target protein to predict binding affinity and orientation. Molecular dynamics simulations further refine these interactions by accounting for the flexibility of both ligand and protein over time. Recent advancements in geometric deep learning have enhanced the accuracy of binding site predictions and affinity estimations, facilitating the design of more effective drugs. Tools like AutoDock, Schrödinger's Glide, and GROMACS are commonly used in SBDD workflows. These methodologies have been instrumental in accelerating the drug discovery process by enabling the rational design of compounds with improved efficacy and specificity.

Ligand-Based Drug Design (LBDD) focuses on the analysis of known active compounds to predict the activity of new molecules. This approach employs techniques such as Quantitative Structure-Activity Relationship (QSAR) modeling, which correlates chemical structure with biological activity, and pharmacophore modeling, which identifies the essential features responsible for a compound's activity. Machine learning algorithms have been increasingly integrated into LBDD to enhance predictive accuracy and handle large datasets. These methods are particularly useful when the three-dimensional structure of the target protein is unknown, allowing for the identification of potential leads based on ligand information alone. Software tools like MOE, LigandScout, and KNIME facilitate various aspects of LBDD, from data preprocessing to model validation.

Hybrid approaches in CADD combine elements of both SBDD and LBDD to leverage the strengths of each methodology. For instance, integrating ligand-based pharmacophore models with structure-based docking can improve the identification of novel compounds with desired biological activity. Recent developments in artificial intelligence and deep learning have further expanded the capabilities of CADD, enabling the analysis of complex biological data and the prediction of drug–target interactions with higher accuracy (Hessler and Baringhaus 2018; Schneider et al. 2020). These emerging techniques are particularly valuable in addressing challenging targets and in the design of drugs with improved pharmacokinetic and pharmacodynamic properties. The continuous evolution of computational tools and algorithms promises to enhance the efficiency and success rate of the drug discovery process.

23.3 Scope of Molecular Modeling in Aquiculture Disease Management

23.3.1 Host-Pathogen Interaction Study

Molecular modeling serves as a critical computational approach for studying the complex biochemical and structural interactions between pathogens and their hosts. Simulating molecular interactions allows researchers to visualize and predict how pathogen-derived molecules, such as proteins or toxins, interact with host cellular components. This understanding is fundamental to elucidating mechanisms of pathogenesis, immune evasion, and cellular entry. With the increasing availability of structural data and powerful simulation tools, molecular modeling has enabled the prediction of key binding sites, structural conformations, and energetic profiles that define pathogen-host relationships at the atomic level. Molecular modeling techniques are extensively used for understanding host-pathogen interactions from mammalian pathogens. Pitard and Malliavin (2019) reviewed the use of structural biology and molecular modeling in understanding the properties of bacterial toxins and virulence factors. Authors described the mechanism of endosomal translocation of diphtheria, tetanus, and botulinum toxins as well as activation of adenylyl cyclase virulence factor inside the cytoplasm. Molecular dynamics were employed in conjunction with structural experiments to understand the molecular mechanism.

Molecular modeling is extensively used to understand conformational dynamics of envelope proteins of flaviviruses to decipher their infective mechanisms (Huber et al. 2021). Several protocols have been developed to simulate the flavivirus envelope in close accordance with biophysical data generated from cryo-electron microscopy, aiming to elucidate the conformational changes associated with adaptation to both host environments and immunological factors, such as antibodies. Abduljalil et al. (2023) reviewed the application of MD simulations in understanding conformational changes in the spike protein of SARS-CoV-2 during infection. Unique mutations in the spike protein were identified that are responsible for higher affinity binding to the host receptor and thus higher spread potential than SARS-CoV. The contributions of glycans to the function of the spike protein and its immune evasion were also revealed. Valério et al. (2025) reviewed the use of molecular dynamics simulation to understand the virus entry mechanism. Simulation studies revealed the function of the viral proteins responsible for attaching to the host receptors and promoting membrane fusion, deciphering interactions between these proteins and receptors, and shedding light on the functional significance of key regions, such as the fusion peptide.

Several reports are also currently available where molecular modeling is used for aquatic animal pathogens. Labra et al. (2016) identified and analyzed the Dot/Icm secretion system of *Piscirickettsia salmonis,* which seriously affects the Chilean salmon industry. Using *P. salmonis* genome, 4 Dot/Icm effectors were identified and analyzed in silico. Homology modeling of these virulence effectors was performed using several servers, including Phyre2, RaptorX, and I-TASSER. The author reported that three of them contain ankyrin repeat domains and the typical conserved 3D structures of this protein family. Additionally, all the potential *P. salmonis* effectors contain a classical Dot/Icm secretion signal in their C-terminus. In this way, they identified Dot/Icm-dependent effectors in *P. salmonis*. Zhang et al. (2017) predicted the virus-host protein interactome using bioinformatics tools of grass carp reovirus (GCRV). JAM-A protein was successfully predicted to interact with motifs of GCRV Sigma1-like protein, and shared a similar binding mode compared with orthoreovirus. In this study, the structure of the grass carp JAM-A and the Sigma-1-like protein of the virus was predicted by using the I-TASSER server with a threading approach. Protein docking was carried out by using the Zdock server in a site-specific manner. In conjunction with the wet lab study, the author identified potential interactions that could be critical for understanding the pathogenesis of GCRV and developing targeted interventions.

Rochat et al. (2019) identified a gene encoding a novel elastinolytic enzyme in *Flavobacterium psychrophilum* using a genome-wide association study. In wet lab experiments, expressed enzyme showed sufficient elastin-degrading capability. The 3D structure was predicted using Swiss-Model based on a homology modeling strategy. The encoded protein is predicted to be a cell-surface-exposed lipoprotein with no homology to previously reported elastases. Based on active site analysis, authors proposed that this protein may belong to the zincin tribe and gluzincin clan of metalloproteases found only in some members of the family Flavobacteriaceae. Bi et al. (2023) identified an emerging pathogen, *Acinetobacter johnsonii*, from Rainbow Trout (*Oncorhynchus mykiss*). Outer membrane protein (OMP) family member proteins, namely, OmpA, Omp34, and OmpTsx, were cloned from the pathogen. The protein structure and topology of the monomer were predicted by AlphaFold 2. The protective antigens, hydrophilicity, flexibility, surface accessibility, and antigenic peptides were determined using immunoinformatic tools. Based on the results, it was confirmed that *A. johnsonii* is an emerging opportunistic pathogen, and the OmpA protein can be used as a subunit vaccine candidate molecule.

23.3.2 Computer-Aided Drug Design for Aquatic Animals

Molecular modeling technique is extensively used for drug discovery not only for mammals but also for aquatic animals. Kumar et al. (2020) identified potential drug targets and validated experimentally against the fungal pathogen *Saprolegnia parasitica*. Drug targets were identified using comparative proteomic analysis of the whole protein of the host and parasite. The candidate drugs were selected from the DrugBank database, which were already FDA approved. Homology modeling was performed using the CPHmodels 3.2 Server based on the principle of homology modeling. Molecular docking with selected lead compounds was done using AutoDock Vina in PyRx. The compounds giving good result were tried in in vitro inhibition test. Authors have identified triclosan as a promising compound that inhibits the morphology of hyphal cells. Sumon et al. (2023) identified potential antiviral drugs against Tilapia Lake Virus (TiLV) from two different tropical mangrove plants: *Heritiera fomes* and *Ceriops candolleana*. A key protein, CRM1, involved in pathogenesis, was targeted using the CADD approach. Homology modeling of the target was performed using I-TASSER and docking using PyRx and AutoDock Vina. Further MD simulation of protein-ligand complexes was performed using Desmond, and the MM-GBSA method was used for binding energy calculation. Authors have identified several potential compounds with low toxicity against the virus, but in vivo trials were not conducted.

Kumar et al. (2023) performed molecular modeling of *M. rosenbergii* nodavirus (MrNV) RNA-dependent RNA polymerase (RdRp), a promising drug target. Sequence analysis confirmed the presence of two enzymatic regions, namely, methyl transferase and RdRp, separated by a transmembrane region. The 3D structure was predicted using AlphaFold 2. Structural comparison with known RdRp structures revealed the highest similarity with Bovine Viral Diarrhea Virus RdRp. Authors have identified the priming loop, substrate binding residues, metal binding residues, as well as the ribonucleotide selection region, thus revealing a detailed catalytic mechanism. Docking of this enzyme with known RNA polymerase inhibitors using AutoDock Vina with UCSF-Chimera interface revealed that Dasabuvir can potentially inhibit the enzyme. Dong et al. (2025) identified potential anthelmintic agents against *Gyrodactylus kobayashii* by integrating CADD and in vivo evaluation. They targeted the well-known antiparasitic drug target β-tubulin. Its three-dimensional structure was generated using homology modeling, and virtual screening was done using FDA-approved drugs and common benzimidazoles. Subsequent in vivo anthelmintic efficacy and acute toxicity assays in goldfish revealed etravirine, flubendazole, and mebendazole as potent anthelmintic agents. Molecular dynamics simulations were also employed to confirm the stability of protein-ligand complexes.

23.3.3 Molecular Modeling for Vaccine Design

Molecular modeling has become an integral component in modern vaccine design, offering a cost-effective and efficient means to combat emerging and re-emerging pathogens. (De Groot et al. 2001; Bazhan et al. 2019; Martinelli 2022). This approach, especially when combined with immunoinformatics, allows researchers to simulate immune responses and identify key antigenic targets without relying on traditional wet-lab experiments. One of the major breakthroughs in this field is the advent of reverse vaccinology, which begins with pathogen genome data instead of whole organisms. This strategy accelerates the identification of vaccine candidates and provides a rational framework for developing new vaccines. With the rise of computational power and curated biological databases, molecular modeling enables in-depth insights into protein structure and function, contributing significantly to the field of computational vaccinology.

The vaccine design pipeline using molecular modeling typically involves several key phases: target selection, epitope prediction, vaccine construct design, and refinement. Tools such as protein structure prediction, molecular docking, and MD simulations are essential throughout these stages (Martinelli 2022). These techniques help in identifying potential B-cell and T-cell epitopes, evaluating their binding affinity to major histocompatibility complex (MHC) molecules, and assessing the stability of vaccine constructs. Success stories from this approach include the development of promising vaccine candidates against human pathogens such as SARS-CoV-2, *Trypanosoma cruzi*, dengue virus, Ebola virus, and avian influenza. In aquatic animal health, similar methodologies can be applied to tackle economically significant diseases in aquaculture. By using immunoinformatics and molecular modeling, researchers can design vaccines against major fish pathogens, helping to safeguard fish stocks and ensure the sustainability of aquaculture systems.

23.3.4 Molecular Modeling for Combating Antimicrobial Resistance

Molecular modeling has emerged as a pivotal tool in addressing the escalating challenge of antimicrobial resistance (AMR; Sunuwar and Azad 2022; Palazzotti et al. 2023; Haque et al. 2024). By simulating and analyzing molecular interactions between antibiotics and bacterial targets, researchers can gain insights into resistance mechanisms and design more effective therapeutics. This computational approach enables the identification of novel drug targets, the prediction of resistance mutations, and the optimization of existing antibiotics, thereby accelerating the development of strategies to combat resistant pathogens. The methodology of molecular modeling in the context of AMR encompasses several computational techniques. Molecular docking allows for the prediction of binding affinities between antibiotics and their bacterial targets, facilitating the identification of potential inhibitors. Molecular dynamics simulations provide insights into the stability and conformational changes of these complexes over time, offering a dynamic perspective on drug–target interactions. Additionally, quantitative structure-activity relationship (QSAR) models can predict the efficacy of new compounds based on their chemical structures. These methods collectively contribute to a more rational and efficient drug design process, reducing reliance on trial-and-error approaches.

Palazzotti et al. (2023) studied NorA efflux pump activity associated with *Staphylococcus aureus* resistance to fluoroquinolone antibiotics. Authors have used cryo-electron microscopy (cryo-EM) structures of the two NorA efflux pumps. In this study, supervised molecular dynamics (SuMD) and molecular docking experiments were conducted to understand molecular mechanisms driving the interaction between NorA and efflux pump inhibitors (EPIs). The detailed picture of NorA-EPI intermolecular interactions is key for the identification of novel molecules to fight antimicrobial resistance. Another study by Haque et al. (2024) employed molecular docking and MD simulation approaches to identify potential inhibitors of the metallo-lactamase-1 (NDM-1) enzyme, a key player in antibiotic resistance, highlighting compounds with promising binding affinities. These examples illustrate how molecular modeling can lead to the discovery of novel therapeutic strategies and inform the design of next-generation antibiotics. In aquaculture, molecular modeling can play a critical role in combating AMR by enabling understanding of AMR resistance mechanisms and design of targeted therapeutics. With the increasing emergence of resistant bacterial strains in aquaculture systems, computational tools such as molecular docking, molecular dynamics simulations, and structure-based drug design offer an efficient solution. This not only improves fish health and productivity but also aligns with global efforts to reduce AMR and ensure environmental safety.

23.4 Conclusion and Future Prospects

Molecular modeling has become an indispensable component of modern aquatic animal health management, offering a wide array of computational tools to unravel complex biological interactions, predict pathogen behavior, and design targeted therapeutics. Techniques such as protein structure prediction, molecular docking, molecular dynamics simulations, and computer-aided drug design (CADD) have revolutionized our approach to diagnostics, drug discovery, vaccine development, and understanding antimicrobial resistance (AMR) mechanisms. The integration of these tools into aquaculture research has significantly enhanced our capacity to address disease challenges in a cost-effective and time-efficient manner. The chapter has demonstrated the diverse applications of molecular modeling—from identifying virulence factors and host-pathogen interactions for designing novel antivirals, antifungals, and antiparasitic agents. Case studies across various aquatic pathogens highlight the practical utility and growing relevance of these tools in real-world aquaculture settings. Moreover, the ability to simulate immune responses and predict antigenic epitopes using immunoinformatics further supports vaccine design efforts against emerging and re-emerging pathogens.

Looking ahead, the future of molecular modeling in aquaculture is highly promising. Continued advancements in artificial intelligence (AI), machine learning, and high-performance computing are expected to further enhance prediction accuracy and simulation capabilities. The integration of omics data (genomics, proteomics, metabolomics) with modeling tools will enable more comprehensive systems biology approaches to aquatic health. Additionally, the development of user-friendly software and cloud-based platforms will democratize access to these technologies for researchers and practitioners worldwide. To fully harness the potential of molecular modeling, interdisciplinary collaboration among bioinformaticians, molecular biologists, veterinarians, and aquaculture specialists is essential. As the aquaculture industry continues to expand, adopting these cutting-edge tools will be vital for ensuring sustainable, resilient, and disease-free aquatic production systems.

References

Abdelsalam M, Elgendy MY, Elfadadny MR, Ali SS, Sherif AH, Abolghait SK (2023) A review of molecular diagnoses of bacterial fish diseases. Aquac Int 31(1):417–434

Abduljalil JM, Elghareib AM, Samir A, Ezat AA, Elfiky AA (2023) How helpful were molecular dynamics simulations in shaping our understanding of SARS-CoV-2 spike protein dynamics? Int J Biol Macromol 242:125153

Bassani D, Moro S (2023) Past, present, and future perspectives on computer-aided drug design methodologies. Molecules 28(9):3906

Bazhan SI, Antonets DV, Karpenko LI, Oreshkova SF, Kaplina ON, Starostina EV, Dudko SG, Fedotova SA, Ilyichev AA (2019) In silico designed ebola virus T-cell multi-epitope DNA vaccine constructions are immunogenic in mice. Vaccine 7(2):34

Bertoline LM, Lima AN, Krieger JE, Teixeira SK (2023) Before and after AlphaFold2: an overview of protein structure prediction. Front Bioinform 3:1120370

Bi B, Yuan Y, Jia D, Jiang W, Yan H, Yuan G, Gao Y (2023) Identification and pathogenicity of emerging fish pathogen Acinetobacter johnsonii from a disease outbreak in rainbow trout (Oncorhynchus mykiss). Aquac Res 2023(1):1995494

Bonvin AM (2006) Flexible protein–protein docking. Curr Opin Struct Biol 16(2):194–200

Brooks BR, Brooks CL III, Mackerell AD Jr, Nilsson L, Petrella RJ, Roux B, Won Y, Archontis G, Bartels C, Boresch S, Caflisch A (2009) CHARMM: the biomolecular simulation program. J Comput Chem 30(10):1545–1614

Chen L, Li Q, Nasif KFA, Xie Y, Deng B, Niu S, Pouriyeh S, Dai Z, Chen J, Xie CY (2024) AI-driven deep learning techniques in protein structure prediction. Int J Mol Sci 25(15):8426

De Groot AS, Bosma A, Chinai N, Frost J, Jesdale BM, Gonzalez MA, Martin W, Saint-Aubin C (2001) From genome to vaccine: in silico predictions, ex vivo verification. Vaccine 19(31):4385–4395

Dong J, Xia L, Liu Y, Yang Q, Xu N, Ai X, Zhou S (2025) Discovery of potential anthelmintic agents against gyrodactylus kobayashii through computer-aided drug design and in vivo evaluation. J Fish Dis 48(6):e14102

Eberhardt J, Santos-Martins D, Tillack AF, Forli S (2021) AutoDock Vina 1.2.0: new docking methods, expanded force field, and python bindings. J Chem Inf Model 61(8):3891–3898

FAO (2024) The state of world fisheries and aquaculture 2024 – blue transformation in action, Rome. https://doi.org/10.4060/cd0683en

Gupta A, Mohanty P, Bhatnagar S (2016) Protein structure prediction using homology modeling: methods and tools. In: Methods and algorithms for molecular docking-based drug design and discovery. IGI Global Scientific Publishing, Hershey, pp 339–359

Haque S, Mathkor DM, Johargy AK, Faidah H, Babalghith AO, Hariri SH, Jalal NA, Ahmad F, Bantun F (2024) Molecular modeling and simulation-based identification of inhibitors against New Delhi Metallo-Lactamase 1: implications for bacterial antibiotic resistance. J King Saud Univ Sci 36:103290

Hénin J, Lelièvre T, Shirts MR, Valsson O, Delemotte L (2022) Enhanced sampling methods for molecular dynamics simulations. arXiv preprint arXiv:2202.04164

Hessler G, Baringhaus KH (2018) Artificial intelligence in drug design. Molecules 23(10):2520

Hollingsworth SA, Dror RO (2018) Molecular dynamics simulation for all. Neuron 99(6):1129–1143

Huber RG, Marzinek JK, Boon PL, Yue W, Bond PJ (2021) Computational modelling of flavivirus dynamics: the ins and outs. Methods 185:28–38

Jaies I, Shah FA, Qadiri SSN, Qayoom I, Bhat BA, Dar SA, Bhat FA (2024) Immunological and molecular diagnostic techniques in fish health: present and future prospectus. Mol Biol Rep 51(1):551

Jumper J, Evans R, Pritzel A, Green T, Figurnov M, Ronneberger O, Tunyasuvunakool K, Bates R, Žídek A, Potapenko A, Bridgland A (2021) Highly accurate protein structure prediction with AlphaFold. Nature 596(7873):583–589

Kardani K, Bolhassani A, Namvar A (2020) An overview of in silico vaccine design against different pathogens and cancer. Expert Rev Vaccines 19(8):699–726

Karkoutly O, Dhasmana A, Dhevan V, Chauhan SC, Tripathi MK (2021) Molecular modelling a key method for potential therapeutic drug discovery. Biomed J Sci Tech Res 37(3):29427

Kumar S, Mandal RS, Bulone V, Srivastava V (2020) Identification of growth inhibitors of the fish pathogen Saprolegnia parasitica using in silico subtractive proteomics, computational modeling, and biochemical validation. Front Microbiol 11:571093

Kumar G, Singh AK, Agarwal D (2023) Structural and functional characterization of RNA dependent RNA polymerase of Macrobrachium rosenbergii nodavirus (MnRdRp). J Biomol Struct Dyn 41(22):12825–12837

Labra Á, Arredondo-Zelada O, Flores-Herrera P, Marshall SH, Gómez FA (2016) In sílico identification and characterization of putative Dot/Icm secreted virulence effectors in the fish pathogen Piscirickettsia salmonis. Microb Pathog 92:11–18

Macalino SJY, Gosu V, Hong S, Choi S (2015) Role of computer-aided drug design in modern drug discovery. Arch Pharm Res 38:1686–1701

Martinelli DD (2022) In silico vaccine design: a tutorial in immunoinformatics. Healthc Anal 2:100044

Moreira IS, Fernandes PA, Ramos MJ (2010) Protein–protein docking dealing with the unknown. J Comput Chem 31(2):317–342

Morris GM, Huey R, Lindstrom W, Sanner MF, Belew RK, Goodsell DS, Olson AJ (2009) AutoDock4 and AutoDockTools4: automated docking with selective receptor flexibility. J Comput Chem 30(16):2785–2791

Nelson MT, Humphrey W, Gursoy A, Dalke A, Kalé LV, Skeel RD, Schulten K (1996) NAMD: a parallel, object-oriented molecular dynamics program. Int J Supercomput Appl High Perform Comput 10(4):251–268

Palazzotti D, Felicetti T, Sabatini S, Moro S, Barreca ML, Sturlese M, Astolfi A (2023) Fighting antimicrobial resistance: insights on how the Staphylococcus aureus NorA efflux pump recognizes 2-phenylquinoline inhibitors by supervised molecular dynamics (SuMD) and molecular docking simulations. J Chem Inf Model 63(15):4875–4887

Páll S, Zhmurov A, Bauer P, Abraham M, Lundborg M, Gray A, Hess B, Lindahl E (2020) Heterogeneous parallelization and acceleration of molecular dynamics simulations in GROMACS. J Chem Phys 153(13):134110

Parvizpour S, Pourseif MM, Razmara J, Rafi MA, Omidi Y (2020) Epitope-based vaccine design: a comprehensive overview of bioinformatics approaches. Drug Discov Today 25(6):1034–1042

Pevsner J (2015) Bioinformatics and functional genomics, 3rd edn. John Wiley & Sons, Newark, pp 589–630

Pitard I, Malliavin TE (2019) Structural biology and molecular modeling to analyze the entry of bacterial toxins and virulence factors into host cells. Toxins 11(6):369

Rochat T, Pérez-Pascual D, Nilsen H, Carpentier M, Bridel S, Bernardet JF, Duchaud E (2019) Identification of a novel elastin-degrading enzyme from the fish pathogen Flavobacterium psychrophilum. Appl Environ Microbiol 85(6):e02535–e02518

Salomon-Ferrer R, Case DA, Walker RC (2013) An overview of the Amber biomolecular simulation package. Wiley Interdiscip Rev Comput Mol Sci 3(2):198–210

Schneider P, Walters WP, Plowright AT, Sieroka N, Listgarten J, Goodnow RA Jr, Fisher J, Jansen JM, Duca JS, Rush TS, Zentgraf M (2020) Rethinking drug design in the artificial intelligence era. Nat Rev Drug Discov 19(5):353–364

Simons KT, Bonneau R, Ruczinski I, Baker D (1999) Ab initio protein structure prediction of CASP III targets using ROSETTA. Proteins Struct Funct Genet 37(S3):171–176

Sumon MAA, Asseri AH, Molla MHR, Aljahdali MO, Hasan MR, Rahman MA, Hasan MT, Sumon TA, Gabr MH, Islam MS, Fakhurji B (2023) Identification of natural antiviral drug candidates against Tilapia Lake Virus: computational drug design approaches. PLoS One 18(11):e0287944

Sunuwar J, Azad RK (2022) Identification of novel antimicrobial resistance genes using machine learning, homology modeling, and molecular docking. Microorganisms 10(11):2102

Tunyasuvunakool K, Adler J, Wu Z, Green T, Zielinski M, Žídek A, Bridgland A, Cowie A, Meyer C, Laydon A, Velankar S (2021) Highly accurate protein structure prediction for the human proteome. Nature 596(7873):590–596

Valério M, Buga CC, Melo MN, Soares CM, Lousa D (2025) Viral entry mechanisms: the role of molecular simulation in unlocking a key step in viral infections. FEBS Open Bio 15(2):269–284

Wu X, Xu LY, Li EM, Dong G (2022) Application of molecular dynamics simulation in biomedicine. Chem Biol Drug Des 99(5):789–800

Xiong J (2006) Essential bioinformatics. Cambridge University Press, Cambridge, pp 187–230

Xu D, Zhang Y (2012) Ab initio protein structure assembly using continuous structure fragments and optimized knowledge-based force field. Proteins Struct Funct Bioinf 80(7):1715–1735

Yalcin-Ozkat G (2021) Molecular modeling strategies of cancer multidrug resistance. Drug Resist Updat 59:100789

Yu W, MacKerell AD Jr (2016) Computer-aided drug design methods. In: Antibiotics: methods and protocols. Springer New York, New York, pp 85–106

Zhang Y (2007) Template-based modeling and free modeling by I-TASSER in CASP7. Proteins Struct Funct Bioinf 69(S8):108–117

Zhang A, He L, Wang Y (2017) Prediction of GCRV virus-host protein interactome based on structural motif-domain interactions. BMC Bioinform 18:1–13

24 Bioinformatics in Aquatic Animal Health Management: Recent Advances and Applications

B. Naveen Rajeshwar, K. B. Kushala, M. S. Nithin, Ashwath Nayak, Anirban Chakraborty, and T. Jawahar Abraham

24.1 Introduction

Fisheries and aquaculture are the fastest-growing sectors in global animal food production and are becoming the main sources of aquatic animal food for human consumption. In 2022, global fisheries and aquaculture production registered an unprecedented 223.2 million tonnes (MT), with a per capita consumption of 20.7 kg (FAO 2024). The total production of aquatic animals accounted for 185 MT (FAO 2024), highlighting its importance in the global food supply. Aquaculture plays a vital role in ensuring global food security, necessitating sustainable practices to meet the growing nutritional demands of the human population. With the increasing global demand for sustainable fish production and improved aquaculture practices, the fisheries sector is undergoing a transformation driven by advanced technologies. Among these, bioinformatics has emerged as a critical tool, enabling the analysis of genomic, transcriptomic, and proteomic data to improve selective breeding, enhance disease resistance, and support effective stock management in aquatic species. Bioinformatics is an interdisciplinary scientific field that combines computer science, information technology, and biology as a single entity (Luscombe et al. 2001). Derived from the term "Biological Informatics," bioinformatics is a modern and expanding science that has quickly emerged as the driving force behind current biological research. The Dutch biologists Paulien Hogeweg and Ben Hesper

B. N. Rajeshwar · K. B. Kushala · M. S. Nithin · A. Nayak · A. Chakraborty
Nitte (Deemed to be University), NITTE-GOK COEI AQUAMARIN, Nitte University Centre for Science Education and Research (NUCSER), Paneer Campus, Deralakatte, Mangaluru, Karnataka, India

T. J. Abraham (✉)
Department of Aquatic Animal Health, Faculty of Fishery Sciences, West Bengal University of Animal and Fishery Sciences, Chakgaria, Kolkata, West Bengal, India

© The Author(s), under exclusive license to Springer Nature Singapore Pte Ltd. 2026
B. K. Das et al. (eds.), *Bioinformatics Tools for Predictive Ecology and Fisheries*, Sustainable Aquatic Food Systems, https://doi.org/10.1007/978-981-95-2573-7_24

described the term "bioinformatics" in 1970, which integrates biological research and information technology (Hesper and Hogeweg 1970). Bioinformatics uses software and algorithms to gather, store, recover, manipulate, analyse, and manage biological data. As a multidisciplinary domain, it combines computer science, mathematics, statistics, and biology (Bayat 2002; Vashisht and Thakral 2024). The main objective of bioinformatics is to employ computational methods and algorithms to comprehend and derive significant facts from extensive datasets produced across several biological fields, thereby elucidating complex biological processes, genetic systems, and molecular interactions (Luscombe et al. 2001). The popularity of bioinformatics as a branch of science increased after the launch of the Human Genome Project in 1990. Bioinformatics development deals with the demands of analysing and interpreting huge amounts of data. The major sub-disciplines of bioinformatics are as follows.

- Development of streamlined algorithms and statistical methods to correlate and analyse large-scale datasets
- Analysis and interpretation of extensive datasets comprising nucleotide, amino acid sequences, protein structures, etc.
- Development of tools to effectively access and manage diverse biological information

The main goal of bioinformatics is to systematically store biological data in specialized databases for easy access and submission, like GenBank for sequences and the Protein Data Bank for structures. Another goal is to develop tools for data analysis, such as BLAST for sequence similarity, ClustalW for alignment, and Primer3 for PCR primer design. Ultimately, bioinformatics enables the interpretation of biological data to better understand molecular processes, identify disease mechanisms, and support drug discovery. Without bioinformatics tools, analysing, interpreting, and storing the massive amount of data available would be impossible, especially after whole-genome sequencing projects. This chapter aims to explore the integration of bioinformatics in fisheries science, with an emphasis on its applications in aquatic animal health management. It seeks to demonstrate how computational tools and data-driven approaches enhance fish health management, genetic improvement, and sustainable aquaculture practices.

24.2 Timeline and Milestones of Bioinformatics

Bioinformatics began in the 1960s with the use of computational approaches to study protein sequences. Pioneering work by Margaret Dayhoff and her team led to the creation of the Protein Information Resource (PIR), one of the first protein sequence databases. The field experienced significant growth with the advent of DNA sequencing technologies and the launch of databases like GenBank, which was significantly propelled by the launch of the Human Genome Project. During the 1980s and 1990s, the development of online protein databases and the introduction of the World Wide Web further accelerated its growth, establishing bioinformatics as a vital scientific discipline. In the present day, bioinformatics plays a central role in various domains, including genomics, personalized medicine, environmental monitoring, and sustainable aquaculture, through advanced tools like artificial intelligence (AI), big data analytics, and multi-omics integration (Fig. 24.1).

24.3 Role of Bioinformatics in Fisheries

The fisheries and aquaculture sectors are increasingly recognized for their critical contributions to global food security and nutrition in the twenty-first century. Fisheries include areas such as aquaculture, nutrition, disease management and prevention, breeding, physiology, and biotechnology, each playing an essential role in the sustainable development of the sector. Bioinformatics serves as the covert backbone for all aspects of fisheries and helps improve the field. Like other areas of life sciences, bioinformatics has been applied to aquaculture, particularly in the management of aquatic animal health, domestication of improved brood stock, development of advanced breeding and aquaculture techniques, cost-effective and sustainable feed development, and maintenance of a congenial marine environment (Lu et al. 2015; Nguyen 2024). Diagnostic tools incorporated with bioinformatics enabled early detection of aquatic diseases, improving health surveillance and response (Altinok and Kurt 2003). By interpreting next-generation sequencing (NGS) data, bioinformatics platforms can assess the environmental variables, host genetics, and pathogen virulence, thus supporting precise disease diagnostics and management strategies (Rana et al. 2020). One of the key implications of bioinformatics in aquatic animal health management is the identification of target proteins for drug development and vaccine design (Banerjee et al. 2017). In addition, it enables large-scale genetic comparisons and phylogenetic reconstruction, allowing researchers to trace species relationships, detect gene duplications or horizontal transfers, and build complex evolutionary models (Roy et al. 2014; Wadas and Domingues 2025). In the area of nutrition, bioinformatics supports the emerging field of nutritional genomics in aquaculture, deciphering how different diets influence gene expression and metabolic pathways in farmed species, thereby guiding the formulation of optimized, efficient, and sustainable feeds (Hakim et al. 2018; Rather et al. 2023). Furthermore, bioinformatics enables genomic selection and quantitative trait locus (QTL) mapping in aquaculture, allowing precise identification of trait-associated genes and accurate prediction of superior broodstock for traits such as enhanced growth, disease resistance, and feed efficiency (Johnston et al. 2024). Bioinformatics also contributes to environmental management through the identification and optimization of microbial genes and metabolic pathways that degrade pollutants, facilitating precision bioremediation strategies to detoxify contaminated environments (Arora and Bae 2014). Moreover, it integrates genomic and environmental data to identify climate-associated genetic variants and predict species' adaptive capacity under future climate scenarios (Sang et al. 2022). Finally, bioinformatics underpins systems biology and network modelling in aquaculture by

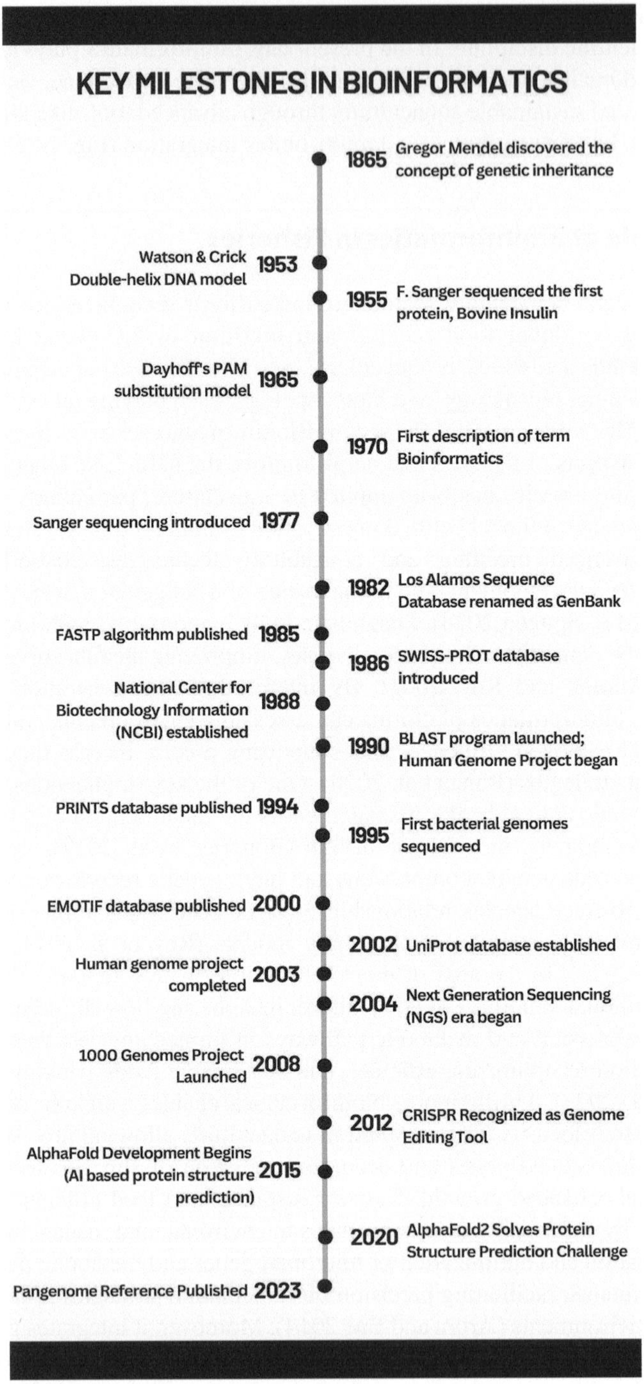

Fig. 24.1 Historical milestones of bioinformatics

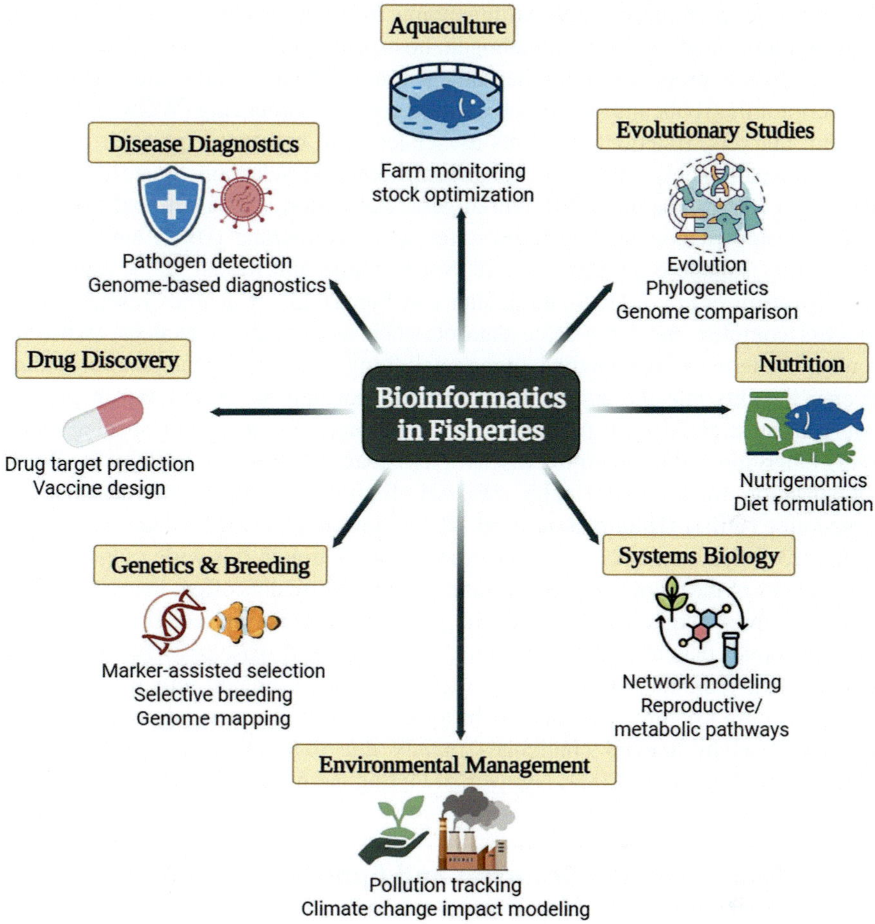

Fig. 24.2 Role of bioinformatics in fisheries

constructing and analysing microbial and metabolic interaction networks, such as Bayesian frameworks for fish gut microbiomes (e.g., SAMBA) and metabolic pathway models like MetaFishNet, for holistic system-level insights and optimization (Li et al. 2010; Soriano et al. 2023) (Fig. 24.2).

24.4 Scope for Bioinformatics in Aquatic Animal Health Management

Aquatic animal health management is growing rapidly with the combination of bioinformatics, with its multidisciplinary approach that combines biology, information science, computer science, and biostatistics used to analyse and interpret the complex biological data (Wani et al. 2018; Baxevanis et al. 2020). In aquaculture and fisheries science, bioinformatics with genomic information has become an essential tool for understanding the host, pathogen, host-pathogen interactions, disease diagnostics, disease progression, biomarkers, genome-wide association study (GWAS), transcriptomics (bulk and single-cell), whole genome sequencing (WGS) to identify genetic markers, functional variants associated with resistance to major bacterial and viral pathogens, vaccine development, genomic selection, quantitative trait locus/loci (QTL) mapping, differential gene expression, and functional prioritisation by using python, R, and high-performance computing (HPC) environments (Liu 2017; Caballero-Solares et al. 2018; Chandhini and Kumar 2019; Rana et al. 2020; Karimi et al. 2023; Shahrajabian and Sun 2023). It enables researchers to process complex and large-scale datasets such as genome sequences, transcriptomes, proteomes, epigenetics, genome editing, and metagenomes, providing deeper insights into the associated molecular mechanisms underlying in growth, overcoming stress, disease resistance and its susceptibility in aquatic species underlying emerging and re-emerging diseases in fish and shellfish as listed by the World Organisation for Animal Health (WOAH, founded as Office International des Epizooties (OIE)) (Hounmanou et al. 2018; Subasinghe 2022; Shahrajabian and Sun 2023; Wijeskara 2023; Kibenge et al. 2024). Through this wide range of tools for analysis of the complex genomic datasets, bioinformatics offers innovative solutions to improve aquatic animal health, productivity, genomic selection integration of multi-omics data to improve breeding programs for disease-resistant stocks, and develop targeted interventions for managing infectious and non-infectious diseases in aquaculture systems, and sustainability in the face of emerging infectious diseases and environmental challenges mainly due to climate change (Allal and Nguyen 2022; Lal et al. 2024; Nguyen 2024; Sagharchi et al. 2025).

24.5 Bioinformatics Branches and Applications in Aquatic Animal Health Management

As aquaculture continues to expand globally, bioinformatics will play a critical role in promoting sustainable and healthy aquatic production (Rise et al. 2019; Houston et al. 2020; Allal and Nguyen 2022; Sundaray et al. 2022; Aly and Fathi 2024; Nguyen 2024; Sahu and Dhimar 2024; Andersen et al. 2025a, b; Nambiar and Banuru 2025). There are several fields involved in bioinformatics, starting from its genomes, metagenomics, transcriptomics, proteomics, cheminformatics, drug discovery, evolutionary studies, waste cleanup, climate change studies, biofuels, and integrated “Multi-omics” technology, with their importance in aquatic animal health management will be discussed in brief in the following sections (Table 24.1).

24.5.1 Genomics

Genomics is the study of an organism’s entire set of DNA (the genome), and it is very important for improving the health of aquatic animals. It helps in identification

Table 24.1 Application of bioinformatic tools in aquatic animal health management with examples

Application area	Description	Examples
Disease diagnostics	Identification of pathogens through genomic tools	LAMP, qPCR, CRISPR-based detection for ISKNV, WSSV
Epidemiology	Tracing the origin and spread of diseases	Phylogenetic tree of viral outbreaks
Vaccine development	In silico design of subunit and DNA vaccines	Reverse vaccinology for iridoviruses
Antimicrobial resistance	Monitoring and prediction of resistance genes	Metagenomics of aquaculture biofilms
Selective breeding	Genetic selection for disease-resistant strains	QTL mapping in tilapia or vannamei shrimp
Environmental monitoring	Detection of harmful pathogens or stressors in water	eDNA-based detection of *Vibrio* spp., harmful algal blooms (HABs)

LAMP loop-mediated isothermal amplification, *qPCR* quantitative PCR, *CRISPR* clustered regularly interspaced short palindromic repeats, *ISKNV* infectious spleen and kidney necrosis virus, *WSSV* white spot syndrome virus, *QTL* quantitative trait locus, *DNA* deoxyribonucleic acid, *eDNA* Environmental DNA,

and discovery and describes the genes that are involved in disease resistance, pathogen virulence, immune responses, biomarker discovery, genome-wide association study (GWAS), WGS to identify important genetic markers, vaccine development focused on its immunogenicity potential, genomic selection to improve growth, QTL mapping, and how pathogens and hosts interact. Genomics is different from traditional genetics because it has a wide range of functions with complex interactions and expressions of its functions of all the genes in an organism, not just focusing on one gene at a time (Yue and Wang 2017; Houston et al. 2020; Sundaray et al. 2022; Andersen et al. 2025a, b). Structural genomics looks at the physical aspects of genomes, like sequencing, mapping, and adding notes, as its first step in figuring out genetic architecture and making tools for diagnosis or selective breeding (Burley 2000; Mohanta et al. 2021). The goals of structural genomics are to understand the molecular basis of host–pathogen interactions, disease resistance, vaccine development, and targeting antimicrobial drugs in aquatic animals.

A genomic sequence of any organism may be eukaryotic or prokaryotic and contains the entire genetic code is a vital initial point for research in the field of molecular genetics and is essential for future genomic applications (Lander et al. 2001). In aquatic organisms, genomics has facilitated the explanation of the molecular basis of host immunity, functional genes, pathogen virulence, antibiotic resistance, phage genes, and stress resilience (Rodrigues et al. 2018; Wenne et al. 2007, 2020). Techniques like single-nucleotide polymorphism (SNP) genotyping are used to find significant genetic polymorphisms that affect health-related traits in fish and shellfish species. In addition, a set of advanced molecular breeding tools used in animal breeding has been used on aquatic species including GWAS, genomic selection (GS), marker-assisted selection (MAS), and gene transfer (GT) with genome editing (GE) which have greatly helped breeding programs in the tilapia and other fish and shellfish to be more disease resistant with good growth potential (Yue and Wang 2017; You et al. 2020; Kumar et al. 2024; Lal et al. 2024; Rasal et al. 2024a, b). In aquaculture, genomics enables the sequencing and analysis of host species as well as the pathogens that pave the way to foundational insights into host–pathogen interactions, enabling targeted disease management through genetic understanding of host and pathogen (Yue and Wang 2017; Natnan et al. 2021; Robinson et al. 2023; Olymon et al. 2024; Nambiar and Banuru 2025).

Applications

- Genome sequencing of Nile tilapia *Oreochromis niloticus* has identified disease-resistance loci useful in marker-assisted selection (Joshi et al. 2018).
- WGS of pathogens like ISKNV and red seabream iridovirus (RSIV) helps reveal virulence genes and epidemiological lineages (Puneeth et al. 2021; Kushala et al. 2025).
- Comparative genomics of different strains of *Vibrio harveyi* in shrimp farms distinguishes pathogenic variants (Fu et al. 2021; Wang et al. 2024a; Wang et al. 2024b).
- Identification of genetic markers (e.g., SNPs) for disease resistance (Das and Sahoo 2014; Zhou et al. 2017; Wenne 2018, Wenne et al. 2020; Uma 2025).

24.5.2 Metagenomics

Metagenomics allows culture-independent profiling of microbial communities, essential in monitoring water quality, gut health, and pathogen emergence in aquaculture environments. It involves the analysis of genetic material recovered directly from environmental samples, offering a culture-independent method to assess microbial communities (Yukgehnaish et al. 2020). It is useful for studying microbial communities in sediments, water, and host microbiomes and indeed helpful in understanding the dominance of a group of microbiomes that facilitate the growth and well-being of the aquatic species (Yukgehnaish et al. 2020; Chunyan et al. 2023). It uses profiling gut microbiota to make new probiotics and finds new or opportunistic pathogens in natural waters and recirculating aquaculture systems [RAS] (Munang'andu 2016; Martínez-Porchas and Vargas-Albores 2017; Johny et al. 2021; Rajeev et al. 2024; Skliros et al. 2024).

Applications

- Gut microbiome analysis in *Penaeus vannamei* showed altered microbial profiles under Early Mortality Syndrome (EMS), indicating potential microbial markers (Rungrassamee et al. 2013).
- Water metagenomics has been used to track *Vibrio* spp. and antibiotic resistance genes in shrimp farms (Patil et al. 2021a, b).
- Detection of eDNA from *Flavobacterium columnare* and *Aeromonas hydrophila* through shotgun metagenomics enabled early-warning surveillance systems (Boutin and Dalpke 2020).

- Environmental DNA (eDNA) analysis for pathogen surveillance in natural waters, especially near areas of aquaculture practice, to continuously monitor (Munang'andu 2016; Yukgehnaish et al. 2020).
- Use of 16S rRNA sequencing to assess gut microbiota in shrimp affected by *Enterocytozoon hepatopenaei* (EHP). 16S rRNA-based profiling of pond water microbiota during a disease outbreak. During shrimp disease outbreaks, metagenomic surveys found higher levels of *Vibrio harveyi* and *Photobacterium damselae*. This information helps implement biosecurity measures and pond cleaning (Deris et al. 2022; Das et al. 2024a; Das et al. 2024b).

24.5.3 Transcriptomics and Applications in Aquatic Animal Health

Transcriptomics deals with the complete set of RNA transcripts produced by the genome under specific conditions to understand the expression of genes altered in response to feed alternatives, disease, vaccination, or environmental stress. Transcriptomics reveals host gene expression changes during stress, infection, or vaccination, offering insights for therapeutic and breeding interventions and helping in identifying the key genes, regulatory genes, and the pathways involved (Fuller et al. 2024; Sankappa et al. 2024; Andersen et al. 2024, 2025a, b).

Applications

- RNA sequencing (RNA-Seq) of *Paralichthys olivaceus* infected with *Edwardsiella tarda* identified immune signalling pathways and cytokine regulation (Sun et al. 2020).
- Gene expression profiling in infected vs. healthy fish to identify the highly expressed genes in bulk transcriptomics, whereas single cell transcriptomics helps in pinpointing the cells involved in a particular type of gene expression related to immune gene identification and cytokine response characterization to interpret the pathways and biomarkers involved for early disease diagnosis (Andersen et al. 2024, 2025a).
- Microarray and RNA-seq were used to analyse gene expression in Atlantic salmon *Salmo salar* challenged with infectious salmon anaemia virus (ISAV) (Dettleff et al. 2017), *P. vannamei* infected with *E. hepatopenaei* infection (Yang et al. 2021), turbot *Scophthalmus maximus* challenged with *V. anguillarum* (Gao et al. 2016), etc.

24.5.4 Proteomics

Proteomics encompasses the large-scale study of proteins present in biological systems, particularly their structure and function in an organism. Since proteins are expressed in the cells have important functional molecules that provide insights into host immune responses to stress and pathogen virulence mechanisms (Zubair et al.

2022). Proteomics uncovers the functional molecules mediating immune response and pathogen invasion, offering direct biomarkers and vaccine targets for disease control (Xiong et al. 2011; Galassie and Link 2015; Long et al. 2015; Rodrigues et al. 2018; Ahmed et al. 2019).

Applications

- Immune response in the skin of carp *Carassius auratus* to *Aeromonas hydrophila* infection revealed by proteomics and multi-omics analysis (Li et al. 2022).
- WSSV envelope proteins VP28 and VP24 as vaccine candidates (Thomas et al. 2022).
- Proteomic comparison of resistant and susceptible aquatic fish species showed protein overexpression linked to resistance (Rout and Behera 2024).
- Protein biomarker discovery for diseases like WSSV (Omotayo 2019) and EHP (Ning et al. 2019).

24.5.5 Cheminformatics and Applications in Aquatic Animal Health

Cheminformatics integrates chemical and biological data for rational drug design and toxicity prediction, contributing to safe and effective treatment protocols in aquaculture. Cheminformatics related to aquaculture and its health is an emerging area in the field of information technology concerned with collecting, retaining, separating, examining, analysing, and reorganising chemical information. In aquatic animal health, it serves a critical role in rational drug design, screening of antiparasitic and antimicrobial agents, and toxicity prediction, enabling safer and more effective treatments while reducing the need for costly and time-consuming wet-lab trials (Pridgeon et al. 2011; Roy 2022; Khang et al. 2025). Cheminformatics involves integrating the principles of information science methodologies and physical chemistry with a computer-based approach, which are commonly referred to as "*in silico* techniques", to address a wide range of descriptive and prescriptive chemistry issues, including applications to drug discovery, biological science, and vast molecular areas. On the other hand, the incorporation of machine learning/artificial intelligence (ML/AI) has been considered to be vital in the field of visualising and understanding the structural complexity for drug design based on the compound databases to develop drugs endowed with significant biological features (Vamathevan et al. 2019; Staszak et al. 2022). Small molecule formulae, characteristics, structures, arrangements, spectra, visualization, and activities in the biological and industrial applications are typical examples of chemical data of interest during different infection studies related to aquatic pathogens (Netzeva et al. 2008; Dias et al. 2016; Chen et al. 2024). This cheminformatics began as a supporting tool in the process of drug discovery and development, from a wide range of biological molecules and synthetic designs can be analysed; however, its significance has grown over the decades and making this tool an essential component in numerous domains of biochemistry, biology, organic, and inorganic chemistry (Wishart 2007) that can be used to correlate with the host and the aquatic pathogens.

Applications

- *Virtual screening for antimicrobials*: Cheminformatics has been used to virtually screen hundreds of small molecules and phytochemicals against aquaculture pathogens. Virtual screening of phytochemicals against *A. hydrophila* virulence proteins identified potent inhibitors (Gupta et al. 2021; Gupta et al. 2023).
- *QSAR modelling and toxicity prediction*: Quantitative Structure-Activity Relationship (QSAR) models are applied to predict the efficacy and environmental toxicity of chemical compounds (Netzeva et al. 2008; Dom et al. 2012; Sharma and Darlami 2025).
- *Target identification and ligand design*: Structure-based cheminformatics helps identify binding sites in pathogen proteins and design specific ligands to inhibit them. For instance, docking studies on WSSV structural proteins such as VP28 have aided in the design of antiviral molecules using natural compound libraries (Thomas et al. 2022). Lipinski's rule and docking simulations assist in pre-selecting drugs for fish pathogens (Singh et al. 2021).
- *Integration with ML/AI*: Machine learning-based cheminformatics is now emerging to predict compound–pathogen interactions, accelerating drug repurposing and minimising experimental failures (Harigua-Souiai et al. 2024).

Cheminformatics thus supports a data-driven approach to aquatic pharmacology and immunotherapy development, paving the way for precision medicine in aquaculture.

24.5.6 Drug Discovery and Applications in Aquatic Animal Health

Bioinformatics accelerates aquaculture drug discovery by identifying target sites, modelling protein-ligand interactions, and reducing dependency on trial-based methods. In addition to speeding up the process of identifying drug targets, screening, and refining drug candidates, bioinformatic analysis can also help with side effect characterization and drug resistance prediction (Xia 2017; Zhang et al. 2025a). Mechanism-based drug discovery and repurposing have greatly benefited from high-throughput data that are correlated with cistromic, transcriptomic, proteomic, ribosome, genomic, epigenetic, and genome architecture in the biological sciences. Realistic protein–ligand docking experiments and virtual screening were made possible by the accumulation of protein and RNA structures, the development of homology modelling and protein structure simulation, and the large structure databases of small molecules and metabolites, enabling in identification of potential and usefulness in drug discovery (Xia 2017).

Applications
- Molecular docking of marine natural compounds with WSSV VP28 protein revealed potential antivirals (Thomas et al. 2022; Tusar et al. 2025).
- Structure-based drug design has proposed quorum-sensing inhibitors for *V. harveyi* to disrupt virulence (Govindan et al. 2024).
- In silico screening of CRISPR-Cas12a systems for pathogen-specific diagnostics is paving the way for theranostics in fish health (Dong et al. 2023; Yu et al. 2023; Wang et al. 2025a; Wang et al. 2025b).

24.5.7 Climate Change Studies, Epigenomics, and Applications in Aquatic Animal Health Management

Bioinformatics and its computational tools support the assessment of how aquatic organisms such as fish and shellfish respond to climate stressors like temperature, salinity, and hypoxia. These climate stressors are directly related to global warming and can be assessed by understanding the epigenetic changes in fish and shellfish when exposed to prolonged duration (Cochrane et al. 2009; Brander et al. 2017; Barange et al. 2018; Galappaththi et al. 2020). Epigenomics studies are mainly connected to modifications of DNA or histone proteins without altering the underlying DNA sequence that affect gene expression and the proteins for proper functioning. Epigenomics focuses on genetic changes over the different generations that involve changes in gene expression by inducing DNA methylation, histone modification, and non-coding RNAs. In aquatic species, epigenomic regulation plays a pivotal role in physiological and chemical stressors, stress adaptation, transgenerational plasticity, and immune responses, which are key components of health management in dynamic aquatic environments (Granada et al. 2018; Kumar et al. 2020; Liu et al. 2022; Roy et al. 2022; Piferrer and Wang 2023; Rasal et al. 2024a, b).

Applications
- Studies on stress and its immunity in zebrafish *Danio rerio* have shown stress-induced methylation of DNA in the immune genes like interleukins (IL-1β), indicating an epigenetic link between stressors in the environment and suppression of immune (Aluru et al. 2015).
- In rainbow trout *Oncorhynchus mykiss*, exposure to the pathogen *A. salmonicida* showed altered methylation patterns in leukocytes, signifying an epigenetic memory after exposure to infection (Morera et al. 2011).
- DNA methylation patterns as biomarkers of heavy metal or antibiotic exposure in molluscs and crustaceans (Kumar et al. 2022; Roy et al. 2022).

24.5.8 Evolutionary Studies

Evolutionary studies, including phylogenetics, molecular evolution, and population genomics, help trace the origin, spread, and adaptive mechanisms of pathogens and hosts in aquatic environments (Olymon et al. 2024). These insights support diagnostics, vaccine development, selective breeding, and outbreak surveillance, core elements of aquatic animal health management. Phylogenetics and comparative genomics help track the evolution of pathogens, host resistance, and virulence trends in aquaculture ecosystems (Eaton et al. 2010).

Applications

- *Pathogen evolution and tracking*: Phylogenetic analysis of ISKNV strains across Asia revealed host-specific evolution and regional transmission routes, supporting biosecurity and control efforts (Dong et al. 2017; Girisha et al. 2021; Kushala et al. 2025).
- *Host-pathogen co-evolution*: Molecular evolution studies of *Myxobolus cerebralis* and its salmonid hosts have revealed co-evolutionary dynamics shaping immune evasion and resistance mechanisms (Hedrick et al. 2001; Baerwald 2013; Akram et al. 2023; Saleh et al. 2024).
- *Antibiotic resistance evolution*: Genomic tracking of *V. harveyi* in shrimp farms has shown the rapid acquisition of antibiotic resistance genes via horizontal gene transfer, informing responsible antimicrobial use (Wang et al. 2025a, b; Zhang et al. 2025a, b).

24.5.9 Waste Cleanup

Bioinformatics plays an important role in the bioremediation of aquatic ecosystems such as aquaculture waste and pollutants threaten aquatic animal health and welfare. The leap of industrialization, urbanization, intensive agricultural and aquaculture practices due to a growing population has led to the release of pollutants ranging from heavy metals, pesticides, and persistent organic contaminants into aquatic environments, resulting in a severe threat to aquatic biodiversity and aquatic animal health. The ecological imbalance due to these activities directly influences the aquatic organisms, affecting their reproductive and immune systems, which tend to be more susceptible to infections and reducing aquaculture productivity (Patil et al. 2021a, b). Bioremediation is a process involving microorganisms to degrade or cleanse harmful contaminants into non-toxic metabolites, which has proved to be a cost-effective and eco-friendly biological approach to mitigate pollution. Modern bioinformatics tools use in silico analyses on microbial genomes, gene–gene interactions, metabolic pathways, and enzyme functions to check the efficacy of bioremediation. This computational approach allows for the identification and engineering of microbial strains with enhanced degradation capabilities for specific pollutants found in aquaculture effluents and aquatic habitats. (Bell et al. 2005; Chávez-Crooker and Obreque-Contreras 2010).

Applications

- Screening in Sundarbans ecosystems for their microbiomes by using a metagenomics approach identified probiotic, denitrifying, and ammonia-oxidising bacteria (Das et al. 2024b).
- Sludge samples revealed the presence of lipases and proteases involved in bio-waste degradation in aquaculture effluents (Dhar et al. 2023).
- Microbial consortia, especially *Bacillus* spp., are being formulated as probiotics and bioremediators to reduce organic load in ponds (Soltani et al. 2019).

24.5.10 Biofuels

Algae-based biofuel production is improved using genome and pathway engineering tools to optimise the lipid biosynthesis and growth efficiency. The development of sustainable green technologies depends on the application of computational tools for the analysis and interpretation of sizable biological datasets, made possible by bioinformatics. Bioinformatics in the field of green technologies uses integrated computational tools crucial to the development and improvement by using a multi-omics approach involving transcriptomics, metabolomics, proteomics, and genome analysis. Bioinformatics can be used to study microbial communities for bioenergy production, for the development of environmentally friendly and sustainable technologies (Govindan et al. 2024; Kruthi et al. 2024; Rajeev et al. 2024).

Applications

- Genome-scale metabolic homology modelling of *Nannochloropsis gaditana* helped in the identification of the vital enzymes for triacylglycerol production (Baroukh et al. 2015; Sharma et al. 2018).
- CRISPR/Cas9 editing of *Chlamydomonas reinhardtii* improved lipid accumulation for biodiesel use (Gupta et al. 2021; Lakhawat et al. 2022).
- Transcriptomic shifts in nutrient-deprived conditions guided algal strain improvement for sustainable biofuel generation (Baroukh et al. 2015; Sharma et al. 2018).

24.5.11 Integrated "Multi-omics" Technology and Applications in Aquatic Animal Health Management

A complex biological system-level understanding of aquatic animal health can be gained by merging the biological information from genomics, which plays an important role as a reference to transcriptomics, proteomics, metagenomics, and metabolomics through the integrated multi-omics approach. Fish and shellfish cultures grown in confined environments are becoming more vulnerable to new emerging and re-emerging infectious diseases, as well as other non-infectious diseases, due to intensive farming leading to the development of environmental stressors. Additionally, many conventional approaches use single data, usually overlooking significant variations in host–pathogen dynamics, environmental interactions, and individual genetics. For these reasons, multi-omics technologies such as transcriptomics, proteomics, metabolomics, genomics, metagenomics, and epigenomics are revolutionising aquatic animal health research and disease management by enabling an understanding of complex biological processes (Li et al. 2022; Shahrajabian and Sun 2023; Sagharchi et al. 2025).

By merging omics techniques, the complex biological relationships between pathogens and their host immune systems at multiple molecular levels, such as genes, gene expression, and their proteins, can be identified. For instance, while genomics and transcriptomics show genetic variations and changes in gene expression, proteomics and metabolomics pinpoint the corresponding functional proteins and metabolic responses during infections. By providing a thorough understanding of the immune system, these technologies aid in the identification of critical immune pathways, novel vaccine candidates, and molecular markers for early diagnosis. (Rise et al. 2019; Natnan et al. 2021; Shivam et al. 2021; Babu and Snyder 2023; Andersen et al. 2025b; Lim et al. 2025).

Recent studies have demonstrated the efficacy of integrated omics in the management of fish diseases. For example, high-throughput transcriptomic analysis has identified immune-regulatory gene networks that are activated during *Streptococcus iniae* infection in tilapia. One of the new vaccine candidates discovered by proteomic methods is Dihydrolipoamide dehydrogenase (DLD), an outer membrane protein that shields the orange-spotted grouper *Epinephelus coioides* from several *Vibrio* species (Pang et al. 2016). In environmental toxicology, conventional single-omics techniques are insufficient to determine the causal link between cellular responses and pollution exposure. Using high-throughput sequencing and high-resolution mass spectrometry, multi-omics integration is a novel approach to mapping the effects of pollutants on gene, protein, and metabolite networks. Early identification of low-dose and long-term exposure effects from contaminants is made possible by the integration of transcriptomic, proteomic, and metabolomic data. In aquatic species, it can also detect molecular pathways linked to immunological suppression, oxidative stress, and apoptosis (Augustyn et al. 2021; Shi et al. 2024).

Despite its potential, large-scale multi-omics data integration and analysis remain difficult as they involve processing huge terabytes of data in high-performance environments. Significant challenges in this field of aquaculture include the requirement for interdisciplinary expertise in aquatic pathology, fish physiology, bioinformatics, information science, computer science, molecular biology, high-performance analytical instruments, biostatistics, and data standardization. However, recent advancements in the field of computer science with AI and ML can play a crucial role in multi-omics analysis with integration are quickly becoming more accessible and actionable to be used in complex biological datasets. By 2030, the routine application of multi-omics in aquaculture is expected to transform the monitoring of water quality and fish health. This requires combining hyperspectral cameras, wearable biosensors, chemical sensors, microbiome analysis, and real-time health tracking systems to provide early warning systems and tailored interventions that reduce the dependency on the use of antibiotics and immunizations in fish health.

Applications

- Integration of multi-omics seabass *Dicentrarchus labrax* infected with *V. anguillarum* revealed coordinated responses across gene, expressed genes, protein, and metabolite levels (Buonocore et al. 2014, 2020).
- Multi-omics integrated with transcriptomics and metabolomics has aided in differentiating immune responses due to acute and chronic WSSV infection in shrimp (Fajardo et al. 2022; Kumar et al. 2022).
- An integrated proteomic and metabolomic approach was employed to investigate the host-pathogen interaction in *P. vannamei* infected with EHP. The study identified key proteomic biomarkers associated with growth hormone disorder in infected shrimp (Ning et al. 2019) (Table 24.2).

24.6 Recent Bioinformatics Advancements in Aquatic Animal Health Management

The in silico genetic analysis of the fish and fish pathogens plays a vital role in managing the diseases in aquaculture systems. The selection of fish traits with potentially high-value characteristics like breeding capacity, disease resistance, and stress tolerance is made possible by the advancement in bioinformatic tools. The emerging tools like transcriptomics, nutrigenomics, and metabolomics reveal the expression of genes towards specific nutrients, making it a vital parameter in nutrient and feed selection (Adebayo et al. 2024). The advancement in transcriptomics, metabolomics, proteomics, and recombinant DNA technology resulted in a large genetic repository, paving the way for future therapeutics and breeding strategies in fish and the aquaculture sector (Arya et al. 2024). The multi-omics systems developed for terrestrial animals can also be effective in bringing therapeutic advancements in aquatic systems. Omics tools are efficiently used to understand the host–pathogen mechanism, which can lead to sufficient data pooling that can be used in the development of vaccines. It can also be used as an effective approach in selecting the right nutrient and feed formulations required for fish growth (Natnan et al. 2021).

Apart from the fish growth and nutritional derivatives, the in-silico bioinformatic tools are effectively exploited for the use of terrestrial animals and for human use. The marine-derived anticancer compounds and their new target are being predicted using bioinformatic tools (Santaniello et al. 2022). The advancement in proteomics paved the way to understand the quality of fish protein and their derivatives and the health benefits in a much better way (Raman 2024). The advancement in genomic progress has led better understanding of the fish breeding technology, helping in the efficient selection of brooders (Wang et al. 2024a, b). The development in the omics, especially in the field of genomics, transcriptomics, proteomics, metabolomics, epigenomics, toxicogenomics, has led scientists to better understand the genetic parameters of fish species, which ultimately helps in fish species selection for the different aquaculture practices and their modulations in the culture systems towards stress and pathogens (Mancia 2018). The advancement in genetic tools and NGS

Table 24.2 Overview of omics technologies and their applications in aquatic animal health management

Omics type	Objective	Key tools/ technologies	Applications in aquatic animal health
Genomics	To decode the complete DNA sequence of aquatic organisms	WGS, NGS, SNP analysis	Disease resistance gene identification, selective breeding, and pathogen detection
Transcriptomics	To study gene expression under specific conditions	RNA-Seq, qRT-PCR, Microarrays	Host–pathogen interaction studies, stress response profiling, vaccine target discovery
Proteomics	To analyse the protein profile and its functional roles	2D-GE, LC-MS/MS, MALDI-TOF	Understanding immune responses, biomarker identification, protein–protein interactions
Metabolomics	To assess metabolic changes during disease or stress	NMR, GC-MS, LC-MS	Disease biomarker discovery, physiological status monitoring, nutritional assessments
Epigenomics	To explore heritable changes without DNA sequence changes	Bisulfite sequencing, ChIP-Seq	Impact of environmental stress on gene regulation, transgenerational immune priming
Microbiomics	To examine microbial communities associated with hosts	16S rRNA sequencing, Metagenomics	Gut microbiome analysis, pathogen surveillance, probiotic development
Metagenomics	To assess total genetic material from environmental samples	Shotgun sequencing, 16S/18S rRNA sequencing	Environmental pathogen detection, water quality monitoring, microbial ecology
Integrative multi-omics	To combine multiple omics layers for systems biology	Bioinformatics platforms (e.g., Cytoscape, OmicsNet)	Holistic disease diagnostics, host–pathogen network modelling, precision aquaculture health management

2D-GE two-dimensional gel electrophoresis, *LC-MS/MS* liquid chromatography-tandem mass spectrometry, *MALDI-TOF* matrix-assisted laser desorption/ionization time-of-flight, *NMR* nuclear magnetic resonance, *GC-MS* gas chromatography-mass spectrometry, *LC-MS* liquid chromatography-mass spectrometry, *ChIP-Seq* chromatin immunoprecipitation sequencing

has revolutionised the molecular detection techniques, which are more rapid and sensitive. The accurate genetic information from the codon to the nucleotide level helps scientists to understand the fish traits more accurately and effectively enhance the species selection in aquaculture systems (Jaies et al. 2024). Proteomics approaches are employed in shrimp culture systems to enhance breeding and spawning. Further, the overview on immunological, disease resistance pattern, and microbiome is well understood using a proteomics approach (Diwan et al. 2022).

24.7 Bioinformatics Databases

There are plenty of databases available for nucleotide and amino acid data analysis.

Database	Function
NCBI	Provide access to biomedical and genomic information, including databases and software tools for analysing molecular and genomic data
MobiDB	Explore the CAID Prediction Portal for accurate protein disorder predictions.
KEGG	KEGG is a database resource for understanding the high-level functions and utilities of the biological system
MetaCyc	MetaCyc contains pathways involved in both primary and secondary metabolism, as well as associated metabolites, reactions, enzymes, and genes.
PDB	Protein 3D structure database
PeptideAtlas	PeptideAtlas is a multi-organism, publicly accessible compendium of peptides identified in a large set of tandem mass spectrometry proteomics experiments.
GeneCards	GeneCards is a comprehensive, user-friendly database that integrates gene-centric data from various sources.
BindingDB	The first public molecular recognition database, BindingDB, supports research, education and practice in drug discovery, pharmacology and related fields.
TIGRFAMs	TIGRFAMs is a collection of manually curated protein families focusing primarily on prokaryotic sequences.
FlyBase	*Drosophila* database
ChEMBL	ChEMBL is a manually curated database of bioactive molecules with drug-like properties.
BRENDA	BRENDA is a database of enzymes, metabolic pathways, and gene functions
ModBase	ModBase is a database of comparative protein structure models, calculated by the ModPipe pipeline.
PhosPhAt 4.0	PhosPhAt 4.0 is a database of phosphorylation sites in Arabidopsis thaliana identified by mass spectrometry
UniPathway	UniPathway is a fully manually curated resource for the representation and annotation of metabolic pathways.
PIRSF	PIRSF is a hierarchical system for clustering UniProtKB sequences based on evolutionary and functional similarity
UniCarbKB	UniCarbKB provides detailed profiles for glycoproteins, including information on glycosylation sites, glycan heterogeneity, and associated biological functions.
HUGE	Large protein database
DrugBank	DrugBank is a vital resource for pharmaceutical research, offering comprehensive and reliable drug data, structured for immediate use or easy software integration.
PATRIC	Bacterial bioinformatics
H-InvDB	H-InvDB is a comprehensive annotation resource of human genes and transcripts, based on human full-length cDNAs and mRNAs.
CATH-Gene3D	CATH-Gene3D provides information on the evolutionary relationships of protein domains through sequence, structure and functional annotation data
HAMAP	HAMAP is a system for the classification and annotation of protein sequences, used to annotate protein records in UniProtKB
SABIO-RK	SABIO-RK is a curated database that contains information about biochemical reactions, their kinetic rate equations with parameters and experimental conditions
EMBL-EBI	EMBL-EBI Expression Atlas, an open public repository of gene expression pattern data under different biological conditions

24.8 Common Bioinformatics Tools

The most common bioinformatic tools are listed below:

Category	Tool	Description	Use
Sequence alignment	BLAST	Basic Local Alignment Search Tool	Find regions of similarity between sequences
	Clustal Omega	Multiple sequence alignment tool	Align multiple DNA/ protein sequences
	MAFFT	Fast multiple sequence alignment	Accurate large-scale sequence alignment
Gene prediction	GENSCAN	Predicts exon-intron structures in genomic DNA sequences	Identify genes and gene structures in eukaryotic DNA
	ORFfinder	Identifies open reading frames (ORFs) in nucleotide sequences	Detect coding regions and start/stop codons
Genome assembly	SPAdes	Genome assembler for single-cell and bacterial data	De novo genome assembly
	Velvet	Sequence assembler for short read sequencing data	De novo assembly from NGS reads
	DNASTAR	Small to medium-sized genome sequencing	Commercial genome assembly software suite
	ABySS	Large genomes (including eukaryotic)	De novo genome assembler
Phylogenetics software	MEGA	Builds evolutionary trees	Phylogenetic tree construction and analysis
	RAxML	Maximum likelihood-based tree construction	Large-scale phylogenetics
Transcriptomics	DESeq2	Statistical analysis for differential gene expression	Identify up/downregulated genes
	HISAT2	Fast alignment program for RNA-Seq reads	Align RNA-Seq data to a genome
	StringTie	Transcript assembly and quantification	Construct and quantify full-length transcripts
Functional annotation	Blast2GO	GO annotation based on BLAST results	Assign Gene Ontology terms to sequences
	InterProScan	Classifies sequences into protein families and predicts domains	Functional annotation of protein sequences
	eggNOG-mapper	Functional annotation using orthology assignments	Gene function prediction and pathway mapping
Protein modelling & docking	SWISS-MODEL Modeller	Homology modelling of protein 3D structures	Predict protein tertiary structures
	AutoDock	Molecular docking software	Simulate ligand–protein interactions
	PyMOL	Molecular visualisation system	View and analyse 3D protein structures
	BIOVIA Discovery Studio	Commercial suite for molecular modelling and simulations	Protein-ligand docking, pharmacophore modelling, ADMET

(continued)

Category	Tool	Description	Use
Metagenomics	QIIME2	Analysis of microbiome data from amplicon sequencing	Microbial diversity analysis
	Kraken2	Taxonomic classification of metagenomic reads	Fast microbial community profiling
	PICRUSt	Predicts functional gene content from 16S rRNA gene sequences	Infer metabolic pathways and gene families in microbiomes
Data visualisation	R (ggplot2)	Statistical computing and graphics	Custom plots and bioinformatics data visualization
	Cytoscape	Visualise molecular interaction networks	Network and pathway analysis
Protein structure validation	PROCHECK	Assesses stereochemical quality using Ramachandran plots	Validate bond angles, torsions, and geometry
	MolProbity	All-atom contact analysis, clash scores, and geometry checks	Validate crystallographic and modelled structures
	QMEAN	Provides global and local quality estimates for protein models	Model validation in homology modelling and predictions
	ERRAT	Analyses non-bonded atomic interactions	Identify regions of poor structural quality

24.9 Key Issues and Challenges

The in silico bioinformatic tools harbour their pros and cons. The variability of genetic make-up between and within the aquatic species makes it very difficult to address any complications related to genetic levels. However, the issue can be overcome by targeting the conserved regions within the fish species. The computationally based homologs or genome-based orthologs are analysed based on the global or local databases available on the databank, which uses nucleotides or amino acid sequences as a data query (Ambrosino et al. 2019). The following approach requires a larger database to be screened for an accurate therapeutic approach. The application of bioinformatic tools in Indian aquaculture systems is limited due to the limited data resources in the databank. Due to the non-similarity of the genetic makeup, due to the attributes like different geographical locations and their growing conditions, the comparison between two species is limited (Bruder et al. 2016). The DNA barcoding to detect new fish species lacks specificity due to its intraspecific variations. The International Barcode of Life (iBOL), which uses Mitochondrial COI, is known to have multiple reference sequences and it makes it less suitable for metabarcoding in aquatic systems (Collins et al. 2019). The bioinformatic tools face the major challenge of data visualization; the current methods being employed are fundamentally inadequate (O'Donoghue 2021)

24.10 Conclusions

The application of bioinformatics in aquaculture and fisheries is a paradigm change in aquatic animal health management and sustainable production. As aquaculture grows to satisfy the world's need for nutritious, high-protein food, health management, genetic enhancement, and environmental sustainability have become core issues. Bioinformatics provides a strong and data-intensive platform to address such issues by analysing and interpreting high-complexity biological data. From genomics to metagenomics and to proteomics, transcriptomics, and integrated multi-omics integration, bioinformatics has enabled a better understanding of host–pathogen interaction, immune response, genetic resistance, and the impact of environmental stressors. Its applications cut across vaccine and drug design, disease diagnosis, selective breeding, nutritional genomics, climate resilience, bioremediation, and even green energy through algal biofuels. Tools like GWAS, QTL mapping, and predictive models with machine learning are revolutionising aquaculture with enhanced precision in disease control, genetic selection, and environmental monitoring. The synergy between AI and bioinformatics with wearable biosensors and real-time health monitoring systems will redefine fish health monitoring and disease prevention strategies. By 2030, these technologies will likely become routine in aquaculture, marking a shift towards precision aquaculture. Despite the existing limitations, additional investment in bioinformatics research, capacity building, and global data sharing will drive innovation, sustainability, and resilience in aquatic animal health systems, facilitating a better, more secure, and sustainable aquaculture future.

References

Adebayo OM, Popoola MA, Kuusu DJ, Fanwo RR, Shoyombo AJ, Ndiomu EP et al (2024) Application of bioinformatics in animal breeding and genetics: a review. In: 2024 International conference on science, engineering and business for driving sustainable development goals (SEB4SDG). IEEE, pp 1–7

Ahmed F, Kumar G, Soliman FM, Adly MA, Soliman HAM, El-Matbouli M, Saleh M (2019) Proteomics for understanding pathogenesis, immune modulation and host-pathogen interactions in aquaculture. Comp Biochem Physiol D Genom Proteom 32:100625

Akram N, El-Matbouli M, Saleh M (2023) The immune response to the myxozoan parasite *Myxobolus cerebralis* in salmonids: a review on whirling disease. Int J Mol Sci 24(24):17392

Allal F, Nguyen NH (2022) Genomic selection in aquaculture species. In: Genomic prediction of complex traits: Methods and protocols, Methods in molecular biology, vol. 2467, Ahmadi N, Bartholome J, (eds), Springer, New York, pp 469–491. https://doi.org/10.1007/978-1-0716-2205-6_17

Altinok İ, Kurt İ (2003) Molecular diagnosis of fish diseases: a review. Turk J Fish Aquat Sci 3(2):131–138

Aluru N, Kuo E, Helfrich LW, Karchner SI, Linney EA, Pais JE, Franks DG (2015) Developmental exposure to 2, 3, 7, 8-tetrachlorodibenzo-p-dioxin alters DNA methyltransferase (DNMT) expression in zebrafish (*Danio rerio*). Toxicol Appl Pharmacol 284(2):142–151

Aly SM, Fathi M (2024) Advancing aquaculture biosecurity: a scientometric analysis and future outlook for disease prevention and environmental sustainability. Aquac Int 32(7):8763–8789

Ambrosino L, Tangherlini M, Colantuono C, Esposito A, Sangiovanni M, Miralto M et al (2019) Bioinformatics for marine products: an overview of resources, bottlenecks, and perspectives. Mar Drugs 17(10):576

Andersen LK, Abernathy JW, Farmer BD, Lange MD, McEntire ME, Rawles SD (2024) Gene expression profiles of white bass (*Morone chrysops*) and hybrid striped bass (*M. chrysops x M. saxatilis*) gill tissue following *Flavobacterium covae* infection. Comp Immunol Rep 6:200144

Andersen LK, Abernathy JW, Farmer BD, Lange MD, Sankappa NM, McEntire ME, Rawles SD (2025a) Analysis of striped bass (*Morone saxatilis*) and white bass (*M. chrysops*) splenic transcriptome following *Streptococcus iniae* infection. Mar Biotechnol 27(2):1–19

Andersen LK, Thompson NF, Abernathy JW, Ahmed RO, Ali A, Al-Tobasei R, Beck BH, Calla B, Delomas TA, Dunham RA (2025b) Advancing genetic improvement in the omics era: status and priorities for United States aquaculture. BMC Genomics 26(1):1–24

Arora PK, Bae H (2014) Integration of bioinformatics to biodegradation. Biol Proced Online 16:1–10

Arya D, Gautam P, Goswami R, Palod J, Singh VS, Sharma M (2024) Bioinformatics: an emerging tool to address animal health. Int J Adv Biochem Res 8(6S):260–264

Augustyn DR, Wyciślik Ł, Mrozek D (2021) Perspectives of using Cloud computing in integrative analysis of multi-omics data. Brief Funct Genomics 20(4):198–206

Babu M, Snyder M (2023) Multi-omics profiling for health. Mol Cell Proteomics 22(6):100561

Baerwald MR (2013) Temporal expression patterns of rainbow trout immune-related genes in response to *Myxobolus cerebralis* exposure. Fish Shellfish Immunol 35(3):965–971

Banerjee S, Mahanty A, Mohanty S, Mazumder DG, Cash P, Mohanty BP (2017) Identification of potential biomarkers of hepatotoxicity by plasma proteome analysis of arsenic-exposed carp *Labeo rohita*. J Hazard Mater 336:71–80

Barange M, Bahri T, Beveridge MCM, Cochrane KL, Funge-Smith S, Poulain F (2018) Impacts of climate change on fisheries and aquaculture. Synthesis of current knowledge, adaptation and mitigation options. FAO Fisheries and Aquaculture Technical Paper, 627, FAO, Rome, p. 628

Baroukh C, Muñoz-Tamayo R, Steyer J-P, Bernard O (2015) A state-of-the-art metabolic networks of unicellular microalgae and cyanobacteria for biofuel production. Metab Eng 30:49–60

Baxevanis AD, Bader GD, Wishart DS (2020) Bioinformatics. John Wiley & Sons, Hoboken

Bayat A (2002) Science, medicine, and the future: bioinformatics. BMJ (Clin Res ed) 324(7344):1018–1022

Bell T, Newman JA, Silverman BW, Turner SL, Lilley AK (2005) The contribution of species richness and composition to bacterial services. Nature 436(7054):1157–1160. https://doi.org/10.1038/nature03891

Boutin S, Dalpke AH (2020) The microbiome: A reservoir to discover new antimicrobials agents. Curr Top Med Chem 20(14):1291–1299. https://doi.org/10.2174/1568026620666200320112731

Brander K, Cochrane K, Barange M, Soto D (2017) Climate change implications for fisheries and aquaculture. In: Climate change impacts on fisheries and aquaculture: a global analysis, vol 1. John Wiley & Sons, Inc., Hoboken, pp 45–62

Bruder K, Maiki K, Cooper A, Sible E, Shapiro JW, Watkins SC, Putonti C (2016) Freshwater metaviromics and bacteriophages: a current assessment of the state of the art in relation to bioinformatic challenges: supplementary issue: bioinformatics methods and applications for big metagenomics data. Evol Bioinform 12:EBO-S38549

Buonocore F, Randelli E, Trisolino P, Facchiano A, de Pascale D, Scapigliati G (2014) Molecular characterization, gene structure and antibacterial activity of a g-type lysozyme from the European sea bass (*Dicentrarchus labrax* L.). Mol Immunol 62(1):10–18

Buonocore F, Gerdol M, Pallavicini A, Stocchi V, Randelli E, Belardinelli MC, Miccoli A, Saraceni PR, Secombes CJ, Scapigliati G (2020) Identification, molecular characterization and functional analysis of interleukin (IL)-2 and IL-2like (IL-2L) cytokines in sea bass (*Dicentrarchus labrax* L.). Cytokine 126:154898

Burley SK (2000) An overview of structural genomics. Nat Struct Biol 7(11):932–934

Caballero-Solares A, Xue X, Parrish CC, Foroutani MB, Taylor RG, Rise ML (2018) Changes in the liver transcriptome of farmed Atlantic salmon (*Salmo salar*) fed experimental diets based on terrestrial alternatives to fish meal and fish oil. BMC Genomics 19:1–26

Chandhini S, Kumar VJR (2019) Transcriptomics in aquaculture: current status and applications. Rev Aquac 11(4):1379–1397

Chávez-Crooker P, Obreque-Contreras J (2010) Bioremediation of aquaculture wastes. Curr Opin Biotechnol 21(3):313–317

Chen C, Chen L, Mao C, Jin L, Wu S, Zheng Y, Cui Z, Li Z, Zhang Y, Zhu S (2024) Natural extracts for antibacterial applications. Small 20(9):2306553

Chunyan Z, Xianhui M, Yongji D, Yangen Z, Yichao R (2023) Probiotics mediate intestinal microbiome and microbiota-derived metabolites regulating the growth and immunity of rainbow trout (*Oncorhynchus mykiss*). Microbiol Spectr 11(2):e03980–e03922

Cochrane K, De Young C, Soto D, Bahri T (2009) Climate change implications for fisheries and aquaculture. FAO fisheries and aquaculture technical paper, 530. Rome, FAO, p 212

Collins RA, Bakker J, Wangensteen OS, Soto AZ, Corrigan L, Sims DW et al (2019) Non-specific amplification compromises environmental DNA metabarcoding with COI. Methods Ecol Evol 10(11):1985–2001

Das S, Sahoo PK (2014) Markers for selection of disease resistance in fish: a review. Aquac Int 22:1793–1812

Das SP, Abidin Z, Huang H-T, Lin Y-R, Huang C-Y, Wu Y-S, Hu Y-F, Nan F-H (2024a) Deciphering the influence of dietary synbiotics in white shrimp gut and its effects in regulating immune signaling pathways. Front Mar Sci 10:1342708

Das BK, Chakraborty HJ, Kumar V, Rout AK, Patra B, Das SK, Behera BK (2024b) Comparative metagenomic analysis from Sundarbans ecosystems advances our understanding of microbial communities and their functional roles. Sci Rep 14(1):16218

Deris ZM, Iehata S, Gan HM, Ikhwanuddin M, Najiah M, Asaduzzaman M, Wang M, Liang Y, Danish-Daniel M, Sung YY (2022) Understanding the effects of salinity and *Vibrio harveyi* on the gut microbiota profiles of *Litopenaeus vannamei*. Front Mar Sci 9:974217

Dettleff P, Moen T, Santi N, Martinez V (2017) Transcriptomic analysis of spleen infected with infectious salmon anemia virus reveals distinct pattern of viral replication on resistant and susceptible Atlantic salmon (*Salmo salar*). Fish Shellfish Immunol 61:187–193

Dhar S, Devnath S, Kumar V, Roy S, Rout AK, Mistri A, Parida SN, Bisai K, Jana AK, Behera BK (2023) Bioremediation and its application in aquaculture. In: Biotechnological tools in fisheries and aquatic health management. Behera BK, (ed), Springer, Singapore, pp 197–215. https://doi.org/10.1007/978-981-99-2981-8

Dias DA, Jones OAH, Beale DJ, Boughton BA, Benheim D, Kouremenos KA, Wolfender J-L, Wishart DS (2016) Current and future perspectives on the structural identification of small molecules in biological systems. Metabolites 6(4):46

Diwan AD, Harke SN, Panche AN (2022) Application of proteomics in shrimp and shrimp aquaculture. Comp Biochem Physiol D Genom Proteom 43:101015

Dom N, Knapen D, Blust R (2012) Assessment of aquatic experimental versus predicted and extrapolated chronic toxicity data of four structural analogues. Chemosphere 86(1):56–64

Dong HT, Jitrakorn S, Kayansamruaj P, Pirarat N, Rodkhum C, Rattanarojpong T, Senapin S, Saksmerprome V (2017) Infectious spleen and kidney necrosis disease (ISKND) outbreaks in farmed barramundi (*Lates calcarifer*) in Vietnam. Fish Shellfish Immunol 68:65–73

Dong HT, Chaijarasphong T, Barnes AC, Delamare-Deboutteville J, Lee PA, Senapin S, Mohan CV, Tang KFJ, McGladdery SE, Bondad-Reantaso MG (2023) From the basics to emerging diagnostic technologies: what is on the horizon for tilapia disease diagnostics? Rev Aquac 15:186–212

Eaton HE, Ring BA, Brunetti CR (2010) The genomic diversity and phylogenetic relationship in the family iridoviridae. Viruses 2(7):1458–1475. https://doi.org/10.3390/v2071458

Fajardo C, Martinez-Rodriguez G, Costas B, Mancera JM, Fernandez-Boo S, Rodulfo H, De Donato M (2022) Shrimp immune response: a transcriptomic perspective. Rev Aquac 14(3):1136–1149

FAO (2024) The state of world fisheries and aquaculture 2024. Blue transformation in action. FAO, Rome

Fu S, Ni P, Yang Q, Hu H, Wang Q, Ye S, Liu Y (2021) Delineating the key virulence factors and intraspecies divergence of *Vibrio harveyi* via whole-genome sequencing. Can J Microbiol 67(3):231–248

Fuller SA, Abernathy JW, Sankappa NM, Beck BH, Rawles SD, Green BW, Rosentrater KA, McEntire ME, Huskey G, Webster CD (2024) Hepatic transcriptome analyses of juvenile white bass (*Morone chrysops*) when fed diets where fish meal is partially or totally replaced by alternative protein sources. Front Physiol 14:1308690

Galappaththi EK, Ichien ST, Hyman AA, Aubrac CJ, Ford JD (2020) Climate change adaptation in aquaculture. Rev Aquac 12(4):2160–2176

Galassie AC, Link AJ (2015) Proteomic contributions to our understanding of vaccine and immune responses. Proteomics Clin Appl 9(11–12):972–989

Gao C, Fu Q, Su B, Zhou S, Liu F, Song L et al (2016) Transcriptomic profiling revealed the signatures of intestinal barrier alteration and pathogen entry in turbot (*Scophthalmus maximus*) following *Vibrio anguillarum* challenge. Dev Comp Immunol 65:159–168

Girisha SK, Kushala KB, Nithin MS, Puneeth TG, Naveen Kumar BT, Vinay TN, Suresh T, Ajay SK, Venugopal MN, Ramesh KS (2021) First report of the infectious spleen and kidney necrosis virus (ISKNV) infection in ornamental fishes in India. Transbound Emerg Dis 68(2):964–972. https://doi.org/10.1111/tbed.13793

Govindan R, Gnanasekaran C, Govindan R, Muthuchamy M, Quero F, Jothi A, Chelliah CK, Arunachalam A, Viswanathan MR, Natesan M (2024) Anti-quorum sensing and anti-biofilm effect of *Nocardiopsis synnemataformans* RMN 4 (MN061002) compound 2, 6-Di-tert-butyl, 1, 4-Benzoquinone against biofilm-producing bacteria. Appl Biochem Biotechnol 196(7):3914–3948

Granada L, Lemos MFL, Cabral HN, Bossier P, Novais SC (2018) Epigenetics in aquaculture–the last frontier. Rev Aquac 10(4):994–1013

Gupta A, Gupta SK, Priyam M, Siddik MAB, Kumar N, Mishra PK, Gupta KK, Sarkar B, Sharma TR, Pattanayak A (2021) Immunomodulation by dietary supplements: a preventive health strategy for sustainable aquaculture of tropical freshwater fish, *Labeo rohita* (Hamilton, 1822). Rev Aquac 13(4):2364–2394

Gupta SK, Gupta A, Sarkar B, Gupta R, Kumar M, Kumari A, Foysal MJ (2023) Pomegranate (*Punica granatum*) peel extract supplementation in diet influences growth performance, haemato-immunological responses and cytokine expression in pathogen-aggravated *Labeo rohita* fingerlings. Aquaculture 562:738823

Hakim MM, Ganai NA, Ahmad SM, Asmi O, Akram T, Hussain S, Gora AH (2018) Nutrigenomics: omics approach in aquaculture research to mitigate the deficits in conventional nutritional practices. J Entomol Zool Stud 6:582–587

Harigua-Souiai E, Masmoudi O, Makni S, Oualha R, Abdelkrim YZ, Hamdi S, Souiai O, Guizani I (2024) cidalsDB: an AI-empowered platform for anti-pathogen therapeutics research. J Chem 16(1):134

Hedrick RP, McDowell TS, Mukkatira K, Georgiadis MP, MacConnell E (2001) Salmonids resistant to *Ceratomyxa shasta* are susceptible to experimentally induced infections with *Myxobolus cerebralis*. J Aquat Anim Health 13(1):35–42

Hesper B, Hogeweg P (1970) Bioinformatica: een werkconcept. Kameleon 1(6):28–29

Hounmanou YMG, Mdegela RH, Dougnon TV, Achoh ME, Mhongole OJ, Agadjihouèdé H, Gangbè L, Dalsgaard A (2018) Tilapia lake virus threatens tilapiines farming and food security: socio-economic challenges and preventive measures in Sub-Saharan Africa. Aquaculture 493:123–129

Houston RD, Bean TP, Macqueen DJ, Gundappa MK, Jin YH, Jenkins TL, Selly SLC, Martin SAM, Stevens JR, Santos EM (2020) Harnessing genomics to fast-track genetic improvement in aquaculture. Nat Rev Genet 21(7):389–409

Jaies I, Shah FA, Qadiri SSN, Qayoom I, Bhat BA, Dar SA, Bhat FA (2024) Immunological and molecular diagnostic techniques in fish health: present and future prospectus. Mol Biol Rep 51(1):551

Johnston IA, Kent MP, Boudinot P, Looseley M, Bargelloni L, Faggion S et al (2024) Advancing fish breeding in aquaculture through genome functional annotation. Aquaculture 583:740589

Johny TK, Puthusseri RM, Bhat SG (2021) A primer on metagenomics and next-generation sequencing in fish gut microbiome research. Aquac Res 52(10):4574–4600

Joshi R, Woolliams JA, Meuwissen THE, Gjøen HM (2018) Maternal, dominance and additive genetic effects in Nile tilapia; influence on growth, fillet yield and body size traits. Heredity 120(5):452–462

Karimi S, Ferreira JA, Taherzadeh MJ (2023) Filamentous fungi as animal and fish feed ingredients. In: Current developments in biotechnology and bioengineering: Filamentous fungi biorefinery. Taherzadeh MJ, Ferreira JA, Pandey A, (eds), Elsevier, Amsterdam, pp 399–433

Khang LTP, Dinh-Hung N, Islam SI, Dwinanti SH, Mwamburi SM, Permpoonpattana P, Linh NV (2025) A review of in silico approaches for discovering natural viral protein inhibitors in aquaculture disease control. J Fish Dis 48(7):e14120

Kibenge FSB, Godoy MG, Kibenge MJT (2024) Diagnosis of aquatic animal viral diseases. In: Aquaculture virology. Kibenge FSB, Godoy MG, (eds), Elsevier, San Diego, pp 49–88. https://doi.org/10.1016/B978-0-12-801573-5.00004-8

Kruthi MN, Panja S, Kumar A, Sengupta A, Narad P (2024) Application of bioinformatics: integrated computational tools in green technologies. In: Microbial approaches for sustainable green technologies. Shankar J, Verma P, Shah MP, (eds), CRC Press, Boca Raton, pp 112–135. https://doi.org/10.1201/9781003407683

Kumar H, Chaudhary A, Singh A, Sukhija N, Panwar A, Saravanan K, Bhaladhare A, Kaisa K, Panigrahi M (2020) A review on epigenetics: manifestations, modifications, methods and challenges. J Entomol Zool Stud 8(4):1–6

Kumar V, Roy S, Behera BK, Das BK (2022) Heat shock proteins (Hsps) in cellular homeostasis: a promising tool for health management in crustacean aquaculture. Life 12(11):1777

Kumar R, Das SP, Choudhury BU, Kumar A, Prakash NR, Verma R, Chakraborti M, Devi AG, Bhattacharjee B, Das R (2024) Advances in genomic tools for plant breeding: harnessing DNA molecular markers, genomic selection, and genome editing. Biol Res 57(1):80

Kushala KB, Nithin MS, Girisha SK, Dheeraj SB, Vinay TN, Kumar BTN, Suresh T (2025) First documentation of whole genome sequence analysis for infectious spleen and kidney necrosis virus (ISKNV) isolated from angel fish (*Pterophyllum scalare*) in India. Aquaculture 597:741906

Lakhawat SS, Malik N, Kumar V, Kumar S, Sharma PK (2022) Implications of CRISPR-Cas9 in developing next-generation biofuel: a mini-review. Curr Protein Pept Sci 23(9):574–584

Lal J, Vaishnav A, Singh SK, Meena DK, Biswas P, Mehta NK, Priyadarshini MB (2024) Biotechnological innovation in fish breeding: from marker-assisted selection to genetic modification. Discov Biotechnol 1(1):1–21

Lander ES, Linton LM, Birren B, Nusbaum C, Zody MC, Baldwin J, Devon K, Dewar K, Doyle M, Fitzhugh W (2001) Erratum: Initial sequencing and analysis of the human genome: International Human Genome Sequencing Consortium (Nature (2001) 409 (860–921)). Nature 412(6846):565–566

Li S, Pozhitkov A, Ryan RA, Manning CS, Brown-Peterson N, Brouwer M (2010) Constructing a fish metabolic network model. Genome Biol 11:1–15

Li X, Hu X, Lv A, Guan Z (2022) Skin immune response to Aeromonas hydrophila infection in crucian carp Carassius auratus revealed by multi-omics analysis. Fish Shellfish Immunol 127:866–875

Lim B, Kim D-Y, Seo Y-J, Lee J-Y, Kim J-M (2025) Systems biology and integration of multi-omics data. In: Bioinformatics in veterinary science: vetinformatics. Kim J-M, Pathak, RK, (eds), Springer, Singapore, pp 163–183. https://doi.org/10.1007/978-981-97-7395-4_8

Liu ZJ (2017) Bioinformatics in aquaculture: principles and methods. John Wiley & Sons, Hoboken

Liu Z, Zhou T, Gao D (2022) Genetic and epigenetic regulation of growth, reproduction, disease resistance and stress responses in aquaculture. Front Genet 13:994471

Long M, Zhao J, Li T, Tafalla C, Zhang Q, Wang X, Gong X, Shen Z, Li A (2015) Transcriptomic and proteomic analyses of splenic immune mechanisms of rainbow trout (*Oncorhynchus mykiss*) infected by *Aeromonas salmonicida* subsp. *salmonicida*. J Proteome 122:41–54

Lu JK, Chou HY, Gong HY, Chiou PP, Huang CW, Chen LL, Wu JL (2015) The marine aquaculture biotechnology status and its sustainability in Taiwan. Curr Biotechnol 4(3):311–318

Luscombe NM, Greenbaum D, Gerstein M (2001) What is bioinformatics? A proposed definition and overview of the field. Methods Inf Med 40(04):346–358

Mancia A (2018) New technologies for monitoring marine mammal health. In: Marine mammal ecotoxicology: Impacts of multiple stressors on population health. Fossi MC, Panti C, (eds), Academic Press, London, pp 291–320. https://doi.org/10.1016/B978-0-12-812144-3.00011-5

Martínez-Porchas M, Vargas-Albores F (2017) Microbial metagenomics in aquaculture: a potential tool for a deeper insight into the activity. Rev Aquac 9(1):42–56

Mohanta TK, Mishra AK, Al-Harrasi A (2021) The 3D genome: from structure to function. Int J Mol Sci 22(21):11585

Morera D, Roher N, Ribas L, Balasch JC, Doñate C, Callol A, Boltana S, Roberts S, Goetz G, Goetz FW (2011) RNA-Seq reveals an integrated immune response in nucleated erythrocytes. PLoS One 6(10):e26998

Munang'andu HM (2016) Environmental viral metagenomics analyses in aquaculture: applications in epidemiology and disease control. Front Microbiol 7:1986

Nambiar SP, Banuru SC (2025) A comprehensive review of bioinformatics tools and applications revolutionising aquatic animal health management. J Fish Health 5(2):86–110

Natnan ME, Mayalvanan Y, Jazamuddin FM, Aizat WM, Low C-F, Goh H-H, Azizan KA, Bunawan H, Baharum SN (2021) Omics strategies in current advancements of infectious fish disease management. Biology 10(11):1086

Netzeva TI, Pavan M, Worth AP (2008) Review of (quantitative) structure–activity relationships for acute aquatic toxicity. QSAR Comb Sci 27(1):77–90

Nguyen NH (2024) Genetics and genomics of infectious diseases in key aquaculture species. Biology 13(1):29

Ning M, Wei P, Shen H, Wan X, Jin M, Li X, Shi H, Qiao Y, Jiang G, Gu W (2019) Proteomic and metabolomic responses in hepatopancreas of whiteleg shrimp *Litopenaeus vannamei* infected by microsporidian *Enterocytozoon hepatopenaei*. Fish Shellfish Immunol 87:534–545

O'Donoghue SI (2021) Grand challenges in bioinformatics data visualization. Front Bioinform 1:669186

Olymon K, Yadav M, Teronpi V, Kumar A (2024) Unravelling the genomic secrets of bacterial fish pathogens: a roadmap to aquaculture sustainability. Mol Biol Rep 51(1):364

Omotayo RF (2019) Development of dystrophin-based biomarker for penaeid shrimps. University of Malaya (Malaysia)

Pang H, Chen L, Hoare R, Huang Y, Jian J (2016) Identification of DLD by immunoproteomic analysis and evaluation as a potential vaccine antigen against three *Vibrio* species in *Epinephelus coioides*. Vaccine 34(9):1225–1231

Patil PK, Geetha R, Ravisankar T, Avunje S, Solanki HG, Abraham TJ, Vinoth SP, Jithendran KP, Alavandi SV, Vijayan KK (2021a) Economic loss due to diseases in Indian shrimp farming with special reference to *Enterocytozoon hepatopenaei* (EHP) and white spot syndrome virus (WSSV). Aquaculture 533:736231

Patil PC, Patel DA, Tale VS (2021b) Metagenomic analysis of Dam Reservoirs in Pune city for bacterial fingerprints through BLAST and Kaiju tool. Biosci Biotechnol Res Asia 17(4):839–852

Piferrer F, Wang H (2023) Epigenetics in aquaculture. John Wiley & Sons, Hoboken

Pridgeon JW, Klesius PH, Mu X, Song L (2011) An in vitro screening method to evaluate chemicals as potential chemotherapeutants to control *Aeromonas hydrophila* infection in channel catfish. J Appl Microbiol 111(1):114–124

Puneeth TG, Baliga P, Girisha SK, Shekar M, Nithin MS, Suresh T, Naveen Kumar BT (2021) Complete genome analysis of a red seabream iridovirus (RSIV) isolated from Asian seabass (*Lates calcarifer*) in India. Virus Res 291:198199. https://doi.org/10.1016/j.virusres.2020.198199

Rajeev M, Jung I, Kang I, Cho J-C (2024) Genome-centric metagenomics provides insights into the core microbial community and functional profiles of biofloc aquaculture. Msystems 9(10):e00782–e00724

Raman M (2024) Bioinformatics and computational tools. In: Fish structural proteins and its derivatives: functionality and applications. Raman M, Sasidharan A, Sabu S, Rajan DP, (eds), Springer Nature, Singapore, pp 107–134. https://doi.org/10.1007/978-981-97-2562-5_5

Rana KMS, Ahammad K, Salam MA (2020) Bioinformatics: scope and challenges in aquaculture research of Bangladesh-a review. Int J Agric Res Innov Technol 10(2):137–145

Rasal KD, Asgolkar P, Shinde S, Dey D, Sundaray JK (2024a) Role of epigenetics in fisheries and aquaculture. In: Current trends in fisheries biotechnology. Behara BK (ed), Springer, Singapore, pp 27–44. https://doi.org/10.1007/978-981-97-3165-7_3

Rasal KD, Kumar PV, Risha S, Asgolkar P, Harshavarthini M, Acharya A, Shinde S, Dhere S, Rasal A, Sonwane A (2024b) Genetic improvement and genomic resources of important cyprinid species: status and future perspectives for sustainable production. Front Genet 15:1398084

Rather MA, Agarwal D, Bhat TA, Khan IA, Zafar I, Kumar S et al (2023) Bioinformatics approaches and big data analytics opportunities in improving fisheries and aquaculture. Int J Biol Macromol 233:123549

Rise ML, Martyniuk CJ, Chen M (2019) Comparative physiology and aquaculture: toward omics-enabled improvement of aquatic animal health and sustainable production. Comp Biochem Physiol D Genom Proteom 31:100603

Robinson NA, Robledo D, Sveen L, Daniels RR, Krasnov A, Coates A, Jin YH, Barrett LT, Lillehammer M, Kettunen AH (2023) Applying genetic technologies to combat infectious diseases in aquaculture. Rev Aquac 15(2):491–535

Rodrigues PM, Martin SAM, Silva TS, Boonanuntanasarn S, Schrama D, Moreira M, Raposo C (2018) Proteomics in fish and aquaculture research. In: Proteomics in domestic animals: from farm to systems biology. de Almeida AM, Eckersall D, Miller I, (eds), Springer International Publishing, pp 311–338. https://doi.org/10.1007/978-3-319-69682-9_16

Rout AK, Behera BK (2024) Proteomics in fisheries and aquaculture. In: Current trends in fisheries biotechnology. Behara BK (ed), Springer, Singapore, pp 45–64. https://doi.org/10.1007/978-981-97-3165-7_4

Roy K (2022) Chemometrics and cheminformatics in aquatic toxicology. In: Wiley Online Library

Roy SS, Dasgupta R, Bagchi A (2014) A review on phylogenetic analysis: a journey through modern era. Comput Mol Biosci 4(3):39–45

Roy S, Kumar V, Behera BK, Das BK (2022) Epigenetics: perspectives and potential in aquaculture. In: Advances in fisheries biotechnology. Pandey PK, Parhi J (eds), Springer, Singapore, pp 133–150. https://doi.org/10.1007/978-981-16-3215-0_9

Rungrassamee W, Klanchui A, Chaiyapechara S, Maibunkaew S, Tangphatsornruang S, Jiravanichpaisal P, Karoonuthaisiri N (2013) Bacterial population in intestines of the black tiger shrimp (*Penaeus monodon*) under different growth stages. PLoS One 8(4):e60802

Sagharchi SMAM, Sheykhhasan M, Ghorbani A, Afrazeh E, Poondla N, Kalhor N, Tanzadehpanah H, Mahaki H, Manoochehri H (2025) Genomics: a piece of the multi-omics puzzle. In: Effective techniques for bioinformatic exploration. IGI Global, Hershey, PA, pp 23–68

Sahu BK, Dhimar C (2024) Sustainable innovations for harnessing marine biotechnology through advanced bioinformatics. Agric Biotechnol J 16(1):337–354

Saleh M, Hummel K, Schlosser S, Razzazi-Fazeli E, Bartholomew JL, Holzer A, Secombes CJ, El-Matbouli M (2024) The myxozoans *Myxobolus cerebralis* and *Tetracapsuloides bryosalmonae* modulate rainbow trout immune responses: quantitative shotgun proteomics at the portals of entry after single and co-infections. Front Cell Infect Microbiol 14:1369615

Sang Y, Long Z, Dan X, Feng J, Shi T, Jia C et al (2022) Genomic insights into local adaptation and future climate-induced vulnerability of a keystone forest tree in East Asia. Nat Commun 13(1):6541

Sankappa NM, Lange MD, Yildirim-Aksoy M, Eljack R, Kucuktas H, Beck BH, Abernathy JW (2024) Transcriptome analysis and immune gene expression of channel catfish (*Ictalurus punctatus*) fed diets with inclusion of frass from black soldier fly larvae. Front Physiol 14:1330368

Santaniello G, Nebbioso A, Altucci L, Conte M (2022) Recent advancement in anticancer compounds from marine organisms: approval, use and bioinformatic approaches to predict new targets. Mar Drugs 21(1):24

Shahrajabian MH, Sun W (2023) Survey on multi-omics, and multi-omics data analysis, integration and application. Curr Pharm Anal 19(4):267–281

Sharma S, Darlami J (2025) QSAR modeling and its application in veterinary drug design. In: Bioinformatics in veterinary science: vetinformatics. Kim J-M, Pathak, RK, (eds), Springer, Singapore, pp 367–385. https://doi.org/10.1007/978-981-97-7395-4_15

Sharma PK, Saharia M, Srivastava R, Kumar S, Sahoo L (2018) Tailoring microalgae for efficient biofuel production. Front Mar Sci 5:382

Shi C, Cheng L, Yu Y, Chen S, Dai Y, Yang J, Zhang H, Chen J, Geng N (2024) Multi-omics integration analysis: tools and applications in environmental toxicology. Environ Pollut 360:124675

Shivam S, El-Matbouli M, Kumar G (2021) Development of fish parasite vaccines in the OMICs era: progress and opportunities. Vaccine 9(2):179

Singh R, Prasad KP, Tiwari A, Pathak A, Srivastava P (2021) Molecular docking and simulation study to identify antiviral agent by targeting MX protein against betanodavirus causing viral nervous necrosis in Barramundi. Res J Pharm Technol 14(3):1405–1411

Skliros D, Kostakou M, Kokkari C, Tsertou MI, Pavloudi C, Zafeiropoulos H, Katharios P, Flemetakis E (2024) Unveiling emerging opportunistic fish pathogens in aquaculture: a comprehensive seasonal study of microbial composition in Mediterranean fish hatcheries. Microorganisms 12(11):2281

Soltani M, Ghosh K, Hoseinifar SH, Kumar V, Lymbery AJ, Roy S, Ringø E (2019) Genus *Bacillus*, promising probiotics in aquaculture: aquatic animal origin, bio-active components, bioremediation and efficacy in fish and shellfish. Rev Fish Sci Aquac 27(3):331–379

Soriano B, Hafez A, Naya-Català F, Moldovan RA, Toxqui-Rodríguez S, Piazzon MC et al (2023) SAMBA: structure-learning of aquaculture microbiomes using a Bayesian-Network Approach. Genes 14(8):1650

Staszak M, Staszak K, Wieszczycka K, Bajek A, Roszkowski K, Tylkowski B (2022) Machine learning in drug design: use of artificial intelligence to explore the chemical structure–biological activity relationship. Wiley Interdiscip Rev Comput Mol Sci 12(2):e1568

Subasinghe RP (2022) An introduction to global aquaculture. In: Aquaculture pathophysiology, Volume I. Finfish diseases. Kibenge FSB, Baldisserotto B, Chong RS, (eds), Academic Press, Cambridge, Massachusetts, pp. 45–48. https://doi.org/10.1016/C2016-0-01537-1

Sun B, Li X, Ning X, Sun L (2020) Transcriptome analysis of *Paralichthys olivaceus* erythrocytes reveals profound immune responses induced by *Edwardsiella tarda* infection. Int J Mol Sci 21(9):3094

Sundaray JK, Dixit S, Rather A, Rasal KD, Sahoo L (2022) Aquaculture omics: an update on the current status of research and data analysis. Mar Genomics 64:100967

Thomas A, Sudheer NS, Kiron V, Issac Singh BS, Narayanan RB (2022) Oral administration of white spot syndrome virus surface proteins VP24 and VP28 modulates immune gene expression in *Penaeus monodon* juveniles. Aquac Res 53(8):3069–3077

Tusar MTT, Hossen Z, Gazi HR, Haq N, Jubayer A-A, Islam MM, Lisa AK, Sikdar B, Haque ME (2025) High-throughput screening of natural antiviral drug candidates against white spot syndrome virus targeting VP28 in *Penaeus monodon*: computational drug design approaches. J Genet Eng Biotechnol 23(1):100455

Uma A (2025) Pathogen and disease transmission in aquatic animals. In: Management of fish diseases. Mallik SK, Shahi N, Pandey PK, (eds), Springer, Singapore, pp 19–55. https://doi.org/10.1007/978-981-96-0270-4

Vamathevan J, Clark D, Czodrowski P, Dunham I, Ferran E, Lee G, Li B, Madabhushi A, Shah P, Spitzer M (2019) Applications of machine learning in drug discovery and development. Nat Rev Drug Discov 18(6):463–477

Vashisht R, Thakral G (2024) A comprehensive study of bioinformatics with machine learning. In: Federated learning and privacy-preserving in healthcare AI. Lilhore UK, Simaiya S, Poongodi M, Dutt V, (eds), IGI Global, Hershey, pp 46–57. https://doi.org/10.4018/979-8-3693-1874-4.ch004

Wadas I, Domingues I (2025) Systematic review of phylogenetic analysis techniques for RNA viruses using bioinformatics. Int J Mol Sci 26(5):2180

Wang Y, Ni P, Sturrock M, Zeng Q, Wang B, Bao Z, Hu J (2024a) Deep learning for genomic selection of aquatic animals. Mar Life Sci Technol 6(4):631–650

Wang Z, Wang H, Chen D, Li Y (2024b) Genomic characterisation and comparative genomic analysis of pathogenic *Vibrio* isolated from aquaculture-grown white-leg shrimp (*Penaeus vannamei*) in Guangdong and Jiangsu, China. Aquaculture 580:740302

Wang J, Cheng Y, Su B, Dunham RA (2025a) Genome manipulation advances in selected aquaculture organisms. Rev Aquac 17(1):e12988

Wang K, Zhang C, Munang'andu HM, Xu C, Cai W, Yan X, Tao Z (2025b) Comparative genomic analysis of two *Vibrio harveyi* strains from *Larimichthys crocea* with divergent virulence profiles. Microorganisms 13(5):1129

Wani MY, Ganie NA, Rani S, Mehraj S, Mir MR, Baqual MF, Sahaf KA, Malik FA, Dar KA (2018) Advances and applications of bioinformatics in various fields of life. Int J Fauna Biol Stud 5(2):3–10

Wenne R (2018) Single-nucleotide polymorphism markers with applications in aquaculture and assessment of its impact on natural populations. Aquat Living Resour 31:2

Wenne R, Boudry P, Hemmer-Hansen J, Lubieniecki KP, Was A, Kause A (2007) What role for genomics in fisheries management and aquaculture? Aquat Living Resour 20(3):241–255

Wenne R, Zbawicka M, Bach L, Strelkov P, Gantsevich M, Kukliński P, Kijewski T, McDonald JH, Sundsaasen KK, Árnyasi M (2020) Trans-Atlantic distribution and introgression as inferred from single-nucleotide polymorphism: mussels, *Mytilus* and environmental factors. Genes 11(5):530

Wijeskara N (2023) The need for one health approach at the recent Anthropocene. In: Vithanage M, VaraPrasad MN (eds) One health: human, animal, and environment triad. Wiley, Hoboken, New Jersey, pp 1–18

Wishart DS (2007) Introduction to cheminformatics. Curr Protoc Bioinformatics 18(1):11–14

Xia X (2017) Bioinformatics and drug discovery. Curr Top Med Chem 17(15):1709–1726

Xiong X-P, Dong C-F, Xu X, Weng S-P, Liu Z-Y, He J-G (2011) Proteomic analysis of zebrafish (*Danio rerio*) infected with infectious spleen and kidney necrosis virus. Dev Comp Immunol 35(4):431–440

Yang LG, Wang Y, Wang Y, Fang WH, Feng GP, Ying N et al (2021) Transcriptome analysis of Pacific white shrimp (*Penaeus vannamei*) intestines and hepatopancreas in response to *Enterocytozoon hepatopenaei* (EHP) infection. J Invertebr Pathol 186:107665

You X, Shan X, Shi Q (2020) Research advances in the genomics and applications for molecular breeding of aquaculture animals. Aquaculture 526:735357

Yu H, Jing W, Cheng X (2023) CRISPR-Cas and aptamer-based systems for diagnosing pathogens: a review. Zoonoses 3(1):978

Yue GH, Wang L (2017) Current status of genome sequencing and its applications in aquaculture. Aquaculture 468:337–347

Yukgehnaish K, Kumar P, Sivachandran P, Marimuthu K, Arshad A, Paray BA, Arockiaraj J (2020) Gut microbiota metagenomics in aquaculture: factors influencing gut microbiome and its physiological role in fish. Rev Aquac 12(3):1903–1927

Zhang S, Liu K, Liu Y, Hu X, Gu X (2025a) The role and application of bioinformatics techniques and tools in drug discovery. Front Pharmacol 16:1547131

Zhang B, Qiu Y, Shi C, Zhang J (2025b) Development of multiple real-time fluorescent quantitative PCR for *Vibrio* pathogen detection in aquaculture. Vet Sci 12(4):327

Zhou T, Liu S, Geng X, Jin Y, Jiang C, Bao L, Yao J, Zhang Y, Zhang J, Sun L (2017) GWAS analysis of QTL for enteric septicemia of catfish and their involved genes suggest evolutionary conservation of a molecular mechanism of disease resistance. Mol Gen Genomics 292:231–242

Zubair M, Wang J, Yu Y, Faisal M, Qi M, Shah AU, Feng Z, Shao G, Wang Y, Xiong Q (2022) Proteomics approaches: a review regarding an importance of proteome analyses in understanding the pathogens and diseases. Front Vet Sci 9:1079359

Advances in Shrimp Infectious Disease Research: The Role of Next-Generation Sequencing

25

Roberto Cruz-Flores, Hung Nam Mai, and Arun K. Dhar

25.1 Introduction

25.1.1 Importance of Shrimp Aquaculture

Global shrimp aquaculture continues to play a central role in meeting the world's increasing demand for animal protein, particularly in coastal and developing regions. In 2023, global production of farmed shrimp was six million metric tons, with a projection to maintain this volume in 2024—reflecting a steady annual growth rate of 4–5% (FAO 2023a, b). This expansion is driven primarily by intensification and geographic expansion of farming systems across Latin America and Asia, particularly in Ecuador, India, Vietnam, and Indonesia.

The economic significance of shrimp aquaculture is equally noteworthy. The global shrimp market was valued at USD 75.3 billion in 2023, with forecasts indicating an increase to USD 79.6 billion in 2024, and a long-term projection of reaching USD 124 billion by 2032 (MAR 2023). This consistent upward trajectory in both production and market value underscores the strategic importance of shrimp aquaculture as a pillar of global food security, trade, and rural livelihoods. However, this growth is highly contingent on the industry's capacity to address pervasive

R. Cruz-Flores
Aquaculture Pathology Laboratory, School of Animal and Comparative Biomedical Sciences, The University of Arizona, Tucson, AZ, USA

Centro de Investigación Científica y de Educación Superior de Ensenada (CICESE), Ensenada, Baja California, Mexico

H. N. Mai · A. K. Dhar (✉)
Aquaculture Pathology Laboratory, School of Animal and Comparative Biomedical Sciences, The University of Arizona, Tucson, AZ, USA
e-mail: adhar@arizona.edu

© The Author(s), under exclusive license to Springer Nature Singapore Pte Ltd. 2026

B. K. Das et al. (eds.), *Bioinformatics Tools for Predictive Ecology and Fisheries*, Sustainable Aquatic Food Systems, https://doi.org/10.1007/978-981-95-2573-7_25

biological risks—particularly the ongoing and emerging threat of infectious diseases (Stentiford et al. 2012).

25.1.2 Burden of Infectious Diseases in Shrimp

Shrimp aquaculture has long been challenged by a wide range of infectious agents that have repeatedly disrupted production and caused significant economic losses. Historically, the first major viral pathogen to devastate the shrimp industry was infectious hypodermal and hematopoietic necrosis virus (IHHNV) in the 1980s, with estimated losses reaching into the billions (Lightner et al. 2012). This was soon followed by white spot syndrome virus (WSSV), which remains a critical threat, having caused cumulative global losses exceeding USD 30 billion, with annual damages estimated at over USD 1 billion. Infectious myonecrosis caused by the infectious myonecrosis virus (IMNV), first identified in Brazil in 2002, also resulted in significant setbacks, with early outbreaks alone causing losses of approximately USD 1.2 billion (Lightner et al. 2012). An interesting turning point in bacterial disease impact came with the emergence of Acute Hepatopancreatic Necrosis Disease (AHPND), caused by *Vibrio parahaemolyticus* carrying a plasmid containing *pirAB* toxin genes; between 2010 and 2016 alone, this disease inflicted estimated global losses of USD 44 billion (Tang and Bondad-Reantaso 2019). Most recently, *Enterocytozoon hepatopenaei* (EHP), a microsporidian parasite, has emerged as an important threat, causing growth retardation and serious production losses, with India reporting damages of USD 567.6 million (Patil et al. 2021). Beyond these well-characterized pathogens, there exists a growing list of lesser-known or emerging infectious agents—such as covart mortality syndrome virus (CMNV), *Decapod Hepanhamaparvovirus* (DHPV), Decapod Iridescent Virus 1 (DIV1), *Penaeus monodon* nudivirus (PmNV), Yellow Head Virus (YHV), *Macrobrachium rosenbergii* nodavirus (MrNV), and Laem Singh Virus (LSNV)—for which comprehensive economic impact assessments are still lacking. Their cumulative effects may be substantial, particularly under conditions of co-infection or in high-density farming systems. Without a coordinated, genomic-level understanding of pathogen diversity and dynamics, the full scope of disease burden in shrimp aquaculture will remain underestimated, impeding effective control and preparedness strategies.

25.1.3 Limitations of Conventional Diagnostic and Sanger Sequencing Technologies

Current diagnostic practices for shrimp diseases primarily rely on nucleic acid amplification techniques, typically performed in centralized and certified laboratories. A range of molecular diagnostic assays has been developed for shrimp viral pathogens, including conventional polymerase chain reaction (PCR), nested PCR, loop-mediated isothermal amplification (LAMP), recombinase polymerase amplification (RPA), and quantitative real-time PCR (qPCR) (Lee et al. 2024). Several of these protocols are officially recommended by the World Organization for Animal Health (WOAH, Paris, France) as standard diagnostic tools for shrimp disease diagnostics. While these methods are highly sensitive and have been instrumental in controlling pathogen dissemination across borders, they suffer from number of practical and technical limitations. These methods require highly skilled personnel, strict quality assurance, and specialized laboratory infrastructure, often inaccessible in field or remote aquaculture settings (Ma et al. 2021). The time required for sample transportation, processing, and analysis further delays actionable responses, especially during early outbreak stages.

A significant constraint is their static primer dependency, which requires continuous revision of primers and probes to accommodate the high mutation rates of RNA and single-stranded DNA viruses. Pathogens such as DHPV, IMNV, IHHNV have demonstrated primer mismatches due to viral adaptation, leading to false negatives and diagnostic inaccuracies (Srisala et al. 2021; Andrade et al. 2022; Cruz-Flores and Dhar 2024; Dhar et al. 2024). Moreover, while Sanger sequencing has historically played a vital role in pathogen confirmation and genome characterization, it is inherently limited by its reliance on prior sequence knowledge and low throughput. Sanger-based methods typically target known genomic regions and are not suited for unbiased, de novo pathogen discovery. They also face challenges in resolving large viral genomes, bacterial chromosomes, or fungal pathogens, due to time, cost, and scalability limitations (Gomes and Korf 2018; Cruz-Flores et al. 2020).

While Sanger sequencing is still occasionally used as a confirmatory tool, especially for targeted sequencing of small viral genomes, its application in comprehensive disease surveillance and outbreak investigation has become increasingly obsolete with the rise of next-generation sequencing (NGS) technologies. NGS platforms now offer superior sensitivity, scalability, and the ability to characterize entire genomes, plasmids, and complex microbial communities without the need for prior sequence information—features that are crucial for responding to the dynamic landscape of shrimp pathogens.

25.2 Historical Context of NGS Technologies

25.2.1 Timeline: From Sanger to 454 GS20 and Illumina

The evolution of sequencing technologies has dramatically transformed the field of pathogen detection and genome characterization, particularly in non-model organisms such as shrimp. This transformation began with the development of the Sanger sequencing method in the late 1970s, a technique that dominated genomics for nearly three decades (Sanger and Nicklen 1977). Although slow and labor-intensive, Sanger sequencing was a groundbreaking innovation that enabled early genome projects, including the full genome sequencing of viruses such as IHHNV, WSSV, and TSV (Shike et al. 2000; Van Hulten et al. 2001; Robles-Sikisaka et al. 2001). Its precision made it the gold standard for DNA sequencing, particularly for targeted gene analysis and confirmatory diagnostics.

However, by the early 2000s, the limitations of Sanger sequencing became increasingly apparent—especially its low throughput and high cost per base, which constrained its applicability for large-scale pathogen discovery and host-pathogen interaction studies. These limitations set the stage for the emergence of NGS. Despite the limitation of Sanger sequencing, several shrimp viruses including TSV, IMNV, YHV, and gill associated virus (GAV) were sequenced using this approach (Robles-Sikisaka et al. 2001; Sittidilokratna et al. 2002; Poulos et al. 2006).

The commercial release of the 454 GS20 platform by 454 Life Sciences in 2005 marked an essential moment in sequencing history (Margulies et al. 2005; Rothberg and Leamon 2008). As the first true high-throughput sequencing platform, the 454 system leveraged pyrosequencing technology to generate hundreds of thousands of reads per run, significantly reducing sequencing costs while dramatically increasing output (Margulies et al. 2005). Although the 454 platform had relatively high error rates in homopolymeric regions and was eventually phased out, it laid the foundation for a new era of genomic research (Rothberg and Leamon 2008; Wirawan et al. 2014).

Shortly thereafter, Illumina (originally Solexa) introduced its sequencing-by-synthesis technology in 2006, further revolutionizing the field (Bentley et al. 2008). Illumina platforms offered exponentially greater read depth and scalability, coupled with improved accuracy and decreasing costs. By the early 2010s, Illumina HiSeq and MiSeq became the dominant platforms for aquatic pathogenomics and host transcriptomics. In shrimp infectious disease research, these platforms were instrumental for obtaining and characterizing the complete genome and extrachromosomal elements of *V. parahaemolyticus* that cause AHPND and generating the first *Penaeus vannamei* transcriptome (Li et al. 2012; Han et al. 2015).

The relatively rapid transition from Sanger to 454 and then to Illumina technologies marked a paradigm shift—from single-gene diagnostics to holistic, genome-wide surveillance. This shift not only expanded our understanding of shrimp pathogens but also empowered researchers to explore host responses, microbiome dynamics, and co-infection scenarios with unprecedented resolution and speed (Fig. 25.1).

25.3 First Applications of NGS in Shrimp Pathogen Discovery

25.3.1 Acute Hepatopancreatic Necrosis Disease (AHPND) Case Study

In 2009, a devastating new disease emerged in shrimp farms across southern China, characterized by unusually high mortalities during the early post-stocking phase. Initially termed EMS, the condition was later more precisely defined and renamed as acute hepatopancreatic necrosis disease (AHPND). The infectious nature of the disease was soon confirmed, and (2013) identified the causative agent as a specific

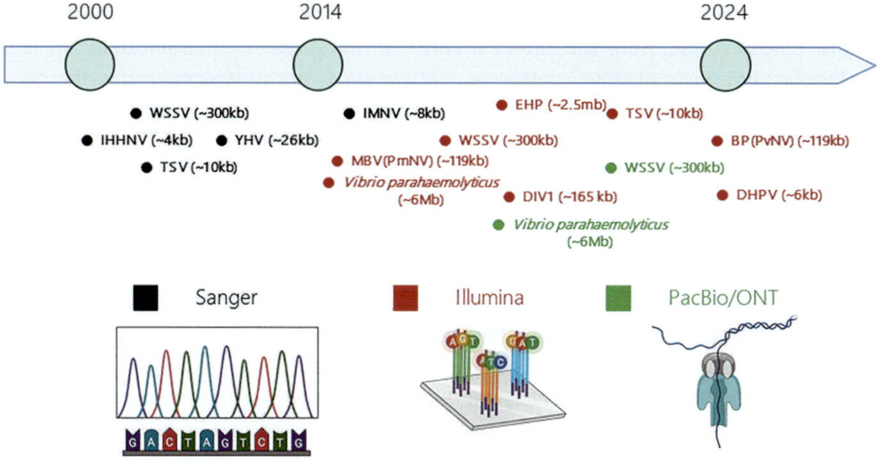

Fig. 25.1 Evolution of sequencing technologies and their application in the complete genome characterization of shrimp pathogens. This timeline illustrates the progression of sequencing platforms used to decode the genomes of key shrimp pathogens. In the early 2000s through 2014, Sanger sequencing was the primary method employed for the genetic characterization of viral pathogens such as WSSV, IHHNV, YHV, and TSV. From 2014 onward, second-generation sequencing technologies (e.g., Illumina) became dominant, enabling broader and deeper genomic analyses of both viruses and bacteria, including DIV1, MBV, *Vibrio parahaemolyticus*, and EHP. Beginning around 2018 and gaining momentum by 2024, third-generation sequencing technologies (e.g., PacBio and Oxford Nanopore Technologies) began to be applied, allowing for the resolution of larger and more complex genomes such as those of WSSV and Vibrio strains. The timeline reflects the increased resolution and throughput achieved with each successive technological advancement

strain of *V. parahaemolyticus*. This pathogen was unique in harboring a virulence-associated plasmid that would later prove critical to understanding the disease's etiology.

An important study by (2014) represents one of the first applications of Illumina technology for genome-level analysis of shrimp pathogens. In their investigation, the authors sequenced the genomes of three AHPND-causing strains and three non-AHPND strains of *V. parahaemolyticus*. While the specific toxin genes were not identified in that study, they noted that all AHPND-associated isolates shared a highly conserved 63-kilobase contig (GenBank: NZ_BAVH01000004.1), later recognized as homologous to the pVPA3-1 plasmid (GenBank: KM067908), now known to carry the virulence determinants of AHPND.

Building on this genomic information, Han et al. (2015) employed Ion Torrent sequencing to perform a complete genome analysis of the AHPND strain 13-028/A3. Their work provided a breakthrough by identifying the *pirA*- and *pirB*-like genes, homologous *to Photorhabdus* insect-related (Pir) toxins, located on the virulence plasmid. These genes encode binary toxins that are now known to be the primary virulence factors responsible for the histopathological damage observed in the hepatopancreas of infected shrimp.

Together, these first two studies marked the beginning of the NGS era in shrimp health research, enabling the direct identification, genomic characterization, and molecular diagnosis of novel pathogens. These tools not only accelerated the pace of discovery but also established a framework for applying high-throughput sequencing technologies to understand host–pathogen interactions, track disease outbreaks, and, most importantly, these tools allowed the development of molecular diagnostics that previously relies mostly on histopathology.

25.3.2 Early Viral Studies: Redefining the Invertebrate RNA Virosphere and Decapod Iridescent Virus 1

A fundamental study in the understanding of RNA viruses in invertebrates was achieved by Shi et al. (2016), who conducted an extensive metatranscriptomic survey of 220 invertebrate species across nine animal phyla, leading to the discovery of 1445 previously uncharacterized RNA viruses. This study, notable for its unprecedented breadth, utilized next-generation sequencing to reveal the true scale and complexity of the invertebrate RNA virosphere, particularly in aquatic organisms.

Importantly, among the invertebrates sampled were commercially significant penaeid shrimp, including *P. vannamei*. The dataset included what are now known as the Wenzhou shrimp viruses (WzSVs)—a group of ten novel RNA viruses. These findings not only extended the evolutionary framework of RNA viruses but also highlighted the previously underappreciated diversity of cryptic and persistent viral infections in shrimp and other crustaceans.

Among these, Wenzhou shrimp virus 1 (WzSV1) has attracted particular attention for its potential role in disease. It appears to represent a distinct genotype of Mourilyan virus (Cowley et al. 2005) and shows phylogenetic proximity to a bunya-like virus associated with disease in *Penaeus chinensis* (Dong et al. 2021). Another member, Wenzhou shrimp virus 8 (WzSV8), has been reclassified as a genotype of *Penaeus vannamei* solinvivirus (PvSV) and has been consistently associated with disease (Cruz-Flores et al. 2022). Furthermore, Laem Singh virus (LSNV) and Wenzhou shrimp virus 9 (WzSV9) have been hypothesized to represent closely related isolates of the same viral species (Taengchaiyaphum et al. 2020), suggesting a potential taxonomic and pathogenic overlap.

The study by Shi et al. (2016) serves as a foundational reference for shrimp virology, emphasizing that many RNA viruses detected via NGS in metagenomic surveys may have clinical relevance yet to be fully uncovered. As surveillance continues and longitudinal health data from shrimp farms are integrated with NGS-based virome screening, it is reasonable to anticipate that more of these newly identified viral entities will be implicated in emerging disease syndromes affecting shrimp aquaculture.

Support for this hypothesis also comes from studies in other crustaceans. For example, Wenling crustacean virus 9, identified in the same metatranscriptomic data, has since been associated with tremor disease in the Chinese mitten crab (*Eriocheir sinensis*), caused by a bunya-like virus (Huang et al. 2019). This case underscores the pathogenic potential of previously uncharacterized viral taxa identified through environmental sequencing.

While *Penaeus monodon* nudivirus (PmNV) was technically the first DNA virus infecting shrimp to be fully characterized at the whole-genome level using next-generation sequencing (NGS), its relevance to contemporary shrimp aquaculture has significantly declined (Yang et al. 2014). In contrast, the first major emerging and economically impactful DNA virus to be comprehensively identified and sequenced using NGS was Decapod Iridescent Virus 1 (DIV1)—originally referred to as Shrimp Hemocyte Iridescent Virus (SHIV) (Qiu et al. 2017). The complete genome of DIV1 was obtained using the Illumina HiSeq 2000 platform, which allowed for high-depth sequencing and detailed molecular characterization. This achievement marked an important moment in shrimp virology, providing rapid insights into the virus's genome architecture and potential virulence factors.

Although whole-genome sequencing could technically have been accomplished using Sanger sequencing, the urgency of the situation—including the rapid spread of the virus, its severe impact on shrimp aquaculture in China, and the pressing need for accurate diagnostic tools—required a faster, high-throughput approach. Illumina's NGS technology offered a practical and timely solution, facilitating not only the complete genome assembly but also downstream applications such as molecular diagnostics, phylogenetic classification, and epidemiological surveillance.

25.3.3 Genetic Characterization of EHP: The First Eukaryotic Shrimp Pathogen to be Sequenced

Enterocytozoon hepatopenaei (EHP), a microsporidian parasite, represents the first eukaryotic pathogen of shrimp to be genetically characterized at the genome level (Wiredu Boakye et al. 2017). Initially described from Thailand in *P. monodon*, EHP has since become a globally distributed and economically devastating pathogen in *P. vannamei* (Tourtip et al. 2009). The inability of EHP to cause acute large-scale mortality, yet induce significant growth retardation, delayed its initial identification as a serious threat to shrimp aquaculture (Chaijarasphong et al. 2020).

A fundamental study by Wiredu Boakye et al. (2017) used Illumina sequencing and comparative genomics to find that EHP had undergone massive gene loss and compaction, hallmarks of microsporidian evolution. Notably, it lacks genes associated with energy production and de novo synthesis of amino acids and nucleotides thereby relying on its host for vegetative growth and proliferation. The loss of canonical fungal metabolic and regulatory genes, combined with a highly reduced genome (~2.5 Mb), would have made EHP nearly impossible for classical PCR characterization or gene-marker-based approaches. Without NGS, it would have been hard to resolve this degree of genomic erosion, or to identify EHP's parasitic dependencies and evolutionary adaptations. The genetic insights provided by NGS not only clarified EHP's phylogenetic placement but also suggested targets for diagnostics, molecular surveillance, and potentially therapeutic intervention. This highlights the importance of application of high-throughput genomics in resolving complex, intracellular eukaryotic pathogens in shrimp aquaculture.

25.4 Host Transcriptomics Using RNA-Seq

Transcriptomics is the large-scale study of RNA transcripts produced by an organism under specific conditions. It offers insights into how gene expression dynamically responds to environmental changes, developmental cues, or disease. In shrimp health research, RNA sequencing (RNA-Seq) has become the standard for transcriptome profiling, enabling identification of differentially expressed genes (DEGs) and key immune pathways. This technology has been crucial for exploring the complexity of host-pathogen interactions, highlighting the molecular mechanisms of disease susceptibility, tolerance, and recovery in shrimp species. In the following section, we highlight representative transcriptomic studies conducted on key shrimp pathogens and the valuable insights they offer into the host's molecular response to infection. These studies illustrate how RNA-Seq has advanced our understanding of shrimp immunity, stress adaptation, and disease progression. We conclude the section with our perspective on how these findings can be translated into practical applications to enhance shrimp health management and improve the sustainability of shrimp aquaculture.

25.4.1 Transcriptomic Responses to White Spot Syndrome Virus (WSSV)

White spot syndrome virus continues and will continue to be one of the most devastating pathogens in shrimp aquaculture until antiviral therapies are available for commercial applications. Transcriptomic studies on WSSV have allowed us to understand how specific organs (hence different cell types) respond to WSSV infection. For an example, Yu et al. 2022 analyzed the transcriptome of hepatopancreas of WSSV-infected shrimp and found differential expression of genes associated with immune regulation, cytoskeletal integrity, apoptosis, and oxidative stress responses. These included heat shock proteins, antioxidant enzymes, and elements of the prophenoloxidase pathway. Metabolomic analysis confirmed that WSSV infection perturbs several metabolic processes, notably amino acid, lipid, and nucleotide metabolism, highlighting a systemic reallocation of resources during immune challenge.

Complementing this findings, Li et al. 2024 examined gene expression in WSSV-challenged shrimp with varying levels of resistance. The study showed that resistant individuals activate immune pathways more effectively, including genes related to pattern recognition, apoptosis modulation, and metabolic regulation. Susceptible shrimp displayed suppressed expression of the same pathways, suggesting that transcriptomic profiling can differentiate between white spot disease resilient and susceptible phenotypes and could serve as a guide for future breeding strategies.

25.4.2 Transcriptomics and Acute Hepatopancreatic Necrosis Disease (AHPND)

AHPND, caused by *V. parahaemolyticus* carrying the *pirAB* toxin genes, severely damages the hepatopancreas. Sun et al. (2023) carried out an integrated transcriptomic and metabolomic study comparing AHPND-resistant and -susceptible shrimp families. The resistant group upregulated genes involved in lysosomal function, phagocytosis, cellular repair, and xenobiotic metabolism. Metabolomic profiling further confirmed active lipid and carbohydrate processing, which supports tissue recovery and immune readiness. In contrast, susceptible shrimp showed suppressed immune and metabolic activity, suggesting an inability to mount sufficient physiological defense during the early stages of infection.

A second study by Velázquez-Lizárraga et al. (2019) applied transcriptomic profiling to identify DEGs in shrimp naturally infected with AHPND under farming conditions. Susceptible shrimp exhibited upregulation of immune-related genes, including C-type lectins, caspases, and antioxidant enzymes, while tolerant individuals maintained expression of genes linked to detoxification, mitochondrial maintenance, and cellular homeostasis. These data suggest that chronic exposure to AHPND favors individuals that invest in sustained metabolic stability over aggressive immune activation.

25.4.3 Transcriptomic Insights into *Enterocytonucelospora hepatopenaei*

Enterocytonucelospora hepatopenaei (formerly known as *Enterocytozoon hepatopenaei*, EHP), the microsporidian parasite infecting the hepatopancreas, has become a major threat in shrimp aquaculture due to its chronic nature and lack of clear treatment. Zhang et al. (2023) analyzed gene expression in shrimp at early and late stages of EHP infection. During the early stage of infection, transcriptomic profiles showed activation of oxidative stress response genes, apoptosis modulators, and immune surveillance elements. In the late stage, however, expression of these genes declined, while transcripts associated with protein degradation, endocytosis, and energy metabolism were altered, indicating that the host gradually shifts from immune response to tissue remodeling and metabolic compensation. The authors concluded that EHP tolerance may depend more on the host's ability to maintain cellular balance and mitigate long-term damage than on direct immune clearance of the parasite. These findings underscore the value of transcriptomics for understanding the subtle, often non-inflammatory nature of host responses to intracellular eukaryotic pathogens like EHP.

25.4.4 Remarks on Applicability of Transcriptomic Data

A substantial body of transcriptomic research has investigated the molecular responses of shrimp to various pathogens, resulting in a wealth of RNA-Seq data.

These studies have uncovered differentially expressed genes, characterized key immune pathways, elucidated host–pathogen interactions, and revealed molecular mechanisms underlying disease susceptibility, tolerance, and recovery across shrimp species. While these findings have greatly advanced our scientific understanding, translating this complex information into practical applications for shrimp aquaculture remains a significant challenge.

One of the most immediate applications is the identification of genetic markers that can support breeding programs aimed at selecting disease-resistant or tolerant phenotypes. However, several promising yet underexplored areas also exist. These include the identification of target genes for therapeutics, the discovery of novel therapeutic interventions, and the characterization of transcriptomic signatures specific to infectious diseases, which could enhance both diagnostics and targeted treatment strategies (Yang et al. 2020; di Iulio et al. 2021). Utilizing the full potential of transcriptomic data will require a concerted effort to integrate these insights into applied breeding, health management, and disease mitigation frameworks within the shrimp aquaculture industry.

25.5 Large Scale Amplicon Sequencing by NGS

Large scale Amplicon sequencing, particularly targeting the 16S rRNA gene, 18S rRNA gene, and ITS, has become a standard tool in microbial ecology to characterize complex microbial communities. This technique selectively amplifies specific hypervariable regions of conserved genes—such as 16S rRNA in bacteria—allowing the profiling of microbial populations without the need for cultivation (Tamaki et al. 2011).

25.5.1 16S rRNA Studies in EHP and White Feces Syndrome

The application of 16S rRNA amplicon sequencing has been instrumental in elucidating microbial dysbiosis associated with important shrimp pathologies such as EHP and White Feces Syndrome (WFS). López-Carvallo et al. (2022) conducted one of the first studies simultaneously examining EHP-infected shrimp by 16S rRNA and ITS. They found distinct shifts and fluctuations in microbial diversity in hepatopancreas of EHP-infected shrimp throughout the time course of an infection cycle. Specifically, EHP infection was associated with increased prevalence of opportunistic bacteria such as *Vibrio* spp. and *Photobacterium*, as well as reduced microbial diversity in shrimp guts.

Similarly, Baliga et al. (2021) reported dynamic changes in bacterial communities in surface and rearing water during EHP outbreaks in farm raised *P. vannamei* shrimp. These authors reported EHP infection creates cellular environment that favors certain taxa such as Vibrio, which were minimally detected in healthy ponds. This shift pointed to the potential role of environmental reservoirs in pathogen persistence and transmission. Munkongwongsiri et al. (2022) further confirmed that gut bacterial profiles differed significantly between healthy and EHP-affected shrimp, noting reductions in beneficial bacteria like *Lactobacillus* and *Bacillus* and overrepresentation of putative pathogens in diseased animals.

Building on this, Huang et al. (2020) used both amplicon sequencing and microbiota transplantation experiments to demonstrate causality between intestinal dysbiosis and WFS. The study introduced the concept of "microecological Koch's postulates" by showing that transplanting dysbiotic microbiota from WFS-affected shrimp into healthy individuals could reproduce clinical signs. This study provided strong evidence that WFS is not caused by a single pathogen but by a microbial community imbalance—a finding with significant implications for diagnosis and management of diseases.

25.5.2 Exploring Eukaryotic Communities with Amplicon Sequencing

While bacterial profiling has dominated microbiome studies, recent work has extended amplicon sequencing to microeukaryotic communities, uncovering potential roles of protists, fungi, and other unicellular eukaryotes in shrimp health and disease. López-Carvallo et al. (2022) included microeukaryotic analysis alongside 16S profiling and found that changes in eukaryotic microbiota were linked with EHP infection, particularly through increases in parasitic protists and reduction in commensal species. These shifts occurred in parallel with bacterial community alterations and further underscored the multi-domain complexity of shrimp dysbiosis.

In a parallel investigation, Song et al. (2022) focused on microeukaryotic dynamics in pond environments and found distinct community structures between healthy and diseased ponds. They reported increased dominance of potentially pathogenic ciliates and a decline in beneficial algae, suggesting that eukaryotic shifts may not only reflect disease states but could actively contribute to pathogenesis or compromise host immunity. Further supporting this, Dai et al. (2019) applied eukaryotic-targeted amplicon sequencing to pond water and shrimp gut samples and observed significant variation in fungal and protistan communities during disease outbreaks. The presence of particular microeukaryotic taxa was strongly associated with poor water quality and host stress, implicating them as both indicators and possible contributors to shrimp disease susceptibility.

25.5.3 Remarks on the Applicability of Amplicon Sequencing Data Infectious Disease Research in Shrimp

The applicability of the vast information generated on microbial community diversity, structure, and interactions within the shrimp host faces limitations similar to those encountered with large-scale RNA-Seq datasets. While the scientific value of these findings is undeniable, translating this complex and high-dimensional data into tangible improvements in shrimp aquaculture remains a daunting challenge. In human medicine, fecal microbiota transplants have been successfully applied to treat diseases associated with microbial dysbiosis (Sahle et al. 2024). A comparable concept was demonstrated by Huang et al. (2020), who showed that transferring microbiota from WFS-affected shrimp to healthy individuals could induce disease symptoms, confirming a causal role of dysbiosis. However, such transplant-based interventions are not practical or scalable for shrimp farming systems due to operational and biosecurity constraints.

Given these limitations, it is compelling to consider the future potential of developing synthetic microbial communities composed of beneficial shrimp-associated microorganisms. Such communities could be rationally designed using amplicon sequencing approaches, as demonstrated in other model organisms by Cheng and colleagues (Cheng et al. 2022), and mass-produced under controlled conditions. These tailored consortia could be administered at early larval stages, promoting stable colonization of the shrimp gut with protective microbiota and enhancing resilience to pathogens and environmental stress. While inoculation at later grow-out stages—especially in large commercial ponds—may be less feasible due to logistical and ecological barriers, the early-life establishment of such microbial communities could represent a promising avenue for disease prevention and performance enhancement. To date, no studies have evaluated the feasibility, stability, or efficacy of such synthetic microbiomes in shrimp, highlighting an exciting and unexplored frontier in shrimp health management.

25.6 Emerging Sequencing Technologies

Next-generation sequencing encompasses a variety of high-throughput platforms, typically categorized into second-generation and third-generation technologies. Second-generation sequencing includes platforms such as Illumina and Ion Torrent, which rely on short-read, clonal amplification and sequencing-by-synthesis approaches. While they offer exceptional throughput and base-calling accuracy, they are limited by short read lengths and complex library preparation. In contrast, third-generation sequencing platforms like Pacific Biosciences (PacBio) and Oxford Nanopore Technologies (ONT) eliminate the need for amplification and can directly sequence long fragments of native DNA or RNA molecules. This transition from short-read to long-read technologies represents a paradigm shift in genomics, especially for complex genomes, structural variant detection, and transcriptomic profiling where context and continuity are critical (Rhoads and Au 2015; Han et al. 2024).

Among third-generation platforms, PacBio has pioneered long-read sequencing through its Single Molecule, Real-Time (SMRT) technology. The original SMRT platform offered reads averaging 10–15 kb in length but had relatively high single-pass error rates. However, the development of HiFi (High-Fidelity) sequencing has significantly enhanced the accuracy of PacBio reads by using circular consensus sequencing (CCS) from the same molecule. According to Han et al. (2024), HiFi reads now routinely achieve accuracies >99.9% while maintaining read lengths of

10–25 kb, making them especially suitable for applications such as de novo genome assembly, isoform discovery, and high-resolution metagenomics.

Oxford Nanopore Technologies represents another major advancement in third-generation sequencing, enabling real-time, portable, and ultra-long-read capabilities. ONT devices, such as the MinION and PromethION, operate by measuring changes in ionic current as DNA or RNA strands pass through nanopores embedded in a membrane. This approach allows the sequencing of entire genomic or transcriptomic fragments exceeding 100 kb in a single read. While ONT reads initially suffered from high error rates (~10–15%), continuous improvements in base calling algorithms and chemistry have improved raw read accuracies to >98% and consensus accuracies even higher (Wang et al. 2021). Importantly, ONT platforms offer significant advantages in scalability, portability, and direct RNA sequencing, enabling field-based applications and rapid pathogen diagnostics in aquaculture.

25.6.1 Application of PacBio and Nanopore Sequencing in AHPND Research

The application of third-generation sequencing technologies—specifically PacBio and Oxford Nanopore—has enabled the complete genome assembly of *V. parahaemolyticus* strains responsible for AHPND in shrimp. Among the earliest studies, Kanrar and Dhar (2018a, b) employed PacBio platform to obtain full-length genome sequences of mutant strains derived from AHPND-causing isolates. These initial reports marked a technical milestone. These studies laid the groundwork for more advanced studies and demonstrated the feasibility of using long-read sequencing to resolve large plasmids like the pVA1 plasmid that carries the *pirAB* toxin genes.

These early sequencing efforts allowed more in-depth investigations on the generated data by Kanrar and Dhar (2018a, b), as exemplified by the study of Aranguren Caro et al. (2020), who conducted a comprehensive analysis of the non-pathogenic nature of the R13 and R14 *V. parahaemolyticus* strains. Their work raised critical questions about the role of functional gene expression and the diversity of virulence plasmids, contributing to a clearer distinction between the mere presence of toxin genes and actual pathogenic potential in certain lineages. Similarly, the recent study by Reyes et al. (2025) applied hybrid sequencing (Illumina + PacBio Revio) to characterize the complete genomes of two virulent strains from South America. Their annotations included virulence factors, AMR genes, CRISPR arrays, and mobile elements, reflecting the value of long-read sequencing in uncovering genomic architecture and accessory features with implications for monitoring and biosecurity. Similarly, Hwang et al. (2020) employed PacBio sequencing to analyze isolates from Taiwan, again emphasizing plasmid structures and resistance genes

Oxford Nanopore sequencing has also been used to sequence complete genomes of AHPND causing *Vibrio* spp. Tapia-Chirinos et al. (2024) used ONT technology to sequence nine *V. parahaemolyticus* isolates from shrimp farms in northern Peru. Their work provided genomic evidence for the presence of pir-like toxin genes, transposable elements, and various antibiotic resistance features. Though the study focused primarily on data generation rather than evolutionary or functional genomics, it substantially contributed to the regional cataloging of pathogenic strains.

Despite the generally descriptive nature of these genome reports, they represent crucial building blocks. They provide foundational data necessary for future comparative genomics, strain typing, and understanding plasmid dynamics. There remains, however, a need for more analytical studies that utilize the full potential of these third-generation sequencing platforms.

25.6.2 Application of PacBio for Studying the Origin of White Spot Syndrome Virus

PacBio long-read sequencing has been important in advancing our understanding of crustacean genomics and the evolutionary origins of white spot syndrome virus (WSSV). Kawato et al. 2019 were among the first to utilize long-read genomic data to explore viral sequences integrated into the genomes of various crustacean species. Their study identified seven distinct phylogenetic groups within the family *Nimaviridae* and revealed the widespread presence of endogenous nimaviruses across multiple crustacean lineages. These findings provided compelling evidence for a long-standing evolutionary relationship between crustacean hosts and WSSV-like viruses and suggested that endogenous viral elements may have played a role in shaping host–virus coevolution. The ability to generate contiguous and complete genome assemblies using PacBio technology was essential for detecting these large and often repetitive viral sequences embedded in host genomes.

Building on this study, Bao et al. (2020) conducted a detailed analysis of one such endogenous nimavirus, Nimav-1_LVa, identified in the genome of *P. vannamei*. Using PacBio sequencing, they reconstructed the entire viral genome—spanning 279,905 base pairs and encoding 117 putative genes, several with high homology to WSSV genes. Importantly, their study revealed that Nimav-1_LVa integrated preferentially into telomeric regions composed of (TAACC/GGTTA)n microsatellites, providing new insights into viral integration mechanisms. These findings not only support the hypothesis of a shared ancestry between WSSV and endogenous Nimaviruses but also underscore the utility of long-read sequencing in resolving complex genomic architecture, enabling the study of virus–host interactions at an unprecedented level of detail. Table 25.1 provides a summarized chronological overview of the most significant studies that have employed NGS for the characterization of shrimp pathogens.

25.6.3 Future Research Harnessing PacBio and Nanopore Sequencing for NHP and EHP

While some progress has been made in the genetic characterization of many shrimp pathogens, several important infectious agents remain incompletely resolved at the genomic level. Notably, Necrotizing Hepatopancreatitis Bacterium (NHP-B), a

Table 25.1 Overview of shrimp pathogen discovery and characterization via sequencing technologies

Year	Pathogen	Technology employed	References
2014	*V. parahaemolyticus* (AHPND)	Illumina	Kondo et al. (2014)
2014	*Penaeus monodon* nudivirus (PmNV)	Illumina	Yang et al. (2014)
2015	*V. parahaemolyticus* (AHPND)	Ion Torrent	Han et al. (2015)
2016	RNA virosphere—Wenzhou Shrimp Viruses	Illumina RNA-Seq	Shi et al. (2016)
2017	*Enterocytozoon hepatopenaei* (EHP)	Illumina	Wiredu Boakye et al. (2017)
2017	*Decapod Iridescent Virus 1*	Illumina	Qiu et al. (2017)
2018	*V. parahaemolyticus* R13 & R14	PacBio	Kanrar and Dhar (2018a, b)
2020	White Spot Syndrome Virus (evolutionary integration)	PacBio	Kawato et al. (2019)
2024	Field strains of AHPND	Nanopore	Tapia-Chirinos et al. (2024)

fastidious intracellular pathogen, has eluded full genome resolution despite several sequencing attempts. Currently, only two near-complete genome assemblies of NHP-B are available on GenBank, yet neither achieves full chromosomal closure. Although NHP-B is expected to possess a relatively small genome, it appears to harbor structural complexities or highly repetitive elements that pose a challenge for short-read second-generation platforms such as Illumina. These limitations have likely contributed to the persistent fragmentation observed in NHP-B assemblies, even after multiple efforts. In this context, third-generation long-read sequencing technologies such as PacBio HiFi and Oxford Nanopore offer an attractive solution. Their ability to produce ultra-long, contiguous reads could help overcome obstacles associated with repetitive regions or secondary structural motifs, enabling complete genome assembly and more accurate annotation of coding and non-coding regions.

A similar opportunity exists for EHP, the first eukaryotic pathogen of shrimp to be studied in detail. Although the initial draft genome of EHP provided valuable insights into its minimal gene set and reductive evolution, the assembly remains fragmented, limiting the resolution of its genome organization and potential regulatory elements. Using third-generation sequencing approaches could drastically improve the completeness and accuracy of the EHP genome by resolving complex or repetitive regions and potentially uncovering key pathogenicity-related loci or structural variants that contribute to virulence. Moreover, improved assemblies would facilitate comparative genomics across microsporidian species and may provide new leads for therapeutic interventions, diagnostic development, or understanding of host-parasite interactions at the genomic level.

25.7 Future Perspectives: Real-Time and Field-Deployable Diagnostics

The integration of third-generation sequencing technologies such as ONT into field diagnostics has revolutionized the way real-time pathogen detection and genomic surveillance is approached. Unlike traditional platforms requiring well-equipped laboratories, stable electricity, and cold chains, nanopore-based sequencing devices such as MinION are highly portable, USB-powered, and operable with minimal infrastructure. These attributes have made them uniquely suited for rapid deployment in resource-limited environments (Wang et al. 2021).

A notable demonstration of this capability was presented during the Ebola outbreak in West Africa, where Quick et al. (2016) successfully established a real-time genomic surveillance system using the MinION platform. The device was transported via standard airline luggage and implemented on-site in Guinea. Their team sequenced Ebola virus genomes directly from clinical samples, generating results within 24 h of receipt—sometimes with sequencing runs as short as 15–60 min. This approach offered high-resolution insights into the virus's evolution and enabled the identification of sub-lineages critical for monitoring transmission during the epidemic. Beyond outbreak scenarios, ONT has been refined for broader diagnostic utility. Furthermore, Quick et al. (2015) demonstrated the application of the MinION platform during a real-time investigation of a hospital outbreak of *Salmonella enterica*. Their work highlighted the capability of nanopore sequencing to deliver species-level identification within 20 min of sample loading, showing that portable sequencing can offer actionable epidemiological data in near real time. This interesting demonstration underscores the immense potential of ONT-based platforms for decentralized, rapid diagnostics — a transformative prospect for pathogen surveillance in shrimp aquaculture.

These advances highlight the potential for nanopore sequencing to be harnessed in aquaculture pathogen surveillance. For diseases like WSSV, AHPND, or microsporidian infections such as EHP, which often break out in remote or rural aquaculture zones, field-deployable diagnostics can revolutionize early detection and control. Future systems could integrate sample preparation, sequencing, and bioinformatics analysis into handheld formats, delivering strain-level resolution on-site—an advancement that would transform disease management in shrimp aquaculture.

25.8 Conclusions and Final Remarks

Next-generation sequencing technologies have greatly transformed our understanding of shrimp pathogens and reshaped the pace at which they can be identified and characterized. The advent of platforms such as Illumina and Ion Torrent enabled the rapid sequencing of complete viral and bacterial genomes that were previously unattainable with traditional molecular tools. These technologies have also allowed for the simultaneous sequencing of multiple samples, generating vast datasets that illuminate the complexity of host-pathogen interactions, microbial community shifts, and gene expression responses across diverse disease contexts.

While RNA-seq and amplicon-based microbiome studies have yielded fundamental insights—revealing key immune pathways, transcriptional responses, and dysbiosis-associated microbial signatures—translating this amount of information into field-applicable solutions remains a challenge. Nonetheless, the scientific value of these datasets is undeniable, and they lay the groundwork for future translational applications, including biomarker development, targeted therapies, and breeding for disease resilience.

Importantly, the field is now entering a new phase with the rise of third-generation sequencing platforms such as PacBio HiFi and ONT. These platforms provide unprecedented opportunities for long-read sequencing, complete genome reconstruction, and structural variant detection—key features needed to resolve complex genomes such as EHP and NHP-B. Additionally, their capacity to deliver portable, real-time diagnostics in field conditions is opening a new opportunity in aquaculture disease management. With portable sequencing devices already deployed in human and animal outbreak scenarios, it is reasonable to expect that in the next 5 years, real-time monitoring of shrimp pathogens using these technologies will become both feasible and routine.

In summary, NGS technologies have initiated a new era in shrimp health research. As sequencing platforms become more accessible and analytic pipelines become more automated and user-friendly, the gap between scientific discovery and practical application will continue to close, bringing in a new generation of shrimp farming that, hopefully, is more resilient, informed, and sustainable.

References

Andrade TPD et al (2022) Novel infectious Myonecrosis virus (IMNV) variant is associated with recent disease outbreaks in Penaeus vannamei shrimp in Brazil. Aquaculture. Elsevier B.V. 554:738159. https://doi.org/10.1016/j.aquaculture.2022.738159

Aranguren Caro LF et al (2020) A mutant of vibrio parahaemolyticus pirABVP (+) that carries binary toxin genes but does not cause acute hepatopancreatic necrosis disease. Microorganisms 8(10):1549. https://doi.org/10.3390/microorganisms8101549

Baliga P et al (2021) Bacterial community changes in penaeus vannamei Boone, 1931 surface and rearing water during enterocytozoon hepatopenaei infection. Asian Fish Sci 34(2):168–180. https://doi.org/10.33997/j.afs.2021.34.2.006

Bao W, Tang KFJ, Alcivar-Warren A (2020) The complete genome of an endogenous nimavirus (Nimav-1_LVa) from the pacific whiteleg shrimp penaeus (Litopenaeus) Vannamei. Genes 11(1):1–20. https://doi.org/10.3390/genes11010094

Bentley DR et al (2008) Accurate whole human genome sequencing using reversible terminator chemistry. Nature 456(7218):53–59. https://doi.org/10.1038/nature07517

Chaijarasphong T et al (2020) The shrimp microsporidian Enterocytozoon hepatopenaei (EHP): biology, pathology, diagnostics and control. J Invertebr Pathol. Elsevier:107458. https://doi.org/10.1016/j.jip.2020.107458

Cheng AG et al (2022) Design, construction, and in vivo augmentation of a complex gut microbiome. Cell:1–20. https://doi.org/10.1016/j.cell.2022.08.003

Cowley JA et al (2005) Preliminary molecular and biological characterization of Mourilyan virus (MoV): a new bunya-related virus of peaneid prawns. In: Diseases in Asian aquaculture V, V(May 2014), pp 113–124. Available at: http://www.genome.ad.jp

Cruz-Flores R, Dhar AK (2024) Tracking the emergence of a novel genotype of Decapod hepanhamaparvovirus in shrimp using laser microdissection and next generation sequencing. PLoS One 19(10):e0311592. https://doi.org/10.1371/journal.pone.0311592

Cruz-Flores R et al (2020) Genome reconstruction of white spot syndrome virus (WSSV) from archival Davidson's-fixed paraffin embedded shrimp (Penaeus vannamei) tissue. Sci Rep. Nature Publishing Group UK 10(1):13425. https://doi.org/10.1038/s41598-020-70435-x

Cruz-Flores R et al (2022) Identification of a novel solinvivirus with nuclear localization associated with mass mortalities in cultured whiteleg shrimp (Penaeus vannamei). Viruses 14(10):2220. https://doi.org/10.3390/v14102220

Dai W et al (2019) Gut eukaryotic disease-discriminatory taxa are indicative of Pacific white shrimp (Litopenaeus vannamei) white feces syndrome. Aquaculture. Elsevier B.V. https://doi.org/10.1016/j.aquaculture.2019.03.034

Dhar AK et al (2024) Comparison of Polymerase Chain Reaction (PCR) assay performance in detecting Decapod penstylhamaparvovirus 1 in penaeid shrimp. J Virol Methods. Elsevier B.V 323:114840. https://doi.org/10.1016/j.jviromet.2023.114840

di Iulio J et al (2021) Transfer transcriptomic signatures for infectious diseases. Proc Natl Acad Sci U S A 118(22):1–9. https://doi.org/10.1073/pnas.2022486118

Dong X et al (2021) A novel bunyavirus discovered in oriental shrimp (Penaeus chinensis). 12:1–9. https://doi.org/10.3389/fmicb.2021.751112

FAO (2023a) Shrimp-quarterly analysis. Available at: https://openknowledge.fao.org/server/api/core/bitstreams/565f0b9d-1b7a-41e7-b4c8-34514ef4d8dc/content

FAO (2023b) Strong demand in China upholds global shrimp trade in 2023. Available at: https://www.fao.org/in-action/globefish/news-events/news/news-detail/Strong-demand-in-China-upholds-globalshrimp-trade-in-2023/en. Accessed 22 July 2025

Gomes A, Korf B (2018) Genetic testing techniques. In: Sanger sequencing, pp 47–64. https://doi.org/10.1016/B978-0-323-48555-5.00005-3

Han J et al (2015) Photorhabdus insect-related (Pir) toxin-like genes in a plasmid of Vibrio parahaemolyticus, the causative agent of acute hepatopancreatic necrosis disease (AHPND) of shrimp. Dis Aquat Org 113(1):33–40. https://doi.org/10.3354/dao02830

Han Y et al (2024) Unlocking the potential of metagenomics with the PacBio high-fidelity sequencing technology. Microorganisms 12(12):1–14. https://doi.org/10.3390/microorganisms12122482

Huang P et al (2019) Genomic and phylogenetic characterization of a bunya-like virus from the freshwater Chinese mitten crab Eriocheir sinensis. Acta Virol 63(4):433–438. https://doi.org/10.4149/av_2019_410

Huang Z et al (2020) Microecological Koch's postulates reveal that intestinal microbiota dysbiosis contributes to shrimp white feces syndrome. Microbiome 8(1):1–13. https://doi.org/10.1186/s40168-020-00802-3

Hwang SD et al (2020) Genome data of shrimp acute hepatopancreatic necrosis disease causative Vibrio parahaemolyticus strains isolated from South Korea aquaculture farms. Data Brief. Elsevier Inc 31:105697. https://doi.org/10.1016/j.dib.2020.105697

Kanrar S, Dhar AK (2018a) Complete genome sequence of a deletion mutant of Vibrio parahaemolyticus from pacific white shrimp (Penaeus vannamei). Genome Announc 6(25):33–40. https://doi.org/10.1128/genomeA.00544-18

Kanrar S, Dhar AK (2018b) Complete genome sequence of a novel mutant strain of Vibrio parahaemolyticus from Pacific white shrimp (Penaeus vannamei). Genome Announc 6(24):19–20. https://doi.org/10.1128/genomeA.00497-18

Kawato S, Shitara A, Wang YNozaki R, Kondo H, Hirono I (2019) Crustacean Genome Exploration Reveals the Evolutionary Origin of White Spot Syndrome Virus. J Virol 93:10.1128/jvi.01144–18. https://doi.org/10.1128/jvi.01144-18

Kondo H et al (2014) Draft genome sequences of six strains of Vibrio parahaemolyticus isolated from early mortality syndrome/acute hepatopancreatic necrosis disease shrimp in Thailand. Genome Announc 2(2):2–3. https://doi.org/10.1128/genomeA.00221-14
Lee Y et al (2024) Nucleic acid amplification-based methods for diagnosis of shrimp viral diseases. Rev Aquac 16(2):892–922. https://doi.org/10.1111/raq.12873
Li C et al (2012) Analysis of Litopenaeus vannamei transcriptome using the next-generation DNA sequencing technique. PLoS One 7(10). https://doi.org/10.1371/journal.pone.0047442
Li X et al (2024) Comparative transcriptomic analysis of WSSV-challenged Penaeus vannamei with variable resistance levels. Int J Mol Sci 25(9). https://doi.org/10.3390/ijms25094961
Lightner et al (2012) Historic emergence, impact and current status of shrimp pathogens in the Americas. J Invertebr Pathol 110(2):174–183. https://doi.org/10.1016/j.jip.2012.03.006
López-Carvallo JA, Cruz-Flores R, Dhar AK (2022) The emerging pathogen Enterocytozoon hepatopenaei drives a degenerative cyclic pattern in the hepatopancreas microbiome of the shrimp (Penaeus vannamei). Sci Rep. Nature Publishing Group UK 12(1):14766. https://doi.org/10.1038/s41598-022-19127-2
Ma H, Bell KN, Loker RN (2021) qPCR and qRT-PCR analysis: regulatory points to consider when conducting biodistribution and vector shedding studies. Mol Ther Methods Clin Dev. Elsevier Ltd 20:152–168. https://doi.org/10.1016/j.omtm.2020.11.007
MAR (2023) Shrimp market summary. Available at: https://www.grandviewresearch.com/industry-analysis/shrimp-market-report. Accessed 22 July 2025
Margulies M et al (2005) Genome sequencing in microfabricated high-density picolitre reactors. Nature 437(7057):376–380. https://doi.org/10.1038/nature03959
Munkongwongsiri N et al (2022) Propionigenium and Vibrio species identified as possible component causes of shrimp white feces syndrome (WFS) associated with the microsporidian Enterocytozoon hepatopenaei. J Invertebr Pathol. Elsevier Inc 192:107784. https://doi.org/10.1016/j.jip.2022.107784
Patil PK et al (2021) Economic loss due to diseases in Indian shrimp farming with special reference to Enterocytozoon hepatopenaei (EHP) and white spot syndrome virus (WSSV). Aquaculture 533. https://doi.org/10.1016/j.aquaculture.2020.736231
Poulos BT et al (2006) Purification and characterization of infectious myonecrosis virus of penaeid shrimp. J Gen Virol 87(4):987–996. https://doi.org/10.1099/vir.0.81127-0
Qiu L et al (2017) Characterization of a new member of Iridoviridae, Shrimp hemocyte iridescent virus (SHIV), found in white leg shrimp (Litopenaeus vannamei). Sci Rep. Springer US:1–13. https://doi.org/10.1038/s41598-017-10738-8
Quick J et al (2015) Rapid draft sequencing and real-time nanopore sequencing in a hospital outbreak of Salmonella. Genome Biol 16(1):1–14. https://doi.org/10.1186/s13059-015-0677-2
Quick J et al (2016) Real-time, portable genome sequencing for Ebola surveillance. Nature. Nature Publishing Group 530(7589):228–232. https://doi.org/10.1038/nature16996
Reyes G et al (2025) Complete genome sequences of Vibrio parahaemolyticus strains L2171 and L2181 associated with AHPND in Penaeus vannamei postlarvae by hybrid sequencing. Data Brief. Elsevier Inc 61:111819. https://doi.org/10.1016/j.dib.2025.111819
Rhoads A, Au KF (2015) PacBio sequencing and its applications. Genom Proteom Bioinform. Beijing Institute of Genomics, Chinese Academy of Sciences and Genetics Society of China 13(5):278–289. https://doi.org/10.1016/j.gpb.2015.08.002
Robles-Sikisaka R et al (2001) Nucleotide sequence of 3′-end of the genome of Taura syndrome virus of shrimp suggests that it is related to insect picornaviruses. Arch Virol 146(5):941–952. https://doi.org/10.1007/s007050170126
Rothberg JM, Leamon JH (2008) The development and impact of 454 sequencing. Nat Biotechnol 26(10):1117–1124. https://doi.org/10.1038/nbt1485
Sahle Z et al (2024) Fecal microbiota transplantation and next-generation therapies: a review on targeting dysbiosis in metabolic disorders and beyond. SAGE Open Med 12. https://doi.org/10.1177/20503121241257486
Sanger F, Nicklen S (1977) DNA sequencing with chain-terminating. Proc Natl Acad Sci USA 74(12):5463–5467

Shi M et al (2016) Redefining the invertebrate RNA virosphere. Nature. Nature Publishing Group 540(7634):539–543. https://doi.org/10.1038/nature20167

Shike H et al (2000) Infectious hypodermal and hematopoietic necrosis virus of shrimp is related to mosquito brevidensoviruses. Virology 277(1):167–177. https://doi.org/10.1006/viro.2000.0589

Sittidilokratna N et al (2002) Complete ORF1b-gene sequence indicates yellow head virus is an invertebrate nidovirus. Dis Aquat Org 50:87–93. https://doi.org/10.3354/dao050087

Song T et al (2022) Common disease-discriminatory fungal taxa accurately diagnose shrimp white feces syndrome, black gills, and retardation diseases. Aquaculture. Elsevier B.V 561:738643. https://doi.org/10.1016/j.aquaculture.2022.738643

Srisala J et al (2021) Potential universal PCR method to detect decapod hepanhamaparvovirus (DHPV) in crustaceans. Aquaculture. Elsevier B.V 541:736782. https://doi.org/10.1016/j.aquaculture.2021.736782

Stentiford GD et al (2012) Disease will limit future food supply from the global crustacean fishery and aquaculture sectors. J Invertebr Pathol. Elsevier Inc 110(2):141–157. https://doi.org/10.1016/j.jip.2012.03.013

Sun M et al (2023) Integrated application of transcriptomics and metabolomics provides insights into acute hepatopancreatic necrosis disease resistance of Pacific white shrimp Litopenaeus vannamei. *mSystems*. Edited by S L Ishaq American Society for Microbiology. https://doi.org/10.1128/msystems.00067-23

Taengchaiyaphum S, Srisala J, Sanguanrut P (2020) Full genome characterization of Laem Singh virus (LSNV) in shrimp Penaeus monodon. bioRxiv 9:1–16. https://doi.org/10.1101/2020.07.25.221432

Tamaki H et al (2011) Analysis of 16S rRNA amplicon sequencing options on the roche/454 next-generation titanium sequencing platform. PLoS One 6(9):6–8. https://doi.org/10.1371/journal.pone.0025263

Tang KFJ, Bondad-Reantaso MG (2019) Impacts of acute hepatopancreatic necrosis disease on commercial shrimp aquaculture. Rev Sci Tech OIE 38(2):477–490. https://doi.org/10.20506/rst.38.2.2999

Tapia-Chirinos S et al (2024) Whole genome sequencing of nine Vibrio parahaemolyticus strains encoding (Pir) toxin-like genes from shrimp cultures in northern Peru using Oxford Nanopore technology. Microbiol Resour Announc. Edited by F J Stewart American Society for Microbiology 13(3):245–251. https://doi.org/10.1128/mra.00511-23

Tourtip S et al (2009) Enterocytozoon hepatopenaei sp. nov. (Microsporida: Enterocytozoonidae), a parasite of the black tiger shrimp Penaeus monodon (Decapoda: Penaeidae): fine structure and phylogenetic relationships. J Invertebr Pathol. Elsevier Inc 102(1):21–29. https://doi.org/10.1016/j.jip.2009.06.004

Van Hulten MCW et al (2001) The white spot syndrome virus DNA genome sequence. Virology 286(1):7–22. https://doi.org/10.1006/viro.2001.1002

Velázquez-Lizárraga AE et al (2019) Transcriptomic analysis of Pacific white shrimp (Litopenaeus vannamei, Boone 1931) in response to acute hepatopancreatic necrosis disease caused by Vibrio parahaemolyticus. PLoS One 14(8):1–28. https://doi.org/10.1371/journal.pone.0220993

Wang Y et al (2021) Nanopore sequencing technology, bioinformatics and applications. Nat Biotechnol. Springer US 39(11):1348–1365. https://doi.org/10.1038/s41587-021-01108-x

Wirawan A et al (2014) HECTOR: a parallel multistage homopolymer spectrum based error corrector for 454 sequencing data. BMC Bioinformatics 15(1). https://doi.org/10.1186/1471-2105-15-131

Wiredu Boakye D et al (2017) Decay of the glycolytic pathway and adaptation to intranuclear parasitism within Enterocytozoonidae microsporidia. Environ Microbiol 19(5):2077–2089. https://doi.org/10.1111/1462-2920.13734

Yang YT et al (2014) The genome and occlusion bodies of marine Penaeus monodon nudivirus (PmNV, also known as MBV and PemoNPV) suggest that it should be assigned to a new nudivirus genus that is distinct from the terrestrial nudiviruses. BMC Genomics 15(1). https://doi.org/10.1186/1471-2164-15-628

Yang X et al (2020) High-throughput transcriptome profiling in drug and biomarker discovery. Front Genet 11:1–12. https://doi.org/10.3389/fgene.2020.00019

Yu D et al (2022) Comprehensive transcriptomic and metabolomic analysis of the Litopenaeus vannamei hepatopancreas after WSSV challenge. Front Immunol 13:1–13. https://doi.org/10.3389/fimmu.2022.826794

Zhang L et al (2023) Dynamic interplay of metabolic and transcriptional responses in shrimp during early and late infection stages of Enterocytozoon hepatopenaei (EHP). Int J Mol Sci 24(23). https://doi.org/10.3390/ijms242316738

MIX
Papier aus verantwortungsvollen Quellen
Paper from responsible sources
FSC® C105338

If you have any concerns about our products,
you can contact us on
ProductSafety@springernature.com

In case Publisher is established outside the EU,
the EU authorized representative is:
Springer Nature Customer Service Center GmbH
Europaplatz 3, 69115 Heidelberg, Germany

Printed by Libri Plureos GmbH
in Hamburg, Germany